T0073138

PRINCIPLES OF PHYSICS

From Quantum Field Theory
to Classical Mechanics

Second Edition

Tsinghua Report and Review in Physics

ISSN: 2010-1414

Series Editor: Bangfen Zhu *(Tsinghua University, China)*

To my daughter Ruyan

Published by

World Scientific Publishing Co. Pte. Ltd.

5 Toh Tuck Link, Singapore 596224

USA office: 27 Warren Street, Suite 401-402, Hackensack, NJ 07601

UK office: 57 Shelton Street, Covent Garden, London WC2H 9HE

Library of Congress Cataloging-in-Publication Data

Names: Ni, Jun, author.

Title: Principles of physics : from quantum field theory to classical mechanics /
    Jun Ni (Tsinghua University, China).

Other titles: Tsinghua report and review in physics ; vol. 3.

Description: Second edition. | Singapore ; Hackensack, NJ : World Scientific, [2017] |
    Series: Tsinghua report and review in physics ; vol. 3 |
    Includes bibliographical references and index.

Identifiers: LCCN 2017019643| ISBN 9789813227095 (hardcover ; alk. paper) |
    ISBN 9813227095 (hardcover ; alk. paper)

Subjects: LCSH: Physics.

Classification: LCC QC21.3 .N5 2017 | DDC 530--dc23

LC record available at https://lccn.loc.gov/2017019643

British Library Cataloguing-in-Publication Data

A catalogue record for this book is available from the British Library.

Desk Editor: Ng Kah Fee

Typeset by Stallion Press
Email: enquiries@stallionpress.com

Printed in Singapore

TSINGHUA
Report and Review in
Physics Vol. 3

PRINCIPLES OF PHYSICS

From Quantum Field Theory to Classical Mechanics

Second Edition

Jun Ni

Tsinghua University, China

 World Scientific

NEW JERSEY · LONDON · SINGAPORE · BEIJING · SHANGHAI · HONG KONG · TAIPEI · CHENNAI

Preface

During the 20th century, physics experienced a rapid expansion. A general theoretical physics curriculum now consists of a collection of separate courses labeled as classical mechanics, electrodynamics, quantum mechanics, statistical mechanics, quantum field theory, general relativity, etc., with each course taught with a different book. I consider there to be a need to write a book which is compact and merge these courses into one single unified course. This book is an attempt to realize this aim. In writing this book, I focus on two purposes. (1) Historically, physics is established from classical mechanics to quantum mechanics, from quantum mechanics to quantum field theory, from thermodynamics to statistical mechanics, from Newtonian gravity to general relativity. However, a more logical presentation is from quantum field theory to classical mechanics, from the physics principles on microscopic scale to physics on macroscopic scale. In this book, I try to achieve this by elucidating the physics from quantum field theory to classical mechanics from a set of common basic principles in a unified way. (2) Physics is considered as an experimental science. This view, however, is being changed. In the history of physics, there are two epic heroes: Newton and Einstein. They represent two epochs in physics. In the Newtonian epoch, physical laws are deduced from the experimental observations. People were amazed that the observed physical laws can be described accurately by mathematical equations. At the same time, it is reasonable to ask why the nature should obey the physical laws described by the mathematical equations. After wondering how accurately the nature obeys the gravitational law that the gravitational force is proportional to the inverse square of the distance, one would ask why it is not operated in other ways. Einstein created a new epoch by deducing physical laws not merely from experiments but also from principles such as simplicity, symmetry and other understandable credos. From the view of Einstein, physical laws should be natural and simple. It is my belief that all physical laws should be understandable. In this book, I endeavor to establish the physical formalisms based on the basic principles that are as simple and understandable as possible.

The book covers all the disciplines of fundamental physics, including quantum field theory, quantum mechanics, statistical mechanics, thermodynamics, general relativity, electromagnetism, and classical mechanics. Instead of the traditional pedagogic way, the subjects and formalisms are arranged in a logical order, i.e. all the formulas are derived from the formulas before them. Also all the formalisms are kept self-contained, i.e. the derivations of all the physical formulas appeared in this book can be found in this book. Most mathematic tools are also given in the appendices. Although this book covers all the disciplines of fundamental physics, the book is compact and has only about 400 pages because the contents are kept concise and treated as an integrated entity. In this book, the main emphasis is the basic formalisms of physics. The topics on applications and approximation methods are kept minimum and are selected based on the generality and importance. Still it was not easy when some important topics had to be omitted. Since it is impossible to provide an exhaustive bibliography, I list only the related textbooks and monographs that I am familiar with. I apologize to the authors whose books have not been included unintentionally.

This book may be used as an advanced textbook by graduate students. It is also suitable for physicists who wish to have an overview of the fundamental physics.

I am grateful to all my colleagues and students for the inspirations and help. I would also like to express my gratitude to World Scientific for the help in publishing this book.

Jun Ni
August 8, 2013
Tsinghua, Beijing

Preface to the second edition

In this edition, the whole frame and style of the book remain unchanged. However, there are some modifications on the derivations of important equations. For example, I have used the new way to derive the Euler-Lagrange equation. There is also expansion on the contents of the book. I have made significant expansion on the parts of quantum electrodynamics, classical electrodynamics and special relativity, which makes the contents of the book more complete and self-contained. Quantum electrodynamics plays an important part in the bridge from quantum field theory to quantum mechanics. I hope these additions could better achieve the purpose of the book. I have also rewritten a few parts of the book and corrected the errors in the first edition.

Jun Ni
April 15, 2017
Tsinghua, Beijing

Contents

Chapter 1

Basic principles

Since the establishment of physics, physics has heaped up countless laws and equations which enable us to describe the phenomena of the nature. What is the common mechanism behind them? Are there the simple common mechanisms behind all of the laws and equations in physics? These are the questions physicists often ask themselves. In the following, we will stipulate five simple principles which can be cataloged as the common mechanisms behind the physical laws and equations.

The following are the five basic principles from which we will start to construct all other physical laws and equations. These five basic principles are: (1) Constituent principle - the basic constituents of matter in the spacetime are various kinds of identical pointlike particles. This can also be called locality principle; (2) Causality principle - The future state depends only on the present state; (3) Covariance principle - Physics should be invariant under an arbitrary coordinate transformation; (4) Invariance or Symmetry principle - The spacetime is homogeneous; (5) Equi-probability principle - All the microscopic states in an isolated system are expected to be occupied with equal probability. These five basic principles can be considered as physical common senses. It is very natural to have these basic principles. More important is that these five basic principles are consistent with each others. From these five principles, we could derive a vast set of equations which explain or promise putatively to explain all the phenomena of the physical world.

Chapter 2

Quantum fields

2.1 Commutators

2.1.1 *Identical particle principle*

We start from the constituent principle. Matter consists of various kinds of identical point-like particles. Since particles are local identities, this principle can also be considered as the locality principle. A particle is characterized by its position and other internal degrees of freedom which are denoted as λ. Such a particle is called to be in the λ state which is denoted by $|\lambda\rangle$. The symbol $|\ \rangle$ is called ket, which was introduced by Dirac. $|\lambda\rangle$ means that there is a particle characterized by λ. $|\lambda\rangle$ is also called a single-particle state. An N-particle state is denoted as $|\lambda_1 \cdots \lambda_i \cdots \lambda_N\rangle$. Here i labels the ith particle. A state of a system corresponds to a configuration of particles. We denote $|0\rangle$ as the vacuum state, which contains no particles. $|\lambda\rangle$ also means that a particle state is created. When there is creation, there should be annihilation. For a vacuum state $|0\rangle$, we can introduce its dual state $\langle 0|$ by

$$\langle 0|0\rangle = 1. \tag{2.1}$$

Eq.(2.1) means that $\langle 0|$ annihilates the state $|0\rangle$. The symbol $\langle\ |$ is called bra. Similarly, for any state $|\lambda\rangle$, we have its dual state $\langle\lambda|$ defined by

$$\langle\lambda|\lambda\rangle = 1. \tag{2.2}$$

Eq.(2.2) means that $\langle\lambda|$ annihilates the state $|\lambda\rangle$.

2.1.2 *Projection operator*

We define a projection operator for single-particle states by $|\lambda\rangle\langle\lambda|$, which projects any state $|\lambda'\rangle$ onto the state $|\lambda\rangle$, resulting in a state $|\lambda\rangle\langle\lambda|\lambda'\rangle$. λ is the fundamental parameter to distinguish particles. Thus, when the states $|\lambda'\rangle$ and $|\lambda\rangle$ are different ($\lambda \neq \lambda'$), the projection of the state $|\lambda'\rangle$ onto the state $|\lambda\rangle$ will be zero. For example, a particle can not be at both position $\mathbf{x}$ and position $\mathbf{x}'$. Since a point-like particle is local, a particle at position $\mathbf{x}$ is not a part of the particle at position $\mathbf{x}'$. We have

$$\langle\lambda|\lambda'\rangle = \delta_{\lambda\lambda'} \tag{2.3}$$

with

$$\delta_{\lambda\lambda'} = 1, \quad \text{if} \quad \lambda = \lambda' \tag{2.4a}$$

$$\delta_{\lambda\lambda'} = 0, \quad \text{if} \quad \lambda \neq \lambda'. \tag{2.4b}$$

The notation $\delta_{\lambda\lambda'}$ is often called the 'Kronecker' delta. When a particle is in the λ state, the projection operator for the λ state projects the state onto itself. When a particle is in the $\lambda' \neq \lambda$ state, the projection operator filters out this state. Eq.(2.3) is called the orthonormal relation of states. We also call $\langle\lambda|\lambda'\rangle$ as the scalar product of two states. When λ is a continuous variable, the Kronecker delta should be replaced by the delta function.

We can add the projections $|\lambda\rangle\langle\lambda|$ of all the states together. Since a particle at least is in one state, we have

$$\sum_{\lambda} |\lambda\rangle\langle\lambda| = 1. \tag{2.5}$$

Eq.(2.5) is called the completeness relation of singe-particle state.

2.1.3   *Creation and annihilation operators*

We introduce creation and annihilation operators to describe the particle state. We define a creation operator $\hat{a}_\lambda^\dagger$ as the one mapping an N-particle state onto an (N+1)-particle state. For the vacuum state, we can add particles using the creation operator $\hat{a}_\lambda^\dagger$. λ can be position of a particle. When λ is the position, $\hat{a}_\lambda^\dagger$ means creating a local particle at λ position. If we create a particle characterized by λ, we have a state

$$\hat{a}_\lambda^\dagger|0\rangle = |\lambda\rangle. \tag{2.6}$$

$\hat{a}_\lambda^\dagger$ can also be denoted as

$$|\lambda\rangle \otimes . \tag{2.7}$$

$\otimes$ means that $|\lambda\rangle$ merges with other states by tensor product. $\otimes$ is often omitted for simplicity. Thus Eq.(2.6) can be rewritten as

$$\hat{a}_\lambda^\dagger|0\rangle = |\lambda\rangle \otimes |0\rangle = |\lambda\rangle. \tag{2.8}$$

The N-particle state $|\lambda_1 \cdots \lambda_i \cdots \lambda_N\rangle$ can be formed using N creation operators,

$$|\lambda_1 \cdots \lambda_N\rangle = \hat{a}^\dagger_{\lambda_1} \cdots \hat{a}^\dagger_{\lambda_N} |0\rangle$$
$$= |\lambda_1\rangle \otimes \cdots |\lambda_N\rangle \otimes |0\rangle. \tag{2.9}$$

In exchanging the two creation operators, we exchange the labels of the two generated particles. We denote P_{ij} the operator that exchanges the labels of the particles i and j. For example,

$$P_{12}|\lambda_1\lambda_2\rangle = |\lambda_2\lambda_1\rangle. \tag{2.10}$$

Since the particles are identical, $|\lambda_1\lambda_2\rangle$ and $|\lambda_2\lambda_1\rangle$ are the same states. Although we have assumed that the particles are identical, the following is a way to understand that the point-like particles are identical. When λ_1 and λ_2 are the positions of particles. $|\lambda_1\lambda_2\rangle$ means that there is a particle at $\mathbf{x}_1$ position and a particle at $\mathbf{x}_2$ position. $|\lambda_2\lambda_1\rangle$ means that there is a particle at $\mathbf{x}_2$ position and a particle at $\mathbf{x}_1$ position. If the two particles are fundamental, there are no other internal degrees of freedom to distinguish them, which means that λ has all the parameters to characterize a particle. Then the states $|\lambda_1\lambda_2\rangle$ and $|\lambda_2\lambda_1\rangle$ describe the same state, i.e. a state with a particle at $\mathbf{x}_1$ position and a particle at $\mathbf{x}_2$ position. Thus when we exchange the two particles, we have the same state.

When we execute the exchange operator two times, the particles return to their initial labels and we recover the original state. Thus $P^2 = 1$ and $P = \pm 1$. Because $P = \pm 1$, we have two cases. i) The two creation operators $\hat{a}^\dagger_{\lambda_1}$ and $\hat{a}^\dagger_{\lambda_2}$ commute, $\hat{a}^\dagger_{\lambda_1}\hat{a}^\dagger_{\lambda_2} = \hat{a}^\dagger_{\lambda_2}\hat{a}^\dagger_{\lambda_1}$, which corresponds to $P = 1$; ii) The two creation operators $\hat{a}^\dagger_{\lambda_1}$ and $\hat{a}^\dagger_{\lambda_2}$ anti-commute, $\hat{a}^\dagger_{\lambda_1}\hat{a}^\dagger_{\lambda_2} = -\hat{a}^\dagger_{\lambda_2}\hat{a}^\dagger_{\lambda_1}$, which corresponds to $P = -1$.

If $\hat{a}^\dagger_{\lambda_1}$ and $\hat{a}^\dagger_{\lambda_2}$ commute, we call the particles bosons. For bosons, we have the commutation relation

$$[\hat{a}^\dagger_{\lambda_1}, \hat{a}^\dagger_{\lambda_2}]_- \equiv [\hat{a}^\dagger_{\lambda_1}, \hat{a}^\dagger_{\lambda_2}] \equiv \hat{a}^\dagger_{\lambda_1}\hat{a}^\dagger_{\lambda_2} - \hat{a}^\dagger_{\lambda_2}\hat{a}^\dagger_{\lambda_1} = 0. \tag{2.11}$$

If $\hat{a}^\dagger_{\lambda_1}$ and $\hat{a}^\dagger_{\lambda_2}$ anti-commute, we call the particles fermions. For fermions, we have the anti-commutation relation

$$[\hat{a}^\dagger_{\lambda_1}, \hat{a}^\dagger_{\lambda_2}]_+ \equiv \{\hat{a}^\dagger_{\lambda_1}, \hat{a}^\dagger_{\lambda_2}\} \equiv \hat{a}^\dagger_{\lambda_1}\hat{a}^\dagger_{\lambda_2} + \hat{a}^\dagger_{\lambda_2}\hat{a}^\dagger_{\lambda_1} = 0. \tag{2.12}$$

Thus any two creation operators $\hat{a}^\dagger_{\lambda_1}$ and $\hat{a}^\dagger_{\lambda_2}$ commute or anti-commute depending on the types of particles. For fermions, in the case of $\lambda_1 = \lambda_2 = \lambda$, the anti-commutation relation Eq.(2.12) becomes $2\hat{a}^\dagger_\lambda\hat{a}^\dagger_\lambda = 0$, i.e.

$\hat{a}^\dagger_\lambda \hat{a}^\dagger_\lambda = 0$. Thus two fermions can not be accommodated in the same state, which is known as the *Pauli exclusion principle*.

Now we introduce annihilation operator $\hat{a}_\lambda$. An annihilation operator maps an N-particle state onto an (N-1)-particle state. The annihilation operator $\hat{a}_\lambda$ thus annihilates the particle characterized by λ. In the simplest situation, we have

$$\hat{a}_\lambda|\lambda\rangle = |0\rangle, \tag{2.13}$$

which means that after annihilating a single-particle state, the state turns into the vacuum state.

Similar to the creation operators, we have the following two types of commutation relations for the annihilation operators. For boson, the annihilation operators commute,

$$[\hat{a}_{\lambda_1}, \hat{a}_{\lambda_2}]_- = 0. \tag{2.14}$$

For fermions, the annihilation operators anti-commute,

$$[\hat{a}_{\lambda_1}, \hat{a}_{\lambda_2}]_+ = 0. \tag{2.15}$$

Since $\langle\lambda|$ annihilates the state $|\lambda\rangle$, we can denote $\hat{a}_\lambda$ as

$$\hat{a}_\lambda = \langle\lambda| = (\langle 0|\hat{a}^\dagger_\lambda). \tag{2.16}$$

The bracket means that $\hat{a}^\dagger_\lambda$ acts on the left. Then Eq.(2.13) can be rewritten as

$$\begin{aligned}
\hat{a}_\lambda|\lambda\rangle &= \hat{a}_\lambda|\lambda\rangle \otimes |0\rangle \\
&= \langle\lambda|\lambda\rangle \otimes |0\rangle \\
&= |0\rangle.
\end{aligned} \tag{2.17}$$

Since $\langle\lambda|0\rangle = 0$, we have

$$\hat{a}_\lambda|0\rangle = 0. \tag{2.18}$$

Eq.(2.18) means that when there is no particle for annihilation, the effect of the annihilation operator is same as zero. Eq.(2.18) has a more general version

$$\hat{a}_{\lambda_1}|\lambda_2\rangle = 0, \quad \lambda_1 \neq \lambda_2. \tag{2.19}$$

From Eq.(2.16), we have

$$\langle 0|\hat{a}_\lambda|\lambda'\rangle = \langle\lambda|\lambda'\rangle = \delta_{\lambda\lambda'} = \langle\lambda'|\lambda\rangle = \langle\lambda'|\hat{a}^\dagger_\lambda|0\rangle. \tag{2.20}$$

Thus $\hat{a}_\lambda$ can be considered as the adjoint operator of $\hat{a}^\dagger_\lambda$.

All the states $|\lambda\rangle$ form the Hilbert space $\mathcal{H}$. We call $|\lambda\rangle$ as an orthonormal basis of $\mathcal{H}$. The N-particle states are described in the Hilbert space $\mathcal{H}_N$, which is the N tensor product of the single-particle Hilbert space $\mathcal{H}$.

$$\mathcal{H}_N = \mathcal{H} \otimes \mathcal{H} \otimes \cdots \otimes \mathcal{H}. \qquad (2.21)$$

The N-particle state $|\lambda_1 \cdots \lambda_N\rangle$ is the tensor product of the single-particle states.

$$|\lambda_1 \cdots \lambda_N\rangle = |\lambda_1\rangle \otimes |\lambda_2\rangle \otimes \cdots \otimes |\lambda_N\rangle. \qquad (2.22)$$

Since the particles are elemental and no particle is a part of other particles, the state $|\lambda_1 \cdots \lambda_N\rangle$ are orthonormal. $|\lambda_1 \cdots \lambda_N\rangle$ form the canonical orthonormal basis of $\mathcal{H}_N$. It should be noted that the states with different particle number are also orthonormal. All particle states form the Fock space.

2.1.4 *Symmetrized and anti-symmetrized states*

In order to describe the symmetry properties of the states of bosons and fermions, we introduce the symmetrization operator P_B and the anti-symmetrization operator P_F.

$$P_B|\lambda_1 \cdots \lambda_N\rangle = \frac{1}{N!} \sum_P |\lambda_{P_1} \cdots \lambda_{P_N}\rangle \qquad (2.23a)$$

$$P_F|\lambda_1 \cdots \lambda_N\rangle = \frac{1}{N!} \sum_P (-1)^{S_P} |\lambda_{P_1} \cdots \lambda_{P_N}\rangle, \qquad (2.23b)$$

where P is the permutation of $(1, 2 \cdots, N)$, which brings $(1, 2 \cdots, N)$ to $(P_1, P_2 \cdots, P_N)$. S_P is the number of the transpositions of two elements in the permutation P that brings $(1, 2 \cdots, N)$ to $(P_1, P_2 \cdots, P_N)$. For example, for two particles,

$$P_B|\lambda_1\lambda_2\rangle = \frac{1}{2}(|\lambda_1\lambda_2\rangle + |\lambda_2\lambda_1\rangle), \qquad (2.24a)$$

$$P_F|\lambda_1\lambda_2\rangle = \frac{1}{2}(|\lambda_1\lambda_2\rangle - |\lambda_2\lambda_1\rangle). \qquad (2.24b)$$

The states of bosons are symmetric. We can use $P_B|\lambda_1 \cdots \lambda_N\rangle$ to describe the states of bosons regardless of the symmetry of $|\lambda_1 \cdots \lambda_N\rangle$. The states of fermions are antisymmetric. We can use $P_F|\lambda_1 \cdots \lambda_N\rangle$ to describe the states of fermions regardless of the symmetry of $|\lambda_1 \cdots \lambda_N\rangle$. The states of bosons form the Hilbert space of bosons B_N, while the states of fermions make up the Hilbert space of fermions F_N. Eq.(2.23) can be rewritten in a compact form as

$$P_{\{{}^B_F\}}|\lambda_1 \cdots \lambda_N\rangle = \frac{1}{N!} \sum_P \xi^{S_P}|\lambda_{P_1} \cdots \lambda_{P_N}\rangle, \qquad (2.25)$$

where $\xi = 1$ for P_B and $\xi = -1$ for P_F.

$P_{\{{}^B_F\}}$ can be shown to be the projections that project $\mathcal{H}_N$ onto the Hilbert space of bosons $\mathcal{B}_N$ and the Hilbert space of fermions $\mathcal{F}_N$, respectively. For any N-particle state of $\mathcal{H}_N$, we have

$$P^2_{\{{}^B_F\}}|\lambda_1 \cdots \lambda_N\rangle = \frac{1}{N!}\frac{1}{N!} \sum_P \sum_{P'} \xi^{S_{P'}} \xi^{S_P}|\lambda_{P_1'P_1} \cdots \lambda_{P_N'P_N}\rangle. \quad (2.26)$$

We introduce $Q = P'P$. Since $\xi^{S_{P'}+S_P} = \xi^{S_{P'P}}$ and Q corresponds to P' one by one, we have

$$\begin{aligned}
P^2_{\{{}^B_F\}}&|\lambda_1 \cdots \lambda_N\rangle \\
&= \frac{1}{N!} \sum_P \frac{1}{N!} \sum_{P'} \xi^{S_Q}|\lambda_{Q_1} \cdots \lambda_{Q_N}\rangle \\
&= \frac{1}{N!} \sum_P \left(\frac{1}{N!} \sum_Q \xi^{S_Q}|\lambda_{Q_1} \cdots \lambda_{Q_N}\rangle \right) \qquad (2.27) \\
&= \frac{1}{N!} \sum_P P_{\{{}^B_F\}}|\lambda_1 \cdots \lambda_N\rangle \\
&= P_{\{{}^B_F\}}|\lambda_1 \cdots \lambda_N\rangle.
\end{aligned}$$

Eq.(2.27) holds for any state. Thus $P_{\{{}^B_F\}}$ are the projection operators projecting $\mathcal{H}_N$ onto $\{{}^{\mathcal{B}_N}_{\mathcal{F}_N}\}$.

Using these projection operators, we can define the symmetrized or antisymmetrized states as

$$\begin{aligned}
|\lambda_1, \lambda_2, &\cdots \lambda_N\rangle_S \\
&\equiv \sqrt{N!}P_{\{{}^B_F\}}|\lambda_1, \lambda_2, \cdots \lambda_N\rangle \qquad (2.28) \\
&= \frac{1}{\sqrt{N!}} \sum_P \xi^{S_P}|\lambda_{P_1}\rangle \otimes |\lambda_{P_2}\rangle \otimes \cdots \otimes |\lambda_{P_N}\rangle.
\end{aligned}$$

It is usually convenient to use the normalized symmetrized or antisymmetrized states. The scalar product of the two same symmetrized or anti-symmetrized states is given by

$$
\begin{aligned}
&{}_S\langle\lambda_1,\lambda_2,\cdots\lambda_N|\lambda_1,\lambda_2,\cdots\lambda_N\rangle_S\\
&= N!\langle\lambda_1,\lambda_2,\cdots\lambda_N|P^2_{\{{}^B_F\}}|\lambda_1,\lambda_2,\cdots\lambda_N\rangle\\
&= N!\langle\lambda_1,\lambda_2,\cdots\lambda_N|P_{\{{}^B_F\}}|\lambda_1,\lambda_2,\cdots\lambda_N\rangle\\
&= \sum_P \xi^{S_P}\langle\alpha_1|\alpha_{P_1}\rangle\langle\alpha_2|\alpha_{P_2}\rangle\cdots\langle\alpha_N|\alpha_{P_N}\rangle.
\end{aligned}
\tag{2.29}
$$

According to Eq.(2.3), the only non-vanishing terms in the summation of Eq.(2.29) are the ones with

$$
\lambda_1 = \lambda_{P_1}, \lambda_2 = \lambda_{P_2}, \cdots, \lambda_N = \lambda_{P_N}.
\tag{2.30}
$$

For fermions, there is at most one particle with the same λ. λ_i in the set $(\lambda_1,\cdots,\lambda_N)$ are all different. There is only one nonzero term which corresponds to $S_P = 0$. Thus we have

$$
{}_S\langle\lambda_1,\lambda_2,\cdots\lambda_N|\lambda_1,\lambda_2,\cdots\lambda_N\rangle_S = 1 \quad \text{for fermions,}
\tag{2.31}
$$

which means that $|\lambda_1,\lambda_2,\cdots\lambda_N\rangle_S$ is already normalized.

For bosons, particles with the same λ are allowed. Any permutation which interchanges the particles with the same λ contributes to the sum in Eq.(2.29). If the state $|\lambda_1,\lambda_2,\cdots\lambda_N\rangle$ contains n_{α_1} bosons with $\lambda = \alpha_1$, n_{α_2} bosons with $\lambda = \alpha_2$, $\cdots$, n_{α_p} bosons with $\lambda = \alpha_p$, where all the α_i are different, the scalar product Eq.(2.29) is given by

$$
{}_S\langle\lambda_1,\lambda_2,\cdots\lambda_N|\lambda_1,\lambda_2,\cdots\lambda_N\rangle_S = n_{\alpha_1}!n_{\alpha_2}!\cdots n_{\alpha_p}!
\tag{2.32}
$$

with

$$
\sum_i n_{\alpha_i} = N.
\tag{2.33}
$$

Since $n_\alpha = 1$ for fermions, Eq.(2.32) is also applicable for fermions. Thus we obtain the normalized symmetrized or anti-symmetrized states defined by

$$
|\lambda_1,\lambda_2,\cdots\lambda_N\rangle_{SN} \equiv \frac{1}{\sqrt{\prod_\alpha n_\alpha!}}|\lambda_1,\lambda_2,\cdots\lambda_N\rangle_S.
\tag{2.34}
$$

To simplify the notation, we use $|n_\lambda\rangle$ to denote $|\lambda_1 = \lambda,\cdots\lambda_{n_\lambda} = \lambda\rangle$. For bosons, N-particle state should have the following normalized symmetric form,

$$
\begin{aligned}
&|n_{\alpha_1}n_{\alpha_2}\cdots n_{\alpha_p}\rangle\\
&= \frac{1}{\sqrt{n_{\alpha_1}!n_{\alpha_2}!\cdots n_{\alpha_p}!}}(\hat{a}^\dagger_{\alpha_1})^{n_{\alpha_1}}(\hat{a}^\dagger_{\alpha_2})^{n_{\alpha_2}}\cdots(\hat{a}^\dagger_{\alpha_p})^{n_{\alpha_p}}|0\rangle.
\end{aligned}
\tag{2.35}
$$

2.1.5   *Commutators between creation and annihilation operators*

When we apply $\hat{a}_\lambda^\dagger$ to the symmetrical N-particle boson state $|n_\lambda\rangle$, we have

$$\begin{aligned}
\hat{a}_\lambda^\dagger|n_\lambda\rangle &= \frac{1}{\sqrt{n_\lambda!}}(\hat{a}_\lambda^\dagger)^{n_\lambda+1}|0\rangle \\
&= \frac{\sqrt{n_\lambda+1}}{\sqrt{(n_\lambda+1)!}}(\hat{a}_\lambda^\dagger)^{n_\lambda+1}|0\rangle \\
&= \sqrt{n_\lambda+1}\,|n_\lambda+1\rangle.
\end{aligned} \qquad (2.36)$$

The annihilation operator $\hat{a}_\lambda$ is the adjoint operator of the creation operator. Thus we have

$$\langle n_\lambda - 1|\hat{a}_\lambda|n_\lambda\rangle = \langle n_\lambda|\hat{a}_\lambda^\dagger|n_\lambda - 1\rangle = \sqrt{n_\lambda}\langle n_\lambda|n_\lambda\rangle = \sqrt{n_\lambda}. \qquad (2.37)$$

Therefore, we have

$$\hat{a}_\lambda|n_\lambda\rangle = \sqrt{n_\lambda}\,|n_\lambda - 1\rangle. \qquad (2.38)$$

Using Eqs.(2.36) and (2.38), we have

$$\hat{a}_\lambda^\dagger\hat{a}_\lambda|n_\lambda\rangle = n_\lambda|n_\lambda\rangle, \qquad (2.39)$$

$$\hat{a}_\lambda\hat{a}_\lambda^\dagger|n_\lambda\rangle = (n_\lambda+1)|n_\lambda\rangle. \qquad (2.40)$$

Subtracting the two equations, we obtain

$$[\hat{a}_\lambda, \hat{a}_\lambda^\dagger] = 1. \qquad (2.41)$$

Now we derive the commutator of $\hat{a}_\lambda$ and $\hat{a}_{\lambda'}^\dagger$ with $\lambda \neq \lambda'$. Using Eqs.(2.36) and (2.38), we have

$$\begin{aligned}
&\hat{a}_\lambda\hat{a}_{\lambda'}^\dagger|\cdots n_\lambda \cdots n_{\lambda'} \cdots\rangle \\
&= \sqrt{n_\lambda}\sqrt{n_{\lambda'}+1}|\cdots(n_\lambda-1)\cdots(n_{\lambda'}+1)\cdots\rangle
\end{aligned} \qquad (2.42)$$

and

$$\begin{aligned}
&\hat{a}_{\lambda'}^\dagger\hat{a}_\lambda|\cdots n_\lambda \cdots n_{\lambda'} \cdots\rangle \\
&= \sqrt{n_\lambda}\sqrt{n_{\lambda'}+1}|\cdots(n_\lambda-1)\cdots(n_{\lambda'}+1)\cdots\rangle.
\end{aligned} \qquad (2.43)$$

This leads to

$$[\hat{a}_\lambda, \hat{a}_{\lambda'}^\dagger] = 0, \quad \lambda \neq \lambda'. \qquad (2.44)$$

Thus we obtain the commutation relation for the annihilation operator $\hat{a}_{\lambda_1}$ and the creation operator $\hat{a}_{\lambda_2}^\dagger$

$$[\hat{a}_{\lambda_1}, \hat{a}_{\lambda_2}^\dagger] = \delta_{\lambda_1\lambda_2}. \qquad (2.45)$$

Now we consider fermions. Fermions obey the anti-commutation relations. Thus $\hat{a}_\lambda^\dagger \hat{a}_\lambda^\dagger = 0$ and $\hat{a}_\lambda \hat{a}_\lambda = 0$. n_λ can only be one or zero. Therefore, we have the following relations:

$$\hat{a}_\lambda^\dagger |0\rangle = |(1)_\lambda\rangle, \ \hat{a}_\lambda^\dagger |(1)_\lambda\rangle = 0,$$
$$\hat{a}_\lambda |(1)_\lambda\rangle = |0\rangle, \ \hat{a}_\lambda |0\rangle = 0. \tag{2.46}$$

In order to deduce the commutator $[\hat{a}_\lambda, \hat{a}_{\lambda'}^\dagger]$, we consider the following state

$$(\hat{a}_{\lambda_1}^\dagger)^{n_{\lambda_1}} (\hat{a}_{\lambda_2}^\dagger)^{n_{\lambda_2}} \cdots (\hat{a}_{\lambda_\infty}^\dagger)^{n_{\lambda_\infty}} |0\rangle \equiv |n_{\lambda_1} n_{\lambda_1} \cdots n_{\lambda_\infty}\rangle. \tag{2.47}$$

If $n_{\lambda_s} = 0$, the direct evaluation of $\hat{a}_{\lambda_s}^\dagger |n_{\lambda_1} n_{\lambda_1} \cdots n_{\lambda_\infty}\rangle$ gives

$$\hat{a}_{\lambda_s}^\dagger |n_{\lambda_1} n_{\lambda_2} \cdots n_{\lambda_\infty}\rangle = (-1)^{S_s} (\hat{a}_{\lambda_1}^\dagger)^{n_{\lambda_1}} \cdots (\hat{a}_{\lambda_s}^\dagger) \cdots (\hat{a}_{\lambda_\infty}^\dagger)^{n_{\lambda_\infty}} |0\rangle, \tag{2.48}$$

where the factor S_s is defined by

$$S_s \equiv n_{\lambda_1} + n_{\lambda_2} + \cdots + n_{\lambda_{s-1}}. \tag{2.49}$$

Thus, we have

$$\hat{a}_{\lambda_s}^\dagger |n_{\lambda_1} n_{\lambda_2} \cdots n_{\lambda_\infty}\rangle$$
$$= (-1)^{S_s} |n_{\lambda_1} \cdots (n_{\lambda_s} + 1) \cdots n_{\lambda_\infty}\rangle \quad (\text{if } n_{\lambda_s} = 0). \tag{2.50}$$

When $n_{\lambda_s} = 1$, we can exchange $\hat{a}_{\lambda_s}^\dagger$ to the position λ_s and get a factor $\hat{a}_{\lambda_s}^\dagger \hat{a}_{\lambda_s}^\dagger$, which leads to

$$\hat{a}_{\lambda_s}^\dagger |n_{\lambda_1} n_{\lambda_1} \cdots n_{\lambda_\infty}\rangle = 0. \ (\text{if } n_{\lambda_s} = 1). \tag{2.51}$$

Now we consider the annihilating operator $\hat{a}_{\lambda_s}$. When $n_{\lambda_s} = 1$, since $\hat{a}_{\lambda_s}$ is the adjoint of operator $\hat{a}_{\lambda_s}^\dagger$, we have

$$\langle n_{\lambda_1} \cdots (n_{\lambda_s} - 1) \cdots n_{\lambda_\infty} | \hat{a}_{\lambda_s} | n_{\lambda_1} \cdots n_{\lambda_s} \cdots n_{\lambda_\infty}\rangle$$
$$= \langle n_{\lambda_1} \cdots n_{\lambda_s} \cdots n_{\lambda_\infty} | \hat{a}_{\lambda_s}^\dagger | n_{\lambda_1} \cdots (n_{\lambda_s} - 1) \cdots n_{\lambda_\infty}\rangle \tag{2.52}$$
$$= \langle n_{\lambda_1} \cdots n_{\lambda_s} \cdots n_{\lambda_\infty} | (-1)^{S_s} | n_{\lambda_1} \cdots n_{\lambda_s} \cdots n_{\lambda_\infty}\rangle = (-1)^{S_s}.$$

Thus, we have

$$\hat{a}_{\lambda_s} |n_{\lambda_1} \cdots n_{\lambda_s} \cdots n_{\lambda_\infty}\rangle$$
$$= (-1)^{S_s} |n_{\lambda_1} \cdots (n_{\lambda_s} - 1) \cdots n_{\lambda_\infty}\rangle \quad (\text{if } n_{\lambda_s} = 1). \tag{2.53}$$

If $n_{\lambda_s} = 0$, we can similarly get

$$\hat{a}_{\lambda_s} |n_{\lambda_1} \cdots n_{\lambda_s} \cdots n_{\lambda_\infty}\rangle = 0 \ (\text{if } n_{\lambda_s} = 0). \tag{2.54}$$

In summary of the results given by Eqs.(2.50),(2.51),(2.53) and (2.54), we can easily obtain

$$\{\hat{a}_{\lambda_1}, \hat{a}_{\lambda_2}^\dagger\} = \delta_{\lambda_1 \lambda_2}. \tag{2.55}$$

The above commutation relations are for the operators at the same time and are called the equal-time commutation relations (ETCR). There are also the commutation relations at different times $[\hat{a}_{\lambda_1}(t), \hat{a}^\dagger_{\lambda_2}(t')]_\pm$. In order to calculate the commutation relations at different times, we need to know the equations of motion. We will discuss the commutation relations at different times $[\hat{a}_{\lambda_1}(t), \hat{a}^\dagger_{\lambda_2}(t')]_\pm$ after we derive the equations of motion.

We introduce $\hat{a}(\mathbf{x}, t)$ and $\hat{a}^\dagger(\mathbf{x}, t)$ by taking λ in $\hat{a}^\dagger_\lambda$ and $\hat{a}_\lambda$ as position $\mathbf{x}$. Then $\hat{a}^\dagger_\lambda$ takes the meaning of creating a particle at position $\mathbf{x}$ and $\hat{a}_\lambda$ annihilating a particle at position $\mathbf{x}$. $\hat{a}^\dagger_\lambda$ and $\hat{a}_\lambda$ become $\hat{a}^\dagger(\mathbf{x}, t)$ and $\hat{a}(\mathbf{x}, t)$ respectively. Since $\lambda = \mathbf{x}$ as position is a continuous variable. $\delta_{\lambda_1 \lambda_2}$ in Eq.(2.55) should be replaced by a delta function $\delta^3(\mathbf{x}_1 - \mathbf{x}_2)$. Then we have

$$[\hat{a}(\mathbf{x}, t), \hat{a}^\dagger(\mathbf{x}', t)]_\pm = \delta^3(\mathbf{x} - \mathbf{x}'), \tag{2.56a}$$

$$[\hat{a}^\dagger(\mathbf{x}, t), \hat{a}^\dagger(\mathbf{x}', t)]_\pm = [\hat{a}(\mathbf{x}, t), \hat{a}(\mathbf{x}', t)]_\pm = 0. \tag{2.56b}$$

With the help of the creation and annihilation operators, we can define the particle-number density operator

$$\hat{n}(\mathbf{x}, t) \equiv \hat{a}^\dagger(\mathbf{x}, t)\hat{a}(\mathbf{x}, t) \tag{2.57}$$

and the total particle-number operator

$$\hat{N}(t) \equiv \int d^3x\, \hat{n}(\mathbf{x}, t) = \int d^3x\, \hat{a}^\dagger(\mathbf{x}, t)\hat{a}(\mathbf{x}, t). \tag{2.58}$$

2.2   Equations of motion

2.2.1   *Field operators*

Now we discuss the particle dynamics. For bosons, we define two field operators

$$\hat{\phi}(\mathbf{x}, t) \equiv \frac{1}{\sqrt{2}}(\hat{a}^\dagger(\mathbf{x}, t) + \hat{a}(\mathbf{x}, t)), \tag{2.59a}$$

$$\hat{\pi}(\mathbf{x}, t) \equiv \frac{i}{\sqrt{2}}(\hat{a}^\dagger(\mathbf{x}, t) - \hat{a}(\mathbf{x}, t)). \tag{2.59b}$$

We have for their commutators

$$
\begin{aligned}
[\hat{\phi}(\mathbf{x}, t), \hat{\pi}(\mathbf{x}', t)] =& \frac{i}{2}[(\hat{a}(\mathbf{x}, t)\hat{a}^\dagger(\mathbf{x}', t) - \hat{a}^\dagger(\mathbf{x}', t)\hat{a}(\mathbf{x}, t)) \\
&+ (\hat{a}(\mathbf{x}', t)\hat{a}^\dagger(\mathbf{x}, t) - \hat{a}^\dagger(\mathbf{x}, t)\hat{a}(\mathbf{x}', t))] \\
=& \frac{i}{2}([\hat{a}(\mathbf{x}, t), \hat{a}^\dagger(\mathbf{x}', t)] + [\hat{a}(\mathbf{x}', t), \hat{a}^\dagger(\mathbf{x}, t)]) \\
=& i\delta^3(\mathbf{x} - \mathbf{x}')
\end{aligned}
\tag{2.60}
$$

and also

$$[\hat{\phi}(\mathbf{x},t),\hat{\phi}(\mathbf{x}',t)] = [\hat{\pi}(\mathbf{x},t),\hat{\pi}(\mathbf{x}',t)] = 0. \tag{2.61}$$

For fermions, we can not use the definition Eq.(2.59), which will lead to $\{\hat{\phi},\hat{\pi}\} = 0$. If we define $\hat{\pi} \equiv \frac{i}{\sqrt{2}}(\hat{a}^\dagger + \hat{a}) = i\hat{\phi}^\dagger$, we can have $\{\hat{\phi}(\mathbf{x},t),\hat{\pi}(\mathbf{x}',t)\} = i\delta^3(\mathbf{x} - \mathbf{x}')$. However, $\hat{\phi}$ and $\hat{\pi}$ should be independent. Thus $\hat{\phi}$ should not be a real operator. We can use two real field operators $\hat{\phi}_1$ and $\hat{\phi}_2$ corresponding to a doublet of particles to form a complex field. We define

$$\hat{\phi} = \frac{1}{\sqrt{2}}(\hat{\phi}_1 + i\hat{\phi}_2), \tag{2.62a}$$

$$\hat{\phi}^* = \frac{1}{\sqrt{2}}(\hat{\phi}_1 - i\hat{\phi}_2) \tag{2.62b}$$

with

$$\hat{\phi}_i(\mathbf{x},t) = \frac{1}{\sqrt{2}}(\hat{a}_i^\dagger(\mathbf{x},t) + \hat{a}_i(\mathbf{x},t)). \tag{2.63}$$

Then we have two independent complex field operators and we can treat $\hat{\phi}$ and

$$\hat{\pi} \equiv i\hat{\phi}^\dagger = \frac{1}{\sqrt{2}}(i\hat{\phi}_1^\dagger - \hat{\phi}_2^\dagger). \tag{2.64}$$

as independent field operators. The field operators $\hat{\phi}$ and $\hat{\pi}$ for fermions obey the following commutation relations

$$\{\hat{\phi}(\mathbf{x},t),\hat{\pi}(\mathbf{x}',t)\} = i\delta^3(\mathbf{x} - \mathbf{x}') \tag{2.65}$$

and

$$\{\hat{\phi}(\mathbf{x},t),\hat{\phi}(\mathbf{x}',t)\} = \{\hat{\pi}(\mathbf{x},t),\hat{\pi}(\mathbf{x}',t)\} = 0. \tag{2.66}$$

$\hat{\phi}$ and $\hat{\pi}$ are paired as the canonical field operators. $\hat{\pi}$ is called the conjugate field operator of $\hat{\phi}$. The commutation relations of the canonical field operators $\hat{\phi}$ and $\hat{\pi}$ are called the canonical commutation relations. The canonical commutation relations of the canonical field operators $\hat{\phi}$ and $\hat{\pi}$ for both bosons and fermions can be summarized as follows

$$[\hat{\phi}(\mathbf{x},t),\hat{\pi}(\mathbf{x}',t)]_\pm = i\delta^3(\mathbf{x} - \mathbf{x}') \tag{2.67a}$$

$$[\hat{\phi}(\mathbf{x},t),\hat{\phi}(\mathbf{x}',t)]_\pm = [\hat{\pi}(\mathbf{x},t),\hat{\pi}(\mathbf{x}',t)]_\pm = 0. \tag{2.67b}$$

$\hat{\pi}$ is called as the conjugate field operator of $\hat{\phi}$ because it is equivalent to

$$\hat{\pi} = -i\frac{\delta}{\delta\hat{\phi}}. \tag{2.68}$$

From above relation, we have

$$\langle\phi|\pi\rangle = \frac{1}{(2\pi C)^{\frac{1}{2}}} \exp\left[i\int d^3x\pi(x)\phi(x)\right], \tag{2.69}$$

where $|\phi\rangle$ and $|\pi\rangle$ are the eigenstates of $\hat{\phi}$ and $\hat{\pi}$ defined by

$$\hat{\phi}(\mathbf{x})|\phi\rangle = \phi(\mathbf{x})|\phi\rangle, \tag{2.70a}$$

$$\hat{\pi}(\mathbf{x})|\pi\rangle = \pi(\mathbf{x})|\pi\rangle. \tag{2.70b}$$

C is a factor in the following functional δ-function expression

$$\int d^3x\delta(\phi(\mathbf{x})) = \frac{1}{2\pi C}\int D\pi(\mathbf{x})\exp\left[i\int d^3x\phi(\mathbf{x})\pi(\mathbf{x})\right]. \tag{2.71}$$

We can derive Eq.(2.69) directly from the commutation relations Eqs.(2.60) and (2.65). For bosons, using the commutation relation Eq.(2.60), we have

$$[\hat{\phi}(\mathbf{x}), \hat{\pi}^n(\mathbf{x}')] = in\hat{\pi}^{n-1}(\mathbf{x})\delta^3(\mathbf{x} - \mathbf{x}'), \tag{2.72a}$$

$$[\hat{\pi}(\mathbf{x}), \hat{\phi}^n(\mathbf{x}')] = -in\hat{\phi}^{n-1}(\mathbf{x})\delta^3(\mathbf{x} - \mathbf{x}'). \tag{2.72b}$$

In the derivation of Eq.(2.72), we have used the operator identity

$$[\hat{A}\hat{B}, \hat{C}] = \hat{A}[\hat{B}, \hat{C}] + [\hat{A}, \hat{C}]\hat{B}. \tag{2.73}$$

For fermions, we have a similar operator identity

$$[\hat{A}\hat{B}, \hat{C}] = \hat{A}[\hat{B}, \hat{C}]_+ - [\hat{A}, \hat{C}]_+\hat{B}. \tag{2.74}$$

Using the Taylor expansion, we have

$$\hat{\phi}(\mathbf{x})\exp\left[-i\int d^3x'\phi(\mathbf{x}')\hat{\pi}(\mathbf{x}')\right]|0\rangle_\phi$$

$$= [\hat{\phi}(\mathbf{x}), \exp\left[-i\int d^3x'\phi(\mathbf{x}')\hat{\pi}(\mathbf{x}')\right]]|0\rangle_\phi \tag{2.75}$$

$$= \phi(\mathbf{x})\exp\left[-i\int d^3x'\phi(\mathbf{x}')\hat{\pi}(\mathbf{x}')\right]|0\rangle_\phi.$$

Thus the eigenstate of $\hat{\phi}$ for bosons is given by

$$|\phi\rangle = \exp\left[-i\int d^3x\phi(\mathbf{x})\hat{\pi}(\mathbf{x})\right]|0\rangle_\phi. \tag{2.76}$$

Similarly, we can show that the eigenstate of $\hat{\pi}$ is given by

$$|\pi\rangle = \exp\left[i\int d^3x\pi(\mathbf{x})\hat{\phi}(\mathbf{x})\right]|0\rangle_\pi. \tag{2.77}$$

Then we can calculate $\langle\phi|\pi\rangle$.

$$
\begin{aligned}
\langle\phi|\pi\rangle &= \langle\phi|\exp\left[i\int d^3x\pi(\mathbf{x})\hat{\phi}(\mathbf{x})\right]|0\rangle_\pi \\
&= \exp\left[i\int d^3x\pi(\mathbf{x})\phi(\mathbf{x})\right]\langle\phi|0\rangle_\pi \\
&= \exp\left[i\int d^3x\pi(\mathbf{x})\phi(\mathbf{x})\right]{}_\phi\langle 0|\exp\left[i\int d^3x\phi(\mathbf{x})\hat{\pi}(\mathbf{x})\right]|0\rangle_\pi \\
&= \exp\left[i\int d^3x\pi(\mathbf{x})\phi(\mathbf{x})\right]{}_\phi\langle 0|0\rangle_\pi.
\end{aligned}
\tag{2.78}
$$

${}_\phi\langle 0|0\rangle_\pi$ is just a constant for normalization, which we will take as $1/(2\pi C)^{\frac{1}{2}}$. Thus we get Eq.(2.69). We express the orthonormal relation in terms of the functional δ-function.

$$
\begin{aligned}
\langle\phi'|\phi\rangle &= {}_\phi\langle 0|\exp\left[-i\int d^3x(\phi(\mathbf{x})-\phi'(\mathbf{x}))\hat{\pi}(\mathbf{x})\right]|0\rangle_\phi \\
&= {}_\phi\langle 0|(\phi'-\phi)\rangle_\phi \\
&= \int d^3x\delta(\phi'(\mathbf{x})-\phi(\mathbf{x})),
\end{aligned}
\tag{2.79}
$$

where ${}_\phi\langle 0|0\rangle_\phi$ is normalized as

$$
{}_\phi\langle 0|0\rangle_\phi = \int d^3x\delta(0).
\tag{2.80}
$$

Since

$$
\begin{aligned}
\int D\pi\langle\phi'|\pi\rangle\langle\pi|\phi\rangle &= \int D\pi\frac{1}{2\pi C}\exp\left[i\int d^3x\pi(\mathbf{x})(\phi'(\mathbf{x})-\phi(\mathbf{x}))\right] \\
&= \int d^3x\delta(\phi'(\mathbf{x})-\phi(\mathbf{x})) \\
&= \langle\phi'|\phi\rangle,
\end{aligned}
\tag{2.81}
$$

we have the completeness relation

$$
\int D\pi|\pi\rangle\langle\pi| = 1.
\tag{2.82}
$$

Similarly we have

$$
\int D\phi|\phi\rangle\langle\phi| = 1.
\tag{2.83}
$$

We can obtain the similar results for fermions.

$$
\langle\phi|\pi\rangle = \frac{1}{(2\pi C)^{\frac{1}{2}}}\exp\left[i\int d^3x\pi(x)\phi(x)\right].
\tag{2.84}
$$

Generally the particle could have internal degrees of freedom. The particle number is a scalar. Then $\hat{a}^\dagger(\mathbf{x},t)\hat{a}(\mathbf{x},t)$ should be a scalar. However, the field operator $\hat{\phi}$ can be, for example, vector or spinor, in addition to scalar.

2.2.2   *Generator of time translation transformation*

In order to consider the dynamics of particles, we introduce the operator of time translation transformation

$$\hat{O}(\hat{\pi}, \hat{\phi}) \equiv e^{i\hat{G}_t(\hat{\pi}, \hat{\phi})t}, \tag{2.85}$$

where $\hat{G}_t(\hat{\pi}, \hat{\phi})$ is called the generator of time translation transformation. The field operator $\hat{\phi}(\mathbf{x}, t)$ has a set of time-dependent eigenstates satisfying

$$\hat{\phi}(\mathbf{x}, t)|\phi, t\rangle = \phi(\mathbf{x}, t)|\phi, t\rangle. \tag{2.86}$$

Using the operator of time translation transformation $\hat{O}$, the time dependence of the state vector $|\phi, t\rangle$ can be expressed in terms of the constant state vector $|\phi\rangle$ (also called the *Heisenberg vector*). By definition of the operator of time translation transformation, we have

$$|\phi, t\rangle = e^{i\hat{G}_t t}|\phi, 0\rangle \equiv e^{i\hat{G}_t t}|\phi\rangle. \tag{2.87}$$

Since

$$\hat{\phi}(\mathbf{x}, t)|\phi', t\rangle = e^{i\hat{G}_t t}\hat{\phi}(\mathbf{x}, 0)|\phi', 0\rangle = e^{i\hat{G}_t t}\hat{\phi}(\mathbf{x}, 0)e^{-i\hat{G}_t t}|\phi', t\rangle, \tag{2.88}$$

we have

$$\hat{\phi}(\mathbf{x}, t) = e^{i\hat{G}_t t}\hat{\phi}(\mathbf{x}, 0)e^{-i\hat{G}_t t}. \tag{2.89}$$

Eq.(2.89) is the transformation for a finite translation of time t. Similarly we have

$$\hat{\pi}(\mathbf{x}, t) = e^{i\hat{G}_t t}\hat{\pi}(\mathbf{x}, 0)e^{-i\hat{G}_t t}. \tag{2.90}$$

Eqs.(2.89) and (2.90) are equivalent to the following equations

$$[\hat{\phi}, \hat{G}_t] = i\partial_t\hat{\phi}, \tag{2.91a}$$

$$[\hat{\pi}, \hat{G}_t] = i\partial_t\hat{\pi}. \tag{2.91b}$$

The equations in Eq.(2.91) looks like the *equations of motion*. However, they are just a reformulation of the transformation of time translation. We can prove that the solutions of Eq.(2.91) are Eqs.(2.89) and (2.90) using the general operator identity

$$e^{\hat{A}}\hat{B}e^{-\hat{A}} = \hat{B} + [\hat{A}, \hat{B}] + \frac{1}{2!}[\hat{A}, [\hat{A}, \hat{B}]] + \cdots. \tag{2.92}$$

From the commutation relations for the generator of time translation transformation Eq.(2.91), it can be seen that the right-hand side of Eq.(2.89) is the Taylor expansion of the operator function $\hat{\phi}(\mathbf{x}, t)$ for t.

$$e^{i\hat{G}_t t}\hat{\phi}(\mathbf{x}, t_0)e^{-i\hat{G}_t t}$$

$$= \hat{\phi}(t_0) + [i\hat{G}_t t, \hat{\phi}(t_0)] + \frac{1}{2!}[i\hat{G}_t t, [i\hat{G}_t t, \hat{\phi}(t_0)] + \cdots$$

$$= \hat{\phi}(t_0) + t\frac{\partial}{\partial t}\hat{\phi}\bigg|_{t_0} + t^2\frac{1}{2!}\frac{\partial^2}{\partial t^2}\hat{\phi}\bigg|_{t_0} + \cdots \qquad (2.93)$$

$$= \hat{\phi}(t + t_0),$$

which shows that $\hat{G}_t$ is indeed the generator of the transformation of time translation.

2.2.3 *Transition amplitude*

Using the operator of time translation transformation, we can determine the scalar product of two state vectors taken at different times $\langle\phi', t'|\phi, t\rangle$, which is also called the *transition amplitude* between the two state vectors. Using Eq.(2.87), we have

$$\langle\phi', t'|\phi, t\rangle = \langle\phi'|e^{-i(t'-t)\hat{G}_t}|\phi\rangle. \qquad (2.94)$$

This amplitude is also named as the *Feynman kernel*. This is the amplitude for making a transition from the field configuration $\phi(\mathbf{x})$ at time t to the field configuration $\phi(\mathbf{x}')$ at time t'.

2.2.4 *Causality principle*

Now let us discuss the properties of the generator of time translation transformation $\hat{G}_t$. All the time evolution processes should obey the causality principle, which is the most basic principle of physics. The *causality principle* can be expressed as follows: The future state is only determined by the present state, which is determined by the field operators at the present time t. Therefore, the generator of time translation transformation $\hat{G}_t$ can be expressed solely as a function of the field operators at t without any time derivatives of $\hat{\phi}$ and $\hat{\pi}$ because the time derivatives depend on the quantities in the future. This statement does not mean that one can not have an expression of $\hat{G}_t$ with time derivatives of $\hat{\phi}$ and $\hat{\pi}$ in it. It says that one can find an expression of $\hat{G}_t$ without time derivatives of $\hat{\phi}$ and $\hat{\pi}$ in it. Now we express $\hat{G}_t$ as a functional of the field operators

$$\hat{G}_t = \int d^3x\hat{\mathcal{G}}_t(\hat{\pi}, \hat{\phi}). \qquad (2.95)$$

$\hat{\mathcal{G}}_t(\hat{\pi}, \hat{\phi})$ does not contain any time derivatives of $\hat{\phi}$ and $\hat{\pi}$, while spatial derivatives are allowed.

2.2.5   *Euler-Lagrange equation in operator form*

In the following, we will derive the first equation of motion in this book, i.e. the Euler-Lagrange equation in operator form, using the canonical commutation relations of the canonical field operators $\hat{\phi}$ and $\hat{\pi}$ and the causality principle. The Euler-Lagrange equation in operator form can be considered as the source equation of all the equations of motion. We have derived the canonical commutation relations of the canonical field operators $\hat{\phi}$ and $\hat{\pi}$ in Eq.(2.67). The commutation relations Eq.(2.67) are the commutation relations for the operators at the same time. The dynamic information are contained in the commutation relations for the operators at different times. Therefore, we need to derive the equations of motion to obtain the dynamic information of a system.

We introduce an operator $\hat{\mathcal{L}}$ defined as

$$\hat{\mathcal{L}} \equiv \hat{\pi} \partial_t \hat{\phi} - \hat{\mathcal{G}}_t(\hat{\pi}, \hat{\phi}) \tag{2.96}$$

$\hat{\mathcal{L}}$ is called the Lagrangian density operator. From the Lagrangian density operator $\hat{\mathcal{L}}$, we can define the Lagrangian operator

$$\hat{L} \equiv \int d^3x \hat{\mathcal{L}} \tag{2.97}$$

and the action operator as

$$\hat{S} \equiv \int dt \hat{L} = \int d^4x \hat{\mathcal{L}}. \tag{2.98}$$

Since $\hat{\mathcal{G}}_t(\hat{\pi}, \hat{\phi})$ does not contain the time derivatives of $\hat{\phi}$ and $\hat{\pi}$, we do not need to deal with $[\hat{\phi}, \dot{\hat{\phi}}]$, $[\hat{\phi}, \dot{\hat{\pi}}]$, $[\hat{\pi}, \dot{\hat{\phi}}]$ and $[\hat{\pi}, \dot{\hat{\pi}}]$ in $[\hat{\phi}, \hat{G}_t(\hat{\pi}, \hat{\phi})]$ and $[\hat{\pi}, \hat{G}_t(\hat{\pi}, \hat{\phi})]$ because $[\hat{\phi}, \hat{G}_t(\hat{\pi}, \hat{\phi})]$ and $[\hat{\pi}, \hat{G}_t(\hat{\pi}, \hat{\phi})]$ contain none of them. For bosons, using Eq.(2.72), we have

$$[\hat{\phi}, \hat{G}_t(\hat{\pi}, \hat{\phi})] = i\frac{\delta \hat{G}_t(\hat{\pi}, \hat{\phi})}{\delta \hat{\pi}}, \tag{2.99a}$$

$$[\hat{\pi}, \hat{G}_t(\hat{\pi}, \hat{\phi})] = -i\frac{\delta \hat{G}_t(\hat{\pi}, \hat{\phi})}{\delta \hat{\phi}}. \tag{2.99b}$$

Similarly, one can obtain Eq.(2.99) for the fermion fields. Comparing Eq.(2.91) with Eq.(2.99), we have

$$\partial_t \hat{\phi} = \frac{\delta \hat{G}_t(\hat{\pi}, \hat{\phi})}{\delta \hat{\pi}}, \tag{2.100a}$$

$$\partial_t \hat{\pi} = -\frac{\delta \hat{G}_t(\hat{\pi}, \hat{\phi})}{\delta \hat{\phi}}. \tag{2.100b}$$

The total functional differential of the Lagrangian operator $\hat{L} \equiv \int d^3x \hat{\mathcal{L}}$ as an operator functional of $\hat{\phi}$ and $\dot{\hat{\phi}}$ is

$$\delta \hat{L} = \int d^3x \left(\frac{\delta \hat{L}}{\delta \hat{\phi}} \delta \hat{\phi} + \frac{\delta \hat{L}}{\delta \dot{\hat{\phi}}} \delta \dot{\hat{\phi}} \right), \tag{2.101}$$

Using Eq.(2.96), we have

$$\delta \hat{L} = \int d^3x [\hat{\pi} \delta \dot{\hat{\phi}} + (\delta \hat{\pi}) \dot{\hat{\phi}}] - \delta \hat{G}_t$$

$$= \int d^3x \left[\hat{\pi} \delta \dot{\hat{\phi}} + (\delta \hat{\pi}) \dot{\hat{\phi}} - \frac{\delta \hat{G}_t}{\delta \hat{\phi}} \delta \hat{\phi} - (\delta \hat{\pi}) \frac{\delta \hat{G}_t}{\delta \hat{\pi}} \right]. \tag{2.102}$$

We have used the expression of $\hat{G}_t$ with $\hat{\pi}$ being ordered in front of $\hat{\phi}$, which can always be realized using the commutation relations. Inserting Eq.(2.100), we obtain

$$\delta \hat{L} = \int d^3x [\hat{\pi} \delta \dot{\hat{\phi}} + (\delta \hat{\pi}) \dot{\hat{\phi}} + \dot{\hat{\pi}} \delta \hat{\phi} - (\delta \hat{\pi}) \dot{\hat{\phi}}]$$

$$= \int d^3x (\hat{\pi} \delta \dot{\hat{\phi}} + \dot{\hat{\pi}} \delta \hat{\phi}). \tag{2.103}$$

Comparing Eq.(2.101) with Eq.(2.103), we find

$$\dot{\hat{\pi}} = \frac{\delta \hat{L}}{\delta \hat{\phi}}, \quad \hat{\pi} = \frac{\delta \hat{L}}{\delta \dot{\hat{\phi}}}, \tag{2.104}$$

which is equivalent to the following equation

$$\frac{d}{dt} \left(\frac{\delta \hat{L}}{\delta \dot{\hat{\phi}}} \right) - \frac{\delta \hat{L}}{\delta \hat{\phi}} = 0. \tag{2.105}$$

Eq.(2.105) is called the *Euler-Lagrange equation* in operator form. We can reform Eq.(2.105) into the covariant form. The functional differential of $\hat{L}$ can be written as

$$\delta \hat{L} = \int d^3x \left(\frac{\partial \hat{\mathcal{L}}}{\partial \hat{\phi}} \delta \hat{\phi} + \frac{\partial \hat{\mathcal{L}}}{\partial \nabla \hat{\phi}} \delta \nabla \hat{\phi} + \frac{\partial \hat{\mathcal{L}}}{\partial \dot{\hat{\phi}}} \delta \dot{\hat{\phi}} \right)$$

$$= \int d^3x \left[\left(\frac{\partial \hat{\mathcal{L}}}{\partial \hat{\phi}} - \nabla \frac{\partial \hat{\mathcal{L}}}{\partial \nabla \hat{\phi}} \right) \delta \hat{\phi} + \frac{\partial \hat{\mathcal{L}}}{\partial \dot{\hat{\phi}}} \delta \dot{\hat{\phi}} \right]. \tag{2.106}$$

In the derivation, we have used an integration by parts. Comparing Eq.(2.106) with Eq.(2.101), we obtain

$$\frac{\delta \hat{L}}{\delta \hat{\phi}} = \frac{\partial \hat{\mathcal{L}}}{\partial \hat{\phi}} - \nabla \frac{\partial \hat{\mathcal{L}}}{\partial \nabla \hat{\phi}}, \tag{2.107a}$$

$$\frac{\delta \hat{L}}{\delta \dot{\hat{\phi}}} = \frac{\partial \hat{\mathcal{L}}}{\partial \dot{\hat{\phi}}}. \tag{2.107b}$$

Expressed in terms of the Lagrangian density, the Euler-Lagrange equation Eq.(2.105) in operator form becomes

$$\frac{\partial \hat{\mathcal{L}}}{\partial \hat{\phi}} - \nabla \frac{\partial \hat{\mathcal{L}}}{\partial \nabla \hat{\phi}} - \frac{\partial}{\partial t} \frac{\partial \hat{\mathcal{L}}}{\partial \dot{\hat{\phi}}} = 0 \tag{2.108}$$

or in covariant form

$$\frac{\partial \hat{\mathcal{L}}}{\partial \hat{\phi}} - \frac{\partial}{\partial x^{\mu}} \frac{\partial \hat{\mathcal{L}}}{\partial (\partial_{\mu} \hat{\phi})} = 0. \tag{2.109}$$

The Euler-Lagrange equation Eq.(2.109) is general. In the derivation, we have not assumed the form of Lagrangian. The Lagrangian density operator and the time translation operator can be any form. The only condition is that the operator of time translation transformation $\hat{G}_t(\hat{\pi}, \hat{\phi})$ does not contain the time derivatives of field operators, which is required by the causality principle.

2.2.6 *Path integral formulas*

We can construct the path integral formulas to calculate the transition amplitude. We divide the time interval (t, t') into many small slices with equal length.

$$t_n = t + n\epsilon \tag{2.110}$$

with

$$\epsilon = \frac{t' - t}{N}. \tag{2.111}$$

We insert a complete set of basis states $|\phi, t\rangle$ at each of the grid points $t_n (n = 1, \cdots, N-1)$ in the Feynman kernel.

$$\langle \phi', t' | \phi, t \rangle = \int D\phi_{N-1} \cdots \int D\phi_2 \int D\phi_1$$
$$\langle \phi', t' | \phi_{N-1}, t_{N-1} \rangle \cdots \langle \phi_2, t_2 | \phi_1, t_1 \rangle \langle \phi_1, t_1 | \phi, t \rangle. \tag{2.112}$$

Using Eq.(2.94), each of the kernel elements under the integral can be rewritten as

$$\langle \phi_{n+1}, t_{n+1} | \phi_n, t_n \rangle = \langle \phi_{n+1} | e^{-i\hat{G}_t(\hat{\pi}, \hat{\phi})\epsilon} | \phi_n \rangle. \tag{2.113}$$

When ϵ is small, the time evolution operator can be approximated by a Taylor expansion

$$\langle \phi_{n+1}, t_{n+1} | \phi_n, t_n \rangle = \langle \phi_{n+1} | [1 - i\hat{G}_t(\hat{\pi}, \hat{\phi})\epsilon] | \phi_n \rangle + \mathcal{O}(\epsilon^2). \tag{2.114}$$

Since the generator $\hat{G}_t$ depends on $\hat{\pi}$ and $\hat{\phi}$, we also insert a complete set of state $|\pi_n\rangle$. Using the completeness relation Eq.(2.82), we have

$$\langle \phi_{n+1} | \hat{G}_t(\hat{\pi}, \hat{\phi}) | \phi_n \rangle = \int D\pi_n \langle \phi_{n+1} | \pi_n \rangle \langle \pi_n | \hat{G}_t(\hat{\pi}, \hat{\phi}) | \phi_n \rangle. \tag{2.115}$$

Due to the causality principle, $\hat{G}_t$ contains only the operators $\hat{\pi}$ and $\hat{\phi}$. The operators $\hat{\pi}$ and $\hat{\phi}$ can act to the left or to the right on their eigenstates. We have

$$\langle \pi_n | \hat{G}_t(\hat{\pi}, \hat{\phi}) | \phi_n \rangle = \langle \pi_n | \phi_n \rangle G_t(\pi_n, \phi_n). \tag{2.116}$$

One might use a more symmetric prescription, so-called Weyl's operator ordering. $\langle \pi_n | \phi_n \rangle G_t(\pi_n, \phi_n)$ in Eq.(2.116) can be replaced by $\langle \pi_n | \phi_n \rangle G_t(\pi_n, \frac{1}{2}(\phi_{n+1} + \phi_n))$. We will use the notation $G_t(\pi_n, \bar{\phi}_n)$ in the following so that we can choose $\bar{\phi}_n = \phi_n$ or $\bar{\phi}_n = \frac{1}{2}(\phi_{n+1} + \phi_n)$ for the convenience of usage.

Using Eq.(2.69), we have

$$\langle \phi_{n+1}, t_{n+1} | \phi_n, t_n \rangle$$
$$= \int \frac{D\pi_n}{2\pi C} \exp\left[i \int d^3x \pi_n(x)(\phi_{n+1}(x) - \phi_n(x))\right] \tag{2.117}$$
$$\times [1 - iG_t(\pi_n, \bar{\phi}_n)\epsilon] + \mathcal{O}(\epsilon^2).$$

Taking the limit $\epsilon \to 0$ or $N \to \infty$, we have

$$\langle \phi', t' | \phi, t \rangle = \lim_{N \to \infty} \int \prod_{n=1}^{N-1} D\phi_n \prod_{n=0}^{N-1} \frac{D\pi_n}{2\pi C}$$
$$\exp\left[\sum_{n=0}^{N-1} i \int d^3x \epsilon \pi_n(x) \frac{\phi_{n+1}(x) - \phi_n(x)}{\epsilon}\right] \tag{2.118}$$
$$\prod_{n=0}^{N-1} [1 - iG_t(\pi_n, \bar{\phi}_n)\epsilon].$$

We can reform Eq.(2.118) by using the representation of the exponential function

$$\lim_{N\to\infty} \prod_{n=0}^{N-1}\left(1 + \frac{x_n}{N}\right) = \exp\left(\lim_{N\to\infty}\frac{1}{N}\sum_{n=0}^{N-1}x_n\right). \tag{2.119}$$

Then Eq.(2.118) becomes

$$\langle\phi', t'|\phi, t\rangle = \lim_{N\to\infty}\int\prod_{n=1}^{N-1}D\phi_n\prod_{n=0}^{N-1}\frac{D\pi_n}{2\pi C}$$
$$\exp\left[i\epsilon\sum_{n=0}^{N-1}\left(\int d^3x\pi_n(x)\frac{\phi_{n+1}(x) - \phi_n(x)}{\epsilon} - G_t(\pi_n, \bar{\phi}_n)\right)\right]. \tag{2.120}$$

In the limit $N \to \infty$, the sample values become continues. The summation is then replaced by the integral. We introduce the notation of path integral

$$\int\prod_{n=1}^{N-1}D\phi_n \to \int D\phi \quad\text{and}\quad \int\prod_{n=0}^{N-1}D\phi_n \to \int D\pi. \tag{2.121}$$

In the limit $\epsilon \to 0$,

$$\frac{\phi_{n+1}(x) - \phi_n(x)}{\epsilon} \to \dot{\phi}(t_n) \quad\text{and}\quad \epsilon\sum_{n=0}^{N-1}f(t_n) \to \int_t^{t'}d\tau f(\tau). \tag{2.122}$$

Then we obtain the path integral expression for the Feynman kernel (the transition amplitude) in Eq.(2.94).

$$Z \equiv \langle\phi', t'|\phi, t\rangle$$
$$= \mathcal{N}\int D\phi\int D\pi\exp\left[i\int_t^{t'}d\tau\int d^3x(\pi\partial_t\phi - \mathcal{G}_t(\pi, \phi))\right] \tag{2.123}$$

with the boundary condition

$$\phi(\mathbf{x}, t') = \phi'(\mathbf{x}), \tag{2.124a}$$
$$\phi(\mathbf{x}, t) = \phi(\mathbf{x}), \tag{2.124b}$$

where $\mathcal{N}$ is a constant factor, which is generally omitted for the simplicity of expression.

It should be noted that we need to use Grassmann algebra (a brief introduction on Grassmann algebra is shown in Appendix D) in the path integration for fermions.

2.2.7 Lagrangian and action

Using the Lagrangian density $\mathcal{L}$ defined by

$$\mathcal{L} = \pi \partial_t \phi - \mathcal{G}_t(\pi, \phi) \tag{2.125}$$

and the action S

$$S = \int d^4 x \mathcal{L}, \tag{2.126}$$

Eq.(2.123) becomes

$$Z = \int D\phi \int D\pi \exp\left[i \int d^4 x \mathcal{L}(\pi, \phi)\right] = \int D\phi \int D\pi \exp\left(iS\right). \tag{2.127}$$

From Eq.(2.123), after integrating over $\int D\pi$, we have

$$\langle \phi', t' | \phi, t \rangle = \mathcal{N} \int D\phi \int D\pi \exp\left[i \int d^4 x (\pi \partial_t \phi - \mathcal{G}_t(\pi, \phi)\right]$$

$$\equiv \mathcal{N}' \int D\phi \exp\left[i \int d^4 x \mathcal{L}'(\phi, \dot{\phi})\right]. \tag{2.128}$$

Instead of $\mathcal{L}(\pi, \phi)$, we have the function $\mathcal{L}'(\phi, \dot{\phi})$ as Lagrangian density. Using the Lagrangian density $\mathcal{L}'(\phi, \dot{\phi})$, we can define the action S' of the field by

$$S' \equiv \int d^4 x \mathcal{L}'(\phi, \dot{\phi}). \tag{2.129}$$

The Lagrangian density $\mathcal{L}'(\phi, \dot{\phi})$ is usually used for bosons to get a covariant Lagrangian density. Thus we have two types of formulas for the Lagrangian density of bosons. We will show that the Lagrangian density $\mathcal{L}'(\phi, \dot{\phi})$ is equal to $\mathcal{L}(\pi, \phi)$ for bosons and the two definitions are equivalent.

2.2.8 Covariance principle

In the following, we assume that the path integral should satisfy the principle of general covariance stating that the physics, as embodied in the path integral, must be invariant under an arbitrary coordinate transformation. Generally, we shall consider any curved spacetime. First we discuss the flat spacetime, which is applicable to the case of vacuum state. For a Riemann metric, we can always find a local Minkowski metric. We will show in later section that when the field is weak, as in the case of near vacuum state, we can use Minkowski metric. In order to satisfy the causality principle, time can only be one-dimensional. We have assumed that space is three-dimensional. There are several reasons for a three dimensional space as we will see in later sections. At present stage, we assume that the space is three-dimensional. Matter, space, and time should be considered as an integrated entity, as Einstein proposed. If time and space are independent, the interaction between particles will be instantaneous, which is not consistent in concept with the causality principle. Because of the causality principle, a flat spacetime can only be Minkowski type. An Euclidean type spacetime will not be consistent with the causality principle because it extends time into four-dimensional. The Lagrangian density $\mathcal{L}$ or $\mathcal{L}'$ should be scalar in the Minkowski metric. We use a Minkowski metric $\eta^{\mu\nu}$ with signature $[+, -, -, -]$ in this chapter. Only a few forms of Lagrangian densities are found to satisfy both the causality principle and the covariance principle. Because $\mathcal{G}_t(\pi, \phi)$ depends only on time locally, it does not depend on the time derivatives of field functions. It can depend on the spatial derivatives of field functions. As we have shown, there are two cases: 1) $\mathcal{L}$ is Lorentz covariant; 2) $\mathcal{L}'$ is Lorentz covariant. For the first case, from Eq.(2.123), we can see that $\mathcal{L}$ depends on $\dot{\phi}$ linearly. In order to get a covariant Lagrangian density $\mathcal{L}$, $\mathcal{G}_t(\pi, \phi)$ should depend on the spatial derivatives linearly. We will show that this case corresponds to the spinor fermion field in the later section. For the second case, we need to carry out the integration over field function π. When $\mathcal{G}_t(\pi, \phi)$ is a quadratic function of π, we can get a $\dot{\phi}^2$ term in $\mathcal{L}'(\phi, \dot{\phi})$ after completing the Gaussian integration over π in the path integral in Eq.(2.128). The $\dot{\phi}^2$ term can match with other spatial derivative terms to form a covariant Lagrangian density. Thus $\mathcal{G}_t(\pi, \phi)$ should also contain the quadratic spatial derivatives of field functions. After integrating out the field function π, we obtain the Lorentz-covariant Lagrangian $\mathcal{L}'(\phi, \dot{\phi})$ in the Minkowski spacetime. We will show that one can get two types of covariant Lagrangians in this way. They correspond to the scalar and vector bosons. For $\mathcal{G}_t(\pi, \phi)$ with other orders of spatial derivative of field functions or power functions of π , we can not find any covariant constructions of Lagrangian. Although this is not a strict proof, it is plausible that there are no other types of $\mathcal{G}_t(\pi, \phi)$ that can lead to covariant Lagrangian $\mathcal{L}$ or $\mathcal{L}'$. In addition, we will show later that the energy is conserved due to the homogeneity of spacetime. Then the Hamiltonian operator should commute with the generator of time translation transformation, which also excludes other possibility. From Eq.(2.125), we can see that there is only first order derivative $\dot{\phi}$ in the Lagrangian $\mathcal{L}$. Therefore, Lagrangian can only depend on the first order derivative $\dot{\phi}$. In the $\mathcal{L}'(\phi, \dot{\phi})$, there is only $\dot{\phi}^2$ term. $\dot{\phi}^2$ may be transformed into $\ddot{\phi}$ through an integration by parts. Therefore, Lagrangian can only contain $\dot{\phi}$ or $\dot{\phi}^2$ (or equivalently

$\ddot{\phi}$) terms linearly. This constrains the form of covariant Lagrangians stiffly and leads to very limited forms of the covariant Lagrangians.

Inserting all the possible forms of the covariant Lagrangian into the Euler-Lagrange equation, we obtain the equations of motion for various types of particles. Therefore, we do not need to postulate the equations of motion in the basic principles.

2.3 Scalar field

The Lagrangian should be a scalar in the Minkowski spacetime due to the covariance principle. Since the simplest field is the scalar field, we first consider the scalar field. It should be noted that the underlining principle is independent of the types of the fields contained in the Lagrangian.

2.3.1 *Lagrangian*

We use the second definition of the Lagrangian density. Since we will show $\mathcal{L}' = \mathcal{L}$, we will not use $'$ to distinguish them. As we have discussed in the previous section, the derivatives in $\mathcal{L}$ can only be quadratic. Thus the general form of the covariant Lagrangian density for a scalar field is given by

$$\mathcal{L} = \frac{1}{2}f(\phi)\partial_\mu\phi\partial^\mu\phi - U(\phi). \qquad (2.130)$$

In Eq.(2.130), the differential operators ∂_μ and ∂^μ are defined as

$$\partial_\mu = \frac{\partial}{\partial x^\mu} = (\partial_0, \partial_1, \partial_2, \partial_3) = (\partial_t, \partial_1, \partial_2, \partial_3) = (\frac{\partial}{\partial t}, \boldsymbol{\nabla}), \qquad (2.131\text{a})$$

$$\partial^\mu = \frac{\partial}{\partial x_\mu} = g^{\mu\nu}\partial_\nu. \qquad (2.131\text{b})$$

$f(\phi)$ can be put into the metric $g_{\mu\nu}$ when we use the curved spacetime formalism, which we will discuss in detail in the section on the curved spacetime. $U(\phi)$ is generally divided into the mass term $\frac{1}{2}m^2\phi^2$ and interaction term $V(\phi)$.

$$U(\phi) = \frac{1}{2}m^2\phi^2 + V(\phi), \qquad (2.132)$$

where m is called the *mass* and $V(\phi)$ is the *self-interaction*. Thus the general form of Lagrangian density in the Minkowski spacetime for a scalar field is given by

$$\mathcal{L} = \frac{1}{2}\partial_\mu\phi\partial^\mu\phi - \frac{1}{2}m^2\phi^2 - V(\phi). \qquad (2.133)$$

We have chosen the proper unit of field function such that the first term in Eq.(2.133) has the form without any parameter. We can also put a constant factor $\hbar^2$ in the first term and reformulate the first term as $\frac{\hbar^2}{2}\partial_\mu\phi\partial^\mu\phi$ to make the unit transformation easier, where $\hbar$ is called the *Planck constant*. All the terms in Eq.(2.133) are scalars in the spacetime. Thus the Lagrangian density in Eq.(2.133) is Lorentz covariant. The corresponding function $\mathcal{G}_t(\pi,\phi)$ relating to the Lagrangian density in Eq.(2.133) is given by

$$\mathcal{G}_t = \frac{1}{2}\pi^2 + \frac{1}{2}(\nabla\phi)^2 + \frac{1}{2}m^2\phi^2 + V(\phi), \qquad (2.134)$$

which does not contain the time derivative terms. We can get the Lagrangian density in Eq.(2.133) by inserting Eq.(2.134) into Eq.(2.128) and integrating over π using the Gaussian integral formula Eq.(C.21) in Appendix C. Thus the generator of time translation transformation corresponding to the Lagrangian density in Eq.(2.133) is given by

$$\hat{G}_t = \int d^3x \left[\frac{1}{2}\hat{\pi}^2 + \frac{1}{2}(\nabla\hat{\phi})^2 + \frac{1}{2}m^2\hat{\phi}^2 + V(\hat{\phi})\right]. \qquad (2.135)$$

2.3.2 Klein-Gordon equation

Now we consider the scalar field as boson field so that $\hat{\phi}$ and $\hat{\pi}$ satisfy the commutation relations for bosons in Eqs.(2.60) and (2.61). We can not construct a consistent formulation for scalar fermions with anti-commutation relations. If we use the anti-commutation relations for $\hat{\phi}$ and $\hat{\pi}$, we get $[\hat{\phi}(\mathbf{x},t),\hat{G}_t(\hat{\pi},\hat{\phi})] = 0$. $\hat{G}_t$ given by Eq.(2.135) can not be the generator of time translation transformation in this case. Therefore, the Lagrangian density Eq.(2.133) can only be used to describe the scalar bosons.[1]

Calculating the commutator $[\hat{\phi}(\mathbf{x},t),\hat{G}_t(\hat{\pi},\hat{\phi})]$ of $\hat{\phi}(\mathbf{x},t)$ with $\hat{G}_t$, we have

$$[\hat{\phi}(\mathbf{x},t),\hat{G}_t(\hat{\pi},\hat{\phi})] = i\hat{\pi}. \qquad (2.136)$$

Comparing Eq.(2.136) with Eq.(2.91a), we can see that

$$\partial_0\hat{\phi} = \hat{\pi} = -i[\hat{\phi}(\mathbf{x},t),\hat{G}_t(\hat{\pi},\hat{\phi})]. \qquad (2.137)$$

---

[1]One can also use microcausality to prove that only boson field can be used in the Lagrangian density Eq.(2.133) (W. Pauli, Phys. Rev. **58**, 716 (1940); M. Fierz, Helv. Phys. Acta **12**, 3 (1939).). Here we consider microcausality is a result of causality principle. One can prove that the microcausality is satisfied by the scalar boson field with the Lagrangian density Eq.(2.133).

Using commutation relations Eq.(2.91b), we have

$$\dot{\hat{\pi}} = -i[\hat{\pi}(\mathbf{x},t), \hat{G}_t(\hat{\pi}, \hat{\phi})] = (\nabla^2 - m^2)\hat{\phi}(\mathbf{x},t) - V'(\hat{\phi}). \qquad (2.138)$$

In deriving Eq.(2.138), we have used the relation

$$[\hat{\pi}(\mathbf{x},t), \nabla'\hat{\phi}(\mathbf{x}',t)] = \nabla'[\hat{\pi}(\mathbf{x},t), \hat{\phi}(\mathbf{x}',t)] = -i\nabla'\delta^3(\mathbf{x} - \mathbf{x}') \quad (2.139)$$

and also an integration by parts. Neglecting the interaction term and combining Eqs.(2.137) and (2.138), we find that the field operator for free scalar bosons satisfies the following equation

$$\ddot{\hat{\phi}}(\mathbf{x},t) = (\nabla^2 - m^2)\hat{\phi}(\mathbf{x},t). \qquad (2.140)$$

Eq.(2.140) is called the *Klein-Gordon equation*. The Klein-Gordon equation Eq.(2.140) can be expressed in the following covariant form

$$\partial_\mu \partial^\mu \hat{\phi} + m^2 \hat{\phi} = \Box \hat{\phi} + m^2 \hat{\phi} = 0. \qquad (2.141)$$

The derivation of the Klein-Gordon equation is based only on the causality principle and the covariance principle.

Using $\partial_0 \hat{\phi} = \hat{\pi}$, Eqs.(2.133) and (2.134), we have $\mathcal{L} = \pi \partial_t \phi - \mathcal{G}_t(\pi, \phi)$. Thus the two definitions of the Lagrangian density given by Eqs.(2.125) and (2.128) are equivalent.

We can also derive the Klein-Gordon equation directly using the Euler-Lagrange equation in operator form Eq.(2.109). We introduce the Lagrangian density operator by replacing the field function ϕ with the field operator $\hat{\phi}$.

$$\hat{\mathcal{L}} = \frac{1}{2}\partial_\mu \hat{\phi} \partial^\mu \hat{\phi} - \frac{1}{2}m^2 \hat{\phi}^2 - V(\hat{\phi}). \qquad (2.142)$$

We can also introduce the action operator $\hat{S}$ by replacing the field function ϕ with the field operator $\hat{\phi}$.

Since

$$\frac{\partial \hat{\mathcal{L}}}{\partial \hat{\phi}} = -m^2 \hat{\phi} - V'(\hat{\phi}) \qquad (2.143)$$

and

$$\frac{\partial \hat{\mathcal{L}}}{\partial(\partial_\mu \hat{\phi})} = \partial^\mu \hat{\phi}, \qquad (2.144)$$

we have

$$\frac{\partial \hat{\mathcal{L}}}{\partial \hat{\phi}} - \frac{\partial}{\partial x^\mu}\left[\frac{\partial \hat{\mathcal{L}}}{\partial(\partial_\mu \hat{\phi})}\right] = -(\partial_\mu \partial^\mu \hat{\phi} + m^2 \hat{\phi} + V'(\hat{\phi})) = 0, \qquad (2.145)$$

which gives the Klein-Gordon equation. The Euler-Lagrange equation in operator form is the general form of the equation of motion. It leads to various equations of motion for different types of particles when we use the corresponding forms of the Lagrangian densities of particles.

We also call the Euler-Lagrange equation in operator form as quantum Euler-Lagrange equation. In contrast, when the field function is used, the corresponding Euler-Lagrange equation

$$\frac{\partial \mathcal{L}}{\partial \phi} - \frac{\partial}{\partial x^\mu}\left[\frac{\partial \mathcal{L}}{\partial(\partial_\mu \phi)}\right] = 0 \qquad (2.146)$$

is called the classical Euler-Lagrange equation or simply Euler-Lagrange equation.

2.3.3   *Solutions of the Klein-Gordon equation*

Eq.(2.140) is a wave equation. Thus we have the particle-wave duality for the scalar bosons. We can solve the operator equation (2.140) by expanding $\hat{\phi}(\mathbf{x}, t)$ with respect to a basis. We usually use the set of plane waves

$$u_\mathbf{p}(\mathbf{x}) = N_\mathbf{p} e^{i\mathbf{p}\cdot\mathbf{x}} \qquad (2.147)$$

for solving the wave equations, where $N_\mathbf{p}$ is the normalization constant. Then we have

$$\hat{\phi}(\mathbf{x}, t) = \int d^3p\, u_\mathbf{p}(\mathbf{x})\hat{a}_\mathbf{p}(t) = \int d^3p\, N_\mathbf{p} e^{i\mathbf{p}\cdot\mathbf{x}}\hat{a}_\mathbf{p}(t). \qquad (2.148)$$

Inserting Eq.(2.148) into Eq.(2.140), we get the equation of motion for the operators $\hat{a}_\mathbf{p}(t)$:

$$\ddot{\hat{a}}_\mathbf{p}(t) = -(\mathbf{p}^2 + m^2)\hat{a}_\mathbf{p}(t). \qquad (2.149)$$

The solution of Eq.(2.149) is given by

$$\hat{a}_\mathbf{p}(t) = \hat{a}_\mathbf{p}^{(1)} e^{-i\omega_\mathbf{p} t} + \hat{a}_\mathbf{p}^{(2)} e^{i\omega_\mathbf{p} t}, \qquad (2.150)$$

where $\hat{a}_\mathbf{p}^{(1)}$ and $\hat{a}_\mathbf{p}^{(2)}$ are the constant operators in time. $\omega_\mathbf{p}$ is given by the dispersion relation

$$\omega_\mathbf{p} = \sqrt{\mathbf{p}^2 + m^2}. \qquad (2.151)$$

According to Eq.(2.59a), the field operator $\hat{\phi}$ is hermitian, $\hat{\phi}^\dagger = \hat{\phi}$. The constraint gives

$$\hat{a}_\mathbf{p}^{(1)\dagger} = \hat{a}_{-\mathbf{p}}^{(2)}. \qquad (2.152)$$

Then the basis expansion Eq.(2.148) becomes

$$\hat{\phi}(\mathbf{x}, t) = \int d^3 p N_{\mathbf{p}} [\hat{a}_{\mathbf{p}}^{(1)} e^{i(\mathbf{p} \cdot \mathbf{x} - \omega_{\mathbf{p}} t)} + \hat{a}_{\mathbf{p}}^{(1)\dagger} e^{-i(\mathbf{p} \cdot \mathbf{x} - \omega_{\mathbf{p}} t)}]. \quad (2.153)$$

Denoting $\hat{a}_{\mathbf{p}}^{(1)}$ simply by $\hat{a}_{\mathbf{p}}$, we have

$$\hat{\phi}(\mathbf{x}, t) = \int d^3 p N_{\mathbf{p}} [\hat{a}_{\mathbf{p}} e^{i(\mathbf{p} \cdot \mathbf{x} - \omega_{\mathbf{p}} t)} + \hat{a}_{\mathbf{p}}^{\dagger} e^{-i(\mathbf{p} \cdot \mathbf{x} - \omega_{\mathbf{p}} t)}]. \quad (2.154)$$

Because $\hat{\pi} = \dot{\hat{\phi}}$, the basis expansion of the conjugate field is given by

$$\hat{\pi}(\mathbf{x}, t) = -i \int d^3 p N_{\mathbf{p}} \omega_{\mathbf{p}} [\hat{a}_{\mathbf{p}} e^{i(\mathbf{p} \cdot \mathbf{x} - \omega_{\mathbf{p}} t)} - \hat{a}_{\mathbf{p}}^{\dagger} e^{-i(\mathbf{p} \cdot \mathbf{x} - \omega_{\mathbf{p}} t)}], \quad (2.155)$$

which is consistent with Eq.(2.136).

2.3.4 *Commutators for creation and annihilation operators in p-space*

The operators $\hat{a}_{\mathbf{p}}$ and $\hat{a}_{\mathbf{p}}^{\dagger}$ can be shown to fulfill the commutators for the creation and annihilation operators, i.e.

$$[\hat{a}_{\mathbf{p}}, \hat{a}_{\mathbf{p}'}^{\dagger}] = \delta^3(\mathbf{p} - \mathbf{p}') \quad (2.156)$$

and

$$[\hat{a}_{\mathbf{p}}, \hat{a}_{\mathbf{p}'}] = [\hat{a}_{\mathbf{p}}^{\dagger}, \hat{a}_{\mathbf{p}'}^{\dagger}] = 0. \quad (2.157)$$

The commutation relations Eqs.(2.156) and (2.157) for $\hat{a}_{\mathbf{p}}$ and $\hat{a}_{\mathbf{p}}^{\dagger}$ in the expansion of the field operators $\hat{\phi}$ and $\hat{\pi}$ can be derived as follows. We introduce $p = (\omega_{\mathbf{p}}, \mathbf{p})$ and define the normalized plane waves as

$$u_{\mathbf{p}}(\mathbf{x}, t) = N_{\mathbf{p}} e^{-ip \cdot x} \equiv \frac{1}{\sqrt{2\omega_{\mathbf{p}}(2\pi)^3}} e^{-i(\omega_{\mathbf{p}} t - \mathbf{p} \cdot \mathbf{x})}, \quad (2.158)$$

where we have used the normalization factor

$$N_{\mathbf{p}} = \frac{1}{\sqrt{2\omega_{\mathbf{p}}(2\pi)^3}}. \quad (2.159)$$

Then Eqs.(2.154) and (2.155) become

$$\hat{\phi}(\mathbf{x}, t) = \int d^3 p [\hat{a}_{\mathbf{p}} u_{\mathbf{p}}(\mathbf{x}, t) + \hat{a}_{\mathbf{p}}^{\dagger} u_{\mathbf{p}}^*(\mathbf{x}, t)], \quad (2.160a)$$

$$\hat{\pi}(\mathbf{x}, t) = -i \int d^3 p \, \omega_{\mathbf{p}} [\hat{a}_{\mathbf{p}} u_{\mathbf{p}}(\mathbf{x}, t) - \hat{a}_{\mathbf{p}}^{\dagger} u_{\mathbf{p}}^*(\mathbf{x}, t)]. \quad (2.160b)$$

Projection of the field operator $\hat{\phi}(\mathbf{x}, t)$ on $u_{\mathbf{p}}$ and $u_{\mathbf{p}}^*$ gives

$$\hat{a}_{\mathbf{p}} = i \int d^3 x u_{\mathbf{p}}^*(\mathbf{x}, t) \overleftrightarrow{\partial_0} \hat{\phi}(\mathbf{x}, t) \equiv (u_{\mathbf{p}}, \hat{\phi}) \quad (2.161)$$

and

$$\hat{a}_{\mathbf{p}}^\dagger = -i \int d^3 x u_{\mathbf{p}}(\mathbf{x}, t) \overset{\leftrightarrow}{\partial_0} \hat{\phi}(\mathbf{x}, t) \equiv -(u_{\mathbf{p}}^*, \hat{\phi}). \tag{2.162}$$

We have defined the scalar product of two Klein-Gordon wave functions ϕ_1 and ϕ_2 as

$$(\phi_1, \phi_2) \equiv i \int d^3 x \phi_1^*(\mathbf{x}, t) \overset{\leftrightarrow}{\partial_0} \phi_2(\mathbf{x}, t), \tag{2.163}$$

where

$$A \overset{\leftrightarrow}{\partial_0} B \equiv A(\partial_0 B) - (\partial_0 A)B. \tag{2.164}$$

We can easily verify that the plane waves form an orthonormal set with respect to the scalar product Eq.(2.163) as follows:

$$\begin{aligned} (u_{\mathbf{p}'}, u_{\mathbf{p}}) &= i \int d^3 x u_{\mathbf{p}'}^*(\mathbf{x}, t) \overset{\leftrightarrow}{\partial_0} u_{\mathbf{p}}(\mathbf{x}, t) \\ &= \delta^3(\mathbf{p} - \mathbf{p}') \end{aligned} \tag{2.165}$$

and

$$(u_{\mathbf{p}'}^*, u_{\mathbf{p}}^*) = -\delta^3(\mathbf{p} - \mathbf{p}'). \tag{2.166}$$

Similarly,

$$(u_{\mathbf{p}'}, u_{\mathbf{p}}^*) = (u_{\mathbf{p}'}^*, u_{\mathbf{p}}) = 0. \tag{2.167}$$

Now we evaluate the commutator

$$[\hat{a}_{\mathbf{p}}, \hat{a}_{\mathbf{p}'}] = i^2 \int d^3 x \int d^3 x' [u_{\mathbf{p}}^*(\mathbf{x}, t) \overset{\leftrightarrow}{\partial_0} \hat{\phi}(\mathbf{x}, t), u_{\mathbf{p}'}^*(\mathbf{x}', t) \overset{\leftrightarrow}{\partial_0} \hat{\phi}(\mathbf{x}', t)]. \tag{2.168}$$

The functions $u_{\mathbf{p}}(\mathbf{x}, t)$ are c numbers and commute with the field operators. We have

$$\begin{aligned} &[u_{\mathbf{p}}^*(\mathbf{x}, t) \overset{\leftrightarrow}{\partial_0} \hat{\phi}(\mathbf{x}, t), u_{\mathbf{p}'}^*(\mathbf{x}', t) \overset{\leftrightarrow}{\partial_0} \hat{\phi}(\mathbf{x}', t)] \\ =& u_{\mathbf{p}}^*(\mathbf{x}, t) u_{\mathbf{p}'}^*(\mathbf{x}', t)[\dot{\hat{\phi}}(\mathbf{x}, t), \dot{\hat{\phi}}(\mathbf{x}', t)] \\ &- u_{\mathbf{p}}^*(\mathbf{x}, t) \dot{u}_{\mathbf{p}'}^*(\mathbf{x}', t)[\dot{\hat{\phi}}(\mathbf{x}, t), \hat{\phi}(\mathbf{x}', t)] \\ &- \dot{u}_{\mathbf{p}}^*(\mathbf{x}, t) u_{\mathbf{p}'}^*(\mathbf{x}', t)[\hat{\phi}(\mathbf{x}, t), \dot{\hat{\phi}}(\mathbf{x}', t)] \\ &+ \dot{u}_{\mathbf{p}}^*(\mathbf{x}, t) \dot{u}_{\mathbf{p}'}^*(\mathbf{x}', t)[\hat{\phi}(\mathbf{x}, t), \hat{\phi}(\mathbf{x}', t)]. \end{aligned} \tag{2.169}$$

Using the commutation relations of $\hat{\phi}$ and $\hat{\pi}$, we get

$$\begin{aligned} [\hat{a}_{\mathbf{p}}, \hat{a}_{\mathbf{p}'}] &= -i \int d^3 x [u_{\mathbf{p}}^*(\mathbf{x}, t) \dot{u}_{\mathbf{p}'}^*(\mathbf{x}, t) - \dot{u}_{\mathbf{p}}^*(\mathbf{x}, t) u_{\mathbf{p}'}^*(\mathbf{x}, t)] \\ &= -i \int d^3 x [u_{\mathbf{p}}^*(\mathbf{x}, t) \overset{\leftrightarrow}{\partial_0} u_{\mathbf{p}'}^*(\mathbf{x}, t) = -(u_{\mathbf{p}}, u_{\mathbf{p}'}^*) = 0. \end{aligned} \tag{2.170}$$

Similar calculations give

$$[\hat{a}_{\mathbf{p}}^{\dagger}, \hat{a}_{\mathbf{p}'}^{\dagger}] = 0. \tag{2.171}$$

Now we calculate the commutator $[\hat{a}_{\mathbf{p}}, \hat{a}_{\mathbf{p}'}^{\dagger}]$. Using the projection formulas Eqs.(2.161) and (2.162), the commutator becomes

$$[\hat{a}_{\mathbf{p}}, \hat{a}_{\mathbf{p}'}^{\dagger}] = -i^2 \int d^3x \int d^3x' [u_{\mathbf{p}}^*(\mathbf{x}, t) \overset{\leftrightarrow}{\partial_0} \hat{\phi}(\mathbf{x}, t), u_{\mathbf{p}'}(\mathbf{x}', t) \overset{\leftrightarrow}{\partial_0} \hat{\phi}(\mathbf{x}', t)]. \tag{2.172}$$

According to Eq.(2.164), we have

$$[u_{\mathbf{p}}^*(\mathbf{x}, t) \overset{\leftrightarrow}{\partial_0} \hat{\phi}(\mathbf{x}, t), u_{\mathbf{p}'}(\mathbf{x}', t) \overset{\leftrightarrow}{\partial_0} \hat{\phi}(\mathbf{x}', t)]$$
$$= u_{\mathbf{p}}^*(\mathbf{x}, t) u_{\mathbf{p}'}(\mathbf{x}', t)[\dot{\hat{\phi}}(\mathbf{x}, t), \dot{\hat{\phi}}(\mathbf{x}', t)]$$
$$- u_{\mathbf{p}}^*(\mathbf{x}, t) \dot{u}_{\mathbf{p}'}(\mathbf{x}', t)[\dot{\hat{\phi}}(\mathbf{x}, t), \hat{\phi}(\mathbf{x}', t)] \tag{2.173}$$
$$- \dot{u}_{\mathbf{p}}^*(\mathbf{x}, t) u_{\mathbf{p}'}(\mathbf{x}', t)[\hat{\phi}(\mathbf{x}, t), \dot{\hat{\phi}}(\mathbf{x}', t)]$$
$$+ \dot{u}_{\mathbf{p}}^*(\mathbf{x}, t) \dot{u}_{\mathbf{p}'}(\mathbf{x}', t)[\hat{\phi}(\mathbf{x}, t), \hat{\phi}(\mathbf{x}', t)].$$

Using the commutation relations of $\hat{\phi}$ and $\hat{\pi}$, we obtain

$$[\hat{a}_{\mathbf{p}}, \hat{a}_{\mathbf{p}'}^{\dagger}] = i \int d^3x [u_{\mathbf{p}}^*(\mathbf{x}, t) \overset{\leftrightarrow}{\partial_0} u_{\mathbf{p}'}(\mathbf{x}, t) = (u_{\mathbf{p}}, u_{\mathbf{p}'}) = \delta^3(\mathbf{p} - \mathbf{p}'). \tag{2.174}$$

Thus we get all the commutators for the creation operator $\hat{a}_{\mathbf{p}}^{\dagger}$ and the annihilation operator $\hat{a}_{\mathbf{p}}$.

2.3.5 *Symmetry and conservation law*

In the following, we will show that a symmetry in the field is related to a conservation law. Suppose we have an infinitesimal transformation defined by a transformation in coordinates

$$x'_{\mu} = x_{\mu} + \delta x_{\mu} \tag{2.175}$$

and a transformation in the field operators $\hat{\phi}_a(x)$

$$\hat{\phi}'_a(x') = \hat{\phi}_a(x) + \delta \hat{\phi}_a(x). \tag{2.176}$$

where the subscript a in $\hat{\phi}_a$ is introduced to denote different fields so that our formula have a more generalized form for later use.

If the transformations Eqs.(2.175) and (2.176) leave the action operator invariant, we say that the system possesses the symmetry defined by the transformations Eqs.(2.175) and (2.176). We introduce a variation that keeps the value of the coordinates x fixed

$$\tilde{\delta} \hat{\phi}_a(x) = \hat{\phi}'_a(x) - \hat{\phi}_a(x). \tag{2.177}$$

$\tilde{\delta}\hat{\phi}_a(x)$ is also called the total variation, while the variation $\delta\hat{\phi}_a(x) = \hat{\phi}'_a(x') - \hat{\phi}_a(x)$ is called the local variation. The two types of variations have the following relation

$$
\begin{aligned}
\tilde{\delta}\hat{\phi}_a(x) &= \hat{\phi}'_a(x) - \hat{\phi}'_a(x') + \hat{\phi}'_a(x') - \hat{\phi}_a(x) \\
&= \delta\hat{\phi}_a(x) - (\hat{\phi}'_a(x') - \hat{\phi}'_a(x)) \\
&= \delta\hat{\phi}_a(x) - \frac{\partial\hat{\phi}'_a}{\partial x^\mu}\delta x^\mu \\
&= \delta\hat{\phi}_a(x) - \frac{\partial\hat{\phi}_a}{\partial x^\mu}\delta x^\mu.
\end{aligned}
\tag{2.178}
$$

In the derivation of Eq.(2.178), $\frac{\partial\hat{\phi}'_a}{\partial x^\mu}$ is approximated by $\frac{\partial\hat{\phi}_a}{\partial x^\mu}$ because their difference contributes only higher order terms. According to the definition Eq.(2.177), we have

$$
\frac{\partial}{\partial x^\mu}\tilde{\delta}\hat{\phi}_a(x) = \tilde{\delta}\left(\frac{\partial\hat{\phi}_a}{\partial x^\mu}\right).
\tag{2.179}
$$

Thus $\tilde{\delta}$ commutes with the differentiation $\frac{\partial}{\partial x^\mu}$. The symmetry transformation leaves the action operator invariant.

$$
\begin{aligned}
\delta\hat{S} &= \int d^4x'\hat{\mathcal{L}}'(x') - \int d^4x\hat{\mathcal{L}}(x) \\
&= \int d^4x'\delta\hat{\mathcal{L}}(x) + \int d^4x'\hat{\mathcal{L}}(x) - \int d^4x\hat{\mathcal{L}}(x) \\
&= 0,
\end{aligned}
\tag{2.180}
$$

where

$$
\delta\hat{\mathcal{L}}(x) = \hat{\mathcal{L}}'(x') - \hat{\mathcal{L}}(x).
\tag{2.181}
$$

The transformation of the volume element in Eq.(2.180) is determined by the Jacobian determinant

$$
\begin{aligned}
d^4x' &= \left|\frac{\partial(x'^\mu)}{\partial(x^\mu)}\right| d^4x \\
&= \begin{vmatrix}
1 + \frac{\partial(\delta x^0)}{\partial x^0} & \frac{\partial(\delta x^0)}{\partial x^1} & \cdots & \cdots \\
\frac{\partial(\delta x^1)}{\partial x^0} & 1 + \frac{\partial(\delta x^1)}{\partial x^1} & \cdots & \cdots \\
\vdots & \vdots & \ddots & \cdots \\
\vdots & \vdots & \vdots & 1 + \frac{\partial(\delta x^3)}{\partial x^3}
\end{vmatrix} d^4x \\
&= \left(1 + \frac{\partial(\delta x^\mu)}{\partial x^\mu}\right) d^4x.
\end{aligned}
\tag{2.182}
$$

The terms of higher orders have been neglected in Eq.(2.182). Thus Eq.(2.180) becomes

$$
\begin{aligned}
\delta \hat{S} &= \int d^4x \, \delta \hat{\mathcal{L}}(x) + \int d^4x \hat{\mathcal{L}}(x) \frac{\partial(\delta x^\mu)}{\partial x^\mu} \\
&= \int d^4x \left(\tilde{\delta}\hat{\mathcal{L}}(x) + \frac{\partial \hat{\mathcal{L}}(x)}{\partial x^\mu} \delta x^\mu \right) + \int d^4x \hat{\mathcal{L}}(x) \frac{\partial(\delta x^\mu)}{\partial x^\mu} \\
&= \int d^4x \left[\tilde{\delta}\hat{\mathcal{L}}(x) + \frac{\partial}{\partial x^\mu}(\hat{\mathcal{L}}(x)\delta x^\mu) \right] \\
&= 0.
\end{aligned}
\tag{2.183}
$$

$\tilde{\delta}\hat{\mathcal{L}}(x)$ is given by the following chain rule.

$$
\begin{aligned}
\tilde{\delta}\hat{\mathcal{L}}(x) =& \frac{\partial \hat{\mathcal{L}}(x)}{\partial \hat{\phi}_a} \tilde{\delta}\hat{\phi}_a(x) + \frac{\partial \hat{\mathcal{L}}(x)}{\partial(\partial_\mu \hat{\phi}_a)} \tilde{\delta} \left(\frac{\partial \hat{\phi}_a}{\partial x^\mu} \right) \\
=& \left[\frac{\partial \hat{\mathcal{L}}(x)}{\partial \hat{\phi}_a} \tilde{\delta}\hat{\phi}_a(x) - \frac{\partial}{\partial x^\mu} \left(\frac{\partial \hat{\mathcal{L}}(x)}{\partial(\partial_\mu \hat{\phi}_a)} \right) \tilde{\delta}\hat{\phi}_a(x) \right] \\
&+ \frac{\partial}{\partial x^\mu} \left(\frac{\partial \hat{\mathcal{L}}(x)}{\partial(\partial_\mu \hat{\phi}_a)} \right) \tilde{\delta}\hat{\phi}_a(x) + \frac{\partial \hat{\mathcal{L}}(x)}{\partial(\partial_\mu \hat{\phi}_a)} \frac{\partial}{\partial x^\mu} \left(\tilde{\delta}\hat{\phi}_a(x) \right) \\
=& \left[\frac{\partial \hat{\mathcal{L}}(x)}{\partial \hat{\phi}_a} - \frac{\partial}{\partial x^\mu} \left(\frac{\partial \hat{\mathcal{L}}(x)}{\partial(\partial_\mu \hat{\phi}_a)} \right) \right] \tilde{\delta}\hat{\phi}_a(x) \\
&+ \frac{\partial}{\partial x^\mu} \left[\frac{\partial \hat{\mathcal{L}}(x)}{\partial(\partial_\mu \hat{\phi}_a)} \tilde{\delta}\hat{\phi}_a(x) \right],
\end{aligned}
\tag{2.184}
$$

Inserting Eq.(2.184) into Eq.(2.183), we have

$$
\begin{aligned}
\delta \hat{S} = \int d^4x \Bigg\{ & \left[\frac{\partial \hat{\mathcal{L}}(x)}{\partial \hat{\phi}_a} - \frac{\partial}{\partial x^\mu} \left(\frac{\partial \hat{\mathcal{L}}(x)}{\partial(\partial_\mu \hat{\phi}_a)} \right) \right] \tilde{\delta}\hat{\phi}_a(x) \\
&+ \frac{\partial}{\partial x^\mu} \left[\frac{\partial \hat{\mathcal{L}}(x)}{\partial(\partial_\mu \hat{\phi}_a)} \tilde{\delta}\hat{\phi}_a(x) + \hat{\mathcal{L}}(x)\delta x^\mu \right] \Bigg\} \\
=& \, 0.
\end{aligned}
\tag{2.185}
$$

Since the range of the integration can be chosen arbitrarily, the integrand of Eq.(2.183) should be zero when $\delta\hat{S} = 0$. Thus we obtain

$$
\begin{aligned}
& \left[\frac{\partial \hat{\mathcal{L}}(x)}{\partial \hat{\phi}_a} - \frac{\partial}{\partial x^\mu} \left(\frac{\partial \hat{\mathcal{L}}(x)}{\partial(\partial_\mu \hat{\phi}_a)} \right) \right] \tilde{\delta}\hat{\phi}_a(x) \\
& + \frac{\partial}{\partial x^\mu} \left[\frac{\partial \hat{\mathcal{L}}(x)}{\partial(\partial_\mu \hat{\phi}_a)} \tilde{\delta}\hat{\phi}_a(x) + \hat{\mathcal{L}}(x)\delta x^\mu \right] = 0.
\end{aligned}
\tag{2.186}
$$

Since the field operators satisfy the quantum Euler-Lagrange equation Eq.(2.109), we have

$$\frac{\partial}{\partial x^\mu}\left[\frac{\partial\hat{\mathcal{L}}(x)}{\partial(\partial_\mu\hat{\phi}_a)}\tilde{\delta}\hat{\phi}_a(x) + \hat{\mathcal{L}}(x)\delta x^\mu\right]$$

$$= \frac{\partial}{\partial x^\mu}\left[\frac{\partial\hat{\mathcal{L}}(x)}{\partial(\partial_\mu\hat{\phi}_a)}\left(\delta\hat{\phi}_a(x) - \frac{\partial\hat{\phi}_a}{\partial x^\nu}\delta x^\nu\right) + \hat{\mathcal{L}}(x)\delta x^\mu\right] \qquad (2.187)$$

$$= 0.$$

We define the current density operator $\hat{j}^\mu$ by

$$\hat{j}^\mu \equiv \frac{\partial\hat{\mathcal{L}}(x)}{\partial(\partial_\mu\hat{\phi}_a)}\delta\hat{\phi}_a(x) - \left(\frac{\partial\hat{\mathcal{L}}(x)}{\partial(\partial_\mu\hat{\phi}_a)}\frac{\partial\hat{\phi}_a}{\partial x^\nu} - \eta^\mu_\nu\hat{\mathcal{L}}(x)\right)\delta x^\nu. \qquad (2.188)$$

$\hat{j}^\mu$ is also called the *Noether current*. Then we have the equation of continuity

$$\frac{\partial}{\partial x^\mu}\hat{j}^\mu = 0. \qquad (2.189)$$

Eq.(2.189) is called *Noether's theorem*, which states that each continuous symmetry transformation corresponds to a conservation law. Expressing it in terms of the time and space components, Eq.(2.189) becomes

$$\frac{\partial}{\partial t}\hat{j}^0(x) + \nabla\cdot\hat{\mathbf{j}}(x) = 0. \qquad (2.190)$$

Then

$$\hat{J} \equiv \int d^3x j^0(x) \qquad (2.191)$$

is a conserved quantity because of Gauss's theorem.

In Eq.(2.185), $\delta\hat{S} = 0$ is a mathematical identity due to the invariance of symmetry transformation. We have shown that $\delta\hat{S}$ can be divided into two parts. One part contains

$$\frac{\partial\hat{\mathcal{L}}}{\partial\hat{\phi}} - \frac{\partial}{\partial x^\mu}\left[\frac{\partial\hat{\mathcal{L}}}{\partial(\partial_\mu\hat{\phi})}\right], \qquad (2.192)$$

which vanishes due to the equations of motion. The other part has the form $\frac{\partial}{\partial x^\mu}\hat{j}^\mu$, which has to be vanish when the field satisfies the equations of motion. Thus the conservation law is equivalent to the equations of motion. They are the same essentially but manifested in different ways. It should be noted that without the quantum Euler-Lagrange equation, we would have no conservation law. The equivalence of the Euler-Lagrange equation with the conservation laws is related to the simple form of the Lagrangian dictated by the causality and symmetry principles. If we do not have the causality and symmetry principles, there would be no conservation laws. Then we would not have stable states and everything would have no lifetime.

2.3.6 *Homogeneity of spacetime*

2.3.6.1 *Conservation of energy-momentum*

Now we use the symmetry principle that demands the homogeneity of spacetime. The action operator should possess the symmetry of spacetime translation. We transform the field operator via $x'_\mu = x_\mu + a_\mu$, where a_μ is a constant four-vector. For an infinitesimal translation δa^ν, we have the total variation $\tilde{\delta}\hat{\phi}(x) = \delta a^\nu \partial_\nu \hat{\phi}(x)$. Under an infinitesimal spacetime translation, the variation of the action operator should be zero. We have $\delta\hat{S} = 0$.

Since $\hat{\phi}'(x') = \hat{\phi}(x)$ under a translation $x'_\mu = x_\mu + \delta a_\mu$, we have $\delta\hat{\phi} = 0$. In this case, Eq.(2.188) becomes

$$\hat{j}^\mu = -\left(\frac{\partial\hat{\mathcal{L}}(x)}{\partial(\partial_\mu\hat{\phi}_a)} \frac{\partial\hat{\phi}_a}{\partial x^\nu} - \eta^\mu_\nu \hat{\mathcal{L}}(x) \right) \delta a^\nu. \tag{2.193}$$

Then the conservation of the Noether current Eq.(2.189) becomes

$$\partial_\mu \hat{\Theta}^\mu_\nu(x) = 0 \tag{2.194}$$

with

$$\hat{\Theta}^\mu_\nu(x) \equiv \frac{\partial\hat{\mathcal{L}}(x)}{\partial(\partial_\mu\hat{\phi}(x))} \partial_\nu\hat{\phi}(x) - \eta^\mu_\nu\hat{\mathcal{L}}(x). \tag{2.195}$$

$\hat{\Theta}^\mu_\nu(x)$ is called the *energy-momentum tensor operator*. Eq.(2.194) is the *Noether's theorem* for the symmetry of spacetime translation. The conservation law Eq.(2.194) is called the *conservation of energy-momentum*. $\hat{\Theta}^0_0$ also has the name of the Hamiltonian density operator and is usually denoted as $\hat{\mathcal{H}}$. The integral of the Hamiltonian density operator

$$\hat{H} = \int d^3x\hat{\mathcal{H}} \tag{2.196}$$

is called the Hamiltonian operator.

2.3.6.2 *Heisenberg's equations of motion*

Since $\frac{\partial\hat{\mathcal{L}}}{\partial\dot{\hat{\phi}}} = \hat{\pi}$ from Eq.(2.104), we have

$$\hat{\mathcal{H}}(\hat{\pi}, \hat{\phi}) \equiv \hat{\Theta}^0_0 = \frac{\partial\hat{\mathcal{L}}(x)}{\partial(\partial_t\hat{\phi}(x))} \partial_t\hat{\phi}(x) - \eta^0_0\hat{\mathcal{L}}(x) = \hat{\mathcal{G}}_t(\hat{\pi}, \hat{\phi}). \tag{2.197}$$

Inserting $\hat{\mathcal{G}}_t(\hat{\pi}, \hat{\phi}) = \hat{\mathcal{H}}(\hat{\pi}, \hat{\phi})$ into Eq.(2.96), we have

$$\hat{\mathcal{L}} = \hat{\pi}\partial_t\hat{\phi} - \hat{\mathcal{H}}(\hat{\pi}, \hat{\phi}), \tag{2.198}$$

which shows that $\hat{\mathcal{L}}$ is the Lagrangian density operator as defined originally by Lagrange. Since $\hat{\mathcal{H}} = \hat{\mathcal{G}}_t$, we have $\hat{H} = \hat{G}_t$. Inserting $\hat{G}_t = \hat{H}$ into Eq.(2.91), we have

$$i\partial_t \hat{\phi} = [\hat{\phi}, \hat{H}], \tag{2.199a}$$

$$i\partial_t \hat{\pi} = [\hat{\pi}, \hat{H}], \tag{2.199b}$$

which is Heisenberg's equations of motion.

The equivalence of the conservation law Eq.(2.194) with the equations of motion Eq. (2.141) can be easily verified by multiplying the two sides of Eq. (2.141) with $\partial_\nu \hat{\phi}$, which gives

$$\partial_\nu \hat{\phi}(\partial_\mu \partial^\mu \hat{\phi} + m^2 \hat{\phi}) = 0. \tag{2.200}$$

The above equation is equivalent to

$$\partial_\mu \left[\partial^\mu \hat{\phi} \partial_\nu \hat{\phi} - \eta_\nu^\mu \left(\frac{1}{2} \partial^\sigma \hat{\phi} \partial_\sigma \hat{\phi} - \frac{1}{2} m^2 \hat{\phi}^2 \right) \right] = 0, \tag{2.201}$$

which is Eq.(2.194) for the conservation of Noether current.

Now we look at the physical meaning of $\hat{\Theta}_\nu^\mu$. We define

$$\hat{P}_\nu \equiv \int d^3 x \hat{\Theta}_\nu^0(x) \tag{2.202}$$

as the energy-momentum vector operator. $\hat{\Theta}_\nu^0$ is called the Poynting vector operator. Using Eq.(2.195), we have

$$\hat{P}_\nu = \int d^3 x \left[\frac{\hat{\mathcal{L}}(x)}{\partial \partial_0 \hat{\phi}(x)} \partial_\nu \hat{\phi}(x) - \eta_\nu^0 \hat{\mathcal{L}}(x) \right]. \tag{2.203}$$

Expressing Eq.(2.194) in terms of the time and space components, We obtain

$$\frac{\partial \hat{\Theta}_\nu^0(x)}{\partial t} + \nabla_i \hat{\Theta}_\nu^i = 0. \tag{2.204}$$

Using Gauss's theorem, we have

$$\frac{d\hat{P}_\nu}{dt} = 0. \tag{2.205}$$

Thus $\hat{P}_\nu$ is the conserved four-vector. One can easily verify that Eq.(2.204) is equivalent to

$$i\frac{\partial \hat{\Theta}_\nu^0(x)}{\partial t} = [\hat{\Theta}_\nu^i, \hat{G}_t]. \tag{2.206}$$

2.3.6.3 *Hamiltonian operator*

For the scalar bosons, inserting the Lagrangian density in Eq.(2.133) into Eq.(2.203), we have

$$\hat{H} \equiv \hat{P}_0 = \int d^3x \left[\frac{1}{2}(\partial_0\hat{\phi})^2 + \frac{1}{2}(\nabla\hat{\phi})^2 + \frac{1}{2}m^2\hat{\phi}^2 + V(\hat{\phi}) \right]. \quad (2.207)$$

$\hat{P}_0 = \hat{H}$ is the Hamiltonian operator of the field and is also called as the energy operator of the field ϕ. When we replace the field operator $\hat{\phi}$ with the field function ϕ, we call the corresponding function H as *Hamiltonian*.

$$H = \int d^3x \left[\frac{1}{2}(\partial_0\phi)^2 + \frac{1}{2}(\nabla\phi)^2 + \frac{1}{2}m^2\phi^2 + V(\phi) \right]. \quad (2.208)$$

Since we have

$$\pi = \dot{\phi} = \frac{\partial \mathcal{L}}{\partial \dot{\phi}}, \quad (2.209)$$

the Hamiltonian density can be written as

$$\mathcal{H}(x) = \pi(x)\dot{\phi}(x) - \mathcal{L}(x) = \frac{1}{2}\pi^2 + \frac{1}{2}(\nabla\phi)^2 + \frac{1}{2}m^2\phi^2 + V(\phi). \quad (2.210)$$

2.3.6.4 *Hamilton's equations*

Inserting $\hat{H} = \hat{G}_t$ into Eq.(2.100), we have

$$\hat{\pi} = \partial_t\hat{\phi} = -i[\hat{\phi}, \hat{H}] = \frac{\partial\hat{\mathcal{H}}}{\partial\hat{\pi}}, \quad (2.211a)$$

$$\partial_t\hat{\pi} = -i[\hat{\pi}, \hat{H}] = -\frac{\partial\hat{\mathcal{H}}}{\partial\hat{\phi}}. \quad (2.211b)$$

This is the operator form of Hamilton's equations, which is equivalent to the operator form of the Euler-Lagrange equation. Using $\hat{\pi} = \partial_t\hat{\phi}$, Eq.(2.211) can also be written as

$$\frac{\partial}{\partial t}\left(\frac{\partial\hat{\mathcal{H}}}{\partial\hat{\pi}}\right) + \frac{\partial\hat{\mathcal{H}}}{\partial\hat{\phi}} = 0. \quad (2.212)$$

2.3.6.5 *Hamiltonian operator of free scalar bosons*

We can express the Hamiltonian of free scalar bosons in terms of the creation and annihilation operators $\hat{a}_{\mathbf{p}}^\dagger$ and $\hat{a}_{\mathbf{p}}$. Inserting the expansion formula for the field operators $\hat{\phi}$ and $\hat{\pi}$, we have

$$
\begin{aligned}
\hat{H} =& \frac{1}{2} \int d^3x \left[\hat{\pi}^2 + (\nabla\hat{\phi})^2 + m^2\hat{\phi}^2 \right] \\
=& \frac{1}{2} \int d^3x \left[-\int d^3p'\omega_{\mathbf{p}'}(\hat{a}_{\mathbf{p}'}u_{\mathbf{p}'} - \hat{a}_{\mathbf{p}'}^\dagger u_{\mathbf{p}'}^*) \int d^3p\omega_{\mathbf{p}}(\hat{a}_{\mathbf{p}}u_{\mathbf{p}} - \hat{a}_{\mathbf{p}}^\dagger u_{\mathbf{p}}^*) \right. \\
& - \int d^3p'p'(\hat{a}_{\mathbf{p}'}u_{\mathbf{p}'} - \hat{a}_{\mathbf{p}'}^\dagger u_{\mathbf{p}'}^*) \int d^3pp(\hat{a}_{\mathbf{p}}u_{\mathbf{p}} - \hat{a}_{\mathbf{p}}^\dagger u_{\mathbf{p}}^*) \\
& \left. + \int d^3p'm(\hat{a}_{\mathbf{p}'}u_{\mathbf{p}'} + \hat{a}_{\mathbf{p}'}^\dagger u_{\mathbf{p}'}^*) \int d^3pm(\hat{a}_{\mathbf{p}}u_{\mathbf{p}} + \hat{a}_{\mathbf{p}}^\dagger u_{\mathbf{p}}^*) \right].
\end{aligned}
\tag{2.213}
$$

The integration over x can be carried out, which gives the delta function.

$$
\int d^3x u_{\mathbf{p}'}^*(\mathbf{x},t)u_{\mathbf{p}}(\mathbf{x},t) = \frac{1}{2\omega_{\mathbf{p}}}\delta^3(\mathbf{p}-\mathbf{p}'), \tag{2.214a}
$$

$$
\int d^3x u_{\mathbf{p}'}(\mathbf{x},t)u_{\mathbf{p}}(\mathbf{x},t) = \frac{1}{2\omega_{\mathbf{p}}}e^{-2i\omega_{\mathbf{p}}t}\delta^3(\mathbf{p}+\mathbf{p}'). \tag{2.214b}
$$

Using Eq.(2.214), we get

$$
\begin{aligned}
\hat{H} =& \frac{1}{2}\left[-\int d^3p \frac{\omega_{\mathbf{p}}^2}{2\omega_{\mathbf{p}}}(\hat{a}_{-\mathbf{p}}\hat{a}_{\mathbf{p}}e^{-2i\omega_{\mathbf{p}}t} - \hat{a}_{\mathbf{p}}^\dagger\hat{a}_{\mathbf{p}} - \hat{a}_{\mathbf{p}}\hat{a}_{\mathbf{p}}^\dagger + \hat{a}_{-\mathbf{p}}^\dagger\hat{a}_{\mathbf{p}}^\dagger e^{2i\omega_{\mathbf{p}}t}) \right. \\
& - \int d^3p \frac{p^2}{2\omega_{\mathbf{p}}}(-\hat{a}_{-\mathbf{p}}\hat{a}_{\mathbf{p}}e^{-2i\omega_{\mathbf{p}}t} - \hat{a}_{\mathbf{p}}^\dagger\hat{a}_{\mathbf{p}} - \hat{a}_{\mathbf{p}}\hat{a}_{\mathbf{p}}^\dagger - \hat{a}_{-\mathbf{p}}^\dagger\hat{a}_{\mathbf{p}}^\dagger e^{2i\omega_{\mathbf{p}}t}) \\
& \left. + \int d^3p \frac{m^2}{2\omega_{\mathbf{p}}}(\hat{a}_{-\mathbf{p}}\hat{a}_{\mathbf{p}}e^{-2i\omega_{\mathbf{p}}t} + \hat{a}_{\mathbf{p}}^\dagger\hat{a}_{\mathbf{p}} + \hat{a}_{\mathbf{p}}\hat{a}_{\mathbf{p}}^\dagger + \hat{a}_{-\mathbf{p}}^\dagger\hat{a}_{\mathbf{p}}^\dagger e^{2i\omega_{\mathbf{p}}t}) \right].
\end{aligned}
\tag{2.215}
$$

The terms involving $\hat{a}\hat{a}$ and $\hat{a}^\dagger\hat{a}^\dagger$ are multiplied by a factor $(-\omega_{\mathbf{p}}^2 + p^2 + m^2)$ which is zero. The remaining expression for the Hamiltonian is given by

$$
\begin{aligned}
\hat{H} &= \frac{1}{2} \int d^3p\omega_{\mathbf{p}}(\hat{a}_{\mathbf{p}}^\dagger\hat{a}_{\mathbf{p}} + \hat{a}_{\mathbf{p}}\hat{a}_{\mathbf{p}}^\dagger) \\
&= \int d^3p\omega_{\mathbf{p}} \left[\hat{a}_{\mathbf{p}}^\dagger\hat{a}_{\mathbf{p}} + \frac{1}{2}\delta^3(0) \right] \\
&= \int d^3p\omega_{\mathbf{p}}\hat{a}_{\mathbf{p}}^\dagger\hat{a}_{\mathbf{p}} + E_0
\end{aligned}
\tag{2.216}
$$

with

$$
E_0 = \frac{1}{2}\int d^3p\omega_{\mathbf{p}}\delta^3(0) = \frac{1}{2}\int \frac{d^3x}{(2\pi)^3}d^3p\omega_{\mathbf{p}}. \tag{2.217}
$$

E_0 is called the *vacuum energy*. Since there is an infinite number of modes, the vacuum energy E_0 is divergent. Because physical observables involve energy differences rather than the absolute value of the energy, the divergent zero-point energy E_0 can be dropped out. Then the Hamiltonian can be rewritten as

$$\hat{\hat{H}} = \hat{H} - E_0 = \int d^3p \omega_{\mathbf{p}} \hat{a}_{\mathbf{p}}^\dagger \hat{a}_{\mathbf{p}}. \tag{2.218}$$

In the Hamiltonian Eq.(2.218), the creation operator is on the left of the annihilation operator. We call this arrangement of operators as *normal ordering* or *normal product*. We denote a normal product of the operators $\hat{A}$ and $\hat{B}$ by $: \hat{A}\hat{B} :$ or $N(\hat{A}\hat{B})$. Thus Eq.(2.218) has the form

$$\begin{aligned}\hat{\hat{H}} &= \frac{1}{2} \int d^3x : \left[\hat{\pi}^2 + (\nabla\hat{\phi})^2 + m^2\hat{\phi}^2\right] : \\ &= \int d^3p \omega_{\mathbf{p}} \hat{a}_{\mathbf{p}}^\dagger \hat{a}_{\mathbf{p}}. \end{aligned} \tag{2.219}$$

Since $\hat{a}_{\mathbf{p}}^\dagger$ and $\hat{a}_{\mathbf{p}}$ are the creation and annihilation operators, respectively, $\hat{n}_{\mathbf{p}} = \hat{a}_{\mathbf{p}}^\dagger \hat{a}_{\mathbf{p}}$ is the particle-number operator. We can see that $\omega_{\mathbf{p}}$ is the energy of a plane wave containing one single-particle, which is also called single-particle state.

2.3.6.6 *Momentum operator of free scalar bosons*

Now let us turn to the momentum of the field.

$$P_\mu = \int d^3x \Theta_\mu^0 = \int d^3x \left(\pi \frac{\partial \phi}{\partial x^\mu} - \eta_\mu^0 \mathcal{L} \right). \tag{2.220}$$

P^i is defined as the momentum.

$$P^i = -P_i = -\int d^3x \pi \frac{\partial \phi}{\partial x^i}. \tag{2.221}$$

In vector notation,

$$\mathbf{P} = -\int d^3x \pi \nabla\phi. \tag{2.222}$$

The momentum operator of the field is given by

$$\hat{\mathbf{P}} = -\int d^3x \hat{\pi}(\mathbf{x}, t) \nabla \hat{\phi}(\mathbf{x}, t). \tag{2.223}$$

It is more natural to use a symmetric form of momentum operator

$$\hat{\mathbf{P}} = -\frac{1}{2} \int d^3x (\hat{\pi}\nabla\hat{\phi} + \nabla\hat{\phi}\hat{\pi}), \tag{2.224}$$

which guarantees that $\hat{\mathbf{P}}$ is a hermitian operator.

Using the commutator of $\hat{\phi}$ and $\hat{\pi}$, we have

$$[\hat{\phi}, \hat{P}_k] = i\partial_k \hat{\phi}, \tag{2.225a}$$

$$[\hat{\pi}, \hat{P}_k] = i\partial_k \hat{\pi}. \tag{2.225b}$$

Therefore, the momentum operator $\hat{P}_k$ is the generator of space translation transformation. Using Eq.(2.92), we have

$$e^{i\hat{P}_i x^i} \hat{\phi}(\mathbf{x}_0, t_0) e^{-i\hat{P}_i x^i} = \hat{\phi}(\mathbf{x}_0 + \mathbf{x}, t_0), \tag{2.226a}$$

$$e^{i\hat{P}_i x^i} \hat{\pi}(\mathbf{x}_0, t_0) e^{-i\hat{P}_i x^i} = \hat{\pi}(\mathbf{x}_0 + \mathbf{x}, t_0). \tag{2.226b}$$

Thus $e^{i\hat{P}_i x^i}$ is the operator of space translation.

We use the expansion formulas for the field operators and get

$$\begin{aligned}
\hat{\mathbf{P}} &= -\frac{1}{2} \int d^3x \Big[-i \int d^3p' \omega_{\mathbf{p}'} (\hat{a}_{\mathbf{p}'} u_{\mathbf{p}'} - \hat{a}_{\mathbf{p}'}^\dagger u_{\mathbf{p}'}^*) \int d^3p(-i\mathbf{p})(\hat{a}_{\mathbf{p}} u_{\mathbf{p}} - \hat{a}_{\mathbf{p}}^\dagger u_{\mathbf{p}}^*) \\
&\quad + \int d^3p(-i\mathbf{p})(\hat{a}_{\mathbf{p}} u_{\mathbf{p}} - \hat{a}_{\mathbf{p}}^\dagger u_{\mathbf{p}}^*)(-i) \int d^3p' \omega_{\mathbf{p}'} (\hat{a}_{\mathbf{p}'} u_{\mathbf{p}'} - \hat{a}_{\mathbf{p}'}^\dagger u_{\mathbf{p}'}^*) \Big] \\
&= -\frac{1}{2} \int d^3p \frac{1}{2\omega_{\mathbf{p}}} \omega_{\mathbf{p}} (\mathbf{p}\hat{a}_{-\mathbf{p}} \hat{a}_{\mathbf{p}} e^{-2i\omega_{\mathbf{p}}t} - \mathbf{p}\hat{a}_{\mathbf{p}}^\dagger \hat{a}_{\mathbf{p}} - \mathbf{p}\hat{a}_{\mathbf{p}} \hat{a}_{\mathbf{p}}^\dagger \\
&\quad + \mathbf{p}\hat{a}_{-\mathbf{p}}^\dagger \hat{a}_{\mathbf{p}}^\dagger e^{2i\omega_{\mathbf{p}}t} + \mathbf{p}\hat{a}_{-\mathbf{p}} \hat{a}_{\mathbf{p}} e^{-2i\omega_{\mathbf{p}}t} - \mathbf{p}\hat{a}_{\mathbf{p}}^\dagger \hat{a}_{\mathbf{p}} - \mathbf{p}\hat{a}_{\mathbf{p}} \hat{a}_{\mathbf{p}}^\dagger + \mathbf{p}\hat{a}_{-\mathbf{p}}^\dagger \hat{a}_{\mathbf{p}}^\dagger e^{2i\omega_{\mathbf{p}}t}) \\
&= \frac{1}{2} \int d^3p \, \mathbf{p} (\hat{a}_{\mathbf{p}}^\dagger \hat{a}_{\mathbf{p}} + \hat{a}_{\mathbf{p}} \hat{a}_{\mathbf{p}}^\dagger).
\end{aligned} \tag{2.227}$$

It should be noted that the contribution involving $\hat{a}_{-\mathbf{p}} \hat{a}_{\mathbf{p}}$ and $\hat{a}_{-\mathbf{p}}^\dagger \hat{a}_{\mathbf{p}}^\dagger$ are dropped out since the integrand is an odd function of $\mathbf{p}$. We can see that $\mathbf{p}$ has the meaning of the momentum of single-particle state. The particle states generated by $\hat{a}_{\mathbf{p}}^\dagger$ are also called the field quanta, which carry the momentum $\mathbf{p}$ and energy $\omega_{\mathbf{p}} = (\mathbf{p}^2 + m^2)^{1/2}$, and are counted by the number operator $\hat{n}_{\mathbf{p}} = \hat{a}_{\mathbf{p}}^\dagger \hat{a}_{\mathbf{p}}$.

A quanta is a quantum state. We can also consider this quantum state as a quasi-particle with an energy $\omega_{\mathbf{p}} = \sqrt{\mathbf{p}^2 + m^2}$ and momentum $\mathbf{p}$. The point-like particles generated by $\hat{a}^\dagger(x)$ do not have definite energy and momentum and thus are not stable. Only quantum states with definite energy or confined by other symmetry are stable. The quantum states generated by $\hat{a}_{\mathbf{p}}^\dagger$ can be consider as energy excitations. Thus quasi-particles are the energy excitations of quantum fields.

From the vacuum state $|0\rangle$, one can construct the quasi-particle states using the creation operator $\hat{a}_{\mathbf{p}}^\dagger$. One-quasi-particle states are the linear superposition of $\hat{a}_{\mathbf{p}}^\dagger|0\rangle$. Two-quasi-particle states are the linear superposition of $\hat{a}_{\mathbf{p}}^\dagger \hat{a}_{\mathbf{p}'}^\dagger|0\rangle$ for $\mathbf{p}' \neq \mathbf{p}$ and $\frac{1}{\sqrt{2}}[\hat{a}_{\mathbf{p}}^\dagger]^2|0\rangle$. Other quasi-particle states can be similarly constructed. The $\frac{1}{\sqrt{2}}$ factor comes in a similar way as that in Eq.(2.35). Usually we do not distinguish the particles and quasi-particles. We only use the name of quasi-particles when it is necessary.

2.4 Complex scalar field

2.4.1 *Lagrangian of complex boson field*

We have discussed the scalar boson field with one component. This field describes the simplest boson particles. The equation of motion for the field is the Klein-Gordon equation. For the scalar boson field with one component, there is a problem in its solution Eq.(2.153) of the equations of motion. The terms corresponding to the annihilation operator $\hat{a}(\mathbf{x}, t)$ and creation operator $\hat{a}^\dagger(\mathbf{x}, t)$ are complex, which is not consistent with the properties of particles without internal degrees of freedom. This problem can be solved by introducing the complex scalar field with the internal degrees of freedom. The equations of motion for the complex scalar field have the solutions with the real annihilation operators $\hat{a}_i(\mathbf{x}, t)$ and creation operators $\hat{a}_i^\dagger(\mathbf{x}, t)$. Although the real scalar boson field is not real because of the above reasons, it can be considered as a toy model useful to introduce the basic concepts and formalisms. For a realistic filed we consider the scalar boson field with two components. This field is equivalent to the complex field $\phi \neq \phi^*$, which corresponds to a doublet of particles and antiparticles.

The covariant Lagrangian density, which is a real-valued function, should be given by

$$\mathcal{L} = \frac{\partial \phi^*}{\partial x_\mu} \frac{\partial \phi}{\partial x^\mu} - m^2 \phi^* \phi, \qquad (2.228)$$

where ϕ and ϕ^* can be treated as independent fields. This can be seen by transforming ϕ and ϕ^* into two real field functions ϕ_1 and ϕ_2 with

$$\phi_1 = \frac{1}{\sqrt{2}}[\phi + \phi^*], \qquad (2.229a)$$

$$\phi_2 = -\frac{i}{\sqrt{2}}[\phi - \phi^*]. \qquad (2.229b)$$

Then we can go to the real valued fields and use the same procedure to derive G_t as in the last section. The two real fields describe two kinds of particles integrated. These two kinds of particles have the same mass

m and the Lagrangian density exhibits an internal symmetry under phase transformation.

$$\phi' = \phi e^{-i\alpha}, \tag{2.230a}$$

$$\phi^{*\prime} = \phi^* e^{i\alpha} \tag{2.230b}$$

with real phase α. The complex scalar boson field is important because it is the basic constituent to construct boson fields with the $SU(N)$ symmetry and we can introduce interaction terms with other types of fields with the gauge invariance.

We have shown that the symmetry of spacetime translation is related to the conservation of energy-momentum. In the following, we will show that the continuous symmetry transformation Eq.(2.230) leads to another conserved quantity.

2.4.2   *Charge conservation*

For the complex scalar boson field, the Lagrangian density has an internal symmetry under the transformation Eq.(2.230). The infinitesimal form of the transformation Eq.(2.230) is given by

$$\phi'(x) = \phi(x) - i\alpha\phi(x), \tag{2.231a}$$

$$\phi^{*\prime}(x) = \phi^*(x) + i\alpha\phi^*(x), \tag{2.231b}$$

where α is an infinitesimal parameter. It is conventional to scale the infinitesimal parameter α out of the current j.

Noether's theorem for this continuous symmetry transformation leads to a conserved quantity we now call the charge

$$
\begin{aligned}
Q &= \int d^3x\, j^0(x) = -i \int d^3x \left(\frac{\partial \mathcal{L}}{\partial \partial_0 \phi} \phi - \frac{\partial \mathcal{L}}{\partial \partial_0 \phi^*} \phi^* \right) \\
&= -i \int d^3x (\dot{\phi}^* \phi - \dot{\phi}\phi^*) = i \int d^3x (\phi^* \overset{\leftrightarrow}{\partial_0} \phi).
\end{aligned}
\tag{2.232}
$$

The transformation Eq.(2.230) is a global one because α is a constant and the transformation is same for all the spacetime positions. Later we will see that we should have a local transformation with $\alpha = \alpha(x)$ when the scalar bosons are coupled with the massless vector bosons because the local gauge symmetry should be guaranteed, which demands the local transformation.

2.5 Spinor fermions

2.5.1 *Lagrangian*

Now we turn to another type of the covariant Lagrangian. We use Eq.(2.127) directly without carrying out the integration over π. The Lagrangian contains only linear time derivative term. Dirac had found out the covariant form for this type of Lagrangian in a genius way. The covariant Lagrangian density has been found to be

$$\mathcal{L} = \bar{\psi}(i\gamma^\mu \partial_\mu - m)\psi = i\psi^\dagger \dot{\psi} + i\psi^\dagger \boldsymbol{\alpha} \cdot \boldsymbol{\nabla}\psi - m\psi^\dagger \beta \psi, \qquad (2.233)$$

where we have used the conventional ψ, instead of ϕ, to represent the field function for this field. The field function ψ has four components and satisfies the transformation laws of a relativistic spinor. The adjoint spinor is defined as $\bar{\psi} \equiv \psi^\dagger \gamma^0$. γ^μ ($\mu = 0, 1, 2, 3$) are the four Dirac's matrices or gamma matrices, satisfying the algebra

$$\gamma^\mu \gamma^\nu + \gamma^\nu \gamma^\mu = 2\eta^{\mu\nu} \qquad (2.234)$$

and

$$\gamma^{0\dagger} = \gamma^0, \qquad (2.235a)$$

$$\gamma^{i\dagger} = -\gamma^i. \qquad (2.235b)$$

We have also introduced $\boldsymbol{\alpha}$ and β defined by $\boldsymbol{\alpha} \equiv \gamma^0 \boldsymbol{\gamma}$ and $\beta \equiv \gamma^0$. A set of objects obeying the relations Eqs.(2.234) and (2.235) is said to construct a Clifford algebra. The field ψ is called spinor field. m in Eq.(2.233) is a constant called the mass of particle.

According to Eq.(2.125), we have

$$\pi = \frac{\partial \mathcal{L}}{\partial \dot{\psi}} = i\psi^\dagger \qquad (2.236)$$

and

$$\mathcal{G}_t = \pi \partial_0 \psi - \mathcal{L}(\pi, \psi). \qquad (2.237)$$

Eq.(2.236) gives $\pi = i\psi^\dagger$, which is consistent with Eq.(2.64) for fermions. Since π and ψ should be independent field functions, ψ should not be a real function. Similar to complex scalar field, we need two independent real field functions ϕ_1 and ϕ_2. We define

$$\psi \equiv \frac{1}{\sqrt{2}}(\phi_1 + i\phi_2), \qquad (2.238a)$$

$$\psi^* \equiv \frac{1}{\sqrt{2}}(\phi_1 - i\phi_2). \qquad (2.238b)$$

Then we have two independent complex field functions and we can treat ψ and $\pi = i\psi^\dagger$ as independent fields. This is why the wave functions of electrons are complex functions.

2.5.2   Generator of time translation transformation

The spinor ψ and $\psi^\dagger = -i\pi$ are treated as independent fields, each having four components. Using Eqs.(2.233) and (2.236), Eq.(2.237) for $\mathcal{G}_t$ becomes

$$\begin{aligned}
\mathcal{G}_t &= \psi^\dagger(-i\boldsymbol{\alpha}\cdot\boldsymbol{\nabla} + \beta m)\psi \\
&= \pi(-\boldsymbol{\alpha}\cdot\boldsymbol{\nabla} - i\beta m)\psi.
\end{aligned} \tag{2.239}$$

Transforming ψ and π in Eq.(2.239) into operators, we get the generator of time translation transformation

$$\hat{G}_t(\hat{\pi}, \hat{\psi}) = \int d^3x\hat{\pi}(-\boldsymbol{\alpha}\cdot\boldsymbol{\nabla} - i\beta m)\hat{\psi}, \tag{2.240}$$

which does not contain the time derivative term and fulfills the causality principle.

2.5.3   Dirac equation

For the spinors with internal variables, when we write out the indices explicitly, the commutators Eqs.(2.65) and (2.66) become

$$\{\hat{\psi}_\alpha(\mathbf{x},t), \hat{\pi}_\beta(\mathbf{x}',t)\} = i\delta_{\alpha\beta}\delta^3(\mathbf{x} - \mathbf{x}'), \tag{2.241}$$

$$\{\hat{\psi}_\alpha(\mathbf{x},t), \hat{\psi}_\beta(\mathbf{x}',t)\} = \{\hat{\pi}_\alpha(\mathbf{x},t), \hat{\pi}_\beta(\mathbf{x}',t)\} = 0. \tag{2.242}$$

We have chosen the anti-commutation relations for the fermions. We will show later that the alternate choice of boson commutators would lead to inconsistencies in the formulation.

Using Eq.(2.91), we can derive the equations of motion

$$\begin{aligned}
\dot{\hat{\psi}}_\sigma(\mathbf{x},t) =&i\int d^3x'\Big[\{\hat{\psi}_\sigma(\mathbf{x},t), \hat{\pi}_\alpha(\mathbf{x}',t)\}\boldsymbol{\alpha}_{\alpha\beta}\cdot\boldsymbol{\nabla}'\hat{\psi}_\beta(\mathbf{x}',t) \\
&- \hat{\pi}_\alpha(\mathbf{x}',t)\boldsymbol{\alpha}_{\alpha\beta}\cdot\boldsymbol{\nabla}'\{\hat{\psi}_\sigma(\mathbf{x},t), \hat{\psi}_\beta(\mathbf{x}',t)\} \\
&+ im\{\hat{\psi}_\sigma(\mathbf{x},t), \hat{\pi}_\alpha(\mathbf{x}',t)\}\beta_{\alpha\beta}\hat{\psi}_\beta(\mathbf{x}',t) \\
&- im\hat{\pi}_\alpha(\mathbf{x}',t)\beta_{\alpha\beta}\{\hat{\psi}_\sigma(\mathbf{x},t), \hat{\psi}_\beta(\mathbf{x}',t)\}\Big] \\
=&\int d^3x'\Big[-\delta_{\sigma\alpha}\delta^3(\mathbf{x} - \mathbf{x}')\boldsymbol{\alpha}_{\alpha\beta}\cdot\boldsymbol{\nabla}'\hat{\psi}_\beta(\mathbf{x}',t) \\
&- im\delta_{\sigma\alpha}\delta^3(\mathbf{x} - \mathbf{x}')\beta_{\alpha\beta}\hat{\psi}_\beta(\mathbf{x}',t)\Big] \\
=&(-\boldsymbol{\alpha}\cdot\boldsymbol{\nabla} - im\beta)_{\sigma\beta}\hat{\psi}_\beta(\mathbf{x},t)
\end{aligned} \tag{2.243}$$

or, in compact form,

$$i\frac{\partial\hat{\psi}}{\partial t} = [\hat{\psi}, \hat{G}_t] = -i\boldsymbol{\alpha}\cdot\boldsymbol{\nabla}\hat{\psi} + \beta m\hat{\psi}. \tag{2.244}$$

Similarly we have

$$i\frac{\partial \hat{\pi}}{\partial t} = [\hat{\pi}, \hat{G}_t] = -i\boldsymbol{\nabla}\hat{\pi} \cdot \boldsymbol{\alpha} - m\hat{\pi}\beta. \tag{2.245}$$

Thus

$$i\frac{\partial \hat{\psi}^{\dagger}}{\partial t} = -i\boldsymbol{\nabla}\hat{\psi}^{\dagger} \cdot \boldsymbol{\alpha} - m\hat{\psi}^{\dagger}\beta. \tag{2.246}$$

We can see that Eqs.(2.244) and (2.246) are consistent. If one takes hermitian conjugate operation on both sides of Eq.(2.244), Eq.(2.244) becomes Eq.(2.246). Multiplying Eq.(2.244) by $\gamma^0 = \beta$, we obtain the *Dirac equation* in operator form

$$(i\gamma^{\mu}\partial_{\mu} - m)\hat{\psi} = 0. \tag{2.247}$$

Introducing the adjoint field operator

$$\hat{\bar{\psi}} \equiv \hat{\psi}^{\dagger}\gamma^0, \tag{2.248}$$

we have for Eq.(2.246)

$$\hat{\bar{\psi}}(i\gamma^{\mu}\overleftarrow{\partial}_{\mu} + m) = 0. \tag{2.249}$$

The arrow indicates that the partial derivative acts on the left function.

For spinor fermions, we have the Lagrangian density operator

$$\begin{aligned}\hat{\mathcal{L}} &= \hat{\bar{\psi}}(i\gamma^{\mu}\partial_{\mu} - m)\hat{\psi} \\ &= i\hat{\bar{\psi}}\gamma^{\mu}\partial_{\mu}\hat{\psi} - m\hat{\bar{\psi}}\hat{\psi}.\end{aligned} \tag{2.250}$$

Using the derivatives of the Lagrangian density operator

$$\frac{\partial \hat{\mathcal{L}}}{\partial \hat{\bar{\psi}}} = -m\hat{\psi} \tag{2.251}$$

and

$$\frac{\partial \hat{\mathcal{L}}}{\partial(\partial_{\mu}\hat{\bar{\psi}})} = -i\gamma^{\mu}\hat{\psi}, \tag{2.252}$$

we have

$$\frac{\partial \hat{\mathcal{L}}}{\partial \hat{\bar{\psi}}} - \partial_{\mu}\frac{\partial \hat{\mathcal{L}}}{\partial(\partial_{\mu}\hat{\bar{\psi}})} = (i\gamma^{\mu}\partial_{\mu} - m)\hat{\psi} = 0. \tag{2.253}$$

Eq.(2.253) is the Euler-Lagrange equation in operator form for spinor fermions. The above equation also shows that the Dirac equation Eq.(2.247) is equivalent to the Euler-Lagrange equation in operator form.

Using the Lagrangian density operator in Eq.(2.250), we obtain

$$\frac{\partial \hat{\mathcal{L}}}{\partial \hat{\bar{\psi}}} = i\hat{\psi}^{\dagger}(x) = \hat{\pi}(x). \tag{2.254}$$

2.5.4   *Dirac's matrices*

From Eq.(2.234), we have

$$(\gamma^0)^2 = 1 \quad \text{and} \quad (\gamma^i)^2 = -1, \tag{2.255}$$

which shows that the eigenvalues of the matrix γ^0 are ± 1 and those of γ^i are $\pm i$. In order to be consistent with the condition that the eigenvalues of γ^0 are ± 1 and those of γ^i are $\pm i$, we take γ^0 as hermitian and γ^i as anti-hermitian. This selection is consistent with the condition that the Hamiltonian operator is hermitian and has real eigenvalues, which can be seen easily when we obtain the Hamiltonian for Dirac fermions later.

Eq.(2.234) also gives

$$\gamma^0 = \gamma^i \gamma^0 \gamma^i, \tag{2.256a}$$
$$\gamma^i = -\gamma^0 \gamma^i \gamma^0. \tag{2.256b}$$

Taking the trace on both sides of Eq.(2.256), we obtain

$$\text{Tr}\gamma^0 = \text{Tr}(\gamma^i \gamma^0 \gamma^i) = -\text{Tr}\gamma^0, \tag{2.257a}$$
$$\text{Tr}\gamma^i = -\text{Tr}(\gamma^0 \gamma^i \gamma^0) = -\text{Tr}\gamma^i, \tag{2.257b}$$

which leads to

$$\text{Tr}\gamma^\mu = 0. \tag{2.258}$$

The trace of a matrix is the sum of its eigenvalues. $\text{Tr}\gamma^\mu = 0$ means that $\gamma^0(\gamma^i)$ shall have as many eigenvalues of $+1(+i)$ as those of $-1(-i)$ so that the sum of them is zero. Therefore, the order N of the matrix γ^μ should be an even number. For $N = 2$, we have the unit matrix

$$I = \begin{pmatrix} 1 & 0 \\ 0 & 1 \end{pmatrix} \tag{2.259}$$

and three Pauli's matrices

$$\sigma^1 = \begin{pmatrix} 0 & 1 \\ 1 & 0 \end{pmatrix}, \quad \sigma^2 = \begin{pmatrix} 0 & -i \\ i & 0 \end{pmatrix}, \quad \sigma^3 = \begin{pmatrix} 1 & 0 \\ 0 & -1 \end{pmatrix} \tag{2.260}$$

as a set of independent basis. However, they are not enough to construct γ^μ. Thus the smallest possible order is $N = 4$. There are 16 independent 4×4 matrices. The representation of γ^μ in 4×4 complex matrices is called the spinor representation and correspondingly ψ is the column matrix with four components, which is called the *Dirac spinor*.

2.5.5 *Dirac-Pauli representation*

The 16 independent matrices Γ^i as a set of independent basis can be constructed using γ^μ and the unit matrix I in the following way. We consider all the possible ways of multiplying γ^μ together. Since $(\gamma^\mu)^2$ is equal to $+1$ or -1, we need only consider the multiplications $\gamma^\mu\gamma^\nu$, $\gamma^\mu\gamma^\nu\gamma^\lambda$ and $\gamma^\mu\gamma^\nu\gamma^\lambda\gamma^\rho$ with $\mu \neq \nu \neq \lambda \neq \rho$. There is only one product of four matrices, which we denoted as γ^5

$$\gamma^5 \equiv i\gamma^0\gamma^1\gamma^2\gamma^3. \tag{2.261}$$

γ^5 anti-commutes with γ^μ ($\mu = 0,1,2,3$),

$$\{\gamma^5, \gamma^\mu\} = 0. \tag{2.262}$$

There are four different products of three gamma matrices. They are $\gamma^\mu\gamma^5$ ($\mu = 0,1,2,3$). Since γ^μ anti-commutes with each other, we have six products of two gamma matrices

$$\sigma^{\mu\nu} \equiv \frac{i}{2}[\gamma^\mu, \gamma^\nu]. \tag{2.263}$$

Together with the unit matrix and four γ^μ, we have the complete set of 16 matrices

$$\{I, \gamma^\mu, \sigma^{\mu\nu}, \gamma^\mu\gamma^5, \gamma^5\}. \tag{2.264}$$

The most used representation of γ_μ is so-called *Dirac-Pauli representation* which has the form

$$\gamma^0 = \begin{pmatrix} I & 0 \\ 0 & -I \end{pmatrix}, \quad \gamma^i = \begin{pmatrix} 0 & \sigma^i \\ -\sigma^i & 0 \end{pmatrix}. \tag{2.265}$$

In this representation, γ^0 is diagonal. There are also other representations which are equivalent to each other. If we choose γ^5 as diagonal matrix, it is called the *Weyl representation*.

Comparing Eq.(2.235) with Eq.(2.256), we have

$$\gamma^{\mu\dagger} = \gamma^0\gamma^\mu\gamma^0. \tag{2.266}$$

Thus the hermitian conjugate of $\sigma^{\mu\nu}$ is

$$\sigma^{\mu\nu\dagger} = \gamma^0\sigma^{\mu\nu}\gamma^0. \tag{2.267}$$

The explicit form of $\sigma^{\mu\nu}$ in the standard representation is given by

$$\sigma^{0j} = i\gamma^0\gamma^j = i\begin{pmatrix} 0 & \sigma^j \\ \sigma^j & 0 \end{pmatrix} = i\alpha^j, \tag{2.268a}$$

$$\sigma^{ij} = i\gamma^i\gamma^j(1 - \delta_{ij}) = \epsilon^{ijk}\begin{pmatrix} \sigma^k & 0 \\ 0 & \sigma^k \end{pmatrix} = \epsilon^{ijk}\Sigma^k \tag{2.268b}$$

with

$$\Sigma^k \equiv \begin{pmatrix} \sigma^k & 0 \\ 0 & \sigma^k \end{pmatrix}, \tag{2.269}$$

where ϵ^{ijk} is the antisymmetric Levi-Civita symbol, which is totally anti-symmetric with $\epsilon^{123} = 1$. Σ^k is the double Pauli's matrix, which can be expressed in a vector form

$$\Sigma = \begin{pmatrix} \sigma & 0 \\ 0 & \sigma \end{pmatrix}. \tag{2.270}$$

From Eq.(2.268a), we have

$$\alpha^j = \begin{pmatrix} 0 & \sigma^j \\ \sigma^j & 0 \end{pmatrix}. \tag{2.271}$$

2.5.6   *Lorentz transformation for spinors*

It is not easy to see that the Lagrangian density of spinor fermions in Eq.(2.233) is covariant.[2] Before we prove the covariance of the Lagrangian density of spinor fermions given by Eq.(2.233), we inspect first the Lorentz transformation of spinor fields.

A Lorentz transformation is expressed as

$$(x')^\mu = \Lambda^\mu{}_\nu x^\nu. \tag{2.274}$$

The Lagrangian density of spinor fermions and Dirac equation should be covariant for any Lorentz transformation. Since Dirac equation is linear, the transformation relation between $\psi'(x')$ and $\psi(x)$ should be linear. Then we have

$$\psi'(x') = S(\Lambda)\psi(x), \tag{2.275}$$

where $S(\Lambda)$ is a 4×4 matrix. The components form of Eq.(2.275) is given by

$$\psi'_a(x') = S_{ab}(\Lambda)\psi_b(x). \tag{2.276}$$

---

[2]The reason that Dirac demanded γ^μ obeying Eq.(2.234) is as follows: Since it is not easy to see directly that the Dirac equation is covariant, it is natural to do some trying. Multiplying the operator $(i\gamma^\mu \partial_\mu + m)$ on the Dirac equation gives

$$-(\gamma^\mu \gamma^\nu \partial_\mu \partial_\nu + m^2)\psi = -\left[\frac{1}{2}(\gamma^\mu \gamma^\nu + \gamma^\nu \gamma^\mu)\partial_\mu \partial_\nu + m^2\right]\psi = 0. \tag{2.272}$$

If γ^μ obey Eq.(2.234), Eq.(2.272) becomes

$$(\partial_\mu \partial^\mu + m^2)\psi = 0, \tag{2.273}$$

which is Lorentz covariant.

Covariance requires $\psi'(x')$ to be a solution of the Dirac equation.

$$(i\gamma^\mu \partial'_\mu - m)\psi'(x') = 0. \tag{2.277}$$

Multiplying the Dirac equation Eq.(2.247) from the left by S

$$\begin{aligned}
S(i\gamma^\mu \partial_\mu - m)\psi(x) &= S(i\gamma^\mu \partial_\mu S^{-1}S - m)\psi(x) \\
&= (iS\gamma^\mu S^{-1}\partial_\mu - m)\psi'(x') \\
&= (iS\gamma^\mu S^{-1}\Lambda^\nu{}_\mu \partial'_\nu - m)\psi'(x') \\
&= 0.
\end{aligned} \tag{2.278}$$

Comparing Eq.(2.278) with Eq.(2.277), we have

$$S\gamma^\mu S^{-1}\Lambda^\nu{}_\mu = \gamma^\nu. \tag{2.279}$$

Eq.(2.279) can be rewritten as

$$S^{-1}\gamma^\mu S = \Lambda^\mu{}_\nu \gamma^\nu. \tag{2.280}$$

An infinitesimal proper Lorentz transformation is given by

$$\Lambda^\mu{}_\nu = \delta^\mu{}_\nu + \Delta\omega^\mu{}_\nu, \tag{2.281}$$

where $\Delta\omega^\mu{}_\nu$ is antisymmetric,

$$\Delta\omega^\mu{}_\nu = -\Delta\omega_\nu{}^\mu. \tag{2.282}$$

Eq.(2.282) can be easily derived using the relation

$$\Lambda^\lambda{}_\mu \Lambda_\lambda{}^\nu = \delta^\nu_\mu. \tag{2.283}$$

Inserting Eq.(2.281), we have

$$\begin{aligned}
\Lambda^\lambda{}_\mu \Lambda_\lambda{}^\nu &= (\delta^\lambda{}_\mu + \Delta\omega^\lambda{}_\mu)(\delta_\lambda{}^\nu + \Delta\omega_\lambda{}^\nu) \\
&= \delta^\lambda{}_\mu \delta_\lambda{}^\nu + \delta^\lambda{}_\mu \Delta\omega_\lambda{}^\nu + \delta_\lambda{}^\nu \Delta\omega^\lambda{}_\mu \\
&= \delta^\nu_\mu + \Delta\omega_\mu{}^\nu + \Delta\omega^\nu{}_\mu \\
&= \delta^\nu_\mu,
\end{aligned} \tag{2.284}$$

which gives

$$\Delta\omega_\mu{}^\nu + \Delta\omega^\nu{}_\mu = 0 \tag{2.285}$$

or

$$\Delta\omega^{\mu\nu} = -\Delta\omega^{\nu\mu}. \tag{2.286}$$

Under an infinitesimal Lorentz transformation, $S(\Lambda)$ should have the form

$$S(\Lambda) = 1 - \frac{i}{4}\Delta\omega^{\mu\nu}\sigma_{\mu\nu}, \tag{2.287}$$

where $\sigma_{\mu\nu}$ is a 4×4 antisymmetric matrix.

$$\sigma_{\mu\nu} = -\sigma_{\nu\mu}. \tag{2.288}$$

The factor $-\frac{i}{4}$ in Eq.(2.287) is introduced for simplicity in notation, which will become clear later. From Eq.(2.287), we have

$$S^{-1}(\Lambda) = 1 + \frac{i}{4}\Delta\omega^{\mu\nu}\sigma_{\mu\nu}. \tag{2.289}$$

Inserting Eqs.(2.287) and (2.289) into Eq.(2.280), we have

$$(1 + \frac{i}{4}\Delta\omega^{\alpha\beta}\sigma_{\alpha\beta})\gamma^{\mu}(1 - \frac{i}{4}\Delta\omega^{\alpha\beta}\sigma_{\alpha\beta}) = (\delta_{\nu}{}^{\mu} + \Delta\omega^{\mu}{}_{\nu})\gamma^{\nu}. \tag{2.290}$$

Neglecting the quadratic terms in $\Delta\omega^{\mu\nu}$, we have

$$\begin{aligned}
\frac{i}{4}\Delta\omega^{\alpha\beta}(\sigma_{\alpha\beta}\gamma^{\mu} - \gamma^{\mu}\sigma_{\alpha\beta}) &= \Delta\omega^{\mu}{}_{\nu}\gamma^{\nu} \\
&= \delta^{\mu}{}_{\sigma}\Delta\omega^{\sigma}{}_{\nu}\gamma^{\nu} \\
&= -\Delta\omega_{\beta}{}^{\alpha}\delta^{\mu}{}_{\alpha}\gamma^{\beta} \\
&= -\Delta\omega^{\beta\alpha}\delta^{\mu}{}_{\alpha}\gamma_{\beta} \\
&= -\frac{1}{2}(\Delta\omega^{\beta\alpha}\delta^{\mu}{}_{\alpha}\gamma_{\beta} - \Delta\omega^{\alpha\beta}\delta^{\mu}{}_{\alpha}\gamma_{\beta}) \\
&= -\frac{1}{2}\Delta\omega^{\beta\alpha}(\delta^{\mu}{}_{\alpha}\gamma_{\beta} - \delta^{\mu}{}_{\beta}\gamma_{\alpha}) \\
&= \frac{1}{2}\Delta\omega^{\alpha\beta}(\delta^{\mu}{}_{\alpha}\gamma_{\beta} - \delta^{\mu}{}_{\beta}\gamma_{\alpha}).
\end{aligned} \tag{2.291}$$

Thus we obtain the relation

$$2i(\delta^{\mu}{}_{\alpha}\gamma_{\beta} - \delta^{\mu}{}_{\beta}\gamma_{\alpha}) = [\gamma^{\mu}, \sigma_{\alpha\beta}]. \tag{2.292}$$

The solution of Eq.(2.292) is given by

$$\sigma_{\alpha\beta} = \frac{i}{2}[\gamma_{\alpha}, \gamma_{\beta}]. \tag{2.293}$$

This solution of Eq.(2.292) for $\sigma_{\alpha\beta}$ is the same as that defined by Eq.(2.263), where we have intentionally used the same symbol. Thus the operator $S(\Lambda)$ for an infinitesimal proper Lorentz transformation has the form

$$S(\Delta\omega^{\mu\nu}) = 1 + \frac{1}{8}[\gamma_{\mu}, \gamma_{\nu}]\Delta\omega^{\mu\nu}. \tag{2.294}$$

We can also introduce the infinitesimal generators given by

$$(I_{\mu\nu})_{\alpha\beta} = -\frac{i}{2}(\sigma_{\mu\nu})_{\alpha\beta} = \frac{1}{4}[\gamma_{\mu}, \gamma_{\nu}], \tag{2.295}$$

where $\mu, \nu = 0, \cdots, 3$ are the Lorentz indices and $\alpha, \beta = 1, \cdots, 4$ are the Dirac indices. Then Eq.(2.294) becomes

$$S(\Delta\omega^{\mu\nu}) = 1 + \frac{1}{2}I_{\mu\nu}\Delta\omega^{\mu\nu}. \tag{2.296}$$

2.5.7 Covariance of spinor fermion Lagrangian

From Eq.(2.267), we have

$$S^\dagger = \gamma^0 S^{-1} \gamma^0. \tag{2.297}$$

Since $\psi'^\dagger(x') = \psi^\dagger(x) S^\dagger$, we have

$$\bar{\psi}'^\dagger(x') = \psi^\dagger(x) S^\dagger \gamma^0 = \psi^\dagger(x) \gamma^0 \gamma^0 S^\dagger \gamma^0 = \bar{\psi}(x) S^{-1}. \tag{2.298}$$

Using Eq.(2.297), we have

$$\begin{aligned}
\bar{\psi}'(x')\psi'(x') &= \psi'^\dagger(x')\gamma^0\psi'(x') \\
&= \psi^\dagger(x) S^\dagger \gamma^0 S\psi(x) \\
&= \psi^\dagger(x)\gamma^0 S^{-1} S\psi(x) \\
&= \bar{\psi}(x)\psi(x),
\end{aligned} \tag{2.299}$$

which shows that $\bar{\psi}\psi$ is a scalar. Similarly we can show that $\bar{\psi}\gamma^\mu\psi$ is a Lorentz vector. Using Eqs.(2.280) and (2.297), we have

$$\begin{aligned}
\bar{\psi}'(x')\gamma^\mu\psi'(x') &= \psi'^\dagger(x')\gamma^0\gamma^\mu\psi'(x') \\
&= \psi^\dagger(x) S^\dagger \gamma^0 \gamma^\mu S\psi(x) \\
&= \bar{\psi}(x) S^{-1}\gamma^\mu S\psi(x) \\
&= \Lambda^\mu{}_\nu \bar{\psi}(x)\gamma^\nu\psi(x),
\end{aligned} \tag{2.300}$$

which shows that $\bar{\psi}(x)\gamma^\mu\psi(x)$ is a vector. Since $\bar{\psi}(x)\psi(x)$ is a scalar and $\bar{\psi}(x)\gamma^\mu\psi(x)$ is a vector, the Lagrangian given in Eq.(2.233) is a Lorentz scalar and thus Lorentz covariant.

2.5.8 Spatial reflection

In addition to the scalar $\bar{\psi}\psi$ and Lorentz vector $\bar{\psi}\gamma^\mu\psi$, we can also define pseudo scalar and pseudo vector related to the spatial reflection.

A spatial reflection is defined by the following transformation

$$\mathbf{x}' = -\mathbf{x}, \tag{2.301a}$$

$$t' = t. \tag{2.301b}$$

The corresponding transformation matrix is

$$\Lambda^\mu{}_\nu = \begin{pmatrix} 1 & 0 & 0 & 0 \\ 0 & -1 & 0 & 0 \\ 0 & 0 & -1 & 0 \\ 0 & 0 & 0 & -1 \end{pmatrix}. \tag{2.302}$$

The spatial reflection is one of the improper Lorentz transformation because it can not be generated by means of infinitesimal rotations. We denote the corresponding spinor transformation $S(\Lambda)$ as P (P is for parity). According to Eq.(2.280), we have

$$P^{-1}\gamma^\mu P = \Lambda^\mu{}_\nu \gamma^\nu. \tag{2.303}$$

Comparing with Eq.(2.302), we have

$$\Lambda^\nu{}_\mu = \eta^\nu{}_\mu. \tag{2.304}$$

Then we have

$$\Lambda^\mu{}_\lambda \Lambda^\lambda{}_\nu \gamma^\nu = P\Lambda^\mu{}_\nu \gamma^\nu P^{-1}, \tag{2.305}$$

which gives

$$\delta^\mu{}_\nu \gamma^\nu = P\sum_{\nu=0}^{3} \eta^{\mu\nu}\gamma^\nu P^{-1} \tag{2.306}$$

or equivalently

$$P^{-1}\gamma^\mu P = \eta^{\mu\mu}\gamma^\mu. \tag{2.307}$$

It should be noted that there is no summation on the right hand of Eq.(2.307). The solution of Eq.(2.307) for P is

$$P = e^{i\varphi}\gamma^0, \tag{2.308a}$$

$$P^{-1} = e^{-i\varphi}\gamma^0, \tag{2.308b}$$

where φ is a phase factor. Using Eq.(2.266), we have

$$P^\dagger = e^{-i\varphi}\gamma^0 = P^{-1}. \tag{2.309}$$

The explicit form of the spinor transformation under the spatial reflection is given by

$$\psi'(\mathbf{x}',t) = \psi'(-\mathbf{x},t) = P\psi(x) = e^{i\varphi}\gamma^0\psi(\mathbf{x},t). \tag{2.310}$$

Using Eq.(2.308), we have

$$\begin{aligned}
\gamma^5 P &= \gamma^5 e^{i\varphi}\gamma^0 = -e^{i\varphi}\gamma^0\gamma^5 \\
&= -P\gamma^5 = \det|\Lambda|P\gamma^5
\end{aligned} \tag{2.311}$$

or

$$P^{-1}\gamma^5 P = \det|\Lambda|\gamma^5. \tag{2.312}$$

For a proper Lorentz transformation, $S(\Lambda)$ contains only $\sigma_{\mu\nu}$. We note

$$\gamma^\mu\gamma^5 + \gamma^\mu\gamma^5 = i\gamma^\mu\gamma^0\gamma^1\gamma^2\gamma^3 + i\gamma^0\gamma^1\gamma^2\gamma^3\gamma^\mu = 0, \tag{2.313}$$

which leads to

$$[\gamma^5, \sigma_{\mu\nu}] = \frac{1}{2}[\gamma^5(\gamma^\mu\gamma^\nu - \gamma^\nu\gamma^\mu) - (\gamma^\mu\gamma^\nu - \gamma^\nu\gamma^\mu)\gamma^5] = 0. \tag{2.314}$$

Since $S(\Lambda)$ contains only $\sigma_{\mu\nu}$, we have

$$[S(\Lambda), \gamma^5] = 0, \tag{2.315}$$

which gives

$$S(\Lambda)^{-1}\gamma^5 S(\Lambda) = \gamma^5. \tag{2.316}$$

Combining Eq.(2.312) with Eq.(2.316), we see that γ^5 behaves as a pseudo scalar.

2.5.9   *Energy-momentum tensor and Hamiltonian operator*

The action should possess the symmetry of spacetime translation. Under an infinitesimal spacetime translation, the variation of the action operator should be zero. We have $\delta\hat{S} = 0$. Similar to the derivation of Eq.(2.194), we have the conservation of energy-momentum

$$\partial_\mu \hat{\Theta}^\mu_\nu(x) = 0 \qquad (2.317)$$

with the canonical energy-momentum tensor $\hat{\Theta}_{\mu\nu}$ given by

$$\hat{\Theta}_{\mu\nu} = \frac{\partial\hat{\mathcal{L}}}{\partial\partial^\mu\hat{\psi}}\partial_\nu\hat{\psi} + \frac{\partial\hat{\mathcal{L}}}{\partial\partial^\mu\hat{\psi}^\dagger}\partial_\nu\hat{\psi}^\dagger - \eta_{\mu\nu}\hat{\mathcal{L}}$$
$$= \hat{\bar{\psi}}i\gamma_\mu\partial_\nu\hat{\psi} - \eta_{\mu\nu}\hat{\bar{\psi}}(i\gamma^\sigma\partial_\sigma - m)\hat{\psi}. \qquad (2.318)$$

Using Eq.(2.318), we get the conserved energy-momentum vector operator

$$\hat{P}_\nu = \int d^3x\hat{\Theta}^0_\nu = \int d^3x(\hat{\bar{\psi}}i\gamma^0\partial_\nu\hat{\psi} - \eta^0_\nu\hat{\bar{\psi}}(i\gamma^\sigma\partial_\sigma - m)\hat{\psi}). \qquad (2.319)$$

The time component of this vector operator is the energy operator

$$\hat{P}_0 = \int d^3x\hat{\bar{\psi}}(i\gamma^0\partial_0 - i\gamma^0\partial_0 - i\boldsymbol{\gamma}\cdot\boldsymbol{\nabla} + m)\hat{\psi}$$
$$= \int d^3x\hat{\psi}^\dagger(-i\boldsymbol{\alpha}\cdot\boldsymbol{\nabla} + \beta m)\hat{\psi}. \qquad (2.320)$$

Thus we get the Hamiltonian operator

$$\hat{H} = \hat{P}_0 = \int d^3x\hat{\psi}^\dagger(-i\boldsymbol{\alpha}\cdot\boldsymbol{\nabla} + \beta m)\hat{\psi}. \qquad (2.321)$$

Using Eq.(2.236) and replacing the field functions by the corresponding operators, we can see that Eq.(2.321) becomes Eq.(2.240). Therefore, the generator of time translation transformation $\hat{G}_t$ is the Hamiltonian operator $\hat{H}$.

$$\hat{G}_t = \hat{H}. \qquad (2.322)$$

From Eq.(2.319), we get the momentum vector

$$\mathbf{P} = -i\int d^3x\psi^\dagger\boldsymbol{\nabla}\psi = -\int d^3x\pi\boldsymbol{\nabla}\psi. \qquad (2.323)$$

2.5.10 *Lorentz invariance*

Since the Lagrangian density is a scalar due to the covariance principle, the Lagrangian density and thus the action is invariant under Lorentz transformation because of Eq.(A.20) in Appendix A and $\det(\Lambda) = 1$ for Λ in Eq.(A.36). Under an infinitesimal Lorentz transformation, we have

$$\delta x^\nu = \delta\omega^{\nu\mu}x_\mu \tag{2.324}$$

and

$$\delta\psi = -\frac{i}{4}\delta\omega_{\mu\nu}\sigma^{\mu\nu}\psi(x). \tag{2.325}$$

Eq.(2.325) is the reformulation of Eq.(2.275) with S replaced by the expression in Eq.(2.296). Under an infinitesimal Lorentz transformation, $\delta S = 0$. According to Eq.(2.188), we have the conserved current

$$j_\mu(x) = \frac{\partial\mathcal{L}(x)}{\partial(\partial^\mu\psi)}\left(-\frac{i}{4}\delta\omega_{\nu\lambda}\sigma^{\nu\lambda}\psi(x)\right) - \Theta_{\mu\nu}\delta\omega^{\nu\lambda}x_\lambda, \tag{2.326}$$

where we have omitted the derivative term containing $\frac{\partial\mathcal{L}(x)}{\partial(\partial^\mu\psi^\dagger)}$ because it is zero. Since $\delta\omega^{\nu\lambda}$ is antisymmetric, the last term in Eq.(2.326) can be written as

$$\Theta_{\mu\nu}\delta\omega^{\nu\lambda}x_\lambda = \frac{1}{2}\delta\omega^{\nu\lambda}(\Theta_{\mu\nu}x_\lambda - \Theta_{\mu\lambda}x_\nu). \tag{2.327}$$

Thus Eq.(2.326) becomes

$$j_\mu(x) = \frac{1}{2}\delta\omega^{\nu\lambda}M_{\mu\nu\lambda}(x) \tag{2.328}$$

with

$$M_{\mu\nu\lambda}(x) = \Theta_{\mu\lambda}x_\nu - \Theta_{\mu\nu}x_\lambda + \frac{\partial\mathcal{L}(x)}{\partial(\partial^\mu\psi)}\left(-\frac{i}{2}\sigma_{\nu\lambda}\right)\psi(x). \tag{2.329}$$

The conserved quantity is the antisymmetric tensor

$$\begin{aligned} M_{\nu\lambda} &\equiv \int M_{0\nu\lambda}d^3x \\ &= \int d^3x\left[\Theta_{0\lambda}x_\nu - \Theta_{0\nu}x_\lambda + \frac{\partial\mathcal{L}(x)}{\partial(\partial^0\psi)}\left(-\frac{i}{2}\sigma_{\nu\lambda}\right)\psi(x)\right]. \end{aligned} \tag{2.330}$$

$M_{\nu\lambda}$ is called the *tensor of generalized angular momentum*.

$M_{\nu\lambda}$ consists of two parts

$$M_{\nu\lambda} = L_{\nu\lambda} + S_{\nu\lambda} \tag{2.331}$$

with

$$L_{\nu\lambda} \equiv \int d^3x (\Theta_{0\lambda}x_\nu - \Theta_{0\nu}x_\lambda)$$

$$= \int d^3x \frac{\partial \mathcal{L}}{\partial \partial^0 \psi} \left(x_\nu \frac{\partial}{\partial x^\lambda} - x_\lambda \frac{\partial}{\partial x^\nu} \right) \psi \qquad (2.332)$$

$$= i \int d^3x \psi^\dagger \left(x_\nu \frac{\partial}{\partial x^\lambda} - x_\lambda \frac{\partial}{\partial x^\nu} \right) \psi$$

and

$$S_{\nu\lambda} \equiv \int d^3x \frac{\partial \mathcal{L}}{\partial \partial^0 \psi} \left(-\frac{i}{2}\sigma_{\nu\lambda} \right) \psi = \frac{1}{2} \int d^3x \psi^\dagger \sigma_{\nu\lambda} \psi. \qquad (2.333)$$

The conservation of the angular momentum reflects the spatial rotation invariance in the Minkowski spacetime. For a spatial rotation, the indices take the values $1, 2, 3$ for ν and λ. Since both L_{ij} and S_{ij} are antisymmetric, we can use vectors to represent them. We define

$$L^k \equiv \frac{1}{2}\epsilon^{ijk}L_{ij} \qquad (2.334)$$

and

$$S^k \equiv \frac{1}{2}\epsilon^{ijk}S_{ij}. \qquad (2.335)$$

L^k is called the vector of *orbital angular momentum* and S^k is called the vector of *spin angular momentum*. Using the vector symbol with components $L_i(S_i) = -L^i(S^i)$, we have the three-dimensional vectors of orbital and spin angular momentum

$$\mathbf{L} = -i \int d^3x \psi^\dagger \mathbf{x} \times \boldsymbol{\nabla}\psi, \qquad (2.336a)$$

$$\mathbf{S} = -\frac{1}{2} \int d^3x \psi^\dagger \boldsymbol{\Sigma}\psi. \qquad (2.336b)$$

Since

$$\left(\frac{1}{2}\boldsymbol{\Sigma} \right) \cdot \left(\frac{1}{2}\boldsymbol{\Sigma} \right) = \left(\frac{1}{2}\boldsymbol{\sigma} \right) \cdot \left(\frac{1}{2}\boldsymbol{\sigma} \right) = \frac{3}{4} = \frac{1}{2}\left(1 + \frac{1}{2} \right), \qquad (2.337)$$

we define a spin $S = \frac{1}{2}$ for Dirac fermions

For scalar bosons, ϕ is a scalar and thus $\delta\phi_a = 0$ under an infinitesimal Lorentz transformation. Therefore, there is no spin for scalar bosons or equivalently $S = 0$ for scalar bosons. Then for scalar bosons, Eq.(2.329) is replaced by

$$M_{\mu\nu\lambda}(x) = \Theta_{\mu\lambda}x_\nu - \Theta_{\mu\nu}x_\lambda. \qquad (2.338)$$

The conserved antisymmetric tensor is given by

$$M_{\nu\lambda} = \int M_{0\nu\lambda} d^3x$$
$$= \int d^3x \left[\Theta_{0\lambda} x_\nu - \Theta_{0\nu} x_\lambda\right] \tag{2.339}$$
$$= L_{\nu\lambda}.$$

Thus

$$L_{ij} = \int d^3x \frac{\partial \mathcal{L}}{\partial \partial^0 \phi} \left(x_i \frac{\partial}{\partial x^j} - x_j \frac{\partial}{\partial x^i}\right) \phi, \tag{2.340}$$

which gives

$$\mathbf{L} = -\int d^3x \pi(x)(\mathbf{x} \times \boldsymbol{\nabla})\phi(x). \tag{2.341}$$

When we write the orbital angular momentum in an operator form by the changing of $\phi \to \hat{\phi}$ and $\pi \to \hat{\pi}$ in Eq.(2.341), we have

$$\hat{\mathbf{L}} = -\int d^3x \hat{\pi}(x)(\mathbf{x} \times \boldsymbol{\nabla})\hat{\phi}(x). \tag{2.342}$$

Now we consider the commutation relations for the orbital angular momentum operator $\hat{\mathbf{L}}$. Using Eq.(2.342), we have

$$[\hat{L}_i, \hat{L}_j] = \int d^3x d^3x' [\hat{\pi}(x)(\mathbf{x} \times \boldsymbol{\nabla})_i \hat{\phi}(x), \hat{\pi}(x')(\mathbf{x}' \times \boldsymbol{\nabla}')_j \hat{\phi}(x')]$$
$$= \int d^3x d^3x' \{\hat{\pi}(x)(\mathbf{x} \times \boldsymbol{\nabla})_i [\hat{\phi}(x), \hat{\pi}(x')](\mathbf{x}' \times \boldsymbol{\nabla}')_j \hat{\phi}(x') \tag{2.343}$$
$$+ \hat{\pi}(x')(\mathbf{x}' \times \boldsymbol{\nabla}')_j [\hat{\pi}(x), \hat{\phi}(x')](\mathbf{x} \times \boldsymbol{\nabla})_i \hat{\phi}(x)\}.$$

In Eq.(2.343), $x_0 = x_0'$ because we evaluate the equal-time commutation relations. Then we use the commutation relation

$$[\hat{\phi}, \hat{\pi}] = i\delta^3(\mathbf{x} - \mathbf{x}'), \tag{2.344}$$

which gives

$$[\hat{L}_i, \hat{L}_j] = \int d^3x d^3x' [\hat{\pi}(x)(\mathbf{x} \times \boldsymbol{\nabla})_i i\delta^3(\mathbf{x} - \mathbf{x}')(\mathbf{x}' \times \boldsymbol{\nabla}')_j \hat{\phi}(x')$$
$$- \hat{\pi}(x')(\mathbf{x}' \times \boldsymbol{\nabla}')_j i\delta^3(\mathbf{x} - \mathbf{x}')(\mathbf{x} \times \boldsymbol{\nabla})_i \hat{\phi}(x)]. \tag{2.345}$$

Integrating by parts, we have

$$[\hat{L}_i, \hat{L}_j] = -i \int d^3x d^3x' \delta^3(\mathbf{x} - \mathbf{x}')[(\mathbf{x} \times \boldsymbol{\nabla})_i \hat{\pi}(x)(\mathbf{x}' \times \boldsymbol{\nabla}')_j \hat{\phi}(x')$$
$$- (\mathbf{x}' \times \boldsymbol{\nabla}')_j \hat{\pi}(x')(\mathbf{x} \times \boldsymbol{\nabla})_i \hat{\phi}(x)]$$
$$= i \int d^3x [\hat{\pi}(x)(\mathbf{x} \times \boldsymbol{\nabla})_i (\mathbf{x} \times \boldsymbol{\nabla})_j \hat{\phi}(x) \tag{2.346}$$
$$- \hat{\pi}(x)(\mathbf{x} \times \boldsymbol{\nabla})_j (\mathbf{x} \times \boldsymbol{\nabla})_i \hat{\phi}(x)].$$

Using the mathematical identity

$$(\mathbf{x} \times \boldsymbol{\nabla})_i (\mathbf{x} \times \boldsymbol{\nabla})_j - (\mathbf{x} \times \boldsymbol{\nabla})_j (\mathbf{x} \times \boldsymbol{\nabla})_i$$
$$= x_j \partial_i - x_i \partial_j \tag{2.347}$$
$$= -\epsilon^{ijk} (\mathbf{x} \times \boldsymbol{\nabla})_k,$$

we obtain

$$[\hat{L}_i, \hat{L}_j] = i\epsilon^{ijk} \hat{L}_k. \tag{2.348}$$

Similarly we can prove that Eq.(2.348) holds also for the spinor fermions. Using Eq.(F.2), we can easily obtain

$$[\hat{S}_i, \hat{S}_j] = i\epsilon^{ijk} \hat{S}_k. \tag{2.349}$$

The spin angular momentum operators obey the same commutation relations as the orbital angular momentum operators

2.5.11 *Symmetric energy-momentum tensor*

Noether's theorem leads to conservation law. The density and current obtained in this way are not fixed uniquely because one can add some four dimensional divergence terms without influencing the equation of continuity. For the canonical energy-momentum tensor $\Theta_{\mu\nu}$, we can define a modified tensor through

$$T_{\mu\nu}(x) \equiv \Theta_{\mu\nu}(x) + \partial^\kappa \chi_{\kappa\mu\nu}, \tag{2.350}$$

where $\chi_{\kappa\mu\nu}$ is an arbitrary antisymmetric tensor with respect to the first two indices.

$$\chi_{\kappa\mu\nu} = -\chi_{\mu\kappa\nu}. \tag{2.351}$$

The conservation law remains unchanged for the transformation Eq.(2.350).

$$\begin{aligned}
\partial^\mu T_{\mu\nu} &= \partial^\mu \Theta_{\mu\nu} + \partial^\mu \partial^\kappa \chi_{\kappa\mu\nu} \\
&= \partial^\mu \Theta_{\mu\nu} + \frac{1}{2} \partial^\mu \partial^\kappa (\chi_{\kappa\mu\nu} - \chi_{\mu\kappa\nu}) \\
&= \partial^\mu \Theta_{\mu\nu} \\
&= 0.
\end{aligned} \tag{2.352}$$

Also the total energy and momentum are not affected by the transformation Eq.(2.350).

$$\begin{aligned}
\tilde{P}_\nu &= \int d^3x \, T^0_\nu \\
&= \int d^3x (\Theta^0_\nu + \partial_0 \chi^{00}{}_\nu - \partial_i \chi^{0i}{}_\nu).
\end{aligned} \tag{2.353}$$

Since $\chi_{\kappa\mu\nu}$ is antisymmetric, $\chi^{00}{}_\nu = 0$. Also we use Gauss's theorem and neglect the surface integral terms. Then we obtain

$$\tilde{P}_\nu = \int d^3x \Theta^0_\nu = P_\nu. \tag{2.354}$$

The transformation Eq.(2.350) allows us to construct a symmetric energy-momentum tensor $T_{\mu\nu}$

$$T_{\mu\nu} = T_{\nu\mu}, \tag{2.355}$$

which can be achieved in the following way. Since $\delta\omega^{\nu\lambda}$ is an any anti-symmetric tensor, the equation of continuity Eq.(2.189) for $M^{\mu\nu\lambda}$ can be written as

$$\partial_\mu M^{\mu\nu\lambda} = 0. \tag{2.356}$$

$M^{\mu\nu\lambda}$ in Eq.(2.329) can be written as

$$M^{\mu\nu\lambda}(x) = \Theta^{\mu\lambda}x^\nu - \Theta^{\mu\nu}x^\lambda + \Sigma^{\mu\nu\lambda} \tag{2.357}$$

with

$$\Sigma^{\mu\nu\lambda} = \frac{\partial\mathcal{L}(x)}{\partial(\partial_\mu\psi)}\left(-\frac{i}{2}\sigma_{\nu\lambda}\right)\psi(x). \tag{2.358}$$

We introduce

$$\tau^{\kappa\mu\nu} = \chi^{\kappa\mu\nu} + \chi^{\nu\kappa\mu}. \tag{2.359}$$

Since $\chi^{\kappa\mu\nu}$ is antisymmetric in its fist two indices, $\tau^{\kappa\mu\nu}$ is antisymmetric in its last two indices. Thus we have

$$\tau^{\kappa\mu\nu} + \tau^{\kappa\nu\mu} = 0. \tag{2.360}$$

Then χ can be expressed in terms of τ.

$$\chi^{\kappa\mu\nu} = \frac{1}{2}(\tau^{\kappa\mu\nu} + \tau^{\mu\nu\kappa} - \tau^{\nu\kappa\mu}). \tag{2.361}$$

Thus Eq.(2.350) becomes

$$T^{\mu\nu}(x) = \Theta^{\mu\nu}(x) + \frac{1}{2}\partial_\kappa(\tau^{\kappa\mu\nu} + \tau^{\mu\nu\kappa} - \tau^{\nu\kappa\mu}). \tag{2.362}$$

Since $\Sigma^{\kappa\mu\nu}$ is antisymmetric in its last indices, we can set

$$\tau^{\kappa\mu\nu} = \Sigma^{\kappa\mu\nu}. \tag{2.363}$$

Then

$$T^{\mu\nu}(x) = \Theta^{\mu\nu}(x) + \frac{1}{2}\partial_\kappa(\Sigma^{\kappa\mu\nu} + \Sigma^{\mu\nu\kappa} - \Sigma^{\nu\kappa\mu}). \tag{2.364}$$

Using Eq.(2.360), we have

$$T^{\mu\nu} - T^{\nu\mu} = \Theta^{\mu\nu} - \Theta^{\nu\mu} + \partial_\kappa\Sigma^{\kappa\mu\nu}. \tag{2.365}$$

Inserting Eq.(2.357) into Eq.(2.356), we obtain

$$\partial_\kappa M^{\kappa\mu\nu} = \Theta^{\mu\nu} - \Theta^{\nu\mu} + \partial_\kappa\Sigma^{\kappa\mu\nu} = 0. \tag{2.366}$$

Comparing Eq.(2.365) with Eq.(2.366), we have

$$T^{\mu\nu} = T^{\nu\mu}. \tag{2.367}$$

Thus $T^{\mu\nu}$ given by Eq.(2.364) is the symmetric energy-momentum tensor.

2.5.12 Charge conservation

The Lagrangian density of Eq.(2.233) has an internal symmetry. It is invariant under the phase transformations $\psi \to \psi e^{i\chi}$ and $\psi^\dagger \to \psi^\dagger e^{-i\chi}$. This leads to a conserved current-density j_μ in a similar way that was shown for the charge conservation of complex scalar bosons. Using the Noether's theorem, we have

$$j_\mu^e = -ie(\frac{\partial \mathcal{L}}{\partial \partial^\mu \psi}\psi - \frac{\partial \mathcal{L}}{\partial \partial^\mu \psi^\dagger}\psi^\dagger) = e\bar{\psi}\gamma_\mu\psi. \tag{2.368}$$

We have included a factor e to conform with the conventional definition of electrical current of the Dirac fermion field. e can be considered as a unit factor. The conserved quantity is thus the total charge

$$Q = \int d^3x j_0(\mathbf{x}, t) = e \int d^3x \psi^\dagger \psi. \tag{2.369}$$

2.5.13 Solutions of free Dirac equation

2.5.13.1 Plane wave expansion

The Dirac equation is a wave equation. Thus we have the particle-wave duality for the Dirac spinor fermions. The solutions of the free Dirac equation for the field operator $\hat{\psi}(\mathbf{x}, t)$ can be expanded in a complete set of plane wave functions. First, we consider the solutions of the classical Dirac equation which is the equation obtained by replacing the operators with the field functions in Eq.(2.247). The solutions are given by

$$\psi_{\mathbf{p}}^{(r)}(\mathbf{x}, t) = (2\pi)^{-\frac{3}{2}}\sqrt{\frac{m}{\omega_{\mathbf{p}}}}w_r(\mathbf{p})e^{-i\epsilon_r(\omega_{\mathbf{p}}t - \mathbf{p}\cdot\mathbf{x})}. \tag{2.370}$$

The index r denotes the four independent solutions. $r = 1, 2$ correspond to the solutions with $\epsilon_r = +1$, while $r = 3, 4$ correspond to those with $\epsilon_r = -1$. Inserting the plane wave solutions into the Dirac equation, we have

$$(i\gamma^\mu\partial_\mu - m)\psi_{\mathbf{p}}^{(r)}(\mathbf{x}, t) = 0, \tag{2.371}$$

which gives

$$(\gamma^\mu p_\mu - \epsilon_r m)w_r(\mathbf{p}) = 0, \tag{2.372}$$

where $p_\mu = (\omega_{\mathbf{p}}, \mathbf{p})$. With a special notation $\not{A} \equiv \gamma^\mu A_\mu$ (called A slash) designed for the calculations involving Dirac fermions, Eq.(2.372) can also be expressed as

$$(\not{p} - \epsilon_r m)w_r(\mathbf{p}) = 0. \tag{2.373}$$

The existence condition of a nontrivial solution to Eq.(2.373) is $\det(\not{p} - \epsilon_r m) = 0$, which gives

$$m^2 + \mathbf{p}^2 - p_0^2 = 0. \tag{2.374}$$

Thus we have

$$\omega_{\mathbf{p}} = \sqrt{m^2 + \mathbf{p}^2}. \tag{2.375}$$

2.5.13.2   *Dirac unit spinors*

$w_r(\mathbf{p})$ $(r = 1, 2, 3, 4)$ in Eq.(2.372) are called the *Dirac unit spinors*. In the rest frame of particle, $\mathbf{p} = 0$. Eq.(2.372) becomes

$$m(\gamma^0 - \epsilon_r)w_r(0) = m \begin{pmatrix} (1 - \epsilon_r)I & 0 \\ 0 & -(1 + \epsilon_r)I \end{pmatrix} w_r(0) = 0. \tag{2.376}$$

We can express the Dirac four-component spinor w_r in terms of two two-component spinors ξ and η. The spinors ξ and η are usually called the Pauli spinors. The solution of Eq.(2.376) is given by

$$w_r(0) = \begin{pmatrix} \dfrac{1 + \epsilon_r}{2}\xi \\ \dfrac{1 - \epsilon_r}{2}\eta \end{pmatrix}. \tag{2.377}$$

For $r = 1, 2$, $\epsilon_r = 1$. The solution has the form

$$w_r = \begin{pmatrix} \xi \\ 0 \end{pmatrix}. \tag{2.378}$$

There are two degenerate solutions for ξ. We usually choose two independent Pauli spinors

$$\xi_1 = \begin{pmatrix} 1 \\ 0 \end{pmatrix} \quad \text{and} \quad \xi_2 = \begin{pmatrix} 0 \\ 1 \end{pmatrix}, \tag{2.379}$$

which obeys the normalized condition

$$\xi_s^\dagger \xi_{s'} = \delta_{ss'}. \tag{2.380}$$

For $r = 3, 4$, $\epsilon_r = -1$. The solution has the form

$$w_r = \begin{pmatrix} 0 \\ \eta \end{pmatrix}. \tag{2.381}$$

The two degenerate solutions for η are usually chosen as the following two independent Pauli spinors

$$\eta_1 = \begin{pmatrix} 0 \\ 1 \end{pmatrix} \quad \text{and} \quad \eta_2 = \begin{pmatrix} -1 \\ 0 \end{pmatrix}, \tag{2.382}$$

which obeys the normalized condition

$$\eta_s^\dagger \eta_{s'} = \delta_{ss'}. \tag{2.383}$$

ξ_s and η_s are related conventionally by $\eta_s = -i\sigma^2 \xi_s$. Then we have four unit Dirac spinors in the rest frame

$$w_1(0) = \begin{pmatrix} 1 \\ 0 \\ 0 \\ 0 \end{pmatrix}, \ w_2(0) = \begin{pmatrix} 0 \\ 1 \\ 0 \\ 0 \end{pmatrix}, \ w_3(0) = \begin{pmatrix} 0 \\ 0 \\ 0 \\ 1 \end{pmatrix}, \ w_4(0) = \begin{pmatrix} 0 \\ 0 \\ -1 \\ 0 \end{pmatrix}. \tag{2.384}$$

They are also the eigenfunctions of

$$\Sigma_3 = \sigma_{12} = \begin{pmatrix} \sigma_3 & 0 \\ 0 & \sigma_3 \end{pmatrix} \tag{2.385}$$

with the eigenvalues of ± 1.

$$\Sigma_3 w_r(0) = (\pm 1) w_r(0). \tag{2.386}$$

For $r = 1, 4$, the eigenvalue is $+1$, while for $r = 2, 3$, it is -1.

Now we consider the solutions for $\mathbf{p} \neq 0$. For $r = 1, 2$, Eq.(2.372) has the form

$$(\omega_{\mathbf{p}} - m)\xi - \boldsymbol{\sigma} \cdot \mathbf{p}\eta = 0, \tag{2.387a}$$

$$\boldsymbol{\sigma} \cdot \mathbf{p}\xi - (\omega_{\mathbf{p}} + m)\eta = 0. \tag{2.387b}$$

The solution of Eq.(2.387) is

$$\eta = \frac{\boldsymbol{\sigma} \cdot \mathbf{p}}{\omega_{\mathbf{p}} + m}\xi. \tag{2.388}$$

Then we obtain the Dirac unit spinors in terms of the Pauli spinors

$$w_r(\mathbf{p}) = N \begin{pmatrix} \xi_r \\ \dfrac{\boldsymbol{\sigma} \cdot \mathbf{p}}{\omega_{\mathbf{p}} + m}\xi_r \end{pmatrix}, \tag{2.389}$$

where N is the normalization factor. Similarly, we have the Dirac unit spinors $w_r(\mathbf{p})$ for $r = 3, 4$

$$w_r(\mathbf{p}) = N' \begin{pmatrix} \dfrac{\boldsymbol{\sigma} \cdot \mathbf{p}}{\omega_{\mathbf{p}} + m}\eta_{r'} \\ \eta_{r'} \end{pmatrix}, \tag{2.390}$$

where $r' = r - 2$ and N' is the normalization factor.

With the appropriate choice of the normalization factors, the Dirac unit spinors obey the following orthogonality and completeness relations:

$$w_{r'}^{\dagger}(\epsilon_{r'}\mathbf{p})w_r(\epsilon_r\mathbf{p}) = \frac{\omega_{\mathbf{p}}}{m}\delta_{rr'}, \tag{2.391a}$$

$$\bar{w}_{r'}(\mathbf{p})w_r(\mathbf{p}) = \epsilon_r\delta_{rr'}, \tag{2.391b}$$

$$\sum_{r=1}^{4} w_{r\alpha}(\epsilon_r\mathbf{p})w_{r\beta}^{\dagger}(\epsilon_r\mathbf{p}) = \frac{\omega_{\mathbf{p}}}{m}\delta_{\alpha\beta}, \tag{2.391c}$$

$$\sum_{r=1}^{4} \epsilon_r w_{r\alpha}(\mathbf{p})\bar{w}_{r\beta}(\mathbf{p}) = \delta_{\alpha\beta}. \tag{2.391d}$$

To fulfill Eq.(2.391), the normalization factors should be chosen as

$$N = N' = \sqrt{\frac{\omega_{\mathbf{p}} + m}{2m}}. \tag{2.392}$$

The explicit forms of Eqs.(2.389) and (2.390) are then given by

$$w_1(\mathbf{p}) = \sqrt{\frac{\omega_{\mathbf{p}} + m}{2m}}\begin{pmatrix} 1 \\ 0 \\ \dfrac{p_z}{\omega_{\mathbf{p}} + m} \\ \dfrac{p_+}{\omega_{\mathbf{p}} + m} \end{pmatrix}, \tag{2.393}$$

$$w_2(\mathbf{p}) = \sqrt{\frac{\omega_{\mathbf{p}} + m}{2m}}\begin{pmatrix} 0 \\ 1 \\ \dfrac{p_-}{\omega_{\mathbf{p}} + m} \\ \dfrac{-p_z}{\omega_{\mathbf{p}} + m} \end{pmatrix}, \tag{2.394}$$

$$w_3(\mathbf{p}) = \sqrt{\frac{\omega_{\mathbf{p}} + m}{2m}}\begin{pmatrix} \dfrac{p_-}{\omega_{\mathbf{p}} + m} \\ \dfrac{-p_z}{\omega_{\mathbf{p}} + m} \\ 0 \\ 1 \end{pmatrix}, \tag{2.395}$$

$$w_4(\mathbf{p}) = \sqrt{\frac{\omega_{\mathbf{p}} + m}{2m}}\begin{pmatrix} \dfrac{-p_z}{\omega_{\mathbf{p}} + m} \\ \dfrac{-p_+}{\omega_{\mathbf{p}} + m} \\ -1 \\ 0 \end{pmatrix}, \tag{2.396}$$

where

$$p_{\pm} \equiv p_x \pm i p_y. \tag{2.397}$$

They can also be expressed in a compact form

$$w_r(\mathbf{p}) = \frac{\epsilon_r \not{p} + m}{\sqrt{2m(\omega_{\mathbf{p}} + m)}} w_r(0). \tag{2.398}$$

One can easily check that Eq.(2.391) guarantees the correct normalization to the delta function.

$$\int d^3 x \psi_{\mathbf{p}'}^{(r')\dagger}(\mathbf{x}) \psi_{\mathbf{p}}^{(r)}(\mathbf{x})$$

$$= \int d^3 x \frac{1}{(2\pi)^3} \sqrt{\frac{m^2}{\omega_{\mathbf{p}'}\omega_{\mathbf{p}}}} e^{-i(\epsilon_r \omega_{\mathbf{p}} - \epsilon_{r'}\omega_{\mathbf{p}'})t} e^{i(\epsilon_r \mathbf{p} - \epsilon_{r'}\mathbf{p}')\cdot\mathbf{x}} w_{r'}^{\dagger}(\mathbf{p}') w_r(\mathbf{p})$$

$$= \sqrt{\frac{m^2}{\omega_{\mathbf{p}}\omega_{\mathbf{p}'}}} e^{-i(\epsilon_r \omega_{\mathbf{p}} - \epsilon_{r'}\omega_{\mathbf{p}'})t} \delta^3(\epsilon_r \mathbf{p} - \epsilon_{r'}\mathbf{p}') w_{r'}^{\dagger}(\mathbf{p}') w_r(\mathbf{p}) \tag{2.399}$$

$$= \sqrt{\frac{m^2}{\omega_{\mathbf{p}}\omega_{\mathbf{p}'}}} e^{-i(\epsilon_r \omega_{\mathbf{p}} - \epsilon_{r'}\omega_{\mathbf{p}'})t} \delta^3(\mathbf{p}' - \epsilon_r \epsilon_{r'}\mathbf{p}) w_{r'}^{\dagger}(\epsilon_r \epsilon_{r'}\mathbf{p}) w_r(\mathbf{p})$$

$$= \delta_{rr'} \delta^3(\mathbf{p} - \mathbf{p}'),$$

where $\epsilon_r^2 = 1$ is used.

2.5.13.3 *Plane-wave expansion of field operators*

Since $\epsilon_r = 1$ in $\psi_{\mathbf{p}}^{(r)}(\mathbf{x}, t)$ for $r = 1, 2$ and $\epsilon_r = -1$ for $r = 3, 4$, in comparison with Eq.(2.154), $\psi_{\mathbf{p}}^{(r)}(\mathbf{x}, t)$ for $r = 1, 2$ correspond to the expansion functions for annihilation operators while $\psi_{\mathbf{p}}^{(r)}(\mathbf{x}, t)$ for $r = 3, 4$ correspond to the expansion functions for creation operators. We form the plane-wave expansion of the field operators by

$$\hat{\psi}(\mathbf{x}, t) = \int d^3 p \left[\sum_{r=1}^{2} \hat{b}(\mathbf{p}, r) \psi_{\mathbf{p}}^{(r)}(\mathbf{x}, t) + \sum_{r=3}^{4} \hat{d}^{\dagger}(\mathbf{p}, r) \psi_{\mathbf{p}}^{(r)}(\mathbf{x}, t) \right]$$

$$= \int \frac{d^3 p}{(2\pi)^{\frac{3}{2}}} \sqrt{\frac{m}{\omega_{\mathbf{p}}}} \left[\sum_{r=1}^{2} \hat{b}(\mathbf{p}, r) w_r(\mathbf{p}) e^{-i\epsilon_r p \cdot x} \right. \tag{2.400}$$

$$\left. + \sum_{r=3}^{4} \hat{d}^{\dagger}(\mathbf{p}, r) w_r(\mathbf{p}) e^{-i\epsilon_r p \cdot x} \right].$$

$\hat{b}$ and $\hat{b}^{\dagger}$ are the operators for particles. $\hat{d}$ and $\hat{d}^{\dagger}$ are the operators for antiparticles. The names of particles and antiparticles are just the convention. We have two kinds of particles and we need two names to distinguish them. The hermitian conjugate field operator is given by

$$
\hat{\psi}^\dagger(\mathbf{x}, t) = \int d^3p \left[\sum_{r=1}^{2} \hat{b}^\dagger(\mathbf{p}, r) \psi_{\mathbf{p}}^{(r)\dagger}(\mathbf{x}, t) + \sum_{r=3}^{4} \hat{d}(\mathbf{p}, r) \psi_{\mathbf{p}}^{(r)\dagger}(\mathbf{x}, t) \right]
$$

$$
= \int \frac{d^3p}{(2\pi)^{\frac{3}{2}}} \sqrt{\frac{m}{\omega_{\mathbf{p}}}} \left[\sum_{r=1}^{2} \hat{b}^\dagger(\mathbf{p}, r) \bar{w}_r(\mathbf{p}) \gamma^0 e^{i\epsilon_r p \cdot x} \right. \tag{2.401}
$$

$$
\left. + \sum_{r=3}^{4} \hat{d}(\mathbf{p}, r) \bar{w}_r(\mathbf{p}) \gamma^0 e^{i\epsilon_r p \cdot x} \right].
$$

2.5.13.4 *Creation and annihilation operators in p-space*

We can invert the expansion by projecting on a plane wave using Eq.(2.399)

$$
\int d^3x \psi_{\mathbf{p}}^{(r)\dagger}(\mathbf{x}, t) \hat{\psi}(\mathbf{x}, t)
$$

$$
= \int d^3p' \left[\sum_{r'=1}^{2} \hat{b}(\mathbf{p}', r') + \sum_{r'=3}^{4} \hat{d}^\dagger(\mathbf{p}', r') \right] \int d^3x \psi_{\mathbf{p}}^{(r)\dagger}(\mathbf{x}, t) \psi_{\mathbf{p}'}^{(r')}(\mathbf{x}, t)
$$

$$
= \int d^3p' \left[\sum_{r'=1}^{2} \hat{b}(\mathbf{p}', r') + \sum_{r'=3}^{4} \hat{d}^\dagger(\mathbf{p}', r') \right] \delta_{rr'} \delta^3(\mathbf{p} - \mathbf{p}') \tag{2.402}
$$

$$
= \begin{cases} \hat{b}(\mathbf{p}, r) & \text{for} \quad r = 1, 2 \\ \hat{d}^\dagger(\mathbf{p}, r) & \text{for} \quad r = 3, 4 \end{cases}
$$

or

$$
\int \frac{d^3x}{(2\pi)^{\frac{3}{2}}} \sqrt{\frac{m}{\omega_{\mathbf{p}}}} e^{i\epsilon_r p \cdot x} w_r{}^\dagger(\mathbf{p}) \hat{\psi}(\mathbf{x}, t) = \begin{cases} \hat{b}(\mathbf{p}, r) & \text{for} \quad r = 1, 2 \\ \hat{d}^\dagger(\mathbf{p}, r) & \text{for} \quad r = 3, 4 \end{cases}. \tag{2.403}
$$

Similarly we get

$$
\int d^3x \hat{\psi}^\dagger(\mathbf{x}, t) \psi_{\mathbf{p}}^{(r)}(\mathbf{x}, t)
$$

$$
= \int \frac{d^3x}{(2\pi)^{\frac{3}{2}}} \sqrt{\frac{m}{\omega_{\mathbf{p}}}} e^{-i\epsilon_r p \cdot x} \hat{\psi}^\dagger(\mathbf{x}, t) w_r(\mathbf{p}) \tag{2.404}
$$

$$
= \begin{cases} \hat{b}^\dagger(\mathbf{p}, r) & \text{for} \quad r = 1, 2 \\ \hat{d}(\mathbf{p}, r) & \text{for} \quad r = 3, 4 \end{cases}.
$$

Then the commutation relation of $\hat{b}$ and $\hat{b}^\dagger$ is given by

$$\{\hat{b}(\mathbf{p}, r), \hat{b}^\dagger(\mathbf{p}', r')\}$$

$$= \int d^3x \int d^3x' \psi_{\mathbf{p}\alpha}^{(r)\dagger}(\mathbf{x}, t)\psi_{\mathbf{p}'\beta}^{(r')}(\mathbf{x}', t)\{\hat{\psi}_\alpha(\mathbf{x}, t), \hat{\psi}_\beta^\dagger(\mathbf{x}', t)\}$$

$$= \int d^3x \psi_{\mathbf{p}\alpha}^{(r)\dagger}(\mathbf{x}, t)\psi_{\mathbf{p}'\beta}^{(r')}(\mathbf{x}, t)\delta_{\alpha\beta} \tag{2.405}$$

$$= \delta_{rr'}\delta^3(\mathbf{p} - \mathbf{p}').$$

Also we have

$$\{\hat{d}(\mathbf{p}, r), \hat{d}^\dagger(\mathbf{p}', r')\} = \delta_{rr'}\delta^3(\mathbf{p} - \mathbf{p}'). \tag{2.406}$$

Similarly, other anti-commutation relations can be deduced.

$$\{\hat{b}(\mathbf{p}, r), \hat{b}(\mathbf{p}', r')\} = \{\hat{b}^\dagger(\mathbf{p}, r), \hat{b}^\dagger(\mathbf{p}', r')\} = 0, \tag{2.407a}$$

$$\{\hat{d}(\mathbf{p}, r), \hat{d}(\mathbf{p}', r')\} = \{\hat{d}^\dagger(\mathbf{p}, r), \hat{d}^\dagger(\mathbf{p}', r')\} = 0. \tag{2.407b}$$

2.5.14 *Hamiltonian operator in p-space*

We can express the Hamiltonian operator by $\hat{b}$, $\hat{b}^\dagger$, $\hat{d}$ and $\hat{d}^\dagger$. From Eq.(2.321), we get

$$\hat{H} = \int d^3x \hat{\psi}^\dagger(\mathbf{x}, t)(-i\boldsymbol{\alpha} \cdot \boldsymbol{\nabla} + \beta m)\hat{\psi}(\mathbf{x}, t)$$

$$= \int d^3p \int d^3p' \Big[\sum_{rr'=1,2} \hat{b}^\dagger(\mathbf{p}', r')\hat{b}(\mathbf{p}, r) \int d^3x \psi_{\mathbf{p}'}^{(r')\dagger}(-i\boldsymbol{\alpha} \cdot \boldsymbol{\nabla} + \beta m)\psi_{\mathbf{p}}^{(r)}$$

$$+ \sum_{rr'=3,4} \hat{d}(\mathbf{p}', r')\hat{d}^\dagger(\mathbf{p}, r) \int d^3x \psi_{\mathbf{p}'}^{(r')\dagger}(-i\boldsymbol{\alpha} \cdot \boldsymbol{\nabla} + \beta m)\psi_{\mathbf{p}}^{(r)} \Big]$$

$$= \int d^3p \int d^3p' \Big[\sum_{rr'=1,2} \hat{b}^\dagger(\mathbf{p}', r')\hat{b}(\mathbf{p}, r)\epsilon_r\omega_{\mathbf{p}} \int d^3x \psi_{\mathbf{p}'}^{(r')\dagger}(\mathbf{x})\psi_{\mathbf{p}}^{(r)}(\mathbf{x}) \tag{2.408}$$

$$+ \sum_{rr'=3,4} \hat{d}(\mathbf{p}', r')\hat{d}^\dagger(\mathbf{p}, r)\epsilon_r\omega_{\mathbf{p}} \int d^3x \psi_{\mathbf{p}'}^{(r')\dagger}(\mathbf{x})\psi_{\mathbf{p}}^{(r)}(\mathbf{x}) \Big]$$

$$= \int d^3p \Big(\sum_{r=1,2} \omega_{\mathbf{p}}\hat{b}^\dagger(\mathbf{p}, r)\hat{b}(\mathbf{p}, r) - \sum_{r=3,4} \omega_{\mathbf{p}}\hat{d}(\mathbf{p}, r)\hat{d}^\dagger(\mathbf{p}, r) \Big).$$

In the derivation of Eq.(2.408), we have used Eq.(2.371), which can be rewritten as

$$(-i\boldsymbol{\alpha} \cdot \boldsymbol{\nabla} + \beta m)\psi_{\mathbf{p}}^{(r)}(x) = i\partial_0\psi_{\mathbf{p}}^{(r)}(x) = \epsilon_r\omega_{\mathbf{p}}\psi_{\mathbf{p}}^{(r)}(x). \tag{2.409}$$

The terms involving $\sum_{r=1,2}\sum_{r'=3,4}$ do not contribute due to $\delta_{rr'}$ in Eq.(2.399).

2.5.15　*Vacuum state*

To make Hamiltonian operator a positive-definite, we use the anti-commutator Eq.(2.406) and get

$$
\hat{H} = \int d^3p \Big[\sum_{r=1}^{2} \omega_{\mathbf{p}} \hat{b}^\dagger(\mathbf{p}, r)\hat{b}(\mathbf{p}, r) - \sum_{r=3}^{4} \omega_{\mathbf{p}}(\delta^3(0) - \hat{d}^\dagger(\mathbf{p}, r)\hat{d}(\mathbf{p}, r)) \Big]
$$

$$
= \int d^3p \Big[\sum_{r=1}^{2} \omega_{\mathbf{p}} \hat{b}^\dagger(\mathbf{p}, r)\hat{b}(\mathbf{p}, r) + \sum_{r=3}^{4} \omega_{\mathbf{p}} \hat{d}^\dagger(\mathbf{p}, r)\hat{d}(\mathbf{p}, r) \Big] + E_0,
$$

(2.410)

where

$$
E_0 \equiv -\delta^3(0) \int d^3p \sum_{r=3}^{4} \omega_{\mathbf{p}} = -\int \frac{d^3x}{(2\pi)^3} \int d^3p \sum_{r=3}^{4} \omega_{\mathbf{p}}. \qquad (2.411)
$$

E_0 is the energy of vacuum, which is unobservable and can be subtracted from the Hamiltonian. The physical vacuum is defined to be the state which contains neither particles nor antiparticles

$$
\hat{b}(\mathbf{p}, r)|0\rangle = 0, \qquad\qquad \text{for } r = 1, 2, \qquad\qquad (2.412a)
$$

$$
\hat{d}(\mathbf{p}, r)|0\rangle = 0, \qquad\qquad \text{for } r = 3, 4. \qquad\qquad (2.412b)
$$

For the momentum operator, we have

$$
\hat{\mathbf{P}} = -i \int d^3x \hat{\psi}^\dagger(x)\boldsymbol{\nabla}\hat{\psi}(x)
$$

$$
= \int d^3p\, \mathbf{p}\Big(\sum_{r=1}^{2} \hat{b}^\dagger(\mathbf{p}, r)\hat{b}(\mathbf{p}, r) + \sum_{r=3}^{4} \hat{d}^\dagger(\mathbf{p}, r)\hat{d}(\mathbf{p}, r) \Big),
$$

(2.413)

which means that each quasi-particle created by $\hat{b}^\dagger(\mathbf{p}, r)$ or quasi-antiparticle created by $\hat{d}^\dagger(\mathbf{p}, r)$ carries a momentum $\mathbf{p}$ and energy $\omega_{\mathbf{p}}$.

2.5.16　*Spin state*

We consider the operator $\gamma^5 \slashed{n}$, where n is an arbitrary space-like unit vector $(n_\mu n^\mu = -1)$ being orthogonal to the four-momentum vector p, i.e.

$$
p_\mu n^\mu = 0. \qquad\qquad (2.414)
$$

In the rest frame, $\mathbf{p} = 0$. Eq.(2.414) gives $n^0 = 0$. Then $\mathbf{n} \cdot \mathbf{n} = +1$. We take the z-axis of the rest frame to be in the $\mathbf{n}$ direction. Thus $n^\mu = (0, 0, 0, 1)$. In the standard representation of the γ matrices, we have

$$
\gamma^5 \slashed{n} = \gamma^5 \gamma^3 = \begin{pmatrix} \sigma^3 & 0 \\ 0 & -\sigma^3 \end{pmatrix}. \qquad\qquad (2.415)
$$

The meaning of Eq.(2.415) is that the spin z direction is assigned to the $\mathbf{n}$ direction. Using Eq.(2.415), we have

$$\gamma^5 \not{n} w_r(0) = \begin{cases} w_r(0) & r = 1, 3 \\ -w_r(0) & r = 2, 4 \end{cases}. \tag{2.416}$$

Since $\gamma^5 \not{n}$ is a pseudo scalar, which is Lorentz covariant, Eq.(2.416) should hold in any frame. We have

$$\gamma^5 \not{n} w_r(\mathbf{p}) = \begin{cases} w_r(\mathbf{p}) & r = 1, 3 \\ -w_r(\mathbf{p}) & r = 2, 4 \end{cases}. \tag{2.417}$$

When the momentum $\mathbf{p} \neq 0$, we can choose n as

$$n = \left(\frac{|\mathbf{p}|}{m}, \frac{\omega_{\mathbf{p}}}{m} \frac{\mathbf{p}}{|\mathbf{p}|} \right). \tag{2.418}$$

Using $(\boldsymbol{\gamma} \cdot \mathbf{p})^2 = -\mathbf{p}^2$, we have

$$\begin{aligned} \gamma^5 \not{n} &= \gamma^5 (\gamma^0 n_0 - \boldsymbol{\gamma} \cdot \mathbf{n}) \\ &= \gamma^5 \left(\gamma^0 \frac{|\mathbf{p}|}{m} - \boldsymbol{\gamma} \cdot \frac{\omega_{\mathbf{p}}}{m} \frac{\mathbf{p}}{|\mathbf{p}|} \right) \\ &= -\gamma^5 (\gamma^0)^2 \boldsymbol{\gamma} \cdot \frac{\mathbf{p}}{|\mathbf{p}|} \frac{\omega_{\mathbf{p}}}{m} - \gamma^5 \gamma^0 \frac{1}{|\mathbf{p}|m} (\boldsymbol{\gamma} \cdot \mathbf{p})^2 \\ &= \gamma^5 \gamma^0 \boldsymbol{\gamma} \cdot \frac{\mathbf{p}}{|\mathbf{p}|} \frac{1}{m} (\omega_{\mathbf{p}} \gamma^0 - \mathbf{p} \cdot \boldsymbol{\gamma}) \\ &= \boldsymbol{\Sigma} \cdot \frac{\mathbf{p}}{|\mathbf{p}|} \frac{\not{p}}{m}. \end{aligned} \tag{2.419}$$

In the derivation of Eq.(2.419), we have used the relation

$$\boldsymbol{\Sigma} = \begin{pmatrix} \boldsymbol{\sigma} & 0 \\ 0 & \boldsymbol{\sigma} \end{pmatrix} = \gamma^5 \gamma^0 \boldsymbol{\gamma}. \tag{2.420}$$

According to Eq.(2.373), we have

$$\boldsymbol{\Sigma} \cdot \frac{\mathbf{p}}{|\mathbf{p}|} \frac{\not{p}}{m} w_r(\mathbf{p}) = \boldsymbol{\Sigma} \cdot \frac{\mathbf{p}}{|\mathbf{p}|} \epsilon_r w_r(\mathbf{p}). \tag{2.421}$$

Comparing with Eq.(2.417), we have

$$\boldsymbol{\Sigma} \cdot \frac{\mathbf{p}}{|\mathbf{p}|} w_r(\mathbf{p}) = \begin{cases} w_r(\mathbf{p}) & r = 1, 4 \\ -w_r(\mathbf{p}) & r = 2, 3 \end{cases}. \tag{2.422}$$

Using Eq.(2.422), we can evaluate the operator of the spin projection in the direction of motion.

$$
\begin{aligned}
\hat{\mathbf{S}} \cdot \frac{\mathbf{P}}{|\mathbf{p}|} =& \frac{1}{2} \int d^3 x \hat{\psi}^\dagger(\mathbf{x}, t) \mathbf{\Sigma} \cdot \frac{\mathbf{P}}{|\mathbf{p}|} \hat{\psi}(\mathbf{x}, t) \\
=& \frac{1}{2} \int d^3 p \frac{m}{\omega_\mathbf{p}} \left[\sum_{\substack{r=1,2 \\ r'=1,2}} \hat{b}^\dagger(\mathbf{p}, r') \hat{b}(\mathbf{p}, r) w_{r'}^\dagger(\mathbf{p}) \mathbf{\Sigma} \cdot \frac{\mathbf{P}}{|\mathbf{p}|} w_r(\mathbf{p}) \right. \\
& \left. + \sum_{\substack{r=3,4 \\ r'=3,4}} \hat{d}(\mathbf{p}, r') \hat{d}^\dagger(\mathbf{p}, r) w_{r'}^\dagger(\mathbf{p}) \mathbf{\Sigma} \cdot \frac{\mathbf{P}}{|\mathbf{p}|} w_r(\mathbf{p}) \right] \\
=& \frac{1}{2} \int d^3 p [\hat{b}^\dagger(\mathbf{p}, 1) \hat{b}(\mathbf{p}, 1) - \hat{b}^\dagger(\mathbf{p}, 2) \hat{b}(\mathbf{p}, 2) \\
& - \hat{d}(\mathbf{p}, 3) \hat{d}^\dagger(\mathbf{p}, 3) + \hat{d}(\mathbf{p}, 4) \hat{d}^\dagger(\mathbf{p}, 4)] \\
=& \frac{1}{2} \int d^3 p [\hat{b}^\dagger(\mathbf{p}, 1) \hat{b}(\mathbf{p}, 1) - \hat{b}^\dagger(\mathbf{p}, 2) \hat{b}(\mathbf{p}, 2) \\
& + \hat{d}^\dagger(\mathbf{p}, 3) \hat{d}(\mathbf{p}, 3) - \hat{d}^\dagger(\mathbf{p}, 4) \hat{d}(\mathbf{p}, 4)] + S_0,
\end{aligned}
\tag{2.423}
$$

where S_0 is the total spin of vacuum and can be subtracted. Eq.(2.423) shows that $r = 1, 3$ gives positive sign for spin and $r = 2, 4$ gives negative sign.

To make consistency with the notation using spin s, we introduce a set of new operators

$$\hat{b}(\mathbf{p}, s) = \hat{b}(\mathbf{p}, 1), \tag{2.424a}$$

$$\hat{b}(\mathbf{p}, -s) = \hat{b}(\mathbf{p}, 2), \tag{2.424b}$$

$$\hat{d}(\mathbf{p}, s) = \hat{d}(\mathbf{p}, 3), \tag{2.424c}$$

$$\hat{d}(\mathbf{p}, -s) = \hat{d}(\mathbf{p}, 4). \tag{2.424d}$$

We also introduce $u(p, s)$ and $v(p, s)$ for the unit Dirac spinors

$$u(\mathbf{p}, s) = w_1(\mathbf{p}), \tag{2.425a}$$

$$u(\mathbf{p}, -s) = w_2(\mathbf{p}), \tag{2.425b}$$

$$v(\mathbf{p}, s) = w_3(\mathbf{p}), \tag{2.425c}$$

$$v(\mathbf{p}, -s) = w_4(\mathbf{p}). \tag{2.425d}$$

In the new notation, the anti-commutation relations are given by

$$\{\hat{b}(\mathbf{p}, s), \hat{b}^\dagger(\mathbf{p}', s')\} = \delta_{ss'} \delta^3(\mathbf{p} - \mathbf{p}'), \tag{2.426a}$$

$$\{\hat{d}(\mathbf{p}, s), \hat{d}^\dagger(\mathbf{p}', s')\} = \delta_{ss'} \delta^3(\mathbf{p} - \mathbf{p}'). \tag{2.426b}$$

The solutions for the field operators become

$$\hat{\psi}(\mathbf{x}, t) = \sum_s \int \frac{d^3 p}{(2\pi)^{\frac{3}{2}}} \sqrt{\frac{m}{\omega_{\mathbf{p}}}}$$
$$[\hat{b}(\mathbf{p}, s) u(\mathbf{p}, s) e^{-ip \cdot x} + \hat{d}^\dagger(\mathbf{p}, s) v(\mathbf{p}, s) e^{ip \cdot x}]. \tag{2.427}$$

and

$$\hat{\psi}^\dagger(\mathbf{x}, t) = \sum_s \int \frac{d^3 p}{(2\pi)^{\frac{3}{2}}} \sqrt{\frac{m}{\omega_{\mathbf{p}}}}$$
$$[\hat{b}^\dagger(\mathbf{p}, s) \bar{u}(\mathbf{p}, s) \gamma^0 e^{ip \cdot x} + \hat{d}(\mathbf{p}, s) \bar{v}(\mathbf{p}, s) \gamma^0 e^{-ip \cdot x}]. \tag{2.428}$$

The unit spinors satisfy the following free Dirac equations

$$(\not{p} - m) u(\mathbf{p}, s) = 0, \quad (\not{p} + m) v(\mathbf{p}, s) = 0 \tag{2.429}$$

and

$$\bar{u}(\mathbf{p}, s)(\not{p} - m) = 0, \quad \bar{v}(\mathbf{p}, s)(\not{p} + m) = 0, \tag{2.430}$$

where $\not{p} = p^\mu \gamma_\mu$.

2.5.17 *Helicity*

In terms of u and v, Eq.(2.422) takes the form

$$\mathbf{\Sigma} \cdot \frac{\mathbf{p}}{|\mathbf{p}|} u(\mathbf{p}, s) = s u(\mathbf{p}, s), \tag{2.431a}$$

$$\mathbf{\Sigma} \cdot \frac{\mathbf{p}}{|\mathbf{p}|} v(\mathbf{p}, s) = -s v(\mathbf{p}, s). \tag{2.431b}$$

with $s = \pm 1$. We call $\frac{1}{2} \mathbf{\Sigma} \cdot \frac{\mathbf{p}}{|\mathbf{p}|}$ the helicity operator for a spin $\frac{1}{2}$ particle and $\frac{1}{2} \hat{\mathbf{S}} \cdot \frac{\mathbf{p}}{|\mathbf{p}|}$ in Eq.(2.423) the helicity operator of spinor field. Eq.(2.431) shows that $u(\mathbf{p}, s)$ and $v(\mathbf{p}, s)$ are the eigenstates of the helicity operator. The eigenvalues of the helicity operator are $\pm \frac{1}{2}$.

The eigenstates with the positive ($h = +\frac{1}{2}$) helicity are called the right-handed states and those with the negative ($h = -\frac{1}{2}$) helicity are called the left-handed states. When the spin is oriented opposite to the direction of momentum, we get the opposite helicity. Thus $u(\mathbf{p}, 1)$ and $v(\mathbf{p}, -1)$ (or equivalently $u(-\mathbf{p}, -1)$ and $v(-\mathbf{p}, 1)$) are right-handed states, while $u(\mathbf{p}, -1)$ and $v(\mathbf{p}, 1)$ (or equivalently $u(-\mathbf{p}, 1)$ and $v(-\mathbf{p}, -1)$) are left-handed states.

2.5.18   *Chirality*

γ^5 is called the chirality operator. Since $(\gamma^5)^2 = 1$, the eigenvalues of γ^5 are ± 1. The eigenstate of γ^5 with eigenvalue of $+1$ is said to have a positive chirality(right-handed) and that with eigenvalue of -1 is said to have a negative chirality(left-handed).

Since $\gamma^5[(1+\gamma^5)\psi] = (1+\gamma^5)\psi$ and $\gamma^5[(1-\gamma^5)\psi] = -(1-\gamma^5)\psi$, $(1\pm\gamma^5)\psi$ are the eigenstates of γ^5. We denote

$$\psi_R = \frac{1}{2}(1 + \gamma^5)\psi, \tag{2.432a}$$

$$\psi_L = \frac{1}{2}(1 - \gamma^5)\psi. \tag{2.432b}$$

The Dirac spinor ψ can be decomposed into the left-hand field $\psi_L = \frac{1}{2}(1 - \gamma^5)\psi$ and the right-hand field $\psi_R = \frac{1}{2}(1 + \gamma^5)\psi$. ψ_L and ψ_R are called the *Weyl spinors*.

2.5.19   *Spin statistics relation*

It should be noted that if we use the commutation relations for bosons, through the similar deduction for Eq.(2.406), the commutation relation for $\hat{d}$ and $\hat{d}^\dagger$ becomes $[\hat{d}(\mathbf{p}, r), \hat{d}^\dagger(\mathbf{p}', r')] = -\delta_{rr'}\delta^3(\mathbf{p} - \mathbf{p}')$, which gives a wrong sign. It seems that one can change $\hat{d}$ into creation operator and $\hat{d}^\dagger$ into annihilation operator to eliminate the wrong sign problems. However, it would make $\hat{\psi}$ contain only annihilation operators and $\hat{\psi}^\dagger$ contains only creation operators which contradicts with the definition of $\hat{\psi}$ and $\hat{\psi}^\dagger$ given by Eqs.(2.62) and (2.63). Thus spinor particles can only be fermions.

There is also a positive-definite problem for the Hamiltonian operator if we use the commutation relations for bosons. Hamiltonian

$$\hat{H} = \sum_s \int d^3p\,\omega_{\mathbf{p}}[\hat{b}^\dagger(\mathbf{p}, s)\hat{b}(\mathbf{p}, s) - \hat{d}(\mathbf{p}, s)\hat{d}^\dagger(\mathbf{p}, s)] \tag{2.433}$$

can not be transformed into a positive-definite operator by reordering $\hat{d}$ and $\hat{d}^\dagger$, which is nonphysical in some sense.

2.5.20   *Charge of spinor particles and antiparticles*

From Eq.(2.238), we can see that both spinor particles and antiparticles are composite. We will show that they carry the opposite charge.

The charge of the particles and antiparticles can be calculated using Eq.(2.369). The conserved charge is given by

$$Q = \int d^3x j^0(x) = e \int d^3x \psi^\dagger \psi.$$ (2.434)

Inserting Eqs.(2.427) and (2.428) into Eq.(2.434), we have

$$
\begin{aligned}
Q =& e \int d^3x \sum_s \sum_{s'} \int \frac{d^3p'}{(2\pi)^{\frac{3}{2}}} \int \frac{d^3p}{(2\pi)^{\frac{3}{2}}} \sqrt{\frac{m^2}{\omega_{\mathbf{p}}\omega_{\mathbf{p'}}}} \\
& \times [\hat{b}^\dagger(\mathbf{p'},s')u^\dagger(\mathbf{p'},s')e^{ip'\cdot x} + \hat{d}(\mathbf{p'},s')v^\dagger(\mathbf{p'},s')e^{-ip'\cdot x}] \\
& \times [\hat{b}(\mathbf{p},s)u(\mathbf{p},s)e^{-ip\cdot x} + \hat{d}^\dagger(\mathbf{p},s)v(\mathbf{p},s)e^{ip\cdot x}] \\
=& e \sum_s \sum_{s'} \int d^3p \frac{m}{\omega_{\mathbf{p}}} [\hat{b}^\dagger(\mathbf{p},s')\hat{b}(\mathbf{p},s)u^\dagger(\mathbf{p},s')u(\mathbf{p},s) \\
& + \hat{d}(\mathbf{p},s')\hat{d}^\dagger(\mathbf{p},s)v^\dagger(\mathbf{p},s')v(\mathbf{p},s) \\
& + \hat{b}^\dagger(-\mathbf{p},s')\hat{d}^\dagger(\mathbf{p},s)u^\dagger(-\mathbf{p},s')v(\mathbf{p},s)e^{2i\omega_{\mathbf{p}}t} \\
& + \hat{d}(-\mathbf{p},s')\hat{b}(\mathbf{p},s)v^\dagger(-\mathbf{p},s')u(\mathbf{p},s)e^{-2i\omega_{\mathbf{p}}t}] \\
=& e \sum_s \int d^3p [\hat{b}^\dagger(\mathbf{p},s)\hat{b}(\mathbf{p},s) + \hat{d}(\mathbf{p},s)\hat{d}^\dagger(\mathbf{p},s)] \\
=& e \sum_s \int d^3p [\hat{b}^\dagger(\mathbf{p},s)\hat{b}(\mathbf{p},s) - \hat{d}^\dagger(\mathbf{p},s)\hat{d}(\mathbf{p},s)] + Q_0
\end{aligned}
$$ (2.435)

with

$$Q_0 = e \int \frac{d^3x}{(2\pi)^3} \int d^3p.$$ (2.436)

Q_0 is the charge of vacuum, which is not observable. Eq.(2.435) shows that a spinor particle carries a charge of $+e$ and a spinor antiparticle a charge of $-e$. Their charges are opposite. Historically, the name of antiparticle came from the two properties of antiparticles. It was considered that the charge of an antiparticle is opposite and the energy of an antiparticle is minus to the particle energy. We have avoided the minus energy conception of antiparticle in the formulations.

2.5.21 *Representation in terms of Weyl spinors*

We have seen that the Dirac spinor field ψ with four internal variables can be decomposed into two fields with two internal variables, the left hand field $\psi_L = 1/2(1 - \gamma^5)\psi$ and the right hand field $\psi_R = 1/2(1 + \gamma^5)\psi$. We have

$\psi = \psi_L + \psi_R$. Using the Weyl spinors, the kinetic term in the Lagrangian density in Eq.(2.233) can be expressed as

$$\bar{\psi}i\gamma^\mu\partial_\mu\psi = \bar{\psi}_L i\gamma^\mu\partial_\mu\psi_L + \bar{\psi}_R i\gamma^\mu\partial_\mu\psi_R. \tag{2.437}$$

If we consider only the kinetic term and do not include the mass term, the Lagrangian density in Eq.(2.233) can be expressed as a summation of the terms, with the same form as those expressed in terms of the Dirac spinor field functions, contributed by the two types of independent Weyl spinor field functions ψ_L and ψ_R. All the derivations for Dirac spinor fields can thus be similarly applied to the Weyl spinor fields. Using the Weyl spinor fields, the mass term in the Lagrangian density Eq.(2.233) can be expressed as

$$m\bar{\psi}\psi = m(\bar{\psi}_L\psi_R + \bar{\psi}_R\psi_L). \tag{2.438}$$

The mass term describes the interaction between the left hand field ψ_L and the right hand field ψ_R. Therefore, the mass term should also be considered as the interaction term. The Weyl spinor fermions are the more basic particle units. The Dirac spinors have the order $N = 4$. The spinors with the order $N > 4$ can be consider just as the composite of the Weyl spinors or Dirac spinors.

2.6   Vector bosons

In addition to the scalar and spinor fields, there is another type of fields which are called vector fields that could have covariant Lagrangian satisfying the causality principle. Now we consider the vector fields. First we turn to the massive vector field, which is simpler than the massless vector field.

2.6.1   *Massive vector bosons*

2.6.1.1   *Lagrangian*

There is a possibility of constructing covariant Lagrangian defined in Eq.(2.128) using vector field. The only possible covariant Lagrangian density for massive vector fields without interaction term is given by

$$\mathcal{L} = -\frac{1}{4}F_{\mu\nu}F^{\mu\nu} + \frac{1}{2}m^2 A_\mu A^\mu, \tag{2.439}$$

where A^μ is a vector function in spacetime and

$$F^{\mu\nu} = \partial^\mu A^\nu - \partial^\nu A^\mu. \tag{2.440}$$

Other forms such as

$$\mathcal{L} = -\frac{1}{2}\partial_\mu A_\nu \partial^\mu A^\nu + \frac{1}{2}m^2 A_\mu A^\mu \tag{2.441}$$

is equivalent to the Lagrangian density in Eq.(2.439) because it can be shown that $\partial_\mu A^\mu = 0$ (Eq.(2.463)) for the field described by the Lagrangian density Eq.(2.439). When we add a term $-\frac{1}{2}(\partial_\mu A^\mu)^2$ to the Lagrangian density in Eq.(2.439), the Lagrangian density in Eq.(2.439) becomes

$$\begin{aligned} \mathcal{L} &= -\frac{1}{4}F_{\mu\nu}F^{\mu\nu} - \frac{1}{2}(\partial_\mu A^\mu)^2 + \frac{1}{2}m^2 A_\mu A^\mu \\ &= -\frac{1}{2}\partial_\mu A_\nu \partial^\mu A^\nu + \frac{1}{2}\partial_\mu A_\nu \partial^\nu A^\mu - \frac{1}{2}\partial_\mu A^\mu \partial_\nu A^\nu + \frac{1}{2}m^2 A_\mu A^\mu \\ &= -\frac{1}{2}\partial_\mu A_\nu \partial^\mu A^\nu + \frac{1}{2}\partial_\mu [A_\nu(\partial^\nu A^\mu) - (\partial_\nu A^\nu)A^\mu] + \frac{1}{2}m^2 A_\mu A^\mu \\ &= -\frac{1}{2}\partial_\mu A_\nu \partial^\mu A^\nu + \frac{1}{2}m^2 A_\mu A^\mu. \end{aligned} \tag{2.442}$$

In the derivation of the last line of Eq.(2.442), we have omitted the term $\frac{1}{2}\partial_\mu[A_\nu(\partial^\nu A^\mu) - (\partial_\nu A^\nu)A^\mu]$ because it is a four-divergence and does not contribute to the action integral. Thus the Lagrangian density in Eq.(2.441) is equivalent to the Lagrangian density in Eq.(2.439). One may put a factor $f(A_\mu A^\mu)$ before $F_{\mu\nu}F^{\mu\nu}$. But this factor can be merged into the metric $g_{\mu\nu}$ when we use the curved spacetime.

2.6.1.2 *Generator of time translation transformation*

We will show that the Lagrangian density in Eq.(2.439) is related to the following generator of time translation transformation

$$\hat{G}_t = \int d^3x \frac{1}{2}\left[\hat{\pi}^2 + (\boldsymbol{\nabla} \times \hat{\boldsymbol{\phi}})^2 + m^2\hat{\boldsymbol{\phi}}^2 + \frac{1}{m^2}(\boldsymbol{\nabla} \cdot \hat{\boldsymbol{\pi}})^2\right], \tag{2.443}$$

which does not contain the time derivative term and thus satisfies the causality principle. $\hat{\boldsymbol{\phi}}$ in Eq.(2.443) is a vector operator in three-dimensional space and the field function $\boldsymbol{\phi}$ is $\mathbf{A}$ in Eq.(2.439). It should be noted that $\hat{\boldsymbol{\phi}}$ can not be a four-dimensional vector in spacetime because there is no $\dot{A}_0$ term in the Lagrangian density in Eq.(2.439). Thus we consider $\hat{\boldsymbol{\phi}}$ as a vector in three-dimensional space and construct a four-dimensional vector $\hat{A}^\mu$ in the following way.

First we define a vector $\mathbf{E}$ which satisfies the following equation

$$\dot{\boldsymbol{\phi}} = -\mathbf{E} + \frac{1}{m^2}\boldsymbol{\nabla}(\boldsymbol{\nabla} \cdot \mathbf{E}). \tag{2.444}$$

We then introduce the four dimensional vector

$$A^\mu = (A^0, \mathbf{A}) \equiv (A^0, \phi) \qquad (2.445)$$

with

$$A^0 = -\frac{1}{m^2} \boldsymbol{\nabla} \cdot \mathbf{E}. \qquad (2.446)$$

We have changed the notation $\hat{\phi}$ to $\hat{\mathbf{A}}$ in Eq.(2.445) because the vector field describes the photon field when mass term is zero and $\hat{\mathbf{A}}$ is the notation we usually used. Using notation $\hat{\mathbf{A}}$, we write G_t as

$$G_t = \int d^3x \frac{1}{2} \left[\boldsymbol{\pi}^2 + (\boldsymbol{\nabla} \times \mathbf{A})^2 + m^2 \mathbf{A}^2 + \frac{1}{m^2} (\boldsymbol{\nabla} \cdot \boldsymbol{\pi})^2 \right]. \qquad (2.447)$$

In terms of $\hat{\mathbf{A}}$, Eq.(2.444) has the form

$$\dot{A} = -\mathbf{E} + \frac{1}{m^2} \boldsymbol{\nabla}(\boldsymbol{\nabla} \cdot \mathbf{E}). \qquad (2.448)$$

2.6.1.3   *Deriving Lagrangian from G_t*

Now we prove that $\hat{G}_t$ in Eq.(2.447) leads to the Lagrangian density given by Eq.(2.439). We have three internal variables for the field operators. The commutators for the vector boson field are

$$[\hat{\phi}_i(\mathbf{x}, t), \hat{\pi}_j(\mathbf{x}', t)] = i\delta_{ij}\delta^3(\mathbf{x} - \mathbf{x}'), \qquad (2.449a)$$

$$[\hat{\phi}_i(\mathbf{x}, t), \hat{\phi}_j(\mathbf{x}', t)] = [\hat{\pi}_i(\mathbf{x}, t), \hat{\pi}_j(\mathbf{x}', t)] = 0. \qquad (2.449b)$$

We have used the commutation relations for bosons. If we use the anti-commutation relations for fermions, similar to the scalar field, we can show $[\hat{\phi}_i, \hat{G}_t] = 0$. Then we can not obtain an equation of motion from $\hat{G}_t$. Thus the vector fields can only be boson fields.

We will show that after carrying out the integration over π, we can get a covariant Lagrangian. From Eq.(2.448), we have

$$\mathbf{E} = -\boldsymbol{\nabla} A_0 - \partial_0 \mathbf{A}. \qquad (2.450)$$

We also define

$$\mathbf{B} \equiv \boldsymbol{\nabla} \times \mathbf{A}. \qquad (2.451)$$

We express Lagrangian density $\mathcal{L}$ in terms of $\mathbf{E}$ and $\mathbf{B}$.

$$\mathcal{L} = \frac{1}{2}(\mathbf{E}^2 - \mathbf{B}^2) + \frac{1}{2}m^2(A_0^2 - \mathbf{A}^2). \qquad (2.452)$$

Inserting G_t into Eq.(2.128) and integrating over π using the Gaussian integration formula, we have for L

$$
\begin{aligned}
L &= \int d^4x \left(-\frac{1}{2}\mathbf{E} \cdot \dot{\mathbf{A}} - \frac{1}{2}\mathbf{B}^2 - \frac{1}{2}m^2\mathbf{A}^2 \right) \\
&= \int d^4x \left[-\frac{1}{2}\mathbf{E} \cdot (-\mathbf{E} - \boldsymbol{\nabla}A_0) - \frac{1}{2}\mathbf{B}^2 - \frac{1}{2}m^2\mathbf{A}^2 \right] \quad (2.453) \\
&= \int d^4x \left(\frac{1}{2}\mathbf{E} \cdot \boldsymbol{\nabla}A_0 + \frac{1}{2}\mathbf{E}^2 - \frac{1}{2}\mathbf{B}^2 - \frac{1}{2}m^2\mathbf{A}^2 \right).
\end{aligned}
$$

In the derivation of the first line of Eq.(2.453), we have used Eq.(2.448). Using

$$
\mathbf{E} \cdot \boldsymbol{\nabla}A_0 = \boldsymbol{\nabla} \cdot (\mathbf{E}A_0) - A_0(\boldsymbol{\nabla} \cdot \mathbf{E}) = \boldsymbol{\nabla} \cdot (\mathbf{E}A_0) + m^2 A_0^2, \quad (2.454)
$$

we have

$$
\begin{aligned}
L &= \int d^4x \left[\frac{1}{2}\boldsymbol{\nabla} \cdot (\mathbf{E}A_0) + \frac{1}{2}\mathbf{E}^2 - \frac{1}{2}\mathbf{B}^2 + \frac{1}{2}m^2(A_0^2 - \mathbf{A}^2) \right] \\
&= \int d^4x \left[\frac{1}{2}\mathbf{E}^2 - \frac{1}{2}\mathbf{B}^2 + \frac{1}{2}m^2(A_0^2 - \mathbf{A}^2) \right] \quad (2.455) \\
&= \int d^4x \left(-\frac{1}{4}F_{\mu\nu}F^{\mu\nu} + \frac{1}{2}m^2 A_\mu A^\mu \right).
\end{aligned}
$$

The divergence term $\boldsymbol{\nabla} \cdot (\mathbf{E}A_0)$ has been dropped out because it yields only surface contribution. Therefore, G_t given by Eq.(2.447) leads to the Lagrangian density $\mathcal{L}$ in Eq.(2.439).

2.6.1.4 *Equations of motion*

Using the generator of time translation transformation $\hat{G}_t$ given by Eq.(2.443), we can obtain the equations of motion. Using Eq.(2.91) and $\boldsymbol{\nabla} \times \boldsymbol{\nabla} \times \mathbf{A} = \boldsymbol{\nabla}(\boldsymbol{\nabla} \cdot \mathbf{A}) - \nabla^2\mathbf{A}$, we have

$$
i\frac{\partial \hat{\mathbf{A}}}{\partial t} = [\hat{\mathbf{A}}, \hat{G}_t] = i\left(\hat{\boldsymbol{\pi}} - \frac{1}{m^2}\boldsymbol{\nabla}(\boldsymbol{\nabla} \cdot \hat{\boldsymbol{\pi}}) \right), \quad (2.456a)
$$

$$
i\frac{\partial \hat{\boldsymbol{\pi}}}{\partial t} = [\hat{\boldsymbol{\pi}}, \hat{G}_t] = i(\nabla^2\hat{\mathbf{A}} - \boldsymbol{\nabla}(\boldsymbol{\nabla} \cdot \hat{\mathbf{A}}) - m^2\hat{\mathbf{A}}). \quad (2.456b)
$$

Comparing Eq.(2.456a) with Eq.(2.448), we can see that $\boldsymbol{\pi} = -\mathbf{E}$.

We can define a four-dimensional vector $\pi_\mu = (0, -E_i)$. Then $\pi^\mu = (0, E^i)$. The four dimensional vector $\pi^\mu = (0, E^i)$ is the one usually used in the ordinary field theory on the vector field.

2.6.1.5   Hamiltonian

Now we consider the energy density

$$\mathcal{H} = \Theta_0^0 = \frac{\partial \mathcal{L}}{\partial \partial_0 A_\mu} \dot{A}_\mu - \mathcal{L}$$

$$= -F^{0\mu} \dot{A}_\mu + \frac{1}{4} F_{\mu\nu} F^{\mu\nu} - \frac{1}{2} m^2 A_\mu A^\mu \qquad (2.457)$$

$$= -\mathbf{E} \cdot \dot{\mathbf{A}} - \frac{1}{2}(\mathbf{E}^2 - \mathbf{B}^2) - \frac{1}{2} m^2 (A_0^2 - \mathbf{A}^2).$$

We have used the following formula in the derivation.

$$\frac{\partial \mathcal{L}}{\partial \partial_0 A_i} = -F^{0i} = E^i. \qquad (2.458)$$

Using $\mathbf{E} = -\nabla A_0 - \partial_0 \mathbf{A}$ and Eq.(2.454), we have

$$\mathcal{H} = -\mathbf{E} \cdot (-\mathbf{E} - \nabla A_0) + \frac{1}{2}(-\mathbf{E}^2 + \mathbf{B}^2 + m^2 \mathbf{A}^2) - \frac{1}{2} m^2 A_0^2$$
$$= \frac{1}{2}(\mathbf{E}^2 + \mathbf{B}^2 + m^2 \mathbf{A}^2 + m^2 A_0^2) + \nabla \cdot (\mathbf{E} A_0). \qquad (2.459)$$

Dropping out the divergence term because it yields only surface contribution, we have

$$H = \int d^3 x \frac{1}{2} \left[\mathbf{E}^2 + (\nabla \times \mathbf{A})^2 + m^2 \mathbf{A}^2 + \frac{1}{m^2}(\nabla \cdot \mathbf{E})^2 \right]. \qquad (2.460)$$

Comparing Eq.(2.460) with Eq.(2.447), we can see

$$G_t = H = \int d^3 x \frac{1}{2} \left[\mathbf{E}^2 + (\nabla \times \mathbf{A})^2 + m^2 \mathbf{A}^2 + \frac{1}{m^2}(\nabla \cdot \mathbf{E})^2 \right]. \qquad (2.461)$$

Applying $\nabla \cdot$ on both sides of Eq.(2.456b), we have

$$\nabla \cdot \dot{\mathbf{E}} = \nabla \cdot (m^2 \hat{\mathbf{A}}), \qquad (2.462)$$

which gives

$$\partial_\mu \hat{A}^\mu = 0. \qquad (2.463)$$

2.6.1.6   Fourier decomposition solution

The equations of motion for vector bosons form a wave equation. Thus we have the particle-wave duality for vector bosons. We can use the following plane wave basis to expand the solutions of the equations of motion.

$$A_\mu(\mathbf{k}, \lambda; x) = N_{\mathbf{k}} e^{-i(\omega_k t - \mathbf{k} \cdot \mathbf{x})} \epsilon_\mu(k, \lambda) \qquad (2.464)$$

with

$$N_{\mathbf{k}} = \frac{1}{\sqrt{2\omega_{\mathbf{k}}(2\pi)^3}}. \tag{2.465}$$

where $\mathbf{k}$ is the wave vector and $\omega_{\mathbf{k}} = \sqrt{\mathbf{k}^2 + m^2}$. The four dimensional vector is defined as $k = (\omega_{\mathbf{k}}, \mathbf{k})$. $\epsilon_\mu(k, \lambda)$ denotes a set of four-dimensional polarization vectors that plays a similar role as the unit spinors u and v in the plane-wave decomposition of the spinor field. In the four polarization vector ϵ_μ, there are three space-like and one time-like ones. It is general to define the polarization vectors with respect to the direction of wave vector $\mathbf{k}$.

2.6.1.7 *Polarization vectors*

Without losing generality, we demand that the polarization vectors are orthonormal, satisfying

$$\epsilon_\mu(k, \lambda)\epsilon^\mu(k, \lambda') = \eta_{\lambda\lambda'}. \tag{2.466}$$

We select two space-like transverse polarization vectors

$$\epsilon(k, 1) = (0, \boldsymbol{\epsilon}(\mathbf{k}, 1)), \tag{2.467a}$$

$$\epsilon(k, 2) = (0, \boldsymbol{\epsilon}(\mathbf{k}, 2)) \tag{2.467b}$$

with the condition

$$\boldsymbol{\epsilon}(\mathbf{k}, 1) \cdot \mathbf{k} = \boldsymbol{\epsilon}(\mathbf{k}, 2) \cdot \mathbf{k} = 0 \tag{2.468}$$

and

$$\boldsymbol{\epsilon}(\mathbf{k}, i) \cdot \boldsymbol{\epsilon}(\mathbf{k}, j) = \delta_{ij}. \tag{2.469}$$

We choose the third space-like polarization vector with $\lambda = 3$ to be in parallel to the direction of the wave vector $\mathbf{k}$. To specify the zero component of $\epsilon(\mathbf{k}, 3)$, we impose the condition that the four-vector $\epsilon(\mathbf{k}, 3)$ is orthogonal to the wave four-vector k,

$$k^\mu \epsilon_\mu(k, 3) = 0. \tag{2.470}$$

The components of this longitudinal polarization vector is given by

$$\epsilon(k, 3) = (\frac{|\mathbf{k}|}{m}, \frac{\mathbf{k}}{|\mathbf{k}|}\frac{k_0}{m}). \tag{2.471}$$

For the fourth time-like polarization vector with index $\lambda = 0$, we can use the vector k to construct it by defining

$$\epsilon(k, 0) = \frac{1}{m}k. \tag{2.472}$$

Apparently, $\epsilon(k,0)$ is orthogonal to the other three space-like polarization vectors $\epsilon(k,i)$. The completeness relation for the polarization vectors is given by

$$\sum_{\lambda=0}^{3} \eta_{\lambda\lambda} \epsilon_\mu(k,\lambda)\epsilon_\nu(k,\lambda) = \eta_{\mu\nu}. \tag{2.473}$$

Eq.(2.473) is a tensor equation. Thus we need only show that Eq.(2.473) holds in the rest frame because a transformation would generalize it to any frames.

In the rest frame of particles, the only nonzero component of k^μ is the time component. Only $\epsilon_\mu(k,\lambda)$ with $\lambda = 0$ has a time-like component. Thus

$$\sum_{\lambda=0}^{3} \eta_{\lambda\lambda} \epsilon_\mu(k,\lambda)\epsilon_\nu(k,\lambda) = \begin{pmatrix} 1 \\ 0 \end{pmatrix}_\mu \begin{pmatrix} 1 \\ 0 \end{pmatrix}_\nu - \sum_{l=1}^{3} \begin{pmatrix} 0 \\ \epsilon_l \end{pmatrix}_\mu \begin{pmatrix} 0 \\ \epsilon_l \end{pmatrix}_\nu. \tag{2.474}$$

i) If $\mu = 0$ and $\nu = 0$, the right hand side of Eq.(2.474) is equal to $+1$. ii) If $\mu = 0$ and $\nu = i$ (or $\mu = i$ and $\nu = 0$), the right hand side of Eq.(2.474) is zero. iii) If both indices are spatial ($\mu = i$ and $\nu = j$), the right hand side of Eq.(2.474) becomes $-\sum_{l=1}^{3} \epsilon_i(k,l)\epsilon_j(k,l)$. The ordinary completeness relation for an orthogonal basis in three dimensional space gives

$$\sum_{l=1}^{3} \epsilon_i(k,l)\epsilon_j(k,l) = \delta_{ij}. \tag{2.475}$$

In summary of the results in i),ii),iii), Eq.(2.473) holds for all μ and ν.

For the three physical polarization states, the completeness relation contains an extra term and reads

$$\sum_{l=1}^{3} \epsilon_\mu(k,l)\epsilon_\nu(k,l) = -\left(\eta_{\mu\nu} - \frac{1}{m^2}k_\mu k_\nu\right). \tag{2.476}$$

Using the basis functions $A^\mu(\mathbf{k},\lambda;x)$, the field operator $\hat{A}^\mu$ can be expanded as

$$\hat{A}^\mu(x) = \int d^3k \sum_{l=1}^{3} [\hat{a}_{\mathbf{k}l} A^\mu(\mathbf{k},l;\mathbf{x}) + \hat{a}_{\mathbf{k}l}^\dagger A^{\mu*}(\mathbf{k},l;x)]$$

$$= \int \frac{d^3k}{\sqrt{2\omega_\mathbf{k}(2\pi)^3}} \sum_{l=1}^{3} [\hat{a}_{\mathbf{k}l} \epsilon^\mu(k,l)e^{-ik\cdot x} + \hat{a}_{\mathbf{k}l}^\dagger \epsilon^{\mu*}(k,l)e^{ik\cdot x}]. \tag{2.477}$$

$\hat{A}^\mu(x)$ constructed by Eq.(2.477) is hermitian, which corresponds to a real-valued field. If one wants to describe a vector field containing multi-components with internal symmetry such as charged vector field, we need to replace it by the expansion

$$\hat{A}^\mu(x) = \int d^3k \sum_{l=1}^{3} [\hat{a}_{\mathbf{k}l} A^\mu(\mathbf{k}, l; x) + \hat{b}^\dagger_{\mathbf{k}l} A^{\mu*}(\mathbf{k}, l; x)], \qquad (2.478)$$

where the operators $\hat{a}_{\mathbf{k}l}$ and $\hat{b}^\dagger_{\mathbf{k}l}$ describe particles and antiparticles, respectively. In the following, we will concentrate on the neutral field described by Eq.(2.477). The treatment can be easily applied to the charge field described by Eq.(2.478). The operators as three-dimensional vectors for the vector bosons have the following expansion:

$$\hat{\mathbf{A}}(x) = \int \frac{d^3k}{\sqrt{2\omega_{\mathbf{k}}(2\pi)^3}} \sum_{l=1}^{3} \epsilon(\mathbf{k}, l)(\hat{a}_{\mathbf{k}l} e^{-ik\cdot x} + \hat{a}^\dagger_{\mathbf{k}l} e^{ik\cdot x}), \qquad (2.479)$$

where $\epsilon(\mathbf{k}, l)$ are the polarization vectors described by the spatial part of Eqs.(2.467) and (2.471). the corresponding canonical conjugate field $\hat{\pi}$ is given by

$$\begin{aligned}
\hat{\pi}(x) =& \partial_0 \hat{\mathbf{A}} + \boldsymbol{\nabla} \hat{A}_0 \\
=& \int \frac{d^3k}{\sqrt{2\omega_{\mathbf{k}}(2\pi)^3}} \sum_{l=1}^{3} [-i\omega_{\mathbf{k}}\epsilon(\mathbf{k}, l) + i\mathbf{k}\epsilon^0(\mathbf{k}, l)] \\
& \times (\hat{a}_{\mathbf{k}l} e^{-ik\cdot x} - \hat{a}^\dagger_{\mathbf{k}l} e^{ik\cdot x}) \\
=& -i \int \frac{d^3k}{\sqrt{2\omega_{\mathbf{k}}(2\pi)^3}} \sum_{l=1}^{3} \omega_{\mathbf{k}} \tilde{\epsilon}(\mathbf{k}, l)(\hat{a}_{\mathbf{k}l} e^{-ik\cdot x} - \hat{a}^\dagger_{\mathbf{k}l} e^{ik\cdot x}),
\end{aligned} \qquad (2.480)$$

where $\tilde{\epsilon}(\mathbf{k}, l)$ is the modified polarization vectors given by

$$\begin{aligned}
\tilde{\epsilon}(\mathbf{k}, l) &\equiv \epsilon(\mathbf{k}, l) - \frac{1}{\omega_{\mathbf{k}}} \mathbf{k}\epsilon^0(\mathbf{k}, l) \\
&= \epsilon(\mathbf{k}, l) - \frac{\mathbf{k}}{\omega_{\mathbf{k}}^2} \mathbf{k} \cdot \epsilon(\mathbf{k}, l).
\end{aligned} \qquad (2.481)$$

The relation

$$k \cdot \epsilon(k, l) = \omega_{\mathbf{k}}\epsilon^0 - \mathbf{k} \cdot \epsilon = 0. \qquad (2.482)$$

has been used in the above derivation. Eq.(2.482) is called the transversality condition.

2.6.1.8   *Commutation relations*

Now let us derive the commutation relations of $\hat{a}_{\mathbf{k}l}$ and $\hat{a}_{\mathbf{k}l}^{\dagger}$. We define a scalar product of $A(x)$ by

$$(A(x), A'(x)) \equiv i \int d^3x A^{\mu *}(\mathbf{x}) \overset{\leftrightarrow}{\partial_0} A'_{\mu}(\mathbf{x}), \qquad (2.483)$$

where

$$A \overset{\leftrightarrow}{\partial_0} A' \equiv A(\partial_0 A') - (\partial_0 A)A'. \qquad (2.484)$$

The scalar product of two plane wave components is given by

$$
\begin{aligned}
(A(\mathbf{k}', l'), A(\mathbf{k}, l)) &= i \int d^3x \frac{\epsilon^{\mu}(k', l')}{\sqrt{2\omega_{\mathbf{k}'}(2\pi)^3}} \frac{\epsilon_{\mu}(k, l)}{\sqrt{2\omega_{\mathbf{k}}(2\pi)^3}} e^{ik' \cdot x} \overset{\leftrightarrow}{\partial_0} e^{-ik \cdot x} \\
&= \delta^3(\mathbf{k}' - \mathbf{k})\epsilon^{\mu}(k, l')\epsilon_{\mu}(k, l) \\
&= \delta^3(\mathbf{k}' - \mathbf{k})\eta_{ll'}.
\end{aligned} \qquad (2.485)
$$

Similarly, we have

$$(A^*(\mathbf{k}', l'), A^*(\mathbf{k}, l)) = -\delta^3(\mathbf{k}' - \mathbf{k})\eta_{ll'}, \qquad (2.486)$$

and

$$(A(\mathbf{k}', l'), A^*(\mathbf{k}, l)) = (A^*(\mathbf{k}', l'), A(\mathbf{k}, l)) = 0. \qquad (2.487)$$

Using the above relations and $\eta_{ll} = -1$, we can project out the annihilation and creation operators

$$\hat{a}_{\mathbf{k}l} = (A(\mathbf{k}, l), \hat{A}(x)) = -i \int d^3x A^{\mu *}(\mathbf{k}, l) \overset{\leftrightarrow}{\partial_0} \hat{A}_{\mu}(x) \qquad (2.488)$$

and

$$\hat{a}_{\mathbf{k}l}^{\dagger} = (A^*(\mathbf{k}, l), \hat{A}(x)) = i \int d^3x A^{\mu}(\mathbf{k}, l) \overset{\leftrightarrow}{\partial_0} \hat{A}_{\mu}(x). \qquad (2.489)$$

Inserting the plane wave, we get

$$
\begin{aligned}
\hat{a}_{\mathbf{k}l} &= -i \int d^3x [A^{\mu *}(\mathbf{k}, l)\partial_0 \hat{A}_{\mu}(x) - \partial_0 A^{\mu *}(\mathbf{k}, l)\hat{A}_{\mu}(x)] \\
&= -i \int \frac{d^3x}{\sqrt{2\omega_{\mathbf{k}}(2\pi)^3}} \epsilon^{\mu}(k, l)e^{ik \cdot x}[\partial_0 \hat{A}_{\mu}(x) - i\omega_{\mathbf{k}}\hat{A}_{\mu}(x)].
\end{aligned} \qquad (2.490)
$$

We express the expansion in terms of the three-dimensional field operators $\hat{\mathbf{A}}$ and $\hat{\boldsymbol{\pi}}$:

$$
\begin{aligned}
\hat{a}_{\mathbf{k}l} = -i \int &\frac{d^3x}{\sqrt{2\omega_{\mathbf{k}}(2\pi)^3}} e^{ik \cdot x} \\
&\times (\epsilon^0 \partial_0 \hat{A}_0 - \boldsymbol{\epsilon} \cdot \partial_0 \hat{\mathbf{A}} - i\omega_{\mathbf{k}}\epsilon^0 \hat{A}_0 + i\omega_{\mathbf{k}}\boldsymbol{\epsilon} \cdot \hat{\mathbf{A}}).
\end{aligned} \qquad (2.491)
$$

Using $\partial \hat{A}_0 = -\boldsymbol{\nabla} \cdot \hat{\mathbf{A}}$ and $-\partial_0 \hat{\mathbf{A}} = -\hat{\boldsymbol{\pi}} + \boldsymbol{\nabla} \hat{A}_0$, Eq.(2.491) becomes

$$
\begin{aligned}
\hat{a}_{\mathbf{k}l} = -i \int \frac{d^3 x}{\sqrt{2\omega_{\mathbf{k}}(2\pi)^3}} e^{ik \cdot x} \\
\times (-\epsilon^0 \boldsymbol{\nabla} \cdot \hat{\mathbf{A}} - \boldsymbol{\epsilon} \cdot \hat{\boldsymbol{\pi}} + \boldsymbol{\epsilon} \cdot \boldsymbol{\nabla} \hat{A}_0 - i\omega_{\mathbf{k}}\epsilon^0 \hat{A}_0 + i\omega_{\mathbf{k}}\boldsymbol{\epsilon} \cdot \hat{\mathbf{A}}).
\end{aligned}
\tag{2.492}
$$

Then we integrate Eq.(2.492) by parts, which gives $-\epsilon^0 \boldsymbol{\nabla} \cdot \hat{\mathbf{A}} \to -i\epsilon^0 \mathbf{k} \cdot \hat{\mathbf{A}}$. Since $k \cdot \epsilon = 0$, we have

$$
\boldsymbol{\epsilon} \cdot \boldsymbol{\nabla} \hat{A}_0 - i\omega_{\mathbf{k}}\epsilon^0 \hat{A}_0 \to (i\boldsymbol{\epsilon} \cdot \mathbf{k} - i\omega_{\mathbf{k}}\epsilon^0)\hat{A}_0 = 0.
\tag{2.493}
$$

Thus the expansion components of the field operators become

$$
\begin{aligned}
\hat{a}_{\mathbf{k}l} &= \int \frac{d^3 x}{\sqrt{2\omega_{\mathbf{k}}(2\pi)^3}} e^{ik \cdot x} [(\omega_{\mathbf{k}}\boldsymbol{\epsilon} - \epsilon^0 \mathbf{k}) \cdot \hat{\mathbf{A}} + i\boldsymbol{\epsilon} \cdot \hat{\boldsymbol{\pi}}] \\
&= \int \frac{d^3 x}{\sqrt{2\omega_{\mathbf{k}}(2\pi)^3}} e^{ik \cdot x} [\omega_{\mathbf{k}}\tilde{\boldsymbol{\epsilon}}(\mathbf{k},l) \cdot \hat{\mathbf{A}}(x) + i\boldsymbol{\epsilon}(\mathbf{k},l) \cdot \hat{\boldsymbol{\pi}}(x)].
\end{aligned}
\tag{2.494}
$$

Similarly we have

$$
\hat{a}_{\mathbf{k}l}^\dagger = \int \frac{d^3 x}{\sqrt{2\omega_{\mathbf{k}}(2\pi)^3}} e^{-ik \cdot x} [\omega_{\mathbf{k}}\tilde{\boldsymbol{\epsilon}}(\mathbf{k},l) \cdot \hat{\mathbf{A}}(x) - i\boldsymbol{\epsilon}(\mathbf{k},l) \cdot \hat{\boldsymbol{\pi}}(x)].
\tag{2.495}
$$

Now we can derive the commutation relations of $\hat{a}_{\mathbf{k}l}$ and $\hat{a}_{\mathbf{k}l}^\dagger$ immediately,

$$
\begin{aligned}
[\hat{a}_{\mathbf{k}'l'}, \hat{a}_{\mathbf{k}l}^\dagger] &= \int \frac{d^3 x'}{\sqrt{2\omega_{\mathbf{k}'}(2\pi)^3}} \frac{d^3 x}{\sqrt{2\omega_{\mathbf{k}}(2\pi)^3}} e^{ik' \cdot x'} e^{-ik \cdot x} \\
&\quad \times (\omega_{\mathbf{k}'}\tilde{\boldsymbol{\epsilon}}' \cdot \hat{\mathbf{A}}' + i\boldsymbol{\epsilon}' \cdot \hat{\boldsymbol{\pi}}', \omega_{\mathbf{k}}\tilde{\boldsymbol{\epsilon}} \cdot \hat{\mathbf{A}} - i\boldsymbol{\epsilon} \cdot \hat{\boldsymbol{\pi}}) \\
&= \int d^3 x \frac{1}{2(2\pi)^3 \sqrt{\omega_{\mathbf{k}'}\omega_{\mathbf{k}}}} e^{i(\omega_{\mathbf{k}'} - \omega_{\mathbf{k}})t - i(\mathbf{k}' - \mathbf{k}) \cdot \mathbf{x}} (\omega_{\mathbf{k}'}\tilde{\boldsymbol{\epsilon}}' \cdot \boldsymbol{\epsilon} + \omega_{\mathbf{k}}\boldsymbol{\epsilon}' \cdot \tilde{\boldsymbol{\epsilon}}) \\
&= \frac{1}{2} [\tilde{\boldsymbol{\epsilon}}(\mathbf{k},l') \cdot \boldsymbol{\epsilon}(\mathbf{k},l) + \boldsymbol{\epsilon}(\mathbf{k},l') \cdot \tilde{\boldsymbol{\epsilon}}(\mathbf{k},l)]\delta^3(\mathbf{k}' - \mathbf{k}),
\end{aligned}
\tag{2.496}
$$

where Eq.(2.449) has been used in the derivation of Eq.(2.496). The vectors $\tilde{\boldsymbol{\epsilon}}$ and $\boldsymbol{\epsilon}$ satisfy the following orthogonality relation:

$$
\begin{aligned}
\tilde{\boldsymbol{\epsilon}}(\mathbf{k},l') \cdot \boldsymbol{\epsilon}(\mathbf{k},l) &= \boldsymbol{\epsilon}(\mathbf{k},l') \cdot \boldsymbol{\epsilon}(\mathbf{k},l) - \frac{1}{\omega_{\mathbf{k}}}\epsilon^0(\mathbf{k},l')\mathbf{k} \cdot \boldsymbol{\epsilon}(\mathbf{k},l) \\
&= \boldsymbol{\epsilon}(\mathbf{k},l') \cdot \boldsymbol{\epsilon}(\mathbf{k},l) - \epsilon^0(\mathbf{k},l')\epsilon^0(\mathbf{k},l) \\
&= -\epsilon(\mathbf{k},l') \cdot \epsilon(\mathbf{k},l) = -\eta_{l'l} = \delta_{l'l}.
\end{aligned}
\tag{2.497}
$$

Thus we obtain

$$
[\hat{a}_{\mathbf{k}'l'}, \hat{a}_{\mathbf{k}l}^\dagger] = \delta_{l'l}\delta^3(\mathbf{k}' - \mathbf{k}).
\tag{2.498}
$$

Other commutation relations can be derived similarly, we have

$$
[\hat{a}_{\mathbf{k}'l'}, \hat{a}_{\mathbf{k}l}] = [\hat{a}_{\mathbf{k}'l'}^\dagger, \hat{a}_{\mathbf{k}l}^\dagger] = 0.
\tag{2.499}
$$

2.6.1.9   *Hamiltonian operator in **k**-space*

We can express the Hamiltonian operator in terms of $\hat{a}_{\mathbf{k}l}$ and $\hat{a}_{\mathbf{k}l}^{\dagger}$. The Hamiltonian operator is given by

$$\hat{H} = \int d^3x \frac{1}{2} \left[\hat{\pi}^2 + m^2 \hat{\mathbf{A}} + (\boldsymbol{\nabla} \times \hat{\mathbf{A}})^2 + \frac{1}{m^2} (\boldsymbol{\nabla} \cdot \hat{\pi})^2 \right]. \quad (2.500)$$

The normal ordered form should be used in Eq.(2.500) to eliminate the possible divergent vacuum contribution. Inserting the expansion for $\hat{\mathbf{A}}$ and $\hat{\pi}$, we obtain an expression of $\hat{H}$ in terms of $\hat{a}_{\mathbf{k}l}$ and $\hat{a}_{\mathbf{k}l}^{\dagger}$. The expression of $\hat{H}$ contains various factor combination of $\hat{a}_{\mathbf{k}l}$ and $\hat{a}_{\mathbf{k}l}^{\dagger}$. As an example, we consider the terms containing $\hat{a}^{\dagger}\hat{a}$, which are given by

$$\frac{1}{2} \int d^3x \sum_{ll'} \int \frac{d^3k'}{\sqrt{2\omega_{\mathbf{k}'}(2\pi)^3}} \frac{d^3k}{\sqrt{2\omega_{\mathbf{k}}(2\pi)^3}} e^{i(k'-k)\cdot x} \hat{a}_{\mathbf{k}'l'}^{\dagger} \hat{a}_{\mathbf{k}l}$$

$$\times [\omega_{\mathbf{k}'}\omega_{\mathbf{k}}\tilde{\boldsymbol{\epsilon}}' \cdot \tilde{\boldsymbol{\epsilon}} + (\mathbf{k}' \times \boldsymbol{\epsilon}') \cdot (\mathbf{k} \times \boldsymbol{\epsilon}) + m^2 \boldsymbol{\epsilon}' \cdot \boldsymbol{\epsilon} + \frac{1}{m^2} \omega_{\mathbf{k}'}\omega_{\mathbf{k}}(\mathbf{k}' \cdot \tilde{\boldsymbol{\epsilon}}')(\mathbf{k} \cdot \tilde{\boldsymbol{\epsilon}})]$$
$$(2.501)$$

$$= \frac{1}{2} \sum_{ll'} \int \frac{d^3k}{2\omega_{\mathbf{k}}} \hat{a}_{\mathbf{k}l'}^{\dagger} \hat{a}_{\mathbf{k}l}$$

$$\times [\omega_{\mathbf{k}}^2 \tilde{\boldsymbol{\epsilon}}' \cdot \tilde{\boldsymbol{\epsilon}} + (\mathbf{k} \times \boldsymbol{\epsilon}') \cdot (\mathbf{k} \times \boldsymbol{\epsilon}) + m^2 \boldsymbol{\epsilon}' \cdot \boldsymbol{\epsilon} + \frac{\omega_{\mathbf{k}}^2}{m^2} (\mathbf{k} \cdot \tilde{\boldsymbol{\epsilon}}')(\mathbf{k} \cdot \tilde{\boldsymbol{\epsilon}})].$$

Using the vector identity

$$(\mathbf{k}' \times \boldsymbol{\epsilon}') \cdot (\mathbf{k} \times \boldsymbol{\epsilon}) = \mathbf{k}^2 \boldsymbol{\epsilon}' \cdot \boldsymbol{\epsilon} - (\mathbf{k}' \cdot \boldsymbol{\epsilon}')(\mathbf{k} \cdot \boldsymbol{\epsilon}), \quad (2.502)$$

Eq.(2.501) becomes

$$\frac{1}{2} \sum_{ll'} \int \frac{d^3k}{2\omega_{\mathbf{k}}} \hat{a}_{\mathbf{k}l'}^{\dagger} \hat{a}_{\mathbf{k}l}$$

$$\times \left[\omega_{\mathbf{k}}^2 \tilde{\boldsymbol{\epsilon}}' \cdot \tilde{\boldsymbol{\epsilon}} + \omega_{\mathbf{k}}^2 \boldsymbol{\epsilon}' \cdot \boldsymbol{\epsilon} - (\mathbf{k} \cdot \boldsymbol{\epsilon}')(\mathbf{k} \cdot \boldsymbol{\epsilon}) + \frac{\omega_{\mathbf{k}}^2}{m^2} (\mathbf{k} \cdot \tilde{\boldsymbol{\epsilon}}')(\mathbf{k} \cdot \tilde{\boldsymbol{\epsilon}}) \right]$$

$$= \frac{1}{2} \sum_{ll'} \int \frac{d^3k}{2\omega_{\mathbf{k}}} 2\omega_{\mathbf{k}}^2 (\boldsymbol{\epsilon}' \cdot \boldsymbol{\epsilon} - \epsilon^{0'} \cdot \epsilon^0) \hat{a}_{\mathbf{k}l'}^{\dagger} \hat{a}_{\mathbf{k}l} \quad (2.503)$$

$$= \frac{1}{2} \sum_{ll'} \int \frac{d^3k}{2\omega_{\mathbf{k}}} (-2\omega_{\mathbf{k}}^2) \eta_{l'l} \hat{a}_{\mathbf{k}l'}^{\dagger} \hat{a}_{\mathbf{k}l}$$

$$= \frac{1}{2} \sum_{l=1}^{3} \int d^3k \omega_{\mathbf{k}} \hat{a}_{\mathbf{k}l}^{\dagger} \hat{a}_{\mathbf{k}l}.$$

In the derivation of the second line of Eq.(2.503), we have used Eqs.(2.481) and (2.482). The terms contains $\hat{a}\hat{a}^{\dagger}$ gives the same result as that in

Eq.(2.503). Similarly we can calculate the terms containing $\hat{a}\hat{a}$ and $\hat{a}^\dagger\hat{a}^\dagger$. Both of them vanish. Thus we have

$$\hat{H} = \sum_{l=1}^{3} \int d^3 k \omega_{\mathbf{k}} \hat{a}^\dagger_{\mathbf{k}l} \hat{a}_{\mathbf{k}l}. \tag{2.504}$$

For the momentum vector, we have

$$P^i = \int d^3 x T^{0i} = -\int d^3 x \partial^0 A^\sigma \partial^i A_\sigma. \tag{2.505}$$

Similarly, in terms of $\hat{a}_{\mathbf{k}l}$ and $\hat{a}^\dagger_{\mathbf{k}l}$, the momentum operator $\hat{\mathbf{P}}$ is given by

$$\hat{\mathbf{P}} = \sum_{l=1}^{3} \int d^3 k \mathbf{k} \hat{a}^\dagger_{\mathbf{k}l} \hat{a}_{\mathbf{k}l}. \tag{2.506}$$

The quanta for the vector bosons carry energy $\omega_{\mathbf{k}}$ and the momentum $\mathbf{k}$. Thus we also call $\mathbf{k}$ as the momentum of the vector bosons.

2.6.1.10 *Spin operator*

Now we discuss the angular momentum tensor for vector bosons. The action is Lorentz invariant because it is a scalar. Under an infinitesimal Lorentz transformation,

$$x'^\mu = x^\mu + \delta\omega^{\nu\mu} x_\nu. \tag{2.507}$$

The transformation of a four vector A^μ is given by

$$A'^\mu(x') = A^\mu(x) + \delta\omega^{\mu\nu} A_\nu(x). \tag{2.508}$$

We can also use the general form of an infinitesimal Lorentz transformation given by

$$A'^\mu(x') = A^\mu(x) + \frac{1}{2}\delta\omega_{\alpha\beta}(I^{\alpha\beta})^{\mu\nu} A_\nu(x). \tag{2.509}$$

Comparing Eq.(2.508) with Eq.(2.509), we have

$$\delta\omega_{\alpha\beta}\left[\frac{1}{2}(I^{\alpha\beta})^{\mu\nu} - \eta^{\alpha\mu}\eta^{\beta\nu}\right] = 0. \tag{2.510}$$

Since $\delta\omega_{\alpha\beta}$ is antisymmetric, we can choose $(I^{\alpha\beta})^{\mu\nu}$ to be antisymmetric for α and β, i.e. $(I^{\alpha\beta})^{\mu\nu} = -(I^{\beta\alpha})^{\mu\nu}$ because the symmetric part is cancelled out after contraction with the antisymmetric $\delta\omega_{\alpha\beta}$. Thus the solution of Eq.(2.510) gives

$$(I^{\alpha\beta})^{\mu\nu} = \eta^{\alpha\mu}\eta^{\beta\nu} - \eta^{\alpha\nu}\eta^{\beta\mu}. \tag{2.511}$$

According to Eq.(2.188), we have the conserved current

$$j_\mu(x) = \frac{\partial \mathcal{L}(x)}{\partial(\partial^\mu A^\lambda)}\delta\omega_{\alpha\beta}(I^{\alpha\beta})^{\lambda\nu}A_\nu(x) - \Theta_{\mu\nu}\delta\omega^{\nu\lambda}x_\lambda. \qquad (2.512)$$

Similar to Eq.(2.328), Eq.(2.512) becomes

$$j_\mu(x) = \frac{1}{2}\delta\omega^{\nu\lambda}M_{\mu\nu\lambda}(x) \qquad (2.513)$$

with

$$M_{\mu\nu\lambda}(x) = \Theta_{\mu\lambda}x_\nu - \Theta_{\mu\nu}x_\lambda + \frac{\partial \mathcal{L}(x)}{\partial(\partial^\mu A^\sigma)}(I_{\nu\lambda})^{\sigma\gamma}A_\gamma(x)$$
$$= \Theta_{\mu\lambda}x_\nu - \Theta_{\mu\nu}x_\lambda + F_{\mu\sigma}(\eta_\nu{}^\sigma\eta_\lambda{}^\gamma - \eta_\nu{}^\gamma\eta_\lambda{}^\sigma)A_\gamma(x). \qquad (2.514)$$

Thus, similar to the derivation of Eq.(2.331), we have the following spin matrix of the vector boson field.

$$S_{ij} = \int d^3x(F_{0j}A_i - F_{0i}A_j). \qquad (2.515)$$

We can use a vector to represent it. We define

$$S^k \equiv \frac{1}{2}\epsilon^{ijk}S_{ij}. \qquad (2.516)$$

Using the vector symbol, we have

$$\mathbf{S} = \int d^3x \mathbf{E} \times \mathbf{A}. \qquad (2.517)$$

2.6.1.11 *Spin 1*

In the following, we show that the vector boson field is a spin 1 field. According to Eq.(2.517), the spin operator of vector bosons has the form

$$\hat{\mathbf{S}} = \int d^3x : \hat{\mathbf{E}} \times \hat{\mathbf{A}} : . \qquad (2.518)$$

Inserting Eqs.(2.479) and (2.480) into Eq.(2.518), we have

$$\hat{\mathbf{S}} = \int d^3x \int \frac{d^3k'}{\sqrt{2\omega_{\mathbf{k}'}(2\pi)^3}} \int \frac{d^3k}{\sqrt{2\omega_{\mathbf{k}}(2\pi)^3}} \sum_{ll'=1}^{3}(-i\omega_{\mathbf{k}'})\tilde{\epsilon}(\mathbf{k}',l') \times \epsilon(\mathbf{k},l)$$
$$: \left(\hat{a}_{\mathbf{k}'l'}e^{-ik'\cdot x} - \hat{a}_{\mathbf{k}'l'}^\dagger e^{ik'\cdot x}\right)\left(\hat{a}_{\mathbf{k}l}e^{-ik\cdot x} + \hat{a}_{\mathbf{k}l}^\dagger e^{ik\cdot x}\right) : \qquad (2.519)$$
$$= \frac{i}{2}\int d^3k \sum_{ll'=1}^{3}[\tilde{\epsilon}(\mathbf{k},l') \times \epsilon(\mathbf{k},l)(\hat{a}_{\mathbf{k}l}^\dagger\hat{a}_{\mathbf{k}l'} - \hat{a}_{\mathbf{k}l'}^\dagger\hat{a}_{\mathbf{k}l})$$
$$+ \tilde{\epsilon}(-\mathbf{k},l') \times \epsilon(\mathbf{k},l)(-\hat{a}_{-\mathbf{k}l}\hat{a}_{\mathbf{k}l'}e^{-2i\omega_{\mathbf{k}}t} + \hat{a}_{-\mathbf{k}l'}^\dagger\hat{a}_{\mathbf{k}l}^\dagger e^{2i\omega_{\mathbf{k}}t})].$$

We define the helicity operator

$$\hat{\Lambda} \equiv \hat{\mathbf{S}} \cdot \frac{\mathbf{k}}{|\mathbf{k}|}, \tag{2.520}$$

which gives the projection of the spin in the direction of wave vector. Using Eq.(2.519), Eq.(2.520) becomes

$$\hat{\Lambda} = \frac{i}{2} \int d^3k \sum_{ll'=1}^{2} \frac{\mathbf{k}}{|\mathbf{k}|} \cdot [\tilde{\epsilon}(\mathbf{k}, l') \times \epsilon(\mathbf{k}, l)] \, (\hat{a}_{\mathbf{k}l}^\dagger \hat{a}_{\mathbf{k}l'} - \hat{a}_{\mathbf{k}l'}^\dagger \hat{a}_{\mathbf{k}l}). \tag{2.521}$$

In Eq.(2.521), the summation does not contain the longitudinal polarization term ($l = 3$) because $\tilde{\epsilon}(\mathbf{k}, 3)$ and $\epsilon(\mathbf{k}, 3)$ are parallel to $\mathbf{k}$. For the transverse polarizations ($l = 1, 2$), $\tilde{\epsilon}(\mathbf{k}, l) = \epsilon(\mathbf{k}, l)$. Since the terms with $\hat{a}_{-\mathbf{k}l'}\hat{a}_{\mathbf{k}l}$ and $\hat{a}_{-\mathbf{k}l'}^\dagger \hat{a}_{\mathbf{k}l}^\dagger$ in Eq.(2.519) change their signs when we exchange the labels $l \leftrightarrow l'$ and $\mathbf{k} \leftrightarrow -\mathbf{k}$, they vanish. Thus only the terms containing $\hat{a}_{\mathbf{k}l}^\dagger \hat{a}_{\mathbf{k}l'}$ ($l, l' = 1, 2$) remain.

We choose the unit vectors $\epsilon(\mathbf{k}, 1)$ and $\epsilon(\mathbf{k}, 2)$ in such a way that the unit vectors $\epsilon(\mathbf{k}, 1)$, $\epsilon(\mathbf{k}, 2)$ and $\mathbf{e_k} = \frac{\mathbf{k}}{|\mathbf{k}|}$ form a right-handed orthogonal basis. Thus Eq.(2.521) becomes

$$\hat{\Lambda} = i \int d^3k (\hat{a}_{\mathbf{k}2}^\dagger \hat{a}_{\mathbf{k}1} - \hat{a}_{\mathbf{k}1}^\dagger \hat{a}_{\mathbf{k}2}). \tag{2.522}$$

To diagonalize the operator $\hat{\Lambda}$, we introduce a new set of operators

$$\hat{a}_{\mathbf{k}+} = \frac{1}{\sqrt{2}}(\hat{a}_{\mathbf{k}1} - i\hat{a}_{\mathbf{k}2}), \tag{2.523a}$$

$$\hat{a}_{\mathbf{k}-} = \frac{1}{\sqrt{2}}(\hat{a}_{\mathbf{k}1} + i\hat{a}_{\mathbf{k}2}), \tag{2.523b}$$

$$\hat{a}_{\mathbf{k}0} = \hat{a}_{\mathbf{k}3}. \tag{2.523c}$$

The inverted relations are

$$\hat{a}_{\mathbf{k}1} = \frac{1}{\sqrt{2}}(\hat{a}_{\mathbf{k}+} + \hat{a}_{\mathbf{k}-}), \tag{2.524a}$$

$$\hat{a}_{\mathbf{k}2} = \frac{i}{\sqrt{2}}(\hat{a}_{\mathbf{k}+} - \hat{a}_{\mathbf{k}-}), \tag{2.524b}$$

$$\hat{a}_{\mathbf{k}3} = \hat{a}_{\mathbf{k}0}. \tag{2.524c}$$

The operators $\hat{a}_{\mathbf{k}+}$, $\hat{a}_{\mathbf{k}-}$, $\hat{a}_{\mathbf{k}0}$ and their hermitian conjugate operators satisfy the commutation relations.

$$[\hat{a}_{\mathbf{k}'+}, \hat{a}_{\mathbf{k}+}^\dagger] = [\hat{a}_{\mathbf{k}'-}, \hat{a}_{\mathbf{k}-}^\dagger] = [\hat{a}_{\mathbf{k}'0}, \hat{a}_{\mathbf{k}0}^\dagger] = \delta^3(\mathbf{k} - \mathbf{k}') \tag{2.525}$$

and all the other commutation relations are zero. Thus $\hat{a}_{\mathbf{k}+}$, $\hat{a}_{\mathbf{k}-}$ and $\hat{a}_{\mathbf{k}0}$ are the annihilation operators, while $\hat{a}_{\mathbf{k}+}^\dagger$, $\hat{a}_{\mathbf{k}-}^\dagger$ and $\hat{a}_{\mathbf{k}0}^\dagger$ are the creation operators.

The Hamiltonian operator and momentum operator remain diagonal when they are expressed in terms of the new set of operators $\hat{a}_{\mathbf{k}\sigma}$ and $\hat{a}_{\mathbf{k}\sigma}^{\dagger}$ ($\sigma = +, -, 0$).

$$\hat{H} = \sum_{\sigma=-,0,+} \int d^3 k \omega_{\mathbf{k}} \hat{a}_{\mathbf{k}\sigma}^{\dagger} \hat{a}_{\mathbf{k}\sigma} \tag{2.526}$$

and

$$\hat{\mathbf{P}} = \sum_{\sigma=-,0,+} \int d^3 k \mathbf{k} \hat{a}_{\mathbf{k}\sigma}^{\dagger} \hat{a}_{\mathbf{k}\sigma}. \tag{2.527}$$

Expressed with the new set of operators, the helicity operator in Eq.(2.522) becomes

$$\hat{\Lambda} = \int d^3 k (\hat{a}_{\mathbf{k}+}^{\dagger} \hat{a}_{\mathbf{k}+} - \hat{a}_{\mathbf{k}-}^{\dagger} \hat{a}_{\mathbf{k}-}). \tag{2.528}$$

The quanta created by $\hat{a}_{\mathbf{k}\sigma}^{\dagger}$ are called the circularly polarized quasi-particles with the energy $\omega_{\mathbf{k}}$ and momentum $\mathbf{k}$. Thus the quanta created by $\hat{a}_{\mathbf{k}+}^{\dagger}$ have the helicity of $+1$ and those created by $\hat{a}_{\mathbf{k}-}^{\dagger}$ have the helicity of -1. Since the spin projection in the direction of momentum is ± 1 for the circularly polarized quanta, the vector bosons are spin 1 particles.

Since both scalar bosons and vector bosons have integer spin while spinor fermions have half-integer spin, one may summarize the spin statistics relation as follows: The particles with integer spin are bosons and those with half-integer spin are fermions.

We can also define the circular polarization vectors (also called helicity vectors) by

$$\epsilon^{\mu}(k, \pm) \equiv \frac{1}{\sqrt{2}} [\epsilon^{\mu}(k, 1) \pm i\epsilon^{\mu}(k, 2)], \tag{2.529a}$$

$$\epsilon^{\mu}(k, 0) \equiv \epsilon^{\mu}(k, 3). \tag{2.529b}$$

The field operators can be expanded in terms of the circular polarization vectors defined by Eq.(2.529).

2.6.2   *Massless vector bosons*

2.6.2.1   *Differences between massive boson field and massless boson field*

In the previous treatment of massive spin-1 vector bosons, we have introduced the 0-component of A_{μ} by $A_0 = -1/m^2 \boldsymbol{\nabla} \cdot \mathbf{E}$. Then we can use the four-dimensional vector A_μ to construct a covariant Lagrangian L. However, for massless particles ($m = 0$), this method fails, which leads to some difference. We can not construct the Lagrangian $-\frac{1}{4}F_{\mu\nu}F^{\mu\nu}$ in a similar way used for the massive vector bosons. For a covariant Lagrangian, we need to construct a related G_t satisfying the causality principle, which we can not find for a massless vector boson field with three components. Although we can not construct a consistent covariant Lagrangian for a massless vector bosons with three components, we can construct a covariant Lagrangian for massless vector bosons with two components. In order to understand the way we construct the covariant Lagrangian for massless vector bosons with two components , we will try to construct the Lagrangian $-\frac{1}{2}\partial_\mu A_\nu \partial^\mu A^\nu$ using ordinary procedure to show the problems in the ordinary procedure.

In order to maintain the covariance of Lagrangian L, we use $\partial_\mu A^\mu = 0$ for the introduction of the artificial components of A^μ. We can then do similar deduction as for the massive vector bosons. The fourier expansion of the field operator is given by

$$
\begin{aligned}
\hat{A}^\mu(\mathbf{x}, t) = \int \frac{d^3k}{\sqrt{2\omega_\mathbf{k}(2\pi)^3}} \sum_{\lambda=0}^{3} \Big[& \hat{a}_{\mathbf{k}\lambda}\epsilon^\mu(k, \lambda)e^{-ik\cdot x} \\
& + \hat{a}_{\mathbf{k}\lambda}^\dagger \epsilon^\mu(k, \lambda)e^{ik\cdot x} \Big].
\end{aligned} \tag{2.530}
$$

The difference is that $\omega_\mathbf{k} = k_0 = |k|$ because of the vanishing mass. The field $\hat{\pi}^\mu = \dot{\hat{A}}^\mu$ is given by

$$
\begin{aligned}
\hat{\pi}^\mu(\mathbf{x}, t) = -i \int \frac{d^3k}{\sqrt{2\omega_\mathbf{k}(2\pi)^3}} \omega_\mathbf{k} \sum_{\lambda=0}^{3} \Big[& \hat{a}_{\mathbf{k}\lambda}\epsilon^\mu(k, \lambda)e^{-ik\cdot x} \\
& - \hat{a}_{\mathbf{k}\lambda}^\dagger \epsilon^\mu(k, \lambda)e^{ik\cdot x} \Big].
\end{aligned} \tag{2.531}
$$

The commutation relations for the operators $\hat{a}_{\mathbf{k}\lambda}$ and $\hat{a}_{\mathbf{k}\lambda}^\dagger$ follow from the commutation relations of $\hat{A}^\mu$ and $\hat{\pi}^\mu$.

$$
\begin{aligned}
\hat{a}_{\mathbf{k}\lambda} &= i\eta_{\lambda\lambda} \int d^3x \hat{A}^{\mu*}(\mathbf{k}, \lambda) \overleftrightarrow{\partial_0} \hat{A}^\mu(\mathbf{x}) \\
&= i\eta_{\lambda\lambda} \int \frac{d^3x}{\sqrt{2\omega_\mathbf{k}(2\pi)^3}} e^{ik\cdot x}\epsilon^\mu(k, \lambda)(\dot{\hat{A}}^\mu(\mathbf{x}) - i\omega_\mathbf{k}\hat{A}^\mu(\mathbf{x})).
\end{aligned} \tag{2.532}
$$

Similarly, we have

$$
\begin{aligned}
\hat{a}_{\mathbf{k}\lambda}^\dagger &= -i\eta_{\lambda\lambda} \int d^3x \hat{A}^\mu(\mathbf{k}, \lambda) \overleftrightarrow{\partial_0} \hat{A}^\mu(\mathbf{x}) \\
&= -i\eta_{\lambda\lambda} \int \frac{d^3x}{\sqrt{2\omega_\mathbf{k}(2\pi)^3}} e^{-ik\cdot x}\epsilon^\mu(k, \lambda)(\dot{\hat{A}}^\mu(\mathbf{x}) + i\omega_\mathbf{k}\hat{A}^\mu(\mathbf{x})).
\end{aligned} \tag{2.533}
$$

In order to remain covariant form, we need to choose the artificial component of operators $\hat{A}^\mu$ and $\hat{\pi}^\mu$ to satisfy the following commutation relations.

$$[\hat{A}^\mu(\mathbf{x}, t), \hat{\pi}^\nu(\mathbf{x}', t)] = -i\eta^{\mu\nu}\delta^3(\mathbf{x}' - \mathbf{x}), \qquad (2.534a)$$

$$[\hat{A}^\mu(\mathbf{x}, t), \hat{A}^\nu(\mathbf{x}, t)] = [\hat{\pi}^\mu(\mathbf{x}, t), \hat{\pi}^\nu(\mathbf{x}, t)] = 0. \qquad (2.534b)$$

The commutation relation for $\hat{A}^0(\mathbf{x}, t)$ has the wrong sign. The commutation relation for $\hat{a}_{\mathbf{k}\lambda}$ and $\hat{a}^\dagger_{\mathbf{k}\lambda}$ then becomes

$$[\hat{a}_{\mathbf{k}'\lambda'}, \hat{a}^\dagger_{\mathbf{k}\lambda}] = -\delta^3(\mathbf{k}' - \mathbf{k})\eta_{\lambda\lambda}\eta_{\lambda'\lambda'}\epsilon^\mu(k, \lambda')\epsilon_\mu(k, \lambda). \qquad (2.535)$$

Covariant form in Eq.(2.534) is crucial to get the factor $\epsilon^\mu(k, \lambda')\epsilon_\mu(k, \lambda)$ in Eq.(2.535), which enables us to use the orthogonality relation of the four-dimensional polarization vectors $\epsilon^\mu(k, \lambda')\epsilon_\mu(k, \lambda) = \eta_{\lambda\lambda'}$. Thus we have

$$[\hat{a}_{\mathbf{k}'\lambda'}, \hat{a}^\dagger_{\mathbf{k}\lambda}] = -\delta^3(\mathbf{k}' - \mathbf{k})\eta_{\lambda\lambda'} \qquad (2.536)$$

and

$$[\hat{a}_{\mathbf{k}'\lambda'}, \hat{a}_{\mathbf{k}\lambda}] = [\hat{a}^\dagger_{\mathbf{k}'\lambda'}, \hat{a}^\dagger_{\mathbf{k}\lambda}] = 0. \qquad (2.537)$$

The operators $\hat{a}_{\mathbf{k}0}$ for the polarization $\lambda = 0$ satisfy the commutation relation with the wrong sign. Wrong sign will cause problem if one tries to construct the Fock space for $\hat{a}_{\mathbf{k}0}$. The norm of one-particle state is

$$\begin{aligned}
\langle 1_\mathbf{k} | 1_\mathbf{k} \rangle &= \langle 0 | \hat{a}_{\mathbf{k}0}\hat{a}^\dagger_{\mathbf{k}0} | 0 \rangle \\
&= \langle 0 | (-\eta_{00}\delta^3(\mathbf{k}' - \mathbf{k}) + \hat{a}^\dagger_{\mathbf{k}0}\hat{a}_{\mathbf{k}0}) | 0 \rangle \qquad (2.538) \\
&= -\eta_{00}\delta^3(0)\langle 0 | 0 \rangle.
\end{aligned}$$

Thus the norm of the state for the $\lambda = 0$ case is negative. The number operator for $\lambda = 0$ obtain a wrong minus sign $\hat{n}_{\mathbf{k}0} = -\hat{a}^\dagger_{\mathbf{k}0}\hat{a}_{\mathbf{k}0}$, which is inconsistent with that the particle number should be positive. This also leads to a wrong sign in the Hamiltonian operator $\hat{H} = -\int d^3k\,\omega_\mathbf{k}\hat{a}^\dagger_{\mathbf{k}0}\hat{a}_{\mathbf{k}0}$. These are the problems for covariant Lagrangian of massless spin-1 bosons with three components. However, we can construct a covariant Lagrangian for massless bosons with two components. Since we have two artificial components, one with positive sign and one with negative sign, we can manage them to cancel out each other.

2.6.2.2 *Faddeev-Popov method*

We consider the vector bosons with two internal degrees of freedoms. We have $\hat{A}_1$ and $\hat{A}_2$. We introduce two artificial variables A_0 and A_3 in order to construct a covariant Lagrangian. Then we integrating out the artificial variables and leave only A_1 and A_2 variables using the Faddeev-Popov method. We start with the covariant Lagrangian density

$$\mathcal{L} = -\frac{1}{4}F_{\mu\nu}F^{\mu\nu}. \tag{2.539}$$

There are four variables and we need to integrate out the redundant variables. It should be noted that the covariance should be maintained in the integration of the redundant variables. We note that there is a transformation $A_\mu \to A_\mu - \partial_\mu\Lambda \equiv A_\mu(\Lambda)$ leaving the Lagrangian invariant. Thus we factor out the redundancy by the integration over Λ using the following Faddeev-Popov method. We first define

$$[\Delta(A)]^{-1} \equiv \int D\Lambda\delta[f(A(\Lambda))], \tag{2.540}$$

where $f(x)$ is an auxiliary function. Then

$$\Delta(A)\int D\Lambda\delta[f(A(\Lambda))] = 1. \tag{2.541}$$

Now we consider the path integral

$$Z = \int DAe^{iS(A)}. \tag{2.542}$$

We can multiply $\Delta(A)\int D\Lambda\delta[f(A(\Lambda))]$ on the right side of Eq.(2.542) and get

$$\begin{aligned}
Z &= \int DAe^{iS(A)}\Delta(A)\int D\Lambda\delta[f(A(\Lambda))] \\
&= \int D\Lambda\int DAe^{iS(A)}\Delta(A)\delta[f(A(\Lambda))].
\end{aligned} \tag{2.543}$$

We change $A \to A(-\Lambda) \equiv A + \partial_\mu\Lambda$, which is equivalent to $A(\Lambda) \to A$. $Z = \int D\Lambda\int DAe^{iS(A)}$ is invariant with this transformation. We have also

$$\begin{aligned}
\Delta(A(-\Lambda)) &= \Delta(A + \partial_\mu\Lambda) \\
&= \left[\int D\Lambda'\delta[f(A(\Lambda' - \Lambda))]\right]^{-1} \\
&= \Delta(A).
\end{aligned} \tag{2.544}$$

Then Eq.(2.543) becomes

$$Z = \left(\int D\Lambda\right)\int DAe^{iS(A)}\Delta(A)\delta[f(A)]. \tag{2.545}$$

The integrand does not depend on Λ and the factor $(\int D\Lambda)$ can be thrown away in Eq.(2.545). We then choose $f(A) = \partial A - \sigma$, where σ is a function of x. Eq.(2.540) becomes

$$[\Delta(A)]^{-1} = \int D\Lambda \delta(\partial_\mu A^\mu - \partial^2 \Lambda - \sigma). \tag{2.546}$$

Since $\Delta(A)$ is multiplied by $\delta(f(A))$ in Eq.(2.545) and we use $\Delta(A)$ only in evaluating Eq.(2.545), we can set $f(A) = \partial_\mu A^\mu - \sigma = 0$ in Eq.(2.546) and get $\Delta(A)^{-1} \Rightarrow \int D\Lambda \delta(\partial^2 \Lambda)$. It can be seen that $\Delta(A)$ does not depend on A. Thus we can throw $\Delta(A)$ away in Eq.(2.545). Since Z does not depend on σ, we can integrating Z with an arbitrary functional of σ which we choose as

$$\exp\left[-\frac{i}{2\xi} \int d^4x \sigma^2(x)\right], \tag{2.547}$$

where ξ is a parameter. Thus we have

$$\begin{aligned}
Z &= \int D\sigma \exp\left[-\frac{i}{2\xi} \int d^4x \sigma^2(x)\right] \int DA \exp(iS(A))\delta(\partial_\mu A^\mu - \sigma) \\
&= \int DA \exp\left[iS(A) - \frac{i}{2\xi} \int d^4x (\partial_\mu A^\mu)^2\right].
\end{aligned} \tag{2.548}$$

From Eq.(2.548), we can see that the original action S(A) is replaced by

$$\begin{aligned}
S_{\text{eff}}(A) &= S(A) - \frac{1}{2\xi} \int d^4x (\partial_\mu A^\mu)^2 \\
&= \int d^4x \frac{1}{2} A_\mu \left[\partial^2 \eta^{\mu\nu} - \left(1 - \frac{1}{\xi}\right) \partial^\mu \partial^\nu\right] A_\nu.
\end{aligned} \tag{2.549}$$

Correspondingly we have the new Lagrangian density

$$\mathcal{L} = -\frac{1}{4} F_{\mu\nu} F^{\mu\nu} - \frac{1}{2\xi} (\partial_\mu A^\mu)^2. \tag{2.550}$$

It should be noted that we have used the symmetry that the action S is invariant with the transformation

$$A_\mu \to A'_\mu = A_\mu - \partial_\mu \Lambda \tag{2.551}$$

in the construction of the Lagrangian of the massless vector bosons, which can be seen from the derivation of Eq.(2.545). Eq.(2.551) is called the *gauge transformation*. The symmetry that the action S is invariant under the gauge transformation is called the *gauge symmetry*. Since the Lagrangian for the massless vector bosons is gauge-invariant, we also call the massless vector bosons as *gauge bosons*. The gauge symmetry is a condition imposed on the derivation of the covariant Lagrangian for the massless vector bosons.

Therefore the interaction terms of the massless vector bosons with other particles should also have the gauge symmetry. This is why the gauge symmetry plays the important role to unify the different interactions. Since the mass term breaks the gauge transformation $A_\mu \to A'_\mu = A_\mu - \partial_\mu \Lambda$, we can not have massive bosons with two components. For vector bosons with only one components, there are three virtual components and the way to construct a covariant Lagrangian has not been found.

2.6.2.3 *Coulomb gauge*

Since the action is invariant for the gauge transformation $A_\mu \to A'_\mu = A_\mu - \partial_\mu \Lambda$, we can take $A'_0 = A_0 - \partial_0 \Lambda = 0$. Therefore, with proper gauge transformation, we can take $A_0 = 0$. The action is also invariant for the transformation

$$\partial_\mu A^\mu \to \partial_\mu A'^\mu = \partial_\mu A^\mu - \sigma. \tag{2.552}$$

We can take $\partial_\mu A^\mu = 0$ with proper choice of σ. Then $\nabla \cdot \mathbf{A} = 0$. This is called the Coulomb gauge. The massless vector bosons have only two internal degrees of freedoms. The Coulomb gauge $\nabla \cdot \mathbf{A} = 0$ means that the longitudinal component vanishes and the two transverse components are not zero. Thus we have the massless vector bosons with two transverse freedoms and add two artificial variables, one is the longitudinal component $A_3 = 0$ and another is the fourth component $A_0 = 0$. We have initially the following commutation relations

$$[\hat{A}_i(\mathbf{x}, t), \hat{\pi}_j(\mathbf{x}', t)] = i\delta_{ij}\delta^3(\mathbf{x} - \mathbf{x}'), \tag{2.553a}$$

$$[\hat{A}_i(\mathbf{x}, t), \hat{A}_j(\mathbf{x}', t)] = [\hat{\pi}_i(\mathbf{x}, t), \hat{\pi}_j(\mathbf{x}', t)] = 0 \tag{2.553b}$$

with $i, j = 1, 2$. After introducing the third artificial variable, we have a vector $\hat{\mathbf{A}}$ with three components, which are not independent and constrained by $\nabla \cdot \hat{\mathbf{A}} = 0$. We have only two independent transverse components. We could use the transverse projection operator $\hat{P}_\perp$ to impose the transversality condition. $\hat{P}_\perp$ is defined by

$$(\hat{P}_\perp)_{ij} \equiv \delta_{ij} - \partial_i \frac{1}{\triangle} \partial_j, \tag{2.554}$$

where $\frac{1}{\triangle}$ is the inverse Laplacian. Then we can impose $\nabla \cdot \hat{\mathbf{A}} = 0$ by acting on $\hat{A}_i$ with the projection operator $\hat{P}_\perp$

$$\hat{A}_i(\mathbf{x}) \to \left(\delta_{ij} - \frac{\partial_i \partial_j}{\triangle}\right) \hat{A}_j(\mathbf{x}). \tag{2.555}$$

We can change the commutation relations to the projected commutation relations for a vector with three components. Through projection, the commutation relation Eq.(2.553a) is expressed with the following projected commutation relation

$$[\hat{A}_i(\mathbf{x},t),\hat{\pi}_j(\mathbf{x}',t)] = i\delta^3_{\perp ij}(\mathbf{x}-\mathbf{x}') \tag{2.556}$$

with $i,j = 1,2,3$. $\delta^3_{\perp ij}(\mathbf{x}-\mathbf{x}')$ is the transverse delta function defined by

$$\delta^3_{\perp ij}(\mathbf{x}-\mathbf{x}') \equiv (P_\perp)_{ij}\delta^3(\mathbf{x}-\mathbf{x}')$$
$$= \int \frac{d^3k}{(2\pi)^3} e^{i\mathbf{k}\cdot(\mathbf{x}-\mathbf{x}')} \left(\delta_{ij} - \frac{k_i k_j}{k^2}\right). \tag{2.557}$$

The Coulomb gauge $\nabla \cdot \mathbf{A} = 0$, together with $A_0 = 0$, is called the radiation gauge. Besides the radiation gauge, we can choose other gauges. We have two functions Λ and σ to change A_0 and A_3 using $A_0 = \partial_0\Lambda$ and $\partial_\mu A^\mu = \sigma$.

2.6.2.4 $\hat{G}_t$ of massless vector bosons

We can show that the following generator of time translation transformation $\hat{G}_t$ leads to the covariant Lagrangian Eq.(2.550).

$$\hat{G}_t = \int d^3x \frac{1}{2}\left[\hat{\pi}^2 + (\nabla \times \hat{\mathbf{A}})^2\right], \tag{2.558}$$

which contains no time derivative and satisfies the causality principle. In Eq.(2.558), we have only two transverse components for $\hat{\pi}$ and $\hat{\mathbf{A}}$ as initial components. We introduce the four dimensional vector A_μ with $\nabla \cdot \hat{\mathbf{A}}(\mathbf{x},t) = 0$ and $A_0(\mathbf{x},t)) = 0$. The Lagrangian density in Eq.(2.550) becomes the Lagrangian density given by Eq.(2.539). We then define the *electric field* $\mathbf{E}$ by

$$\mathbf{E} \equiv -\frac{\partial \mathbf{A}}{\partial t} - \nabla A_0. \tag{2.559}$$

We also define

$$\mathbf{B} \equiv \nabla \times \hat{\mathbf{A}}. \tag{2.560}$$

$\mathbf{B}$ is called the *magnetic field*. In the radiation gauge, we can express the Lagrangian density in Eq.(2.550) in terms of $\mathbf{E}$ and $\mathbf{B}$ as

$$\mathcal{L} = -\frac{1}{4}F_{\mu\nu}F^{\mu\nu} = \frac{1}{2}E^2 - \frac{1}{2}B^2. \tag{2.561}$$

After carrying out the integration over π in Eq.(2.128), we can see that $\hat{G}_t$ in Eq.(2.558) leads to the Lagrangian density given by Eq.(2.550). The massless vector bosons described by the Lagrangian density in Eq.(2.561) are also called photons.

2.6.2.5 *Equations of motion for massless vector bosons*

Using the generator of time translation transformation $\hat{G}_t$ given by Eq.(2.558), we obtain the equations of motion

$$i\frac{\partial \hat{\mathbf{A}}}{\partial t} = [\hat{\mathbf{A}}, \hat{G}_t] = iP_\perp \hat{\boldsymbol{\pi}}, \tag{2.562a}$$

$$i\frac{\partial \hat{\boldsymbol{\pi}}}{\partial t} = [\hat{\boldsymbol{\pi}}, \hat{G}_t] = i\nabla^2 P_\perp \hat{\mathbf{A}} - i\boldsymbol{\nabla}(\boldsymbol{\nabla} \cdot P_\perp \hat{\mathbf{A}}). \tag{2.562b}$$

The equations of motion for massless vector bosons form a set of wave equations. Thus we have the particle-wave duality for the massless vector bosons.

2.6.2.6 *Solution of equations of motion*

For the solution of the equations of motion Eq.(2.562), there are only two projected transverse modes. Thus we have the following plane wave expansion of the field operator

$$\hat{\mathbf{A}}(\mathbf{x}, t) = \int \frac{d^3 k}{\sqrt{2\omega_{\mathbf{k}}(2\pi)^3}} \sum_{l=1}^{2} \boldsymbol{\epsilon}(\mathbf{k}, l)(\hat{a}_{\mathbf{k}l} e^{-ik\cdot x} + \hat{a}_{\mathbf{k}l}^{\dagger} e^{ik\cdot x}), \tag{2.563}$$

where $\boldsymbol{\epsilon}(\mathbf{k}, l)$ are the transverse polarization vectors satisfying

$$\mathbf{k} \cdot \boldsymbol{\epsilon}(\mathbf{k}, l) = 0, \tag{2.564a}$$

$$\boldsymbol{\epsilon}(\mathbf{k}, l) \cdot \boldsymbol{\epsilon}(\mathbf{k}, l') = \delta_{ll'} \tag{2.564b}$$

and

$$\omega_{\mathbf{k}} = k_0 = |k|. \tag{2.565}$$

The electric field $\hat{\mathbf{E}}$ is given by

$$\hat{\mathbf{E}}(\mathbf{x}, t) = \int \frac{d^3 k}{\sqrt{2\omega_{\mathbf{k}}(2\pi)^3}} \sum_{l=1}^{2} i\omega_{\mathbf{k}} \boldsymbol{\epsilon}(\mathbf{k}, l)(\hat{a}_{\mathbf{k}l} e^{-ik\cdot x} - \hat{a}_{\mathbf{k}l}^{\dagger} e^{ik\cdot x}). \tag{2.566}$$

Then we get the expansion for $\hat{\boldsymbol{\pi}} = -\hat{\mathbf{E}}$.

$$\hat{\boldsymbol{\pi}}(\mathbf{x}, t) = \int \frac{d^3 k}{\sqrt{2\omega_{\mathbf{k}}(2\pi)^3}} \sum_{l=1}^{2} i\omega_{\mathbf{k}} \boldsymbol{\epsilon}(\mathbf{k}, l)(-\hat{a}_{\mathbf{k}l} e^{-ik\cdot x} + \hat{a}_{\mathbf{k}l}^{\dagger} e^{ik\cdot x}). \tag{2.567}$$

The magnetic field $\hat{\mathbf{B}}$ becomes

$$\hat{\mathbf{B}}(\mathbf{x}, t) = \int \frac{d^3 k}{\sqrt{2\omega_{\mathbf{k}}(2\pi)^3}} \sum_{l=1}^{2} i\mathbf{k} \times \boldsymbol{\epsilon}(\mathbf{k}, l)(\hat{a}_{\mathbf{k}l} e^{-ik\cdot x} - \hat{a}_{\mathbf{k}\lambda}^{\dagger} e^{ik\cdot x}). \tag{2.568}$$

The operators $\hat{a}_{\mathbf{k}l}$ and $\hat{a}^{\dagger}_{\mathbf{k}l}$ have the properties of creation and annihilation operators for the transverse photons. They satisfy the following commutation relations

$$[\hat{a}_{\mathbf{k}'l'}, \hat{a}^{\dagger}_{\mathbf{k}l}] = \delta^3(\mathbf{k}' - \mathbf{k})\delta_{ll'}, \tag{2.569a}$$

$$[\hat{a}_{\mathbf{k}'l'}, \hat{a}_{\mathbf{k}l}] = [\hat{a}^{\dagger}_{\mathbf{k}'l'}, \hat{a}^{\dagger}_{\mathbf{k}l}] = 0. \tag{2.569b}$$

It is easy to check that the commutation relations give the correct results. Using the commutators of $\hat{a}$ and $\hat{a}^{\dagger}$, we have

$$
\begin{aligned}
[\hat{A}^i(\mathbf{x}, t), \hat{\pi}^j(\mathbf{x}', t)] &= \int \frac{d^3k}{\sqrt{2\omega_{\mathbf{k}}(2\pi)^3}} \int \frac{d^3k'}{\sqrt{2\omega_{\mathbf{k}'}(2\pi)^3}} \\
&\quad \times i\omega_{\mathbf{k}'} \sum_{ll'=1}^{2} \epsilon^i(\mathbf{k}, l)\epsilon^j(\mathbf{k}', l')([\hat{a}_{\mathbf{k}l}, \hat{a}^{\dagger}_{\mathbf{k}'l'}]e^{-i(k\cdot x - k'\cdot x')} \\
&\quad - [\hat{a}^{\dagger}_{\mathbf{k}l}, \hat{a}_{\mathbf{k}'l'}]e^{i(k\cdot x - k'\cdot x')}) \\
&= i \int \frac{d^3k}{2(2\pi)^3} \sum_{l=1}^{2} \epsilon^i(\mathbf{k}, l)\epsilon^j(\mathbf{k}, l') \\
&\quad \times (e^{i\mathbf{k}\cdot(\mathbf{x}-\mathbf{x}')} + e^{-i\mathbf{k}\cdot(\mathbf{x}-\mathbf{x}')}).
\end{aligned}
\tag{2.570}
$$

The transverse polarization vectors $\epsilon(\mathbf{k}, 1)$ and $\epsilon(\mathbf{k}, 2)$ are orthogonal to each other. They are also orthogonal to the unit vector $\mathbf{k}/|\mathbf{k}|$ in the direction of momentum $\mathbf{k}$. Thus $\epsilon(\mathbf{k}, 1)$, $\epsilon(\mathbf{k}, 2)$ and $\mathbf{k}/|\mathbf{k}|$ form an orthogonal basis of three dimensional space and satisfy the completeness relation

$$\sum_{l=1}^{2} \epsilon^i(\mathbf{k}, l)\epsilon^j(\mathbf{k}, l) + \frac{k^i k^j}{\mathbf{k}^2} = \delta_{ij}. \tag{2.571}$$

With Eq.(2.571), Eq.(2.570) becomes

$$
\begin{aligned}
[\hat{A}^i(\mathbf{x}, t), \hat{\pi}^j(\mathbf{x}', t)] &= i \int \frac{d^3k}{(2\pi)^3} e^{i\mathbf{k}\cdot(\mathbf{x}-\mathbf{x}')} \left(\delta_{ij} - \frac{k^i k^j}{\mathbf{k}^2}\right) \\
&= i\delta^3_{\perp ij}(\mathbf{x} - \mathbf{x}').
\end{aligned}
\tag{2.572}
$$

2.6.2.7 *Hamiltonian and momentum operators in **k**-space*

Using the expansion expressions Eqs.(2.566) and (2.568), the Hamiltonian operator becomes

$$\hat{H} = \int d^3x \frac{1}{2} : (\hat{\mathbf{E}}^2 + \hat{\mathbf{B}}^2) :$$

$$= \frac{1}{2} \int \frac{d^3k}{2\omega_{\mathbf{k}}} \sum_{ll'=1}^{2} [\omega_{\mathbf{k}}^2 \boldsymbol{\epsilon}(\mathbf{k}, l') \cdot \boldsymbol{\epsilon}(\mathbf{k}, l) + (\mathbf{k} \times \boldsymbol{\epsilon}(\mathbf{k}, l')) \cdot (\mathbf{k} \times \boldsymbol{\epsilon}(\mathbf{k}, l))]$$

$$\times (\hat{a}_{\mathbf{k}l'}^\dagger \hat{a}_{\mathbf{k}l} + \hat{a}_{\mathbf{k}l}^\dagger \hat{a}_{\mathbf{k}l'}) \qquad (2.573)$$

$$= \int d^3k \, \omega_{\mathbf{k}} \sum_{l=1}^{2} \hat{a}_{\mathbf{k}l}^\dagger \hat{a}_{\mathbf{k}l}.$$

In the derivation of the last line of Eq.(2.573), $(\mathbf{k} \times \boldsymbol{\epsilon}') \cdot (\mathbf{k} \times \boldsymbol{\epsilon}) = \mathbf{k}^2 \boldsymbol{\epsilon}' \cdot \boldsymbol{\epsilon} - (\mathbf{k} \cdot \boldsymbol{\epsilon}')(\mathbf{k} \cdot \boldsymbol{\epsilon})$ and $\omega_{\mathbf{k}}^2 - \mathbf{k}^2 = 0$ have been used. We can similarly obtain the momentum operators

$$\hat{\mathbf{P}} = \int d^3x : \hat{\mathbf{E}} \times \hat{\mathbf{B}} := \int d^3k \sum_{l=1}^{2} \mathbf{k} \hat{a}_{\mathbf{k}l}^\dagger \hat{a}_{\mathbf{k}l}. \qquad (2.574)$$

2.6.2.8 *Spin of massless vector bosons*

The spin of the photon field is given by Eq.(2.515). We have

$$\hat{S}_{ij} = \int d^3x (\hat{F}_{0j} \hat{A}_i - \hat{F}_{0i} \hat{A}_j)$$

$$= i \int d^3k \sum_{ll'}^{2} \epsilon_i(\mathbf{k}, l') \epsilon_j(\mathbf{k}, l)(\hat{a}_{\mathbf{k}l}^\dagger \hat{a}_{\mathbf{k}l'} - \hat{a}_{\mathbf{k}l'}^\dagger \hat{a}_{\mathbf{k}l}). \qquad (2.575)$$

Using the vector symbol, the spin operator of the massless vector bosons has the form

$$\hat{\mathbf{S}} = \int d^3x : \hat{\mathbf{E}} \times \hat{\mathbf{A}} :$$

$$= \int d^3x \int \frac{d^3k'}{\sqrt{2\omega_{\mathbf{k}'}(2\pi)^3}} \int \frac{d^3k}{\sqrt{2\omega_{\mathbf{k}}(2\pi)^3}} \sum_{ll'=1}^{2} (i\omega_{\mathbf{k}'}) \boldsymbol{\epsilon}(\mathbf{k}', l') \times \boldsymbol{\epsilon}(\mathbf{k}, l)$$

$$: \left(\hat{a}_{\mathbf{k}'l'} e^{-ik' \cdot x} - \hat{a}_{\mathbf{k}'l'}^\dagger e^{ik' \cdot x} \right) \left(\hat{a}_{\mathbf{k}l} e^{-ik \cdot x} + \hat{a}_{\mathbf{k}l}^\dagger e^{ik \cdot x} \right) : \qquad (2.576)$$

$$= \frac{i}{2} \int d^3k \sum_{ll'=1}^{2} [\boldsymbol{\epsilon}(\mathbf{k}, l') \times \boldsymbol{\epsilon}(\mathbf{k}, l)(\hat{a}_{\mathbf{k}l}^\dagger \hat{a}_{\mathbf{k}l'} - \hat{a}_{\mathbf{k}l'}^\dagger \hat{a}_{\mathbf{k}l})$$

$$+ \boldsymbol{\epsilon}(-\mathbf{k}, l') \times \boldsymbol{\epsilon}(\mathbf{k}, l)(-\hat{a}_{-\mathbf{k}l} \hat{a}_{\mathbf{k}l'} e^{-2i\omega_{\mathbf{k}}t} + \hat{a}_{-\mathbf{k}l'}^\dagger \hat{a}_{\mathbf{k}l}^\dagger e^{2i\omega_{\mathbf{k}}t})].$$

Then the helicity operator is given by

$$\hat{\Lambda} = \hat{\mathbf{S}} \cdot \frac{\mathbf{k}}{|\mathbf{k}|} = i \int d^3k (\hat{a}^\dagger_{\mathbf{k}2} \hat{a}_{\mathbf{k}1} - \hat{a}^\dagger_{\mathbf{k}1} \hat{a}_{\mathbf{k}2}). \tag{2.577}$$

Similar to what we did to diagonalize Eq.(2.522), we can diagonalize Eq.(2.577) using the transformation Eq.(2.523), which gives

$$\hat{\Lambda} = \int d^3k (\hat{a}^\dagger_{\mathbf{k}+} \hat{a}_{\mathbf{k}+} - \hat{a}^\dagger_{\mathbf{k}-} \hat{a}_{\mathbf{k}-}). \tag{2.578}$$

Thus the photons are spin 1 particles.

2.7   Interaction

By now, we have only considered the Lagrangians without interactions. The interactions can be added into the Lagrangians without violating the causality principle when they contain no time derivatives. Since any terms involving field function π in the generator of time translation transformation G_t for bosons will give terms related to time derivative, the interaction terms for bosons should not contain π. The physical mass and interaction terms should achieve the lowest energy for the ground state when the temperature effect is small. By now, we have no good numerical methods to calculate the ground state in the Riemann spacetime. Generally, there are symmetries in the ground state. Some symmetries are related to the Lorentz covariance. These symmetries should always be guaranteed when we add mass and interaction terms. In these symmetries, the most important one is the gauge symmetry, which correlates different types of particles. We have shown that the Lagrangian containing the massless boson field should possess the gauge symmetry in order to fulfill the covariance principle in the previous section. Therefore, any Lagrangian contains the massless boson field should have the gauge symmetry. We will discuss the gauge symmetry in the following section.

2.7.1   *Lagrangian with gauge invariance*

We can couple the vector bosons with the spinor fermions by adding an interaction term $-eA_\mu \bar{\psi} \gamma^\mu \psi$. The Lagrangian with this interaction term is gauge invariant. The Lagrangian for a spinor fermion field interacting with a vector field reads

$$\mathcal{L} = \bar{\psi}(i\gamma^\mu \partial_\mu - m)\psi - eA_\mu \bar{\psi}\gamma^\mu \psi - \frac{1}{4}F_{\mu\nu}F^{\mu\nu}$$
$$= \bar{\psi}[i\gamma^\mu(\partial_\mu + ieA_\mu) - m]\psi - \frac{1}{4}F_{\mu\nu}F^{\mu\nu}. \qquad (2.579)$$

The above Lagrangian density is invariant by the gauge transformation described by

$$A_\mu(x) \to A_\mu(x) - \partial_\mu \Lambda(x) = A_\mu(x) + ie^{-i\Lambda(x)}\partial_\mu e^{i\Lambda(x)} \qquad (2.580)$$

and

$$\psi(x) \to e^{ie\Lambda(x)}\psi(x), \qquad (2.581\text{a})$$
$$\bar{\psi}(x) \to e^{-ie\Lambda(x)}\bar{\psi}(x). \qquad (2.581\text{b})$$

Eq.(2.580) is the gauge transformation for the massless vector boson field. The existence of a massless vector boson field demands that the total Lagrangian should be invariant under the transformation Eq.(2.580). To keep the total Lagrangian in Eq.(2.579) gauge invariant, we have a coupled transformation Eq.(2.581) for the fermion field. We usually call the transformations Eqs.(2.580) and (2.581) together as gauge transformations. We define the *gauge covariant derivative*

$$D_\mu \equiv \partial_\mu + ieA_\mu. \qquad (2.582)$$

Then Eq.(2.579) can be rewritten as follows:

$$\mathcal{L} = \bar{\psi}(i\gamma^\mu D_\mu - m)\psi - \frac{1}{4}F_{\mu\nu}F^{\mu\nu}. \qquad (2.583)$$

Under the gauge transformations Eqs.(2.580) and (2.581), the gauge covariant derivative term $D_\mu\psi$ transforms as

$$D_\mu\psi(x) \to e^{ie\Lambda(x)}D_\mu\psi(x), \qquad (2.584)$$

which leads to the gauge invariance of the first term in Eq.(2.583). The gauge transformation Eq.(2.580) makes $F_{\mu\nu}(x) \to F_{\mu\nu}(x)$, which means that the Lagrangian of photons is invariant. Thus the Lagrangian density in Eq.(2.579) possesses the symmetry of gauge invariance. From Eq.(2.581), we can see that $e\Lambda(x)$ and $e\Lambda(x) + 2\pi$ give exactly the same transformation. The gauge transformation $U = e^{ie\Lambda(x)}$ in Eq.(2.581) forms the abelian group $U(1)$ because $\Lambda(x)$ is a simple function of spacetime coordinates.

2.7.2 *Nonabelian gauge symmetry*

When there are interactions, the field A_μ could become composite. We still need the gauge symmetry to maintain the covariance of the total Lagrangian. Now we consider the gauge transformation for the nonabelian case where the field described by A_μ is composite. In the gauge transformation $\psi(x) \to U\psi(x)$, U can be an element of $SU(N)$ with $U^\dagger U = 1$, which guarantees the term $\psi^\dagger \psi$ for scalar field or $\bar{\psi}\psi$ for fermion field to be invariant. This generalization was introduced by Yang and Mills in 1954.

Similar to the abelian case. We introduce the gauge covariant derivative

$$D_\mu \equiv \partial_\mu - igA_\mu \tag{2.585}$$

to replace the ordinary derivative ∂_μ, where g is called the coupling constant. Under the transformation

$$\psi(x) \to U\psi(x), \tag{2.586}$$

we have

$$D_\mu \to U(x)D_\mu U^\dagger(x). \tag{2.587}$$

Then we have an invariant kinetic term $D_\mu \psi^\dagger D^\mu \psi$ for scalar boson field or $i\bar{\psi}\slashed{D}_\mu \psi$ for spinor fermion field. Now we consider a Lagrangian with N field $\psi_i(x)$ under a continuous $SU(N)$ symmetry transformation $\psi_i(x) \to U_{ij}(x)\psi_j(x)$. The gauge transformation is an $SU(N)$ transformation. An infinitesimal $SU(N)$ transformation has the form

$$U_{jk}(x) = \delta_{jk} - ig\theta^a(x)(T^a)_{jk} + O(\theta^2). \tag{2.588}$$

The indexes j and k run from 1 to N. a runs from 1 to N^2-1. In Eq.(2.588), the summation over a is implied. T^a are the generators of $SU(N)$. Due to the special unitarity of U, T^a are hermitian and traceless. The Lie algebra of group gives the commutation relations

$$[T^a, T^b] = if^{abc}T^c, \tag{2.589}$$

where the real factors f^{abc} are called the structure constant. For $SU(2)$, $f^{abc} = \varepsilon^{abc}$, where ε^{abc} is the antisymmetric Levi-Civita symbol. For the abelian case, $f^{abc} = 0$. The generators obey the normalization condition

$$\text{Tr}(T^a T^b) = \frac{1}{2}\delta^{ab}. \tag{2.590}$$

Under the gauge transformation,

$$\partial_\mu \psi \to \partial_\mu(U\psi) = U\partial_\mu \psi + (\partial_\mu U)\psi = U[\partial_\mu \psi + (U^\dagger \partial_\mu U)\psi]. \tag{2.591}$$

In order to give

$$D_\mu \to U(x)D_\mu U^\dagger(x), \tag{2.592}$$

A_μ should transform as

$$A_\mu(x) \to U(x)A_\mu(x)U^\dagger(x) + \frac{i}{g}U(x)\partial_\mu U^\dagger(x). \tag{2.593}$$

This can be verified directly.

$$D_\mu\psi = \partial_\mu\psi - igA_\mu\psi \to$$
$$U[\partial_\mu\psi + (U^\dagger\partial_\mu U)\psi] - igUA_\mu U^\dagger U\psi + U(\partial_\mu U^\dagger)U\psi \tag{2.594}$$
$$= U[\partial_\mu\psi - igA_\mu]U^\dagger U\psi.$$

We have used $U^\dagger U = 1$ in the derivation of Eq.(2.594). $U(x)$ can be expressed in terms of the generator T^a as

$$U(x) = \exp[-ig\theta(x)T^a]. \tag{2.595}$$

To construct the gauge invariant Lagrangian of vector boson field $A_\mu(x)$, we replace $F_{\mu\nu} = \partial_\mu A_\nu - \partial_\nu A_\mu$ by

$$F_{\mu\nu} \equiv \frac{i}{g}[D_\mu, D_\nu]$$
$$= \partial_\mu A_\nu - \partial_\nu A_\mu - ig[A_\mu, A_\nu]. \tag{2.596}$$

Eqs.(2.592) and (2.596) give

$$F_{\mu\nu}(x) \to U(x)F_{\mu\nu}(x)U^\dagger(x). \tag{2.597}$$

Then we can construct a gauge invariant kinetic term

$$\mathcal{L}_{gb} = -\frac{1}{2}\text{Tr}F^{\mu\nu}F_{\mu\nu}, \tag{2.598}$$

where the subscript 'gb' represents gauge bosons.

We can derive this Lagrangian from the generator of time translation transformation in a similar way as we used for the photon field. We will give the detailed deduction later.

A_μ are N by N matrices. From Eq.(2.588), we have

$$A_\mu \to A_\mu - ig\theta^a[T^a, A_\mu] - \partial_\mu\theta^a T^a. \tag{2.599}$$

Taking the trace of Eq.(2.599), we can see that the trace of A_μ does not transform. Thus A_μ can be traceless and hermitian. Then we can expand A_μ in terms of the generator T^a

$$A_\mu(x) = A_\mu^a(x)T^a. \tag{2.600}$$

Using Eq.(2.590), we have

$$A_\mu^a(x) = 2\text{Tr}A_\mu(x)T^a. \tag{2.601}$$

For $F_{\mu\nu}(x)$, we have

$$\begin{aligned}
F_{\mu\nu} &= \partial_\mu A_\nu - \partial_\nu A_\mu - ig[A_\mu, A_\nu] \\
&= (\partial_\mu A_\nu^c - \partial_\nu A_\mu^c)T^c - igA_\mu^a A_\nu^b[T^a, T^b] \\
&= (\partial_\mu A_\nu^c - \partial_\nu A_\mu^c + gf^{abc}A_\mu^a A_\nu^b)T^c.
\end{aligned} \tag{2.602}$$

Then, we can express $F_{\mu\nu}(x)$ as

$$F_{\mu\nu}(x) = F_{\mu\nu}^c(x)T^c \tag{2.603}$$

with

$$F_{\mu\nu}^c(x) = \partial_\mu A_\nu^c - \partial_\nu A_\mu^c + gf^{abc}A_\mu^a A_\nu^b. \tag{2.604}$$

Thus the kinetic term Eq.(2.598) becomes

$$\mathcal{L}_{gb} = -\frac{1}{4}F^{a\mu\nu}F_{\mu\nu}^a. \tag{2.605}$$

This is the so-called *Yang-Mills Lagrangian density*. From Eq.(2.604), we can see that $\mathcal{L}_{gb}$ contains the self-interactions among the gauge fields. Using Eq.(2.604), we can express the Yang-Mills Lagrangian density Eq.(2.605) by

$$\begin{aligned}
\mathcal{L}_{gb} &= -\frac{1}{4}F_{\mu\nu}^a F^{a\mu\nu} \\
&= -\frac{1}{4}(\partial_\mu A_\nu^a - \partial_\nu A_\mu^a)^2 - \frac{1}{2}g(\partial_\mu A_\nu^a - \partial_\nu A_\mu^a)f^{abc}A^{b\mu}A^{c\nu} \\
&\quad - \frac{1}{4}g^2 f^{abc}f^{ade}A_\mu^b A_\nu^c A^{d\mu}A^{e\nu}.
\end{aligned} \tag{2.606}$$

We use the Faddeev-Popov method to integrating out the redundant components. We consider the path integral

$$Z = \int DA e^{iS(A)}, \tag{2.607}$$

where $S(A) = \int d^4x \mathcal{L}_{gb}$ is the Yang-Mills action. We define a function

$$\Delta(A) \equiv \left\{ \int D\theta \delta[f(A(\theta))] \right\}^{-1}, \tag{2.608}$$

where $f(x)$ is an auxiliary function and

$$A_\mu(\theta) = U(\theta)A_\mu U^\dagger(\theta) + \frac{i}{g}(\partial_\mu U(\theta))U^\dagger(\theta). \tag{2.609}$$

$A(\theta)$ in Eq.(2.609) is the gauge transformation of A and $U(\theta) = e^{-ig\theta(x) \cdot T}$ is the group elements that defines the gauge transformation at x. We use Eq.(2.543) and factor out integration over θ. Then we have

$$Z = \int DA e^{iS(A)} \Delta(A) \delta[f(A)]. \qquad (2.610)$$

We can choose suitable form of function for $f(A)$. A suitable selection is

$$f(A) = \partial A - \sigma. \qquad (2.611)$$

Since $\Delta(A)$ appears in Eq.(2.610) together with $\delta[f(A)]$, only infinitesimal θ is relevant. Under an infinitesimal transformation

$$A_\mu^a \to A_\mu^a + g f^{abc}\theta^b A_\mu^c - \partial_\mu \theta^a, \qquad (2.612)$$

Eq.(2.608) becomes

$$\Delta(A) = \left\{ \int D\theta \delta[\partial A^a - \sigma^a + \partial^\mu(g f^{abc}\theta^b A_\mu^c - \partial_\mu \theta^a)] \right\}^{-1}. \qquad (2.613)$$

Thus

$$\Delta(A)\delta[f(A)] = \left\{ \int D\theta \delta[\partial A^a - \sigma^a + \partial^\mu(g f^{abc}\theta^b A_\mu^c - \partial_\mu \theta^a)] \right\}^{-1}$$
$$\times \delta(f(A)) \qquad (2.614)$$
$$= \left\{ \int D\theta \delta[\partial^\mu(g f^{abc}\theta^b A_\mu^c - \partial_\mu \theta^a)] \right\}^{-1} \delta(f(A)).$$

We define an operator $K^{ab}(x, y)$ by

$$K^{ab}(x, y) \equiv \partial^\mu(g f^{abc} A_\mu^c - \partial_\mu \delta^{ab})\delta^4(x - y). \qquad (2.615)$$

Then we have

$$\partial^\mu(g f^{abc}\theta^b A_\mu^c - \partial_\mu \theta^a) = \int d^4 y K^{ab}(x, y)\theta^b. \qquad (2.616)$$

Since $\int d\theta \delta(K\theta) = 1/\det K$, we have $\Delta(A) = \det K$. We can express the determinant $\det K$ as a functional integral over Grassmann variables by

$$\Delta(A) = \int Dc Dc^\dagger e^{iS_{ghost}(c^\dagger, c)} \qquad (2.617)$$

with

$$S_{ghost}(c^\dagger, c) = \int d^4 x \int d^4 y c_a^\dagger(x) K^{ab}(x, y) c_b(y)$$
$$= \int d^4 x [\partial^\mu c_a^\dagger(x)\partial_\mu c_a(x) - \partial^\mu c_a^\dagger(x) g f^{abc} A_\mu^c c_b(x)]. \qquad (2.618)$$

The field functions c_a and $c_a^\dagger$ are the ghost field functions. They do not correspond to the real particles. Thus scalar fields c_a and $c_a^\dagger$ can be anti-commuting without causing problems.

Since Z does not depend on σ^a, similar to Eq.(2.548), we multiply a Gaussian functional

$$e^{-\frac{i}{2\xi}\int d^4x \sigma^2(x)} \tag{2.619}$$

and integrate over σ. Thus $\delta(f(A))$ is replaced by

$$e^{-\frac{i}{2\xi}\int d^4x (\partial A^a)^2}. \tag{2.620}$$

The final expression for Z is then given by

$$Z = \int DADcDc^\dagger e^{iS(A)-\frac{i}{2\xi}\int d^4x(\partial A^a)^2 + iS_{ghost}(c^\dagger, c)}. \tag{2.621}$$

The gauge Lagrangian density is changed to

$$\begin{aligned}
\mathcal{L}_{gb} = & -\frac{1}{4}(\partial_\mu A_\nu^a - \partial_\nu A_\mu^a)^2 \\
& -\frac{1}{2\xi}(\partial^\mu A_\mu^a)^2 - \frac{1}{2}(\partial_\mu A_\nu^a - \partial_\nu A_\mu^a)gf^{abc}A^{b\mu}A^{c\nu} \\
& -\frac{1}{4}g^2 f^{abc}f^{ade}A_\mu^b A_\nu^c A^{d\mu}A^{e\nu} + \partial^\mu c_a^\dagger \partial_\mu c_a - \partial^\mu c_a^\dagger g f^{abc}A_\mu^c c_b.
\end{aligned} \tag{2.622}$$

Now we construct the generator $\hat{G}_t$ of time translation transformation corresponding to the Lagrangian density $\mathcal{L}_{gb}$. Similar to photons, we have only two internal degrees of freedoms for each A^a. Since the action is invariant for the gauge transformation $A_\mu^a \to A_\mu^a + gf^{abc}\theta^b A_\mu^c - \partial_\mu \theta^a$, we can take $A_0'^a = A_0^a + gf^{abc}\theta^b A_0^c - \partial_\mu\theta^a = 0$ with proper gauge transformation. Therefore, we can choose $A_0^a = 0$. The action is also invariant for the transformation $\partial A \to \partial A'^a = \partial A^a - \sigma^a$. We can take $\partial A^a = 0$ with proper σ^a. Then $\nabla \cdot \mathbf{A}^a = 0$. For the three spatial components of the four-vector A^a, we can choose two nonzero transverse components and one zero longitudinal component to satisfy $\nabla \cdot \mathbf{A}^a = 0$. Thus we have the two components of gauge bosons as two transverse components and add one artificial longitudinal component $A_3^a = 0$ and the fourth component $A_0^a = 0$.

We have initially the following commutation relations

$$[\hat{A}_i^a(\mathbf{x}, t), \hat{\pi}_j^a(\mathbf{x}', t)] = i\delta_{ij}\delta^3(\mathbf{x} - \mathbf{x}'), \tag{2.623a}$$

$$[\hat{A}_i^a(\mathbf{x}, t), \hat{A}_j^a(\mathbf{x}', t)] = [\hat{\pi}_i^a(\mathbf{x}, t), \hat{\pi}_j^a(\mathbf{x}', t)] = 0 \tag{2.623b}$$

with $i, j = 1, 2$. After introducing the third artificial variable, we have three variables of A_i^a which are not independent and constrained by $\nabla \cdot \hat{\mathbf{A}}^a = 0$.

If we use the commutation relations for a vector, we could use the transverse projection operator $(\hat{P}_\perp)_{ij}$ in Eq.(2.554) to impose the transversality condition. Through projection, the commutation relation Eq.(2.623) is expressed with the following projected commutation relation

$$[\hat{A}_i^a(\mathbf{x}, t), \hat{\pi}_j^a(\mathbf{x}', t)] = i\delta_{\perp ij}^3(\mathbf{x} - \mathbf{x}').$$ (2.624)

Using the four-dimensional vector A_μ^a with $\boldsymbol{\nabla} \cdot \hat{\mathbf{A}}^a = 0$ and $A_0^a = 0$, we can express the Yang-Mills Lagrangian density Eq.(2.622) as follows:

$$
\begin{aligned}
\mathcal{L}_{gb} =& \frac{1}{2}\left(\frac{\partial \mathbf{A}^a}{\partial t}\right)^2 - \frac{1}{2}(\boldsymbol{\nabla} \times \mathbf{A}^a)^2 + \frac{\partial A_i^a}{\partial t}f^{abc}A^{bi}A^{c0} \\
&- \frac{1}{2}(\partial_i A_j^a - \partial_j A_i^a)gf^{abc}A^{ai}A^{bj} - \frac{1}{4}g^2 f^{abc}f^{ade}A_\mu^b A_\nu^c A^{d\mu}A^{e\nu} \\
&+ \partial^\mu c_a^\dagger \partial_\mu c_a - \partial^\mu c_a^\dagger g f^{abc}A_\mu^c c_b \\
=& \frac{1}{2}\left(\frac{\partial \mathbf{A}^a}{\partial t}\right)^2 - \frac{1}{2}(\boldsymbol{\nabla} \times \mathbf{A}^a)^2 \\
&- \frac{1}{2}(\partial_i A_j^a - \partial_j A_i^a)gf^{abc}A^{ai}A^{bj} - \frac{1}{4}g^2 f^{abc}f^{ade}A_\mu^b A_\nu^c A^{d\mu}A^{e\nu} \\
&+ \partial^\mu c_a^\dagger \partial_\mu c_a - \partial^i c_a^\dagger f^{abc}A_i^c c_b.
\end{aligned}
$$ (2.625)

We can show that the following generator of time translation transformation $\hat{G}_t$ leads to the Yang-Mills Lagrangian density Eq.(2.625).

$$
\begin{aligned}
\hat{G}_t = \int d^3x \Big[&\frac{1}{2}(\hat{\boldsymbol{\pi}}^a)^2 + \frac{1}{2}(\boldsymbol{\nabla} \times \hat{\mathbf{A}}^a)^2 + \frac{1}{2}(\partial_i A_j^a - \partial_j A_i^a)gf^{abc}A^{ai}A^{bj} \\
&+ \frac{1}{4}g^2 f^{abc}f^{ade}A_\mu^b A_\nu^c A^{d\mu}A^{e\nu} + i\mathrm{Tr}\ln K \Big]
\end{aligned}
$$ (2.626)

with

$$K^{ab}(x, y) = (\partial^i g f^{abc}A_i^c - \partial^\mu \partial_\mu \delta^{ab})\delta^4(x - y).$$ (2.627)

$\hat{G}_t$ in Eq.(2.626) does not contain the time derivative term of the field functions and thus satisfies the causality principle. Inserting G_t in Eq.(2.626) into Eq.(2.128) and carrying out the integration over $\hat{\pi}^a$ and introducing the ghost fields, we get the Yang-Mills Lagrangian density Eq.(2.625). Using the generator of time translation transformation $\hat{G}_t$ given by Eq.(2.626) with the fields replaced by the operators, we obtain the equations of motion

$$i\frac{\partial \hat{\mathbf{A}}^a}{\partial t} = [\hat{\mathbf{A}}^a, \hat{G}_t],$$ (2.628a)

$$i\frac{\partial \hat{\boldsymbol{\pi}}^a}{\partial t} = [\hat{\boldsymbol{\pi}}^a, \hat{G}_t].$$ (2.628b)

Since there are self-interaction terms, we are not be able to solve the equations of motion for the Yang-Mills gauge bosons exactly.

For the photon field, there is only a single gauge field $A^\mu(x)$ and $f^{abc} = 0$. The ghost terms are not coupled to the photon field, which can be integrated out. The Yang-Mills Lagrangian density Eq.(2.622) becomes the Lagrangian density in Eq.(2.550). Similar to the photon field, the nonabelian gauge fields can be coupled with the fermion spinor field by introducing the gauge covariant derivative to replace ∂_μ.

Chapter 3

Quantum fields in the Riemann spacetime

3.1 Lagrangian in Riemann spacetime

Now we turn to the curved spacetime. In order to fulfill the causality principle, the physical spacetime should be the Riemann spacetime as discussed in Appendix A. The action should satisfy the principle of general covariance. Thus, Lagrangian density is a scalar in the Riemann spacetime. Let us construct the Lagrangian density which should be the scalars in the Riemann spacetime. The simplest field is the scalar field. We consider the scalar field as an example. The underlining principle is independent of the types of the fields contained in the Lagrangian.

Let us begin with the Lagrangian density of matter $\mathcal{L}_m$. The general form of the Lagrangian density of matter $\mathcal{L}_m$ with $\dot{\phi}^2$ term for a scalar field in the Riemann spacetime is given by

$$\mathcal{L}_m = \frac{1}{2}(g^{\mu\nu}\partial_\mu\phi_a\partial_\nu\phi_a - m^2\phi_a^2) - V(\phi), \tag{3.1}$$

where m is the mass and $V(\phi)$ is the self-interaction term. $g^{\mu\nu}$ is the metric tensor in the Riemann spacetime. $g^{\mu\nu}$ can be a functional of the field ϕ and its spatial derivatives. It should not contain $\dot{\phi}$. If $g^{\mu\nu}$ is not a function of $\dot{\phi}$, all the procedure of constructing covariant Lagrangian in the Minkowski spacetime can be applied similarly to the Riemann spacetime. The more general form for Eq. (3.1) is

$$\mathcal{L}_m = \frac{1}{2}f(\phi)g^{\mu\nu}\partial_\mu\phi_a\partial_\nu\phi_a - \frac{1}{2}m^2\phi_a^2 - V(\phi). \tag{3.2}$$

$f(\phi)$ can be absorbed into the metric $g^{\mu\nu}$ because the metric $g^{\mu\nu}$ is the functional of the field functions. Therefore the Lagrangian density in Eq. (3.2) is equivalent to that in Eq. (3.1). We will only consider the Lagrangian density in Eq. (3.1). The action of matter for a scalar field in the Riemann spacetime is given by

$$S_m = \int d^4x \sqrt{-g} \left[\frac{1}{2} g^{\mu\nu} \partial_\mu \phi_a \partial_\nu \phi_a - \frac{1}{2} m^2 \varphi_a^2 - V(\phi) \right]. \tag{3.3}$$

For a vector A_μ, we should use a covariant derivative

$$D_\alpha A_\mu = \partial_\alpha A_\mu - \Gamma^\nu_{\alpha\mu} A_\nu. \tag{3.4}$$

where $\Gamma^\nu_{\alpha\mu}$ is the Levi-Civita connection of the metric given by

$$\Gamma^\lambda_{\mu\nu} = \frac{1}{2} g^{\lambda\rho} (\partial_\nu g_{\rho\mu} + \partial_\mu g_{\rho\nu} - \partial_\rho g_{\mu\nu}), \tag{3.5}$$

The first order covariant derivative of a scalar function coincides with the ordinary derivative.

With the Lagrangian density of matter $\mathcal{L}_m$, we can define the energy-momentum tensor

$$T^{\mu\nu} \equiv \frac{1}{\sqrt{-g}} \frac{D(\sqrt{-g}\mathcal{L}_m)}{D(D_\mu \phi_a(x))} D^\nu \phi_a(x) - g^{\mu\nu} \mathcal{L}_m(x). \tag{3.6}$$

Using the energy-momentum tensor $T^{\mu\nu}$, we can also construct a scalar

$$\begin{aligned} S_e &= \alpha_1 \int d^4x \sqrt{-g} g_{\mu\nu} T^{\mu\nu} \\ &\equiv \int d^4x \sqrt{-g} \mathcal{L}_e \end{aligned} \tag{3.7}$$

with

$$\mathcal{L}_e = \alpha_1 g_{\mu\nu} T^{\mu\nu}, \tag{3.8}$$

where α_1 is a constant parameter. This Lagrangian density can be considered as a part divided from the Lagrangian density of matter. With the metric $g^{\mu\nu}$, we can construct a scalar as follows:

$$S_g = \alpha_2 \int d^4x \sqrt{-g} R, \tag{3.9}$$

where α_2 is a constant parameter. R is the Ricci scalar curvature. S_g is the so-called *Einstein-Hilbert action* for gravity. R is defined as $g^{\mu\nu} R_{\mu\nu}$. $R_{\mu\nu}$ is the Ricci tensor defined by $R_{\mu\nu} \equiv R^\kappa_{\mu\kappa\nu}$. $R^\lambda_{\mu\nu\kappa}$ is the Riemann curvature tensor defined by

$$R^\lambda_{\mu\nu\kappa} \equiv \partial_\nu \Gamma^\lambda_{\mu\kappa} - \partial_\kappa \Gamma^\lambda_{\mu\nu} + \Gamma^\sigma_{\mu\kappa} \Gamma^\lambda_{\nu\sigma} - \Gamma^\sigma_{\mu\nu} \Gamma^\lambda_{\kappa\sigma}. \tag{3.10}$$

The total action S_t should be the sum of the above three parts and a constant term.

$$S_t = \int d^4x \sqrt{-g} \mathcal{L}_m + \alpha_1 \int d^4x \sqrt{-g} g_{\mu\nu} T^{\mu\nu}$$

$$+ \alpha_2 \int d^4x \sqrt{-g} R + \int d^4x \sqrt{-g} \Lambda' \qquad (3.11)$$

$$\equiv \int d^4x \sqrt{-g} \mathcal{L}_t$$

with

$$\mathcal{L}_t = \mathcal{L}_m + \alpha_1 g_{\mu\nu} T^{\mu\nu} + \alpha_2 R + \Lambda', \qquad (3.12)$$

where Λ' is a constant. There are other scalar terms such as R^2, which will be discussed later. Now we consider the total action is given by the above three contributions. If all the quantum fields involved are considered in the matter Lagrangian density $\mathcal{L}_m$, then the action S_t contains all kinetic terms and the interactions.

It should be noted that we do not consider the metric $g_{\mu\nu}$ as an independent field. $g_{\mu\nu}$ is a functional of field functions $\phi_a(x)$, i.e. $g_{\mu\nu} = g_{\mu\nu}(\phi(x))$.

3.2   Homogeneity of spacetime

We use the principle that the spacetime is homogeneous. The total action operator should possess the symmetry of spacetime translation. We transform the field operator via $x_\mu \to x_\mu + a^\mu$, where a^μ is a constant four-vector. For an infinitesimal translation δa^ν, we have the total variation $\tilde{\delta}\hat{\phi}(x) = \delta a^\nu \partial_\nu \hat{\phi}(x)$.

For the Minkowski spacetime, $g_{\mu\nu}$ are constants. Thus $\delta \hat{S} = 0$ is an identity. However, for the Riemann spacetime, since $g_{\mu\nu}$ contain x^μ and if there is no constraint on $g_{\mu\nu}$, $\delta \hat{S} = 0$ is not an identity. Thus we need a relation to guarantee $\delta \hat{S} = 0$. This relation is the Einstein equations which have the threefold functions: i) preserving the validation of the Euler-Lagrange equation, which is also equivalent to the equations of motion, ii) guaranteeing the energy-momentum conservation, iii) making the cancellation of the other remaining terms in $\delta \hat{S}$. Thus $\delta \hat{S} = 0$ becomes an identity in the Riemann spacetime. The above procedures can be realized in the following way.

First we consider $\delta \hat{S}_m$, where $\hat{S}_m = \int d^4y \sqrt{-\hat{g}} \hat{\mathcal{L}}_m(y)$ is the action operator for matter. Similar to the derivation of Eq.(2.183), we have

$$\delta \hat{S}_m = \int d^4x [\tilde{\delta}(\sqrt{-\hat{g}}\hat{\mathcal{L}}_m(x)) + D_\mu(\sqrt{-\hat{g}}\hat{\mathcal{L}}_m(x)\delta a^\mu)]. \qquad (3.13)$$

$\tilde{\delta}(\sqrt{-\hat{g}}\hat{\mathcal{L}}_m(x))$ in $\delta\hat{S}_m$ is given by the chain rule

$$\tilde{\delta}(\sqrt{-\hat{g}}\hat{\mathcal{L}}_m(x)) = \frac{D(\sqrt{-\hat{g}}\hat{\mathcal{L}}_m)}{D\hat{\phi}_a(x)}\tilde{\delta}\hat{\phi}_a(x) + \frac{D(\sqrt{-\hat{g}}\hat{\mathcal{L}}_m)}{D(D_\mu\hat{\phi}_a(x))}D_\mu\tilde{\delta}\hat{\phi}_a(x). \quad (3.14)$$

The variation of the total action operator $\hat{S}_t$ becomes

$$\delta\hat{S}_t = \int d^4x \left[\frac{D(\sqrt{-\hat{g}}\hat{\mathcal{L}}_m)}{D\hat{\phi}_a(x)}\tilde{\delta}\hat{\phi}_a(x) - D_\mu \frac{D(\sqrt{-\hat{g}}\hat{\mathcal{L}}_m)}{D(D_\mu\hat{\phi}_a(x))}\tilde{\delta}\hat{\phi}_a(x) \right] \\ + \sqrt{-\hat{g}}D_\mu[\delta a_\nu \hat{T}^{\mu\nu}(x)] + \delta\hat{S}_e + \delta\hat{S}_g + \delta\hat{S}_\Lambda. \quad (3.15)$$

In the Riemann spacetime, the Euler-Lagrange equation in operator form should hold as in the Minkowski spacetime and we will show that Einstein equations would guarantee it. We have

$$\frac{D(\sqrt{-\hat{g}}\hat{\mathcal{L}}_m(x))}{D\hat{\phi}_a(x)} - D_\mu \frac{D(\sqrt{-\hat{g}}\hat{\mathcal{L}}_m(x))}{D(D_\mu\hat{\phi}_a(x))} = 0. \quad (3.16)$$

We would also expect that the energy-momentum conservation holds as in the Minkowski spacetime, which is given by

$$D_\mu \hat{T}^{\mu\nu} = 0. \quad (3.17)$$

The action should possess the symmetry of the spacetime translation. Under an infinitesimal spacetime translation, the variation of the total action operator should be zero. We have

$$\delta\hat{S}_t = 0. \quad (3.18)$$

Now let us consider the equation

$$\delta\hat{S}_e + \delta\hat{S}_g + \delta\hat{S}_\Lambda = \alpha_1\delta(\hat{g}_{\mu\nu}\sqrt{-\hat{g}}\hat{T}^{\mu\nu}) + \alpha_2\delta(\sqrt{-\hat{g}}\hat{R}) + \delta(\sqrt{-\hat{g}}\Lambda') \\ = 0. \quad (3.19)$$

We will show that when Eq.(3.19) holds, both the Euler-Lagrange equation Eq.(3.16) and the conservation of energy-momentum Eq.(3.16) holds. Using the following relation

$$\hat{R} = \frac{1}{1+4\beta}\hat{g}_{\mu\nu}(\hat{R}^{\mu\nu} + \beta\hat{g}^{\mu\nu}\hat{R}), \quad (3.20)$$

we can write Eq.(3.19) in a more symmetric way,

$$\delta\left[\left(\alpha_1\sqrt{-\hat{g}}\hat{T}^{\mu\nu} + \frac{\alpha_2\sqrt{-\hat{g}}}{1+4\beta}(\hat{R}^{\mu\nu} + \beta\hat{g}^{\mu\nu}\hat{R}) + \frac{1}{4}\sqrt{-\hat{g}}\Lambda'\hat{g}^{\mu\nu} \right) \hat{g}_{\mu\nu} \right] \\ = 0. \quad (3.21)$$

3.3   Einstein equations

Since we can use any local coordinate frame, $g_{\mu\nu}$ can be a very general function and we expect the terms in the bracket before $g_{\mu\nu}$ in Eq.(3.21) cancel out except for a constant term. Thus we have

$$\alpha_1 \hat{T}^{\mu\nu} + \frac{\alpha_2}{1 + 4\beta}(\hat{R}^{\mu\nu} + \beta \hat{g}^{\mu\nu} \hat{R}) + \frac{1}{4}\Lambda' \hat{g}^{\mu\nu} = c\hat{g}^{\mu\nu}, \qquad (3.22)$$

where c is a constant. $c\hat{g}^{\mu\nu}$ term can be merged with $\frac{1}{4}\Lambda'\hat{g}^{\mu\nu}$ term and thus we take $c = 0$. Using the Bianchi identity, we can see that $\beta = -1/2$ in order to guarantee the equation of the conservation of energy-momentum $D_\mu \hat{T}^{\mu\nu}(x) = 0$. We introduce the gravitational constant G, which relates the parameters α_1 and α_2 by $\alpha_1 = -8\pi G\alpha_2$. Then Eq.(3.22) becomes

$$\hat{R}^{\mu\nu} - \frac{1}{2}\hat{g}^{\mu\nu}\hat{R} + \hat{g}^{\mu\nu}\Lambda = -8\pi G\hat{T}^{\mu\nu}, \qquad (3.23)$$

where $\Lambda = 2\pi G\Lambda'/\alpha_1$, which is called the cosmological constant. Eq.(3.23) is called the *Einstein equations* or *Einstein field equations* of general relativity.

When we put back the Einstein equations into the total action operator $\hat{S}_t$ and use the parameter relation $\alpha_1 = -8\pi G\alpha_2$, we find the terms $\hat{S}_e$ and $\hat{S}_g$ cancel out. Thus $\delta\hat{S}_e + \delta\hat{S}_g + \delta\hat{S}_\Lambda = 0$. Using Eq.(A.131), we can see that the Einstein equations Eq.(3.23) guarantee Eq.(3.17). Since $\hat{S}_e$ and $\hat{S}_g$ are cancelled, we will show later in Section 3.4 that the Euler-Lagrange equation in operator form is satisfied. Thus Eq.(3.18) is fulfilled automatically and the homogeneity of spacetime is guaranteed. Only the action operator $\hat{S}_m$ for matter remains in the total action operator $\hat{S}_t$. Thus the action becomes

$$S_t = \int d^4x \sqrt{-g}\mathcal{L}_m. \qquad (3.24)$$

The path integral for the action of Eq.(3.24) should be carried out for the field functions with the metric $g^{\mu\nu}$ in the action satisfying the Einstein equations Eq.(3.23). Both $T^{\mu\nu}$ and $g^{\mu\nu}$ are symmetric for the index μ and ν. There are 10 independent functions for $g^{\mu\nu}$. Thus there are 10 equations in Eq.(3.23). Since the energy-momentum conservation equations $D_\mu \hat{T}^{\mu\nu}(x) = 0$ are satisfied automatically. We have 6 independent equations for 10 functions. We have then 4 functions to make the coordinate transformation for $g^{\mu\nu}$. Thus the Einstein equations uniquely determine the metric $g^{\mu\nu}$ from the field functions. On the right hand side of Eq.(3.23), $T^{\mu\nu}$ is determined by Eq.(3.6).

When the Einstein equations hold, the equations of motion Eq.(3.17) and the conservation of energy-momentum Eq.(3.17) hold. Thus we can also call the Einstein equations the equations of motion or the equations of everything.

3.4   Generator of time translation transformation

We will show that the following generator of time translation transformation gives the action of Eq.(3.24)

$$
G_t = \int \sqrt{-g} d^3x \left[\frac{1}{2} \frac{1}{(-gg^{00})} (\pi_a - \sqrt{-g} g^{0i} D_i \phi_a)^2 \right.
$$
$$
\left. - \frac{1}{2} g^{ij} D_i \phi_a D_j \phi_a + \frac{1}{2} m^2 \phi_a^2 + V(\phi) \right],
$$

(3.25)

where the metric $g^{\mu\nu}$ is only the functional of the field functions ϕ_a. $g^{\mu\nu}$ should not contain the conjugate field function π_a and the time derivative terms. Thus the causality principle could be guaranteed. We can get the action of Eq.(3.24) by inserting Eq.(3.25) into Eq.(2.128) and integrating over π_a.

$$
Z = \int D\phi \int D\pi \exp \left[i \int d^4x (\pi \dot{\phi} - \mathcal{G}_t(\pi, \phi)) \right]
$$
$$
= \int D\phi \exp \left\{ i \int \sqrt{-g} d^4x \left[\frac{1}{2} g^{00} (D_0 \phi_a)^2 + g^{0i} D_i \phi_a D_0 \phi_a \right. \right.
$$
$$
\left. \left. + \frac{1}{2} g^{ij} D_i \phi_a D_j \phi_a - \frac{1}{2} m^2 \phi_a^2 - V(\phi) \right] \right\}
$$

(3.26)

$$
= \int D\phi \exp \left\{ i \int \sqrt{-g} d^4x \left[\frac{1}{2} g^{\mu\nu} D_\mu \phi_a D_\nu \phi_a - \frac{1}{2} m^2 \phi_a^2 - V(\phi) \right] \right\}
$$
$$
= \int D\phi \exp \left[i \int \sqrt{-g} d^4x \mathcal{L}_m(\phi) \right].
$$

Thus the generator of time translation transformation corresponding to the action of Eq.(3.24) is given by

$$
\hat{G}_t = \int \sqrt{-\hat{g}} d^3x \left[\frac{1}{2} \frac{1}{(-\hat{g}\hat{g}^{00})} (\hat{\pi}_a - \sqrt{-\hat{g}} \hat{g}^{0i} D_i \hat{\phi}_a)^2 \right.
$$
$$
\left. - \frac{1}{2} \hat{g}^{ij} D_i \hat{\phi}_a D_j \hat{\phi}_a + \frac{1}{2} m^2 \hat{\phi}_a^2 + V(\hat{\phi}) \right].
$$

(3.27)

In the meanwhile, the energy-momentum vector is defined as

$$P_\nu \equiv \int \sqrt{-g}d^3x T_\nu^0$$
$$= \int \sqrt{-g}d^3x \left[\frac{1}{\sqrt{-g}} \frac{D(\sqrt{-g}\mathcal{L}_m)}{D(D_0\phi_a(x))} D_\nu\phi_a(x) - g_\nu^0 \mathcal{L}_m(x) \right]. \tag{3.28}$$

Since $g_{\mu\nu}$ is independent of $\dot{\phi}_a$, we have

$$P_0 = \int \sqrt{-g}d^3x \left[\frac{D(\mathcal{L}_m)}{D(D_0\phi_a(x))} D_0\phi_a(x) - g_0^0 \mathcal{L}_m(x) \right]$$
$$= \int \sqrt{-g}d^3x \left[g^{0\mu}D_\mu\phi_a D_0\phi_a - g_0^0 \left(\frac{1}{2} g^{0\mu}D_\mu\phi_a D_0\phi_a \right. \right.$$
$$\left. \left. + \frac{1}{2}g^{i\mu}D_\mu\phi_a D_i\phi_a - \frac{1}{2}m^2\phi_a^2 - V(\phi) \right) \right] \tag{3.29}$$
$$= \int \sqrt{-g}d^3x \left[\frac{1}{2}g^{00}D_0\phi_a D_0\phi_a - \frac{1}{2}g^{ij}D_i\phi_a D_j\phi_a + \frac{1}{2}m^2\phi_a^2 + V(\phi) \right].$$

P_0 is the energy of the fields ϕ_a and is called the Hamiltonian of the fields. When we replace the field functions ϕ_a with the field operators $\hat{\phi}_a$, the corresponding operator is the Hamiltonian operator.

$$\hat{H} \equiv \hat{P}_0$$
$$= \int \sqrt{-\hat{g}}d^3x \left[\frac{1}{2}\hat{g}^{00}D_0\hat{\phi}_a D_0\hat{\phi}_a - \frac{1}{2}\hat{g}^{ij}D_i\hat{\phi}_a D_j\hat{\phi}_a \right. \tag{3.30}$$
$$\left. + \frac{1}{2}m^2\hat{\phi}_a^2 + V(\hat{\phi}) \right].$$

Similar to the flat spacetime, we can not construct a consistent formalism for the scalar fermions which obey anti-commutation relations. Thus we consider that the scalar field is a boson field that $\hat{\phi}_a$ and $\hat{\pi}_a$ satisfy the commutation relations in Eq.(2.60) for bosons. Inserting $\hat{G}_t(\hat{\pi}, \hat{\phi})$ in Eq.(3.27) into the commutator $[\hat{\phi}_a(\mathbf{x}, t), \hat{G}_t(\hat{\pi}, \hat{\phi})]$, we have

$$[\hat{\phi}_a(\mathbf{x}, t), \hat{G}_t(\hat{\pi}, \hat{\phi})] = \frac{1}{\sqrt{-\hat{g}}\hat{g}^{00}} i\hat{\pi}_a - \frac{\hat{g}^{0i}}{\hat{g}^{00}} iD_i\hat{\phi}_a. \tag{3.31}$$

Using the equations of motion

$$i\partial_t\hat{\phi}_a = [\hat{\phi}_a, \hat{G}_t], \tag{3.32a}$$
$$i\partial_t\hat{\pi}_a = [\hat{\pi}_a, \hat{G}_t], \tag{3.32b}$$

We find

$$i\partial_t\hat{\phi}_a = \frac{1}{\sqrt{-\hat{g}}\hat{g}^{00}} i\hat{\pi}_a - \frac{\hat{g}^{0i}}{\hat{g}^{00}} iD_i\hat{\phi}_a \tag{3.33}$$

Thus we have

$$\hat{\pi}_a = \sqrt{-\hat{g}}\hat{g}^{0\nu}D_\nu\hat{\phi}_a. \tag{3.34}$$

Using Eq.(3.33) to express P_0 in terms of $\hat{\pi}_a$, we have

$$\hat{G}_t = \hat{H}. \tag{3.35}$$

Therefore the generator of time translation transformation $\hat{G}_t$ is equal to the Hamiltonian operator $\hat{H}$ and Eq.(2.91) becomes Heisenberg's equations of motion.

In order to calculate the second equation of motion Eq.(3.32b), we need to know the relation of $g_{\mu\nu}(\hat{\phi})$ with the field operator $\hat{\phi}$. The relation of $g_{\mu\nu}(\hat{\phi})$ with the field operator $\hat{\phi}$ is described by the Einstein equations Eq.(3.23), which we can not solve exactly.

Using the commutation relations, we have

$$i\partial_t\hat{\phi}_a = [\hat{\phi}_a, \hat{H}] = \frac{\partial\hat{\mathcal{H}}}{\partial\hat{\pi}_a}, \tag{3.36a}$$

$$i\partial_t\hat{\pi}_a = [\hat{\pi}_a, \hat{H}] = -\frac{\partial\hat{\mathcal{H}}}{\partial\hat{\phi}_a}. \tag{3.36b}$$

Eq.(3.36) is equivalent to the Euler-Lagrange equation in operator form

$$\frac{D(\sqrt{-\hat{g}}\hat{\mathcal{L}}_m(x))}{D\hat{\phi}_a(x)} - D_\mu\frac{D(\sqrt{-\hat{g}}\hat{\mathcal{L}}_m)}{D(D_\mu\hat{\phi}_a(x))} = 0. \tag{3.37}$$

It should be noted that if we use the anti-commutation relations of fermions for $\hat{\phi}$ and $\hat{\pi}$, we get $[\hat{\phi}(\mathbf{x}, t), \hat{G}_t(\hat{\pi}, \hat{\phi})] = 0$. $\hat{G}_t$ given by Eq.(3.25) can not be the generator of time translation transformation. Therefore, the Lagrangian density in Eq.(3.2) can only be used to describe the scalar bosons.

In the derivation of the Einstein equations, we have only used the R term. We have not considered the high order terms of R because they involve the time derivatives of third order on the left side of the Einstein equations while only the time derivatives of up to the second order are involved on the right side. The terms contributing the third order time derivative should be zero. Thus the terms containing the high order terms of R should vanish and only the linear R term is needed.

The gravitational effect are mainly caused by the mass of particles and the mass effect is involved not only in the range of low energy. In order to calculate the gravitational effect in a general way, we need consider a broad range of energy spectrum. This poses a computational difficulty of gravitational effect due to the failure of ordinary perturbation. Since in the ordinary perturbation, when we consider the Minkowski metric as a starting metric and use plane wave basis, we encounter the integral of $\int d^4k$ that is integrated over the whole range of energy. Thus the ordinary perturbation calculations are failed.

3.5 Relations of terms in the total action

Now let us consider the value of the parameter α_1. Since the action with α_1 has the similar terms as the action for matter, it is natural to choose $\alpha_1 = 1/4$, which gives $\alpha_1 g_\mu^\mu = 1$ because $g_\mu^\mu = 4$. Then the total action becomes

$$
\begin{aligned}
S_t &= \int d^4x \sqrt{-g} \mathcal{L}_t \\
&\equiv \int d^4x \sqrt{-g} (\mathcal{L}_m + \mathcal{L}_i)
\end{aligned}
\tag{3.38}
$$

with

$$
\mathcal{L}_i \equiv \frac{1}{4} g_{\mu\nu} \frac{D(\sqrt{-g}\mathcal{L}_m)}{D(D_\mu \phi_a(x))} D^\nu \phi_a(x) - \mathcal{L}_m(x) - \frac{1}{32\pi G} R + \Lambda'. \tag{3.39}
$$

The terms in $\mathcal{L}_i$ are canceled out except for a constant term due to the symmetry of spacetime translation. We call $\mathcal{L}_i$ the *invariant Lagrangian density*. The invariant Lagrangian density $\mathcal{L}_i$ plays the role of guaranteeing the conservation of energy-momentum. In addition, we find that it implicates a symmetry of the scale invariance for the total action. It can be seen that in the total action, the matter Lagrangian density has a minus counterpart $-\mathcal{L}_m(x)$ in the invariant Lagrangian density. Therefore any mass and interaction terms can be canceled out without changing the total action. In the second term $\mathcal{L}_i$ of Eq.(3.38), the first kinetic term plays a special role. Since the matter action can be canceled out, the remaining action related to matter particles in Eq.(3.38) are the following action

$$
S_r = \int d^4x \sqrt{-g} \mathcal{L}_r = \frac{1}{4} \int d^4x \sqrt{-g} g^{\mu\nu} \frac{D(\sqrt{-g}\mathcal{L}_m)}{D(D_\mu \phi_a(x))} D_\nu \phi_a(x). \tag{3.40}
$$

The action in Eq.(3.40) has the symmetry of scale invariance. It is natural to list the symmetry of scale invariance as a basic principle. Like the symmetry of spacetime translation that no point in spacetime is special and the covariance principle that no frame is special, no scale in spacetime should be special. However, if we list the symmetry of scale invariance as a basic principle, it would make this book too difficult to be understood. Due to the symmetry breaking of the scale invariance, the particles obtain the mass and interaction. Then there is no more scale invariance.

3.6   Interactions

Similar to the Minkowski spacetime, the interaction can be added into the Lagrangian. Any terms involving field function π in the generator of time translation transformation G_t for bosons will give terms containing time derivative. The interaction terms generally should not contain boson field π except for the linear term of π. Although we can add any mass and interaction terms without changing the total Lagrangian, the suitable form of the mass and interaction terms should be that achieves the lowest energy for the ground state when temperature effect is small. Determination of the form of the interaction terms that achieves the lowest energy for the ground state involves the calculations of the ground state in the Riemann spacetime, which is difficult. We note that a sign change of all mass and interaction terms is equivalent to the sign change of kinetic terms and thus the sign change of the gravitational constant G. Therefore, the sign of G is related to the ground state.

Generally, the Lagrangian of matter to achieve the ground state should have high symmetry. There are some symmetries which are related to the covariance of Lagrangians. These symmetries should always be guaranteed when we add interaction terms. In these symmetries, the most important one is the gauge symmetry, which correlates different types of particles. We have shown that massless boson vector field should possess the gauge symmetry. Therefore, any Lagrangian containing the massless boson field should have the gauge symmetry. We can couple the vector bosons with the spinor fermions by adding an interaction term $gA_\mu\bar{\psi}\gamma^\mu\psi$, where g is the coupling constant. We introduce the gauge covariant derivative $\tilde{D}_\mu = D_\mu - igA_\mu$ to replace the spacetime covariant derivative D_μ to include this interaction term. For the Lagrangian of the gauge boson field $A_\mu(x)$, we use $F_{\mu\nu} = \tilde{D}_\mu A_\nu - \tilde{D}_\nu A_\mu$ for the abelian gauge symmetry and use

$$F_{\mu\nu} = \tilde{D}_\mu A_\nu - \tilde{D}_\nu A_\mu - ig[A_\mu, A_\nu]. \tag{3.41}$$

for the nonabelian gauge symmetry. Then we have a gauge invariant kinetic term for the Yang-Mills Lagrangian

$$\mathcal{L}_{gb} = -\frac{1}{2}\mathrm{Tr}F^{\mu\nu}F_{\mu\nu}. \tag{3.42}$$

Since $\Gamma^\nu_{\alpha\mu} = \Gamma^\nu_{\mu\alpha}$, the terms involving $\Gamma^\nu_{\alpha\mu}$ canceled out and we do not need to consider them. We can construct the generator of time translation in a similar way used for the case of the Minkowski metric in the section 2.7.2.

3.7 Vierbein

For spinor fields, we have used γ^μ matrix to obtain the Lorentz covariance of the Lagrangian. γ^μ behaves as a vector in the Minkowski spacetime. However, there is not a direct way to make γ^μ to behave as a vector in a curved spacetime. In order to achieve the general covariance for a spinor field, we need use the vierbein representation of the curved spacetime. For the Riemann spacetime, there is a tangent space which is flat. Spinors can then be defined with the Lorentz covariance at any point in the Riemann spacetime in term of this tangent space. In the following, we will relate the spinor representation in this tangent flat spacetime to that in the Riemann spacetime by means of the vierbein field.

For the tangent space T_P at a point P, we have vector basis $\hat{\mathbf{e}}_{(\mu)}$. The hat on $\hat{\mathbf{e}}_{(\mu)}$ denotes it as a unit basis vector. The four basis vectors are distinguished by the subscript with parentheses. A contravariant four-vector A can be represented as

$$A = A^\mu \hat{\mathbf{e}}_{(\mu)} = (A^0, A^1, A^2, A^3). \tag{3.43}$$

The cotangent space T_P^* is spanned by the basis vectors $\hat{\mathbf{e}}^{(\mu)}$. The bases $\hat{\mathbf{e}}_{(\mu)}$ and $\hat{\mathbf{e}}^{(\mu)}$ are related by the tensor product

$$\hat{\mathbf{e}}^{(\mu)} \otimes \hat{\mathbf{e}}_{(\nu)} = \mathbb{I}^\mu_\nu. \tag{3.44}$$

T_P^* is called the dual space of T_P. A covariant four-vector has the form

$$A = A_\mu \hat{\mathbf{e}}^{(\mu)} = g_{\mu\nu} A^\nu \hat{\mathbf{e}}^{(\mu)}. \tag{3.45}$$

Using Eqs.(3.44) and (3.45), we have

$$\hat{\mathbf{e}}_{(\mu)} \cdot \hat{\mathbf{e}}_{(\nu)} = g_{\mu\nu} \tag{3.46}$$

or

$$\hat{\mathbf{e}}^{(\mu)} \cdot \hat{\mathbf{e}}^{(\nu)} = g^{\mu\nu}. \tag{3.47}$$

Since one is free to choose any basis to span T_P, we introduce non-coordinate unit vectors $\hat{\mathbf{e}}_{(a)}$ as basis vectors. They are orthonormal

$$\hat{\mathbf{e}}^{(a)} \otimes \hat{\mathbf{e}}_{(b)} = \mathbb{I}^a_b. \tag{3.48}$$

Using Eq.(3.48), we have

$$\hat{\mathbf{e}}_{(a)} \cdot \hat{\mathbf{e}}_{(b)} = \eta_{ab}, \tag{3.49a}$$

$$\hat{\mathbf{e}}^{(a)} \cdot \hat{\mathbf{e}}^{(b)} = \eta^{ab}, \tag{3.49b}$$

where η_{ab} is the Minkowski metric of flat spacetime. This orthonormal basis is independent of coordinates and is called a tetrad basis. In order to distinguish with the coordinate basis $\hat{\mathbf{e}}_{(\mu)}$, we use the Latin letters to denote the components of the non-coordinate frame.

Although we can not use this basis to cover all the spacetime, we can express any vector locally using the fixed tetrad basis vector. The coordinate basis $\hat{\mathbf{e}}_{(\mu)}(x)$ can be expressed in terms of the non-coordinate basis $\hat{\mathbf{e}}_{(a)}$ by the following linear combination

$$\hat{\mathbf{e}}_{(\mu)}(x) = e_\mu^a(x)\hat{\mathbf{e}}_{(a)}. \tag{3.50}$$

where $e_\mu^a(x)$ is a 4×4 inversible matrix. $e_\mu^a(x)$ is called the vierbein field (vierbein in Germany means four-legs). We can also express the relations of the bases in the cotangent space T_P^* as follows:

$$\hat{\mathbf{e}}^{(\mu)}(x) = e_a^\mu(x)\hat{\mathbf{e}}^{(a)}. \tag{3.51}$$

Using Eqs.(3.44) and (3.48), we have the following orthonormality conditions

$$e_a^\mu(x)e_\nu^a(x) = \delta_\nu^\mu \tag{3.52}$$

and

$$e_\mu^a(x)e_b^\mu(x) = \delta_b^a. \tag{3.53}$$

Thus the inverse of the vierbein $e_\mu^a(x)$ is $e_a^\mu(x)$.

The inverse vierbein can be used to represent the tetrad basis $\hat{\mathbf{e}}_{(a)}$ in terms of the coordinate basis $\hat{\mathbf{e}}_{(\mu)}(x)$.

$$\hat{\mathbf{e}}_{(a)} = e_a^\mu(x)\hat{\mathbf{e}}_{(\mu)}(x) \tag{3.54}$$

and

$$\hat{\mathbf{e}}^{(a)} = e_\mu^a(x)\hat{\mathbf{e}}^{(\mu)}(x). \tag{3.55}$$

Inserting Eq.(3.50) into Eq.(3.46), we obtain

$$g_{\mu\nu}(x) = e_\mu^a(x)e_\nu^b(x)\eta_{ab} \tag{3.56}$$

or equivalently

$$g_{\mu\nu}(x)e_a^\mu(x)e_b^\nu(x) = \eta_{ab} \tag{3.57}$$

A vector A can be expressed in terms of the coordinate and non-coordinate basis as

$$A = A^\mu \hat{\mathbf{e}}_{(\mu)} = A^a \hat{\mathbf{e}}_{(a)}. \tag{3.58}$$

The components of the vectors in the coordinate frame and non-coordinate frame are related by the vierbein field transformation

$$A^\mu = e^\mu_a A^a, \tag{3.59a}$$

$$A^a = e^a_\mu A^\mu. \tag{3.59b}$$

The tensors can also be cast forth and back between Latin and Greek bases. For example,

$$V^a{}_b = e^a_\mu V^\mu{}_b = e^\mu_b V^a{}_\mu = e^a_\mu e^\nu_b V^\mu{}_\nu. \tag{3.60}$$

The metric tensor can be used to raise and lower the indices of the vierbeins, for example,

$$e^\mu_a = g^{\mu\nu} \eta_{ab} e^b_\nu. \tag{3.61}$$

3.8 Spin connection

A covariant derivative is given by

$$D_\gamma A^\alpha = \partial_\gamma A^\alpha + \Gamma^\alpha_{\beta\gamma} A^\beta. \tag{3.62}$$

In the non-coordinate frame, the ordinary Levi-Civita connection $\Gamma^\alpha_{\beta\gamma}$ is replaced by the tetrad connection $\omega^a_{b\mu}$. The covariant derivative has the form

$$D_\mu A^a = \partial_\mu A^a + \omega^a_{b\mu} A^b. \tag{3.63}$$

$\omega^a_{\mu b}$ is called spin connection because it is used for the covariant derivative of spinors.

The covariant derivative of a vector A expressed in the coordinate basis has the form

$$\begin{aligned} DA &= (D_\mu A^\nu) \hat{\mathbf{e}}^{(\mu)} \otimes \hat{\mathbf{e}}_{(\nu)} \\ &= (\partial_\mu A^\nu + \Gamma^\nu_{\lambda\mu} A^\lambda) \hat{\mathbf{e}}^{(\mu)} \otimes \hat{\mathbf{e}}_{(\nu)}. \end{aligned} \tag{3.64}$$

It can also be expressed in a mixed basis as

$$\begin{aligned} DA &= (D_\mu A^a) \hat{\mathbf{e}}^{(\mu)} \otimes \hat{\mathbf{e}}_{(a)} \\ &= (\partial_\mu A^a + \omega^a_{b\mu} A^b) \hat{\mathbf{e}}^{(\mu)} \otimes \hat{\mathbf{e}}_{(a)}. \end{aligned} \tag{3.65}$$

When DA in Eq.(3.65) is converted into the expression in the coordinated basis, we have

$$
\begin{aligned}
DA &= (\partial_\mu A^a + \omega^a_{b\mu} A^b)\hat{\mathbf{e}}^{(\mu)} \otimes \hat{\mathbf{e}}_{(a)} \\
&= (\partial_\mu(e^a_\nu A^\nu) + \omega^a_{b\mu} e^b_\lambda A^\lambda)\hat{\mathbf{e}}^{(\mu)} \otimes (e^\sigma_a \hat{\mathbf{e}}_{(\sigma)}) \\
&= e^\sigma_a(e^a_\nu \partial_\mu A^\nu + A^\nu \partial_\mu e^a_\nu + \omega^a_{b\mu} e^b_\lambda A^\lambda)\hat{\mathbf{e}}^{(\mu)} \otimes \hat{\mathbf{e}}_{(\sigma)} \\
&= (\partial_\mu A^\sigma + e^\sigma_a(\partial_\mu e^a_\nu)A^\nu + e^\sigma_a e^b_\lambda \omega^a_{b\mu} A^\lambda)\hat{\mathbf{e}}^{(\mu)} \otimes \hat{\mathbf{e}}_{(\sigma)}.
\end{aligned} \tag{3.66}
$$

Relabeling the indexes, we have

$$
\begin{aligned}
DA &= (\partial_\mu A^\nu + e^\nu_a(\partial_\mu e^a_\lambda)A^\lambda + e^\nu_a e^b_\lambda \omega^a_{b\mu} A^\lambda)\hat{\mathbf{e}}^{(\mu)} \otimes \hat{\mathbf{e}}_{(\nu)} \\
&= [\partial_\mu A^\nu + (e^\nu_a \partial_\mu e^a_\lambda + e^\nu_a e^b_\lambda \omega^a_{b\mu})A^\lambda]\hat{\mathbf{e}}^{(\mu)} \otimes \hat{\mathbf{e}}_{(\nu)}.
\end{aligned} \tag{3.67}
$$

Comparing Eq.(3.67) with Eq.(3.64), we have

$$
\Gamma^\nu_{\lambda\mu} = e^\nu_a \partial_\mu e^a_\lambda + e^\nu_a e^b_\lambda \omega^a_{b\mu}, \tag{3.68}
$$

which relates the Levi-Civita connection $\Gamma^\nu_{\lambda\mu}$ with the spin connection $\omega^a_{\mu b}$. Solving for the spin connection, we obtain

$$
\omega^a_{b\mu} = e^a_\nu e^\lambda_b \Gamma^\nu_{\lambda\mu} - e^\lambda_b \partial_\mu e^a_\lambda. \tag{3.69}
$$

The covariant derivative of the vierbein e^a_ν is given by

$$
D_\mu e^a_\nu = \partial_\mu e^a_\nu - e^a_\sigma \Gamma^\sigma_{\mu\nu} + \omega^a_{b\mu} e^b_\nu. \tag{3.70}
$$

Multiplying Eq.(3.69) by e^b_ν gives

$$
\begin{aligned}
\omega^a_{b\mu} e^b_\nu &= e^a_\sigma e^\lambda_b e^b_\nu \Gamma^\sigma_{\lambda\mu} - e^\lambda_b e^b_\nu \partial_\mu e^a_\lambda \\
&= e^a_\sigma \Gamma^\sigma_{\nu\mu} - \partial_\mu e^a_\nu.
\end{aligned} \tag{3.71}
$$

Inserting Eq.(3.71) into Eq.(3.70), we have

$$
\begin{aligned}
D_\mu e^a_\nu &= \partial_\mu e^a_\nu - e^a_\sigma \Gamma^\sigma_{\nu\mu} + e^a_\sigma \Gamma^\sigma_{\nu\mu} - \partial_\mu e^a_\nu \\
&= 0.
\end{aligned} \tag{3.72}
$$

Eq.(3.72) is called the tetrad postulate. Using Eq.(3.56), we have

$$
\begin{aligned}
D_\mu g_{\nu\lambda} &= (D_\mu e^a_\nu)e^b_\lambda \eta_{ab} + e^a_\nu(D_\mu e^b_\lambda)\eta_{ab} \\
&= 0,
\end{aligned} \tag{3.73}
$$

which is consistent with Eq.(A.85).

In the Riemann spacetime, the Lagrangian density for spinor field (see also Eq.(2.233) in the Minkowski spacetime) is expressed as

$$
\mathcal{L} = \bar{\psi}(ie^\mu_a \gamma^a \partial_\mu - m)\psi. \tag{3.74}
$$

The Lorentz transformation is performed on the non-coordinate basis for ψ and γ^a is independent of the coordinate.

Chapter 4

Symmetry breaking

4.1 Scale invariance

4.1.1 *Lagrangian with scale invariance*

Dilatation transformation on spacetime is defined by

$$x \to x' = \lambda x, \tag{4.1}$$

where λ is a real number. With the change in coordinate scale $x \to x' = \lambda x$, if we also define a field transformation of the form

$$\phi(x) \to \phi'(x') = \lambda^{d_\phi} \phi(\lambda x), \tag{4.2}$$

then the transformations Eqs.(4.1) and (4.2) are called the *scale transformation*. d_ϕ is called the *scaling dimension* of the field ϕ. If the action S is invariant with the scale transformations Eqs.(4.1) and (4.2), we say that the system has the symmetry of *scale invariance*.

The Lagrangian without the mass terms and interaction terms has the symmetry of scale invariance, which we call the *plain Lagrangian*. For a d-dimensional space, in order to maintain the scale invariance, the field transformation needs to have the following forms.

i) For scalar bosons,

$$\varphi(x) \to \varphi'(x') = \lambda^{\frac{d-1}{2}} \varphi(\lambda x). \tag{4.3}$$

ii) For spinor fermions,

$$\psi(x) \to \psi'(x') = \lambda^{\frac{d}{2}} \psi(\lambda x). \tag{4.4}$$

iii) For vector bosons,

$$A_\mu(x) \to A'_\mu(x') = \lambda^{\frac{d-1}{2}} A_\mu(\lambda x). \tag{4.5}$$

With these transformation, we have

$$\mathcal{L}(x) \rightarrow \mathcal{L}'(x') = \lambda^{d+1}\mathcal{L}(\lambda x) \tag{4.6}$$

and

$$S = \int d^{d+1}x \mathcal{L}(x) \rightarrow \int d^{d+1}x \lambda^{d+1}\mathcal{L}(\lambda x) = \int d^{d+1}x' \mathcal{L}(x'). \tag{4.7}$$

When S is unchanged with the scale transformation, for a Lagrangian with the gauge symmetry, A_μ should be scaled as ∂_μ because of the presence of the covariant derivative $D_\mu = \partial_\mu - igA_\mu$. We have $\lambda^{\frac{d-1}{2}} = \lambda$ from Eq.(4.5). Then $d = 3$, which means only four-dimensional spacetime can have both scale invariance and gauge invariance. We have shown that the action with massless vector bosons should be gauge invariant. In the three-dimensional space, when we add gauge interaction terms to the plain Lagrangian, we still have the symmetry of scale invariance. For other space dimension, the symmetry of the scale invariance will be broken when we add the gauge interaction terms.

4.1.2   *Conserved current for scale invariance*

In order to consider an infinitesimal transformation, we introduce $e^\alpha \equiv \lambda$. Then the field transformation is expressed as

$$\phi(x) \rightarrow e^{\alpha d_\phi}\phi(e^\alpha x). \tag{4.8}$$

We have

$$\delta x = \alpha x \tag{4.9}$$

and

$$\begin{aligned}
\delta\phi &= e^{\alpha d_\phi}\phi(e^\alpha x) - \phi(x) \\
&= (1 + \alpha d_\phi)\phi(x + \alpha x) - \phi(x) \\
&= \left(\alpha d_\phi + \alpha x_\lambda \frac{\partial}{\partial x_\lambda}\right)\phi.
\end{aligned} \tag{4.10}$$

Then

$$\delta\mathcal{L} = \alpha(4 + x_\lambda \partial^\lambda)\mathcal{L}. \tag{4.11}$$

The variation $\delta S = \int d^4 x \delta\mathcal{L}$ vanishes upon integration by parts. Using Noether's theorem, the invariance of action S leads to the conservation law corresponding to the scale invariance

$$\partial_\mu \theta^\mu = 0 \tag{4.12}$$

with the canonical Noether current θ given by

$$\theta^\mu = \frac{\partial \mathcal{L}}{\partial \partial_\mu \phi_a} (d_\phi \phi_a + \partial_\nu \phi_a x^\nu) - \left(\frac{\partial \mathcal{L}}{\partial \partial_\mu \phi_a} \partial_\nu \phi_a - \delta^\mu_\nu \mathcal{L} \right) x^\nu. \qquad (4.13)$$

In the above formula, the second term can be rewritten in terms of the canonical energy-momentum tensor

$$\Theta^{\mu\nu} = \frac{\partial \mathcal{L}}{\partial \partial_\mu \phi_a} \partial^\nu \phi_a - \eta^{\mu\nu} \mathcal{L}. \qquad (4.14)$$

Then the canonical dilatation or scaling current θ^μ is expressed as

$$\theta^\mu = x_\nu \Theta^{\mu\nu} + \Sigma^\mu, \qquad (4.15)$$

where

$$\Sigma^\mu = d_\phi \frac{\partial \mathcal{L}}{\partial \partial_\mu \phi_a} \phi_a + \frac{\partial \mathcal{L}}{\partial \partial_\mu \phi_a} \partial_\nu \phi_a x^\nu \qquad (4.16)$$

is called the internal part.

It is possible to eliminate the internal part Σ^μ in Eq.(4.15) by redefining an symmetric energy-momentum tensor $T^{\mu\nu}$ in the following way. We use Eq.(2.362)

$$T^{\mu\nu}(x) = \Theta^{\mu\nu}(x) + \frac{1}{2} \partial_\kappa (\tau^{\kappa\mu\nu} + \tau^{\mu\nu\kappa} - \tau^{\nu\kappa\mu}) \qquad (4.17)$$

and set

$$\tau^{\kappa\mu\nu} = \Sigma^{\kappa\mu\nu} + \frac{1}{3} \left(\eta^{\kappa\mu} \partial^\nu f - \eta^{\kappa\nu} \partial^\mu f \right), \qquad (4.18)$$

where f is the solution of the differential equation

$$\Box f = \partial_\kappa (\Sigma^\kappa - \eta_{\mu\nu} \Sigma^{\mu\nu\kappa}). \qquad (4.19)$$

$\Box$ is the d'Alembert operator. Then

$$T^{\mu\nu} - T^{\nu\mu} = \Theta^{\mu\nu} - \Theta^{\nu\mu} + \partial_\kappa \tau^{\kappa\mu\nu}. \qquad (4.20)$$

Using Eq.(2.366), we have

$$\begin{aligned}
T^{\mu\nu} - T^{\nu\mu} &= \partial_\kappa (M^{\kappa\mu\nu} - \Sigma^{\kappa\mu\nu} + \tau^{\kappa\mu\nu}) \\
&= \partial_\kappa (-\Sigma^{\kappa\mu\nu} + \tau^{\kappa\mu\nu}) \\
&= \frac{1}{3} \partial_\kappa (\eta^{\kappa\mu} \partial^\nu f - \eta^{\kappa\nu} \partial^\mu f) \\
&= 0,
\end{aligned} \qquad (4.21)$$

which shows that $T^{\mu\nu}$ is symmetric. Using Eq.(4.17) and $\tau^{\kappa\mu\nu} + \tau^{\kappa\nu\mu} = 0$, we have

$$T^\mu_\mu = \Theta^\mu_\mu + \eta_{\mu\nu} \partial_\kappa \tau^{\mu\nu\kappa}. \qquad (4.22)$$

Using Eq.(4.15), we find

$$\partial_\mu \theta^\mu = \Theta^\mu_\mu + \partial_\mu \Sigma^\mu, \tag{4.23}$$

Inserting Eq.(4.23) into Eq.(4.22), we obtain

$$\begin{aligned}
T^\mu_\mu &= \partial_\kappa (\theta^\kappa - \Sigma^\kappa + \eta_{\mu\nu} \tau^{\mu\nu\kappa}) \\
&= \partial_\kappa (\eta_{\mu\nu} \Sigma^{\mu\nu\kappa} - \Sigma^\kappa) + \frac{1}{3} \eta_{\mu\nu} \partial_\kappa (\eta^{\kappa\mu} \partial^\nu f - \eta^{\kappa\nu} \partial^\mu f) \\
&= \partial_\kappa (\eta_{\mu\nu} \Sigma^{\mu\nu\kappa} - \Sigma^\kappa) + \Box f \\
&= 0.
\end{aligned} \tag{4.24}$$

We introduce T^μ without internal part by

$$T^\mu = x_\nu T^{\mu\nu}. \tag{4.25}$$

Then the conservation law Eq.(4.12) is replaced by

$$\partial_\mu \theta^\mu = T^\mu_\mu = 0, \tag{4.26}$$

which implies that the scale invariance leads to the vanishing of the trace of the energy-momentum tensor.

4.1.3   *Scale invariance for total Lagrangian*

The action given by Eq.(3.40) has a symmetry of scale invariance. When we change the scale of coordinates $x \to x' = \lambda x$ together with the field transformations

$$\phi_a(x) \to \phi'_a(x') = \lambda \phi_a(\lambda x), \tag{4.27a}$$

$$\psi(x) \to \psi'(x') = \lambda^{3/2} \psi(\lambda x), \tag{4.27b}$$

$$A_\mu(x) \to A'_\mu(x') = \lambda A_\mu(\lambda x), \tag{4.27c}$$

where $\phi_a(x)$, $\psi(x)$ and $A_\mu(x)$ are the scalar boson field, spinor fermion field, and vector boson field, respectively. the Lagrangian density $\mathcal{L}_r$ changes as

$$\mathcal{L}_r(x) \to \mathcal{L}'_r(x') = \lambda^4 \mathcal{L}_r(\lambda x) \tag{4.28}$$

Thus the action S_r is unchanged under the scale transformation,

$$\begin{aligned}
S_r &= \int d^4 x \sqrt{-g} \mathcal{L}_r(x) \\
&\to \int d^4 x \sqrt{-g} \lambda^4 \mathcal{L}_r(\lambda x) = \int d^4 x' \sqrt{-g} \mathcal{L}_r(x') = S_r.
\end{aligned} \tag{4.29}$$

Using Noether's theorem, the invariance of action leads to the conservation law corresponding to the scale invariance, $\partial_\mu (x_\nu T^{\mu\nu}) = T^\mu_\mu = 0$, which implies that the scale invariance is equivalent to the vanishing of the trace of the energy-momentum tensor. When we put $T^\mu_\mu = 0$ into the Einstein equations, we have $R = 0$ if the cosmological constant is zero. The spacetime will have zero curvature for a system with the scale invariance.

4.2 Ground state energy

Energy is a physical observable that obeys the conservation law and plays an important role in physics. The energy is excited over a background to generate quasi-particles. There are two special backgrounds which are important. One is the ground state, which is the state having the lowest energy and is thus the state with the highest stability. The other is the vacuum, which does not contain particles. Each ground state has its vacuum. But vacuum is not necessarily equivalent to the ground state. When the ground state does not contain particles, the ground state is then the same with the vacuum. When the ground state contains particles, the ground state is not the vacuum. The universe is not empty. Our universe consists of various types of particles. Because temperature is not zero, we do not know whether the ground state of our universe is a vacuum. When we approximate the ground state as the vacuum. The ground state energy is then equal to the vacuum energy approximately. The vacuum energy is also called the zero-point energy because it is related to $1/2\hbar\omega_\mathbf{p}$ in the case without interactions of particles. As an example, we calculate the vacuum energy of the free scalar boson field in the Minkowski spacetime, which is the expectation value

$$\langle 0|\hat{H}|0\rangle = \frac{1}{2}\int d^4x \langle 0|\hat{\pi}^2 + (\nabla\hat{\phi})^2 + m^2\hat{\phi}^2|0\rangle. \tag{4.30}$$

Let us first calculate

$$\begin{aligned}
\langle 0|\hat{\phi}(\mathbf{x},t)\hat{\phi}(\mathbf{x},t)|0\rangle &= \lim_{\mathbf{x}\to 0}\langle 0|\hat{\phi}(\mathbf{x},0)\hat{\phi}(0,0)|0\rangle \\
&= \lim_{\mathbf{x}\to 0}\int \frac{d^3p}{2\omega_\mathbf{p}(2\pi)^3}e^{-i\mathbf{p}\cdot\mathbf{x}} \\
&= \int \frac{d^3p}{2\omega_\mathbf{p}(2\pi)^3}.
\end{aligned} \tag{4.31}$$

Other terms in Eq.(4.30) can be calculated similarly, we get

$$\begin{aligned}
\langle 0|\hat{H}|0\rangle &= V\int \frac{d^3p}{2\omega_\mathbf{p}(2\pi)^3}\frac{1}{2}(\omega_\mathbf{p}^2 + \mathbf{p}^2 + m^2) \\
&= V\int \frac{d^3p}{(2\pi)^3}\frac{1}{2}\omega_\mathbf{p}.
\end{aligned} \tag{4.32}$$

We can see that the energy of vacuum is the integration of $1/2\omega_\mathbf{p}$ over all momentum mode and over all space. It should be noted that the integration over $\mathbf{p}$ diverges. However, what matters is the difference of energies. These could be i) the energy difference of systems with and without particles; ii) the energy difference between different vacuums corresponding to different ground states.

For the spinor fermions, the energy of vacuum in the Minkowski spacetime is given by Eq.(2.408). We have

$$\langle 0|\hat{H}|0\rangle = \langle 0| \int d^3p \sum_s \omega_{\mathbf{p}}[\hat{b}^\dagger(\mathbf{p},s)\hat{b}(\mathbf{p},s) - \hat{d}(\mathbf{p},s)\hat{d}^\dagger(\mathbf{p},s)]|0\rangle$$

$$= \langle 0| \int d^3p \sum_s \omega_{\mathbf{p}}[\hat{b}^\dagger(\mathbf{p},s)\hat{b}(\mathbf{p},s) + \hat{d}^\dagger(\mathbf{p},s)\hat{d}(\mathbf{p},s) - \delta^3(0)]|0\rangle. \tag{4.33}$$

It is noted that

$$\delta^3(0) = \lim_{\mathbf{p}\to 0} \frac{1}{(2\pi)^3} \int d^3x e^{i\mathbf{p}\cdot\mathbf{x}} = \frac{1}{(2\pi)^3} \int d^3x. \tag{4.34}$$

Then we get

$$\langle 0|\hat{H}|0\rangle = -\frac{1}{(2\pi)^3} \int d^3x \int d^3p \sum_s 2\frac{1}{2}\omega_{\mathbf{p}}. \tag{4.35}$$

The factor 2 comes from the summation of the contributions of particles and antiparticles, such as, electrons and positrons.

Eq.(4.35) has an minus sign with it. The spinor fermion field has a negative vacuum energy, while the scalar boson field has a positive vacuum energy. Since $\omega_{\mathbf{p}} = \sqrt{\mathbf{p}^2 + m^2}$ increases with the increase of m. The mass of particles increases the vacuum energy. A Lagrangian with positive mass term for scalar bosons will increase the vacuum energy. The ground state should have the lowest energy. Thus we can only add a negative mass term, which means that we can not give mass directly to the scalar bosons. Instead, we need a negative mass term and then with a symmetry breaking to transform the negative mass term into positive mass term, which is the reason why we need Higgs mechanism to generate mass for particles. The Higgs mechanism will be discussed in Section 4.4.

4.3   Symmetry breaking

When temperature is high, we have the state with high symmetry due to the entropy effect (see chapter of statistical mechanics). When temperature is low, the high symmetry state will transform into a state, which usually has lower symmetry. The symmetry breaking is related to the phase transitions of second order and critical phenomena. It is one of the most important physical mechanism.

Now we consider the Lagrangian density

$$\mathcal{L} = \frac{1}{2}\partial_\mu\phi\partial^\mu\phi - \frac{1}{2}m^2\phi^2 - \frac{\lambda}{4}(\phi^2)^2, \qquad (4.36)$$

where $\varphi = (\varphi_1, \varphi_2, \cdots, \varphi_N)$. This Lagrangian density corresponds to the scalar bosons with mass m and a self-interaction $\frac{\lambda}{4}\phi^4$. The Lagrangian density exhibits an $O(N)$ symmetry under which φ transfers as an N-component vector and is renormalizable(see Section (5.6)). We can add terms that do not possess the $O(N)$ symmetry to break the $O(N)$ symmetry. For instance, we can add $\varphi_1^2\phi^2$ to break the $O(N)$ symmetry down to $O(N-1)$. However, the Lagrangian does not possess the original $O(N)$ symmetry anymore. There is another way for system to break the symmetry. We keep the Lagrangian with the $O(N)$ symmetry, but the ground state turns out to be a state without the $O(N)$ symmetry. This phenomenon is called as spontaneous symmetry breaking.

4.3.1 *Spontaneous symmetry breaking*

We have explained that for scalar bosons, positive mass term leads to the ground state with higher energy. Thus a Lagrangian density with the correct ground state would have the following form

$$\mathcal{L} = \frac{1}{2}\partial_\mu\phi\partial^\mu\phi + \frac{1}{2}\mu^2\phi^2 - \frac{\lambda}{4}(\phi^2)^2. \qquad (4.37)$$

We have changed the sign of the ϕ^2 term in Eq.(4.37). We can not just say that the field has the particles with mass $\sqrt{-\mu^2} = i\mu$, which is meaningless. $\frac{1}{2}(\partial_i\phi)^2 - \frac{1}{2}\mu^2\phi^2 + \frac{\lambda}{4}(\phi^2)^2$ is the potential term. We notice that $\phi = 0$ is not the position of the minimum for the potential term. It is the maximum position. The minimum is at $\phi \neq 0$. Now we determine the minimum of the potential term. Clearly, any spatial variation in ϕ will increase the energy. Thus we set $\phi(x)$ to be a constant quantity ϕ_0 in spacetime and look for the minimum of potential. We define

$$V(\phi) = -\frac{1}{2}\mu^2\phi^2 + \frac{\lambda}{4}(\phi^2)^2. \qquad (4.38)$$

First we consider the case $N = 1$. The potential is shown in Fig.4.1. The minimum is determined by $\frac{\partial V}{\partial \phi} = 0$ and $\frac{\partial^2 V}{\partial \phi^2} > 0$. The equation

$$\frac{\partial V}{\partial \phi} = -\mu^2\phi + \lambda\phi^3 = \phi(-\mu^2 + \lambda\phi^2) = 0 \qquad (4.39)$$

has three solutions. $\phi = 0$ corresponds to a maximum. $\phi = \pm(\mu^2/\lambda)^{\frac{1}{2}} \equiv \pm v$ are the two minima.

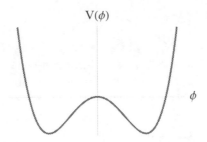

Fig. 4.1   Potential of the field with one component.

There are two possibilities for the ground state: $\phi = +v$ or $\phi = -v$. Physics is equivalent for the two cases. When the nature made the choice, the reflection symmetry $\phi \to -\phi$ of the Lagrangian is broken. It is broken spontaneously. We can choose either two ground states. So we choose the ground state at $+v$ and write $\phi = v + \phi'$. We expand $\mathcal{L}$ in ϕ' and we have

$$\mathcal{L} = \frac{\mu^4}{4\lambda} + \frac{1}{2}\partial_\mu\phi'\partial^\mu\phi' - \mu^2\phi'^2 + O(\phi'^3). \tag{4.40}$$

Now we have a positive mass term for the shifted field ϕ'. The particles corresponding to the field ϕ' with a mass of $\sqrt{2}\mu$. The first term is the constant term, which contributes to the cosmological constant.

4.3.2   *Continuous symmetry*

The case $N \geq 2$ is different with $N = 1$. For $N = 1$, there is a reflection symmetry $\phi \to -\phi$. It is a discrete symmetry with one symmetry transformation. For $N \geq 2$, we have an infinite number of symmetry transformations with continuous parameters. We call it *continuous symmetry*. For $N = 2$, the shape of the potential is shown in Fig.4.2. We have the $O(2)$ symmetry in the Lagrangian density. The potential has the minima at $\phi^2 = \mu^2/\lambda$, which corresponds to an infinite number of equivalent vacua characterized by the direction of ϕ. We can choose any one of them and others are the same with it. So we choose the one with the direction of ϕ to be in 1 direction. In this case, $\phi_1 = v \equiv \sqrt{\mu^2/\lambda}$ and $\phi_2 = 0$.

Now we express the field functions as the fluctuations around the vacuum. We have $\phi_1 = v + \phi_1'$ and $\phi_2 = \phi_2'$ and put them into the Lagrangian density in Eq.(4.37). The Lagrangian density becomes

$$\mathcal{L} = \frac{\mu^4}{4\lambda} + \frac{1}{2}\partial_\mu\phi_1'\partial^\mu\phi_1' + \frac{1}{2}\partial_\mu\phi_2'\partial^\mu\phi_2' - \mu^2\phi_1'^2 + O(\phi'^3). \tag{4.41}$$

In the Lagrangian density given by Eq.(4.41), the particles generated by the field ϕ_1' have a mass $\sqrt{2}\mu$. However, the field ϕ_2' is massless due to the absence of $\phi_2'^2$ term. The emergence of the massless boson field ϕ_2' comes from the symmetry of vacuum. The potential along the bottom of the potential takes the same value. ϕ_2' is the field along the bottom. It costs no addition energy to go around the potential bottom. The mass of particles is zero for the field ϕ_2'. The massless field ϕ_2' is called the *Nambu-Goldstone bosons* or *Goldstone bosons*. For $N > 2$, we have the similar results. After symmetry is broken, the system has the ground state with one massive boson and $N - 1$ Nambu-Goldstone bosons.

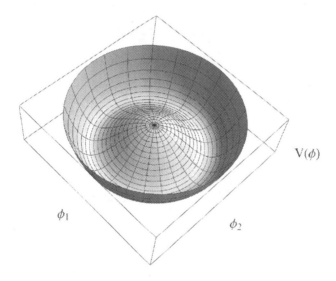

Fig. 4.2 Potential of the field with $O(2)$ symmetry.

4.4 Higgs mechanism

We know that scalar bosons can achieve a positive mass term through symmetry breaking from a negative mass term with self-interaction. Through interaction between different types of particles, other particles can also become massive by the symmetry breaking of the scalar boson field. This mechanism is called the *Higgs mechanism*.

We consider the simplest case: the charged scalar field (complex scalar field) with the local $U(1)$ gauge invariance coupled with the massless vector bosons. We denote the massless vector boson field as A_μ. The vector boson term is given by $-\frac{1}{4}F_{\mu\nu}F^{\mu\nu}$. We add the interaction term $\frac{1}{2}e^2 A_\mu A^\mu \phi^* \phi$ with the gauge invariance. Then the Lagrangian density is given by

$$\mathcal{L} = -\frac{1}{4}F_{\mu\nu}F^{\mu\nu} + \frac{1}{2}[(\partial_\mu - ieA_\mu)\phi^*][(\partial^\mu + ieA^\mu)\phi]$$
$$+ \frac{1}{2}\mu^2\phi^2 - \frac{1}{4}\lambda(\phi^2)^2. \tag{4.42}$$

The Lagrangian is invariant under the local abelian gauge transformation

$$U = e^{-i\theta(x)}. \tag{4.43}$$

The gauge transformation of the field functions has the form

$$\phi(x) \to \phi'(x) = e^{-i\theta(x)}\phi(x), \tag{4.44a}$$

$$\phi^*(x) \to \phi^{*\prime}(x) = e^{i\theta(x)}\phi^*(x), \tag{4.44b}$$

$$A_\mu(x) \to A'_\mu(x) = A_\mu(x) + \frac{1}{e}\partial_\mu\theta(x). \tag{4.44c}$$

As we did in the previous section, we set

$$\phi(x) = v + \xi(x) + i\chi(x), \tag{4.45}$$

where $v = \sqrt{-\mu^2/\lambda}$. We have shifted the field by a value of $\phi_{vac} = v$. Inserting in Eq.(4.42), we have

$$\mathcal{L} = -\frac{1}{4}F_{\mu\nu}F^{\mu\nu} - \frac{e^2 v^2}{2}A_\mu A^\mu + \frac{1}{2}(\partial_\mu\xi)^2 + \frac{1}{2}(\partial_\mu\chi)^2$$
$$- \lambda v^2\xi^2 - evA_\mu\partial^\mu\chi + \cdots. \tag{4.46}$$

A mass term $\frac{e^2 v^2}{2}A_\mu A^\mu$ emerges in Eq.(4.46). The gauge transformation Eq.(4.44) becomes

$$\xi(x) \to \xi'(x) = [v + \xi(x)]\cos\theta(x) + \chi(x)\sin\theta(x) - v, \tag{4.47a}$$

$$\chi(x) \to \chi'(x) = \chi(x)\cos\theta(x) - [v + \xi(x)]\sin\theta(x), \tag{4.47b}$$

$$A_\mu(x) \to A'_\mu(x) = A_\mu(x) + \frac{1}{e}\partial_\mu\theta(x). \tag{4.47c}$$

It seems that the resulted Lagrangian density describes a massive boson field interacted with two scalar boson fields, the massive ξ and massless χ fields. However, it should be noted that the gauge bosons have only two independent transverse components before the symmetry breaking. The third component (longitudinal component) has been gauged out by $\partial_\mu A^\mu - \sigma = 0$ (with σ taken to be zero here). Now longitudinal component A_3 can be nonzero because the vector bosons become massive. We shall use the gauge transformation to gauge out another component of the scale boson field.

Since the Lagrangian density does not change with any choice of the transformation function $\theta(x)$ in Eq.(4.43), we can choose $\theta(x)$ to be equal to the phase of $\phi(x)$ at any spacetime point. In this gauge,

$$\phi'(x) = e^{-i\theta(x)}\phi(x) \tag{4.48}$$

becomes a real field function. We set

$$\phi'(x) = v + \eta(x), \tag{4.49}$$

which shifts $\phi'(x)$ to a new field function $\eta(x)$ in a new vacuum $\phi_{vac} = v$. $\eta(x)$ is a new real function. In the new gauge,

$$A'_\mu(x) = A_\mu(x) + \frac{1}{e}\frac{\partial\theta(x)}{\partial x^\mu}. \tag{4.50}$$

The Lagrangian density in Eq.(4.42) becomes

$$\begin{aligned}
\mathcal{L} = &-\frac{1}{4}F'_{\mu\nu}F'^{\mu\nu} - \frac{e^2v^2}{2}A'_\mu A'^\mu + \frac{1}{2}(\partial_\mu\eta)^2 \\
&- \lambda v^2\eta^2 - \frac{1}{4}\lambda\eta^4 + \frac{1}{2}e^2(A'_\mu)^2(2v\eta + \eta^2),
\end{aligned} \tag{4.51}$$

where

$$F'_{\mu\nu} = \partial_\mu A'_\nu - \partial_\nu A'_\mu. \tag{4.52}$$

The Lagrangian density in Eq.(4.51) now describes a massive vector boson field interacted with a real scalar boson field η. η is called the *Higgs boson field*, which has a mass

$$m_H = \sqrt{2\lambda v^2} = \sqrt{2}\mu. \tag{4.53}$$

All massless particles become massive particles through the Higgs mechanism.

If there is no gauge interaction term, the complex massless scalar bosons become one massive boson and one massless Goldstone boson in the spontaneously broken symmetry. When there is a gauge interaction term between the vector bosons and scalar bosons, the vector bosons acquire mass at the expense of the would-be Goldstone bosons. Vector bosons with two components become massive vector bosons with three components while Goldstone boson disappears. This Higgs mechanism can be applied similarly to the non-abelian gauge bosons.

4.5   Mass and interactions of particles

The mass terms generally are not added purely as an self-interaction term. If we add the mass term as an self-interaction without interaction terms between different types of particles, these massive particles will become dark matters. However, the mass of particles can be generated through interactions by the Higgs mechanism. It is reasonable that the favorable interaction terms are those possessing the scale invariance, which maintain the symmetry of the scale invariance in the total Lagrangian. The most interaction terms included in the standard model of electroweak unification are those possessing the scale invariance and gauge invariance.

The gauge group in the standard model is $U(1) \otimes SU(2)$, for which the gauge invariant Lagrangian density for the gauge bosons has the form

$$\mathcal{L}_{gb} = -\frac{1}{4}W_{\mu\nu}^j W^{j\mu\nu} - \frac{1}{4}B_{\mu\nu}B^{\mu\nu}, \tag{4.54}$$

where

$$B_{\mu\nu} = \partial_\mu B_\nu - \partial_\nu B_\mu, \tag{4.55}$$

and

$$W_{\mu\nu}^j = \partial_\mu W_\nu^j - \partial_\nu W_\mu^j + g\epsilon^{jkl}W_\mu^k W_\nu^l, \tag{4.56}$$

B_μ is the abelian gauge boson field and W_μ^j is the nonabelian gauge boson field. The self-interaction terms for the gauge boson fields are scale invariant. One can add other interaction terms which are both gauge invariant and scale invariant. There are the interaction term of the gauge bosons with the left hand spinor fermions $\mathcal{L}_{gb-lsf}$, the interaction term of the gauge bosons with the right hand spinor fermions $\mathcal{L}_{gb-rsf}$, the interaction term of the scalar bosons with the spinor fermions $\mathcal{L}_{sb-sf}$, and the interaction term of the gauge bosons with the scalar bosons $\mathcal{L}_{gb-lsf}$. They are given by

$$\mathcal{L}_{gb-lsf} = \bar{\psi}_L i\gamma^\mu \left(\partial_\mu - \frac{1}{2}ig'B_\mu + \frac{1}{2}ig\boldsymbol{\tau} \cdot \boldsymbol{W}_\mu\right)\psi_L, \tag{4.57a}$$

$$\mathcal{L}_{gb-rsf} = \bar{\psi}_R i\gamma^\mu(\partial_\mu - ig''B_\mu)\psi_R, \tag{4.57b}$$

$$\mathcal{L}_{gb-sf} = -g_e[(\bar{\psi}_L\phi)\psi_R + \bar{\psi}_R(\phi^\dagger\psi_L)], \tag{4.57c}$$

$$\mathcal{L}_{gb-sb} = \left\{\left(\partial_\mu + \frac{1}{2}ig'B_\mu + \frac{1}{2}ig\boldsymbol{\tau} \cdot \boldsymbol{W}_\mu\right)\phi\right\}^\dagger \tag{4.57d}$$

$$\times \left\{\left(\partial_\mu + \frac{1}{2}ig'B_\mu + \frac{1}{2}ig\boldsymbol{\tau} \cdot \boldsymbol{W}_\mu\right)\phi\right\} - \lambda(\phi^\dagger\phi)^2, \tag{4.57e}$$

where $-\lambda(\phi^\dagger\phi)^2$ is the self-interaction term. When a symmetry breaking term $-\mu^2\phi^\dagger\phi$ in the Lagrangian of matter is generated, the scale invariance symmetry is broken. The particles become massive by the Higgs mechanism.

Before the symmetry breaking of the scale invariance and generating the interactions, we have only three basic types of particles: 1) the scalar bosons with the Lagrangian density

$$\mathcal{L}_{sb} = \frac{1}{2}g^{\mu\nu}\partial_\mu\phi\partial_\nu\phi; \tag{4.58}$$

2) the vector bosons with the Lagrangian density

$$\mathcal{L}_{sb} = -\frac{1}{4}F_{\mu\nu}F^{\mu\nu}; \tag{4.59}$$

3) the spinor fermions with the Lagrangian density

$$\mathcal{L}_{sf} = \bar{\psi}_L i\gamma^\mu\partial_\mu\psi_L + \bar{\psi}_R i\gamma^\mu\partial_\mu\psi_R. \tag{4.60}$$

All these particles are massless. The massless particles moves with the speed of light. When the symmetry of the scale invariance is broken, particles acquire mass through the Higgs mechanism.

Chapter 5

Interacting quantum fields

5.1 Invariant commutation relations for scalar bosons

We have solved the equations of motion for free fields. In order to treat the quantum fields with interactions, we will develop some tools which are useful for the calculations of quantum fields in this chapter.

The commutation relations Eqs.(2.60) and (2.61) are the commutation relations between field operators at two different spatial positions but at equal time. Using the equations of motion, we can calculate the commutation relations between field operators at different times. One of the most important commutation relations is the invariant commutation relation, which possesses the Lorentz-invariance.

5.1.1 *Commutation functions*

In the following, we consider the scalar boson field. The commutation functions for other types of particles will be dealt with in later sections. For scalar bosons, we define the invariant commutation relation between the field operators $\hat{\phi}(\mathbf{x}, x_0)$ and $\hat{\phi}^\dagger(\mathbf{y}, y_0)$ as

$$i\triangle(x - y) \equiv [\hat{\phi}(x), \hat{\phi}^\dagger(y)]. \tag{5.1}$$

$\triangle(x)$ is also called the *Pauli-Jordan function*. We have used the homogeneous character of spacetime to write $\triangle(x-y)$ as a function of $x-y$ instead of x and y separately. It can be seen that $\triangle(x - y)$ is a Lorentz-invariant function from the definition Eq.(5.1) which is not dependent of any specific frame of reference.

For the free complex scalar field, we can calculate the function $\triangle(x-y)$ easily. The generator of time translation transformation $\hat{G}_t$ is given by

$$\hat{G}_t = \hat{H} = \int d^3x \left(\frac{1}{2}\hat{\pi}^\dagger\hat{\pi} + \frac{1}{2}\nabla\hat{\phi}^\dagger\nabla\hat{\phi} + \frac{1}{2}m^2\hat{\phi}^\dagger\hat{\phi} \right). \tag{5.2}$$

The corresponding equations of motion are given by

$$i\partial_0\hat{\phi} = [\hat{\phi}, \hat{G}_t] = i\hat{\pi}^\dagger, \tag{5.3a}$$

$$i\partial_0\hat{\phi}^\dagger = [\hat{\phi}^\dagger, \hat{G}_t] = i\hat{\pi}, \tag{5.3b}$$

$$i\partial_0\hat{\pi} = [\hat{\pi}, \hat{G}_t] = i(\nabla^2 - m^2)\hat{\phi}^\dagger, \tag{5.3c}$$

$$i\partial_0\hat{\pi}^\dagger = [\hat{\pi}^\dagger, \hat{G}_t] = i(\nabla^2 - m^2)\hat{\phi}. \tag{5.3d}$$

The solutions of the equations of motion have the form

$$\hat{\phi}(\mathbf{x}, t) = \int d^3p[\hat{a}_\mathbf{p}u_\mathbf{p}(\mathbf{x}, t) + \hat{b}_\mathbf{p}^\dagger u_\mathbf{p}^*(\mathbf{x}, t)], \tag{5.4a}$$

$$\hat{\phi}^\dagger(\mathbf{x}, t) = \int d^3p[\hat{a}_\mathbf{p}^\dagger u_\mathbf{p}^*(\mathbf{x}, t) + \hat{b}_\mathbf{p}u_\mathbf{p}(\mathbf{x}, t)]. \tag{5.4b}$$

In Eq.(5.4), we have two types of creation and annihilation operators denoted by $(\hat{a}, \hat{a}^\dagger)$ and $(\hat{b}, \hat{b}^\dagger)$, respectively, because we have two components for a complex field. Similar to Eqs.(2.170), (2.171) and (2.174), we can deduce the following commutation relations:

$$[\hat{a}_\mathbf{p}, \hat{a}_{\mathbf{p}'}^\dagger] = [\hat{b}_\mathbf{p}, \hat{b}_{\mathbf{p}'}^\dagger] = \delta^3(\mathbf{p} - \mathbf{p}'), \tag{5.5a}$$

$$[\hat{a}_\mathbf{p}, \hat{a}_{\mathbf{p}'}] = [\hat{b}_\mathbf{p}, \hat{b}_{\mathbf{p}'}] = [\hat{a}_\mathbf{p}^\dagger, \hat{a}_{\mathbf{p}'}^\dagger] = [\hat{b}_\mathbf{p}^\dagger, \hat{b}_{\mathbf{p}'}^\dagger] = 0, \tag{5.5b}$$

$$[\hat{a}_\mathbf{p}, \hat{b}_{\mathbf{p}'}] = [\hat{a}_\mathbf{p}, \hat{b}_{\mathbf{p}'}^\dagger] = [\hat{a}_\mathbf{p}^\dagger, \hat{b}_{\mathbf{p}'}] = [\hat{a}_\mathbf{p}^\dagger, \hat{b}_{\mathbf{p}'}^\dagger] = 0. \tag{5.5c}$$

For the vacuum state, we have

$$\hat{a}_\mathbf{p}|0\rangle = \hat{b}_\mathbf{p}|0\rangle = 0. \tag{5.6}$$

Inserting the expansion Eq.(5.4) into Eq.(5.1), we have

$$\begin{aligned}
i\triangle(x - y) &= \int d^3p \int d^3p'(u_{\mathbf{p}'}(x)u_\mathbf{p}^*(y)[\hat{a}_{\mathbf{p}'}, \hat{a}_\mathbf{p}^\dagger] + u_{\mathbf{p}'}^*(x)u_\mathbf{p}(y)[\hat{b}_{\mathbf{p}'}^\dagger, \hat{b}_\mathbf{p}]) \\
&= \int d^3p(u_\mathbf{p}(x)u_\mathbf{p}^*(y) - u_\mathbf{p}^*(x)u_\mathbf{p}(y)) \\
&= \int \frac{d^3p}{2\omega_\mathbf{p}(2\pi)^3} \left[e^{-ip\cdot(x-y)} - e^{ip\cdot(x-y)} \right] \\
&\equiv i\triangle^{(+)}(x - y) + i\triangle^{(-)}(x - y)
\end{aligned} \tag{5.7}$$

with

$$i\triangle^{(+)}(x - y) = \int \frac{d^3p}{2\omega_\mathbf{p}(2\pi)^3} e^{-ip\cdot(x-y)}, \tag{5.8a}$$

$$i\triangle^{(-)}(x - y) = -\int \frac{d^3p}{2\omega_\mathbf{p}(2\pi)^3} e^{ip\cdot(x-y)}, \tag{5.8b}$$

where we have defined the four-dimensional momentum $p = (p_0, \mathbf{p}) \equiv (\omega_{\mathbf{p}} = \sqrt{\mathbf{p}^2 + m^2}, \mathbf{p})$. $i\triangle^{(+)}(x - y)$ is called the *positive frequency function* and $i\triangle^{(-)}(x - y)$ is the *negative frequency function*. Eq.(5.7) can be expressed as

$$i\triangle(x - y) = -\int \frac{d^3p}{(2\pi)^3} \frac{\sin(p \cdot (x - y))}{\omega_{\mathbf{p}}}. \tag{5.9}$$

We can extend the three-dimensional integration in Eq.(5.9) to a four-dimensional one in order to show the Lorentz invariance explicitly. We denote $z \equiv x - y$ and change p_0 from $\omega_{\mathbf{p}}$ to an independent variable in integration. Then we have

$$
\begin{aligned}
i\triangle(x - y) &= \int \frac{d^3p}{2\omega_{\mathbf{p}}(2\pi)^3} \left[e^{-i(\omega_{\mathbf{p}} z_0 - \mathbf{p} \cdot \mathbf{z})} - e^{i(\omega_{\mathbf{p}} z_0 - \mathbf{p} \cdot \mathbf{z})} \right] \\
&= \int \frac{d^4p}{(2\pi)^3} \frac{1}{2\omega_{\mathbf{p}}} [\delta(p_0 - \omega_{\mathbf{p}}) - \delta(p_0 + \omega_{\mathbf{p}})] e^{-i(p_0 z_0 - \mathbf{p} \cdot \mathbf{z})} \\
&= \int \frac{d^4p}{(2\pi)^3} \frac{\epsilon(p_0)}{2\omega_{\mathbf{p}}} [\delta(p_0 - \omega_{\mathbf{p}}) + \delta(p_0 + \omega_{\mathbf{p}})] e^{-ip \cdot z},
\end{aligned}
\tag{5.10}
$$

where

$$\epsilon(p_0) \equiv \mathtt{Sign}(p_0) = \begin{cases} +1, & \text{for} \quad p_0 > 0 \\ -1, & \text{for} \quad p_0 < 0 \end{cases} \tag{5.11}$$

is the sign function. Using

$$
\begin{aligned}
\frac{1}{2\omega_{\mathbf{p}}} [\delta(p_0 - \omega_{\mathbf{p}}) + \delta(p_0 + \omega_{\mathbf{p}})] &= \delta((p_0 - \omega_{\mathbf{p}})(p_0 + \omega_{\mathbf{p}})) \\
&= \delta(p_0^2 - \omega_{\mathbf{p}}^2) \\
&= \delta(p^2 - m^2),
\end{aligned}
\tag{5.12}
$$

Eq.(5.10) becomes

$$i\triangle(x - y) = \int \frac{d^4p}{(2\pi)^3} \epsilon(p_0)\delta(p^2 - m^2) e^{-ip \cdot z}. \tag{5.13}$$

The sign of p_0 does not change under any Lorentz transformations because the time-like momentum vectors with $p_0 > 0$ ($p^2 = m^2 > 0$) always keep $p_0 > 0$ and thus always lie in the forward light cone while those those with $p_0 < 0$ are always in the backward light cone. Thus $\triangle(x - y)$ is Lorentz-invariant.

We can easily show that other commutation relations are equal to zero.

$$[\hat{\phi}(x), \hat{\phi}(y)] = [\hat{\pi}(x), \hat{\pi}(y)] = 0. \tag{5.14}$$

Since the field operator $\hat{\phi}(x)$ satisfies the Klein-Gordon equation

$$(\Box + m^2)\hat{\phi}(x) = 0, \tag{5.15}$$

the function $\triangle(x)$ also satisfies the Klein-Gordon equation

$$(\Box + m^2)\triangle(x) = 0 \tag{5.16}$$

with the following boundary conditions at vanishing time difference.

$$\triangle(0, \mathbf{x}) = 0 \tag{5.17}$$

and

$$\left. \frac{\partial}{\partial x_0}\triangle(x_0, \mathbf{x})\right|_{x_0=0} = -\delta^3(x). \tag{5.18}$$

Eq.(5.17) comes directly from Eq.(5.9) because the integrand becomes an odd function of $\mathbf{p}$ when $t = 0$. We can verify Eq.(5.18) by differentiating Eq.(5.9). In fact, Eq.(5.18) is just the equal-time commutation relation Eq.(2.60).

$$
\begin{aligned}
\left. i\frac{\partial}{\partial y_0}\triangle(x-y)\right|_{x_0 \to y_0} &= \left. \frac{\partial}{\partial y_0}[\hat{\phi}(x), \hat{\phi}^\dagger(y)]\right|_{x_0 \to y_0} \\
&= \left. [\hat{\phi}(x), \dot{\hat{\phi}}^\dagger(y)]\right|_{x_0 \to y_0} \\
&= \left. [\hat{\phi}(x), \hat{\pi}(y)]\right|_{x_0 \to y_0} \\
&= i\delta^3(\mathbf{x} - \mathbf{y}).
\end{aligned} \tag{5.19}
$$

5.1.2 *Microcausality*

Eq.(5.17) leads to a very important property of quantum field

$$\triangle(x - y) = 0, \quad \text{for} \quad (x - y)^2 < 0. \tag{5.20}$$

The invariant function $\triangle(x - y)$ vanishes when $x - y$ is a space-like four vector. Eq.(5.20) has important implication that two observable quantities can be measured independently when the measurements take place at two points that have a space-like separation. This is the so-called *microcausality*, which states that any disturbances can not propagate faster than the speed of light.

In the following, we will give a deduction that Eq.(5.20) leads to the microcausality of observables. We write the operator for a local observable such as $\hat{P}_\mu$ as

$$\hat{O}(x) = \hat{\phi}^\dagger(x)O(x)\hat{\phi}(x), \tag{5.21}$$

where $O(x)$ is a c-number function or a differential operator. The commutator of two observables is given by

$$
\begin{aligned}
[\hat{O}(x), \hat{O}(y)] &= O(x)O(y)[\hat{\phi}^\dagger(x)\hat{\phi}(x), \hat{\phi}^\dagger(y)\hat{\phi}(y)] \\
&= O(x)O(y)\{\hat{\phi}^\dagger(x)\hat{\phi}^\dagger(y)[\hat{\phi}(x), \hat{\phi}(y)] \\
&\quad + \hat{\phi}^\dagger(x)[\hat{\phi}(x), \hat{\phi}^\dagger(y)]\hat{\phi}(y) + \hat{\phi}^\dagger(y)[\hat{\phi}^\dagger(x), \hat{\phi}(y)]\hat{\phi}(x) \\
&\quad + [\hat{\phi}^\dagger(y), \hat{\phi}^\dagger(x)]\hat{\phi}(y)\hat{\phi}(x)\} \\
&= O(x)O(y)\{\hat{\phi}^\dagger(x)i\triangle(x-y)\hat{\phi}(y) - \hat{\phi}^\dagger(y)i\triangle(y-x)\hat{\phi}(x)\} \\
&= O(x)O(y)(\hat{\phi}^\dagger(x)\hat{\phi}(y) + \hat{\phi}^\dagger(y)\hat{\phi}(x))i\triangle(x-y).
\end{aligned}
\tag{5.22}
$$

From Eq.(5.20), we obtain the microcausality for the scalar boson field.

$$
[\hat{O}(x), \hat{O}(y)] = 0, \quad \texttt{for} \quad (x-y)^2 < 0.
\tag{5.23}
$$

In the frame that two space-like points x and y have the same time, $\hat{O}(x)\hat{O}(y)$ or $\hat{O}(y)\hat{O}(x)$ can be considered as two consecutive measurements made within an infinitesimal time difference. Eq.(5.23) means that the measurement first at x and then at y is equivalent to the measurements first at y and then at x for two space-like points x and y. Thus the observable O can be measured independently at two space-like points. This is also called the no signaling principle in the special relativity.

5.1.3 *Propagator functions*

In addition to the function $\triangle(x-y)$, we can define other invariant functions for the operators, the so-called propagator functions. One of the most important propagator functions is the Feynman propagator $\triangle_F(x-y)$, which is defined as

$$
i\triangle_F(x-y) \equiv \langle 0|T\hat{\phi}(x)\hat{\phi}^\dagger(y)|0\rangle.
\tag{5.24}
$$

The symbol T denotes the time-ordered product of the operators $\hat{\phi}(x)$ and $\hat{\phi}^\dagger(y)$, which is defined by

$$
T\hat{A}(x)\hat{B}(y) \equiv \hat{A}(x)\hat{B}(y)\Theta(x_0 - y_0) \pm \hat{B}(y)\hat{A}(x)\Theta(y_0 - x_0),
\tag{5.25}
$$

where

$$
\Theta(x) = \begin{cases} 1, & \text{for} \quad x > 0 \\ 0, & \text{for} \quad x < 0. \end{cases}
\tag{5.26}
$$

The operator T put the factor of two time-dependent operators $\hat{A}$ and $\hat{B}$ into a chronological order that the operator having the later time argument is put before the other. $\pm$ sign in Eq.(5.25) occurs due to the reordering of operators. The plus(minus) sign is for the boson(fermion) field operators. In the case of the free fields, $\triangle_F(x-y)$ is also called the free propagator.

We can evaluate the Feynman propagator using the solutions of the equations of motion. The solution Eq.(5.4) for the complex scalar bosons consists of the following parts

$$\hat{\phi}^{(+)}(\mathbf{x},t) = \int d^3p\,\hat{a}_{\mathbf{p}}u_{\mathbf{p}}(\mathbf{x},t), \quad \hat{\phi}^{\dagger(+)}(\mathbf{x},t) = \int d^3p\,\hat{b}_{\mathbf{p}}u_{\mathbf{p}}(\mathbf{x},t)), \quad (5.27a)$$

$$\hat{\phi}^{(-)}(\mathbf{x},t) = \int d^3p\,\hat{b}_{\mathbf{p}}^\dagger u_{\mathbf{p}}^*(\mathbf{x},t), \quad \hat{\phi}^{\dagger(-)}(\mathbf{x},t) = \int d^3p\,\hat{a}_{\mathbf{p}}^\dagger u_{\mathbf{p}}^*(\mathbf{x},t)). \quad (5.27b)$$

They have the properties

$$\hat{\phi}^{\dagger(+)}(x)|0\rangle = \langle 0|\hat{\phi}^{(-)}(x) = 0, \quad (5.28a)$$

$$\hat{\phi}^{(+)}(x)|0\rangle = \langle 0|\hat{\phi}^{\dagger(-)}(x) = 0. \quad (5.28b)$$

Thus we have

$$\begin{aligned}
i\triangle_F(x-y) =&\Theta(x_0-y_0)\langle 0|\hat{\phi}^{(+)}(x)\hat{\phi}^{\dagger(-)}(y)|0\rangle \\
&+ \Theta(y_0-x_0)\langle 0|\hat{\phi}^{\dagger(+)}(y)\hat{\phi}^{(-)}(x)|0\rangle \\
=&\Theta(x_0-y_0)\int d^3p\,u_{\mathbf{p}}(x)u_{\mathbf{p}}^*(y) \\
&+ \Theta(y_0-x_0)\int d^3p\,u_{\mathbf{p}}(y)u_{\mathbf{p}}^*(x) \\
=&\Theta(x_0-y_0)\int \frac{d^3p}{(2\pi)^3}\frac{1}{2\omega_{\mathbf{p}}}e^{-ip\cdot(x-y)} \\
&+ \Theta(y_0-x_0)\int \frac{d^3p}{(2\pi)^3}\frac{1}{2\omega_{\mathbf{p}}}e^{ip\cdot(x-y)} \\
=&\Theta(x_0-y_0)i\triangle^{(+)}(x-y) - \Theta(y_0-x_0)i\triangle^{(-)}(x-y).
\end{aligned} \quad (5.29)$$

We can express Eq.(5.29) in a more compact form. Using the following mathematical formula

$$\begin{aligned}
&\frac{1}{2\omega_{\mathbf{p}}}[\Theta(x_0-y_0)e^{-i\omega_{\mathbf{p}}\cdot(x_0-y_0)} + \Theta(y_0-x_0)e^{i\omega_{\mathbf{p}}\cdot(x_0-y_0)}] \\
&= -\int \frac{dp_0}{2\pi i}\frac{e^{-ip_0\cdot(x_0-y_0)}}{p_0^2 - \omega_{\mathbf{p}}^2 + i\epsilon},
\end{aligned} \quad (5.30)$$

where ϵ is an infinitesimal number, we have

$$\triangle_F(x-y) = \int \frac{d^4p}{(2\pi)^4}\frac{e^{-ip\cdot(x-y)}}{p^2 - m^2 + i\epsilon}. \quad (5.31)$$

We can see that the fourier coefficient of $\triangle_F(x-y)$ is

$$\triangle_F(p) = \int d^3x e^{-ip\cdot x}\triangle_F(x) = \frac{1}{p^2 - m^2 + i\epsilon}. \tag{5.32}$$

Since $\triangle_F(x-y)$ satisfies the following relation

$$\begin{aligned}
(\Box_x + m^2)\triangle_F(x-y) &= \int \frac{d^4p}{(2\pi)^4}(-p^2+m^2)\frac{e^{-ip\cdot(x-y)}}{p^2-m^2+i\epsilon} \\
&= -\int \frac{d^4p}{(2\pi)^4}e^{-ip\cdot(x-y)} \\
&= -\delta(x-y),
\end{aligned} \tag{5.33}$$

the function $\triangle_F(x-y)$ is the solution of the inhomogeneous Klein-Gordon equation containing a delta function as a source term. The Feynman propagator has the meaning of the amplitude probability for a process in which a particle created at the point $\mathbf{x}_1, t_1$ in spacetime propagates to the point $\mathbf{x}_2, t_2$ where it is annihilated. Since field operators $\hat{\phi}$ and $\hat{\phi}^\dagger$ contain both the operators for particles and antiparticles, $\triangle_F(x-y)$ describes the processes for both particles and antiparticles depending on the chronological order of $\hat{\phi}$ and $\hat{\phi}^\dagger$.

In contrast, the commutation functions $\triangle(x-y)$, $\triangle^{(+)}(x-y)$, and $\triangle^{(-)}(x-y)$ satisfy the homogeneous Klein-Gordon equation

$$(\Box + m^2)\triangle_i(x-y) = 0, \tag{5.34}$$

where $\triangle_i = \triangle, \triangle^{(+)}$, and $\triangle^{(-)}$.

In addition to $\triangle(x-y)$ and $\triangle_F(x-y)$, there are several other commutation functions and propagator functions. For the propagator functions, in addition to $\triangle_F(x-y)$, we define the *Dyson propagator* as

$$\triangle_D(x) \equiv \Theta(x_0)\triangle^{(-)}(x) - \Theta(-x_0)\triangle^{(+)}(x). \tag{5.35}$$

$\triangle_D(x)$ is also known as anti-causal propagator which has an opposite chronological order as compared to the Feynman propagator.

We can also define two other propagators, the retarded propagator $\triangle_R(x)$ and the advanced propagator $\triangle_A(x)$.

$$\triangle_R(x) \equiv \Theta(x_0)\triangle(x), \tag{5.36a}$$

$$\triangle_A(x) \equiv -\Theta(x_0)\triangle(x). \tag{5.36b}$$

The Pauli-Jordan function $\triangle(x)$ can be written as the difference between the retarded propagator and the advanced propagator

$$\triangle(x) = \triangle_R(x) - \triangle_A(x). \tag{5.37}$$

Using $\triangle_R(x)$ and $\triangle_A(x)$, we can define the principal-part propagator $\bar{\triangle}(x)$ as

$$\bar{\triangle}(x) \equiv \frac{1}{2}[\triangle_R(x) + \triangle_A(x)]. \tag{5.38}$$

Inserting Eq.(5.36) into Eq.(5.38), we have

$$\bar{\triangle}(x) = \frac{1}{2}\epsilon(x_0)\triangle(x). \tag{5.39}$$

The propagator functions $\triangle_F(x)$, $\triangle_D(x)$, $\triangle_R(x)$, $\triangle_A(x)$ and $\bar{\triangle}(x)$ are the solutions of the inhomogeneous Klein-Gordon equation with a delta function as the source term

$$(\Box + m^2)\triangle_i(x) = -\delta^4(x), \tag{5.40}$$

where $\triangle_i = \triangle_F, \triangle_D, \triangle_R(x), \triangle_A(x), \bar{\triangle}(x)$. Since they are the solutions of the inhomogeneous Klein-Gordon equation with delta function source, we also call them the Green's functions. For example, $\triangle_F(x)$ is also called the Feynman Green's function. These propagator functions contain a product of the function $\triangle(x)$ with a unit step function in time such as $\Theta(x_0)$ or $\frac{1}{2}\epsilon(x_0)$. The step function is the one leading to the delta function when the Klein-Gordon operator $\Box + m^2$ is applied.

5.2   n-point Green's function of scalar fields

5.2.1   *Definition of n-point Green's function*

We have calculated the Feynman propagator in the previous section, which is shown to be the Green's function for the equations of motion. The Green's functions are also called the *correlation functions*. They are the useful tools in the calculations of field properties. Now we generalize the two-point Green's function to the n-point Green's function defined as the following time-ordered product.

$$G(x_1, x_2, \cdots, x_n) \equiv \langle 0|T[\hat{\phi}(x_1)\hat{\phi}(x_2)\cdots\hat{\phi}(x_n)]|0\rangle. \tag{5.41}$$

$G(x_1, x_2, \cdots, x_n)$ is also called the *n-point correlation function*. Similar to the two-point Green's function, we can calculate the n-point Green's function using the solution of the equations of motion. Generally, the easiest way to calculate the n-point Green's function is that uses the path integral formalism. Similar to the derivation of Eq.(2.123), we can express the n-point time-ordered product as a path integral. The n-point time-ordered product, which is also called the transition matrix element, has the form in the path integral formalism

$$\langle \phi', t' | T[\hat{\phi}(x_1)\hat{\phi}(x_2) \cdots \hat{\phi}(x_n)] | \phi, t \rangle$$
$$= \int D\phi \phi(x_1)\phi(x_2) \cdots \phi(x_n) e^{i \int_t^{t'} d\tau L[\phi]}. \tag{5.42}$$

5.2.2 *Wick rotation*

Now we consider the evaluation of the two-point function

$$G(x_1, x_2) = \langle 0 | T[\hat{\phi}(x_1)\hat{\phi}(x_2)] | 0 \rangle. \tag{5.43}$$

We can extract the correlation function $G(x_1, x_2)$ from the transition matrix element Eq.(5.42) in the following way. We decompose $|\phi\rangle$ into the eigenstates $|n\rangle$ of $\hat{H}$, which gives

$$|\phi, t\rangle = e^{i\hat{H}t} \sum_n |n\rangle\langle n|\phi\rangle$$
$$= \sum_n e^{iE_n t} |n\rangle\langle n|\phi\rangle \tag{5.44}$$
$$= \sum_n e^{iE_n t} \langle n|\phi\rangle |n\rangle.$$

Using the expansion of Eq.(5.44), we have

$$\langle \phi', t' | T[\hat{\phi}(x_1)\hat{\phi}(x_2)] | \phi, t \rangle$$
$$= \sum_{n,n'} \langle \phi', t' | n' \rangle \langle n' | T[\hat{\phi}(x_1)\hat{\phi}(x_2)] | n \rangle \langle n | \phi, t \rangle \tag{5.45}$$
$$= \sum_{n,n'} e^{-i(E_{n'} t' - E_n t)} \langle \phi', t' | n' \rangle \langle n | \phi, t \rangle \langle n' | T[\hat{\phi}(x_1)\hat{\phi}(x_2)] | n \rangle.$$

The term with $n' = n = 0$ in Eq.(5.45) contains the correlation function $G(x_1, x_2)$. The trick to extract the function $G(x_1, x_2)$ from Eq.(5.45) is to damp out the terms with $n' \neq 0$ or $n \neq 0$. These terms have the oscillatory factor $e^{-i(E_{n'} t' - E_n t)}$. We set the ground state energy $E_0 = 0$. One can introduce an exponentially damping factor by attaching an imaginary part to the time coordinate by $t' \to \tau' e^{-i\delta}$ and $t \to \tau e^{-i\delta}$. When we take the limit $\tau \to -\infty$ and $\tau' \to \infty$, the terms with $n' \neq 0$ or $n \neq 0$ are damped out. Geometrically, this is achieved by a rotation clockwise with an angle $0 > \delta > \pi$ in the complex plane. To calculate the path integral, one can start from any $0 > \delta > \pi$. In terms of the new rotated time coordinates $\tau = e^{i\delta}t$ and $\tau' = e^{i\delta}t'$, the limit of the matrix element has the form

$$\lim_{\substack{t' \to \infty \\ t \to -\infty}} \langle \phi', t' | T[\hat{\phi}(x_1)\hat{\phi}(x_2)] | \phi, t \rangle$$

$$= \lim_{\substack{\tau' \to e^{i\delta}\infty \\ \tau \to -e^{i\delta}\infty}} \langle \phi', e^{-i\delta}\tau' | T[\hat{\phi}(x_1)\hat{\phi}(x_2)] | \phi, e^{-i\delta}\tau \rangle \qquad (5.46)$$

$$\Rightarrow \lim_{\substack{\tau' \to \infty \\ \tau \to -\infty}} \langle \phi', e^{-i\delta}\tau' | T[\hat{\phi}(x_1)\hat{\phi}(x_2)] | \phi, e^{-i\delta}\tau \rangle.$$

In the last line of Eq.(5.46), we have made an analytical continuation by going to real values of the rotated time coordinate τ. This is a mathematical manipulation. If the integral is an analytic function in the time variable, this can also be considered as a procedure that we calculate the well defined limit in the last line of Eq.(5.46) and then make an analytical continuation to $\delta = 0$. Since one can choose any $0 > \delta > \pi$, the most convenient choice is $\delta = \frac{\pi}{2}$, which rotates the time axis into the pure imaginary direction. $t \to -it$. Such a rotation is called a *Wick rotation*.

Using the Wick rotation $t = -i\tau$ with τ being real, Eq.(5.46) becomes

$$\lim_{\substack{t' \to \infty \\ t \to -\infty}} \langle \phi', t' | T[\hat{\phi}(x_1)\hat{\phi}(x_2)] | \phi, t \rangle$$

$$= \lim_{\substack{\tau' \to \infty \\ \tau \to -\infty}} \langle \phi', -i\tau' | T[\hat{\phi}(x_1)\hat{\phi}(x_2)] | \phi, -i\tau \rangle$$

$$= \lim_{\substack{\tau' \to \infty \\ \tau \to -\infty}} \sum_{n,n'} e^{-(E_{n'}\tau' - E_n\tau)} \langle \phi', t' | n' \rangle \langle n | \phi \rangle \langle n' | T[\hat{\phi}(x_1)\hat{\phi}(x_2)] | n \rangle$$

$$= \lim_{\substack{\tau' \to \infty \\ \tau \to -\infty}} e^{-E_0(\tau' - \tau)} \langle \phi', t' | 0 \rangle \langle 0 | \phi \rangle \langle 0 | T[\hat{\phi}(x_1)\hat{\phi}(x_2)] | 0 \rangle.$$

$$(5.47)$$

Similarly, we have

$$\lim_{\substack{t' \to \infty \\ t \to -\infty}} \langle \phi', t' | \phi, t \rangle = \lim_{\substack{\tau' \to \infty \\ \tau \to -\infty}} e^{-E_0(\tau' - \tau)} \langle \phi', t' | 0 \rangle \langle 0 | \phi \rangle. \qquad (5.48)$$

Combining Eq.(5.47) with Eq.(5.48) gives

$$\langle 0 | T[\hat{\phi}(x_1)\hat{\phi}(x_2)] | 0 \rangle = \lim_{\substack{t' \to \infty \\ t \to -\infty}} \frac{\langle \phi', t' | T[\hat{\phi}(x_1)\hat{\phi}(x_2)] | \phi, t \rangle}{\langle \phi', t' | \phi, t \rangle}$$

$$= \lim_{\substack{t' \to \infty \\ t \to -\infty}} \frac{\int D\phi \phi(x_1)\phi(x_2) e^{iS[\phi]}}{\int D\phi e^{iS[\phi]}}, \qquad (5.49)$$

where $S[\phi] = \int d^4x \mathcal{L}(\phi, \dot{\phi}))$ is the action of the field. Eq.(5.49) can be easily extended to more general cases.

$$\langle 0|T[\hat{\phi}(x_1)\hat{\phi}(x_2)\cdots\hat{\phi}(x_n)]|0\rangle$$

$$= \lim_{\substack{t' \to \infty \\ t \to -\infty}} \frac{\langle \phi', t'|T[\hat{\phi}(x_1)\hat{\phi}(x_2)\cdots\hat{\phi}(x_n)]|\phi, t\rangle}{\langle \phi', t'|\phi, t\rangle} \qquad (5.50)$$

$$= \lim_{\substack{t' \to \infty \\ t \to -\infty}} \frac{\int D\phi\phi(x_1)\phi(x_2)\cdots\phi(x_n)e^{iS[\phi]}}{\int D\phi e^{iS[\phi]}}.$$

To make the notations simpler, we generally omit the lim symbol in Eq.(5.50). Then Eq.(5.50) is expressed as

$$G(x_1, x_2, \cdots, x_n) = \langle 0|T[\hat{\phi}(x_1)\hat{\phi}(x_2)\cdots\hat{\phi}(x_n)]|0\rangle$$

$$= \frac{\int D\phi\phi(x_1)\phi(x_2)\cdots\phi(x_n)e^{iS[\phi]}}{\int D\phi e^{iS[\phi]}}. \qquad (5.51)$$

It is noted that on the right hand side of Eq.(5.51), the path integral should be modified slightly in accordance with the transformation in Eq.(5.46).

5.2.3 *Generating functional*

In order to calculate the path integral in Eq.(5.51), we define the *generating functional* of a field by

$$Z[J] \equiv \int D\phi e^{i\int d^4x[\mathcal{L}(\phi, \dot{\phi}) + J\phi]}. \qquad (5.52)$$

Using the generating functional, we can define a normalized functional

$$\mathcal{Z}[J] \equiv \frac{Z[J]}{Z[0]}. \qquad (5.53)$$

Then we have

$$G(x_1, x_2, \cdots, x_n) = \left(\frac{1}{i}\right)^n \frac{\delta^n \mathcal{Z}[J]}{\delta J(x_1)\delta J(x_2)\cdots\delta J(x_n)}\bigg|_{J=0}. \qquad (5.54)$$

$G(x_1, x_2, \cdots, x_n)$ is a symmetric function of its arguments for a scalar field. Eq.(5.54) means

$$\mathcal{Z}[J] = \sum_n \frac{1}{n!} \int d^4x_1 \cdots d^4x_n i^n G(x_1, x_2, \cdots, x_n)$$

$$\times J(x_1)J(x_2)\cdots J(x_n). \qquad (5.55)$$

There is another useful functional $W[J]$ defined by

$$\mathcal{Z}[J] \equiv e^{iW[J]}. \qquad (5.56)$$

$W[J]$ is called the connected generating functional. Using $W[J]$, we can introduce the *connected Green's function G_c* by

$$G_c(x_1, x_2, \cdots, x_n) \equiv \left(\frac{1}{i}\right)^{n-1} \frac{\delta^n W[J]}{\delta J(x_1)\delta J(x_2)\cdots\delta J(x_n)}\bigg|_{J=0}. \quad (5.57)$$

The physical content for the name 'connected' will become clear in the later usage.

5.2.4 *Momentum representation*

For a free field or perturbation calculations based on the free field, it is advantageous to work in the momentum space because the solutions of the equations of motion for the free field can be expanded using plane wave basis. The transformation of the Green's functions into the momentum representation is defined by

$$G(p_1, p_2, \cdots, p_n)(2\pi)^4\delta^4(p_1 + p_2 + \cdots + p_n)$$
$$\equiv \int d^4x_1 \cdots d^4x_n i^n G(x_1, x_2, \cdots, x_n)e^{-i(p_1\cdot x_1 + p_2\cdot x_2 + \cdots p_n\cdot x_n)}. \quad (5.58)$$

The δ-function factor comes from the conservation of energy-momentum due to the translation invariance of spacetime. After evaluating the right hand side of Eq.(5.58), there would be a factor $\delta^4(p_1 + p_2 + \cdots + p_n)$ on the right hand side of the equation so that the factor $\delta^4(p_1 + p_2 + \cdots + p_n)$ would be canceled out.

5.2.5 *Operator representation*

We introduce the operator functional defined by

$$\hat{Z}[J] \equiv Te^{i\int d^4x J(x)\hat{\phi}(x)}, \quad (5.59)$$

where $\hat{\phi}$ is the field operator. It can be seen that $\hat{Z}[0] = 1$. We have for the functional derivatives of $\hat{Z}[J]$

$$\left(\frac{1}{i}\right)^n \frac{\delta^n \hat{Z}[J]}{\delta J(x_1)\delta J(x_2)\cdots\delta J(x_n)} = T[\hat{\phi}(x_1)\hat{\phi}(x_2)\cdots\hat{\phi}(x_n)\hat{Z}[J]]. \quad (5.60)$$

Now we consider the functional derivatives of $\langle 0|\hat{Z}[J]|0\rangle$. Using $\hat{Z}[0] = 1$, we have

$$\left(\frac{1}{i}\right)^n \frac{\delta^n \langle 0|\hat{Z}[J]|0\rangle}{\delta J(x_1)\delta J(x_2)\cdots\delta J(x_n)}\bigg|_{J=0} = \langle 0|T[\hat{\phi}(x_1)\hat{\phi}(x_2)\cdots\hat{\phi}(x_n)]|0\rangle.$$
$$(5.61)$$

According to Eq.(5.54), we have

$$\frac{\delta^n \mathcal{Z}[J]}{\delta J(x_1)\delta J(x_2)\cdots\delta J(x_n)}\bigg|_{J=0} = \frac{\delta^n \langle 0|\hat{\mathcal{Z}}[J]|0\rangle}{\delta J(x_1)\delta J(x_2)\cdots\delta J(x_n)}\bigg|_{J=0}. \qquad (5.62)$$

Inserting Eq.(5.62) into Eq.(5.54), Eq.(5.55) becomes

$$\begin{aligned}
\mathcal{Z}[J] &= \sum_n \frac{1}{n!}\int d^4x_1\cdots d^4x_n \frac{\delta^n \langle 0|\hat{\mathcal{Z}}[J]|0\rangle}{\delta J(x_1)\delta J(x_2)\cdots\delta J(x_n)}\bigg|_{J=0} \\
&\quad \times J(x_1)J(x_2)\cdots J(x_n) \\
&= \langle 0|\hat{\mathcal{Z}}[J]|0\rangle.
\end{aligned} \qquad (5.63)$$

5.2.6 *Free scalar fields*

For a perturbation calculation, we need to define a reference field, whose equations of motion can be solved. Generally, we take the free field as the base field because the equations of motion for the free field can be solved exactly. Now we consider the case of a free scalar field. The generating functional for the free scalar field is given by

$$Z[J] = \int D\phi e^{-i\int d^4x[\frac{1}{2}\phi(\Box+m^2-i\epsilon)\phi - J\phi]}. \qquad (5.64)$$

with $\Box \equiv \partial_\mu\partial^\mu$. Here we have performed an integration by parts for the kinetic term in the Lagrangian density $\mathcal{L} = \frac{1}{2}(\partial_\mu\phi\partial^\mu\phi - m^2\phi^2)$ of a free scalar field to get

$$\int d^4x\partial_\mu\phi\partial^\mu\phi = -\int d^4x\phi\Box\phi. \qquad (5.65)$$

A positive infinitesimal ϵ is introduced in accordance with Eq.(5.46). An alternative method is the Wick rotation, which enable one to evaluate the path integral in the Euclidean space.(see Appendix E)

In order to calculate the generating functional, we introduce a field ϕ_0 satisfying the following equation

$$[\Box + (m^2 - i\epsilon)]\phi_0 = J(x). \qquad (5.66)$$

We take ϕ_0 as the reference field and shift ϕ with respect to ϕ_0. $\phi = \phi_0 + \phi'$.

Then we have

$$
\begin{aligned}
S[\phi, J] &= -\int d^4x \left[\frac{1}{2}\phi(\Box + m^2 - i\epsilon)\phi - J\phi\right] \\
&= -\int d^4x \left[\frac{1}{2}\phi'(\Box + m^2 - i\epsilon)\phi' + \frac{1}{2}\phi_0(\Box + m^2 - i\epsilon)\phi_0 \right.\\
&\qquad \left. + \phi'(\Box + m^2 - i\epsilon)\phi_0 - J\phi_0 - J\phi'\right] \\
&= -\int d^4x \left[\frac{1}{2}\phi'(\Box + m^2 - i\epsilon)\phi' + \frac{1}{2}J\phi_0 \right.\\
&\qquad \left. + \phi'(\Box + m^2 - i\epsilon)\phi_0 - J\phi_0 - J\phi'\right] \\
&= -\frac{1}{2}\int d^4x \left[\phi'(\Box + m^2 - i\epsilon)\phi' - J\phi_0\right].
\end{aligned}
\tag{5.67}
$$

In the derivation of Eq.(5.67), we have used Eq.(5.66).

The solution of Eq.(5.66) is

$$
\phi_0(x) = -\int d^4y \Delta_F(x - y) J(y),
\tag{5.68}
$$

where Δ_F is the Feynman propagator given by

$$
\Delta_F(x - y) = \int \frac{d^4k}{(2\pi)^4} \frac{e^{-ik\cdot(x-y)}}{k^2 - m^2 + i\epsilon},
\tag{5.69}
$$

which is the solution of the equation

$$
(\Box + m^2 - i\epsilon)\Delta_F(x) = -\delta^4(x).
\tag{5.70}
$$

Eq.(5.70) shows that $\Delta_F(x)$ is the minus inverse of $\Box + m^2$. Substituting Eq.(5.68) into $S[\phi, J]$ in Eq.(5.67), we have

$$
\begin{aligned}
S[\phi, J] &= -\frac{1}{2}\int d^4x \phi'(\Box + m^2 - i\epsilon)\phi' \\
&\quad - \frac{1}{2}\int d^4x d^4y J(x)\Delta_F(x - y)J(y).
\end{aligned}
\tag{5.71}
$$

Dependence of J in the path integral is contained in the second term of Eq.(5.71). Thus

$$
\begin{aligned}
Z[J] &= \int D\phi e^{-i\int d^4x\left[\frac{1}{2}\phi(\Box + m^2 - i\epsilon)\phi - J\phi\right]} \\
&= Z[0]e^{-\frac{i}{2}\int d^4x d^4y J(x)\Delta_F(x-y)J(y)}.
\end{aligned}
\tag{5.72}
$$

Then the normalized generating functional $\mathcal{Z}_0[J]$ for the free scalar field is given by

$$
\mathcal{Z}_0[J] = \frac{Z[J]}{Z[0]} = e^{-\frac{i}{2}\int d^4x d^4y J(x)\Delta_F(x-y)J(y)}.
\tag{5.73}
$$

The subscript 0 is used to denote a quantity of the free field. Using Eq.(5.56), we have

$$W_0[J] = -\frac{1}{2} \int d^4x d^4y J(x) \Delta_F(x-y) J(y). \tag{5.74}$$

Usually an integration expression can be considered as a solution of a differential equation. The integration expression Eq.(5.73) for $\mathcal{Z}_0[J]$ can also be considered as a solution of a differential equation for $\mathcal{Z}_0[J]$. This differential equation for $\mathcal{Z}_0[J]$ can be deduced as follows:

A functional derivative of $\mathcal{Z}_0[J]$ leads to

$$\frac{1}{i} \frac{\delta}{\delta J(x)} \mathcal{Z}_0[J] = -\int d^4y \Delta_F(x-y) J(y)$$
$$\times e^{-\frac{i}{2} \int d^4x' d^4y' J(x') \Delta_F(x'-y') J(y')}. \tag{5.75}$$

Multiplying both sides of Eq.(5.75) by $\Box + m^2$, we have

$$(\Box + m^2) \frac{1}{i} \frac{\delta}{\delta J(x)} \mathcal{Z}_0[J] = J(x) e^{-\frac{i}{2} \int d^4x' d^4y' J(x') \Delta_F(x'-y') J(y')} \tag{5.76}$$

Using Eq.(5.73), we find

$$(\Box + m^2) \frac{1}{i} \frac{\delta}{\delta J(x)} \mathcal{Z}_0[J] = J(x) \mathcal{Z}_0[J]. \tag{5.77}$$

Eq.(5.77) is the differential equation for $\mathcal{Z}_0[J]$ and Eq.(5.73) can be considered as the solution of Eq.(5.77).

5.2.7 Wick's theorem

We expand $\mathcal{Z}_0[J]$ in Eq.(5.73)

$$\mathcal{Z}_0[J] = \sum_{n=0}^{\infty} \frac{1}{n!} \left[-\frac{i}{2} \int d^4x d^4y J(x) \Delta_F(x-y) J(y) \right]^n$$
$$= 1 + \sum_{n=1}^{\infty} \frac{1}{n!} \left(\frac{-i}{2} \right)^n \int d^4x_1 \cdots d^4x_{2n} \tag{5.78}$$
$$\times \Delta_{F12} \Delta_{F34} \cdots \Delta_{F\,2n-1\,2n} J_1 J_2 \cdots J_{2n}$$

with $\Delta_{Fij} = \Delta_F(x_i - x_j)$ and $J_k = J(x_k)$. All the n-point functions with odd n vanish because Z contains only even powers of J and an odd functional derivative leaves odd powers of J in the integrand which vanishes at $J = 0$.

For the even powers of functional derivatives of $\mathcal{Z}_0$, we have

$$
\begin{aligned}
G_0(x_1, x_2, \cdots, x_{2k}) &= \left(\frac{1}{i}\right)^{2k} \frac{\delta^{2k} \mathcal{Z}_0(J)}{\delta J(x_1) \delta J(x_2) \cdots \delta J(x_{2k})}\bigg|_{J=0} \\
&= \left(\frac{i}{2}\right)^k \frac{1}{k!} \sum_P \Delta_{Fp_1 p_2} \cdots \Delta_{Fp_{2k-1} p_{2k}}.
\end{aligned}
\tag{5.79}
$$

where the sum runs over all permutations $(p_1, p_2, \cdots, p_{2k})$ of the number $(1, 2, \cdots, 2k)$. Eq.(5.79) shows that the n-point functions of free scalar bosons can be written as a product of two-point functions, which is the so-called *Wick's theorem*. Wick's theorem thus allows us to calculate the n-point functions of the free fields in terms of the Feynman propagator. As an example, we consider the case of $n = 4$, which corresponds to $k = 2$. We have

$$
G_0(x_1, x_2, x_3, x_4) = -\frac{1}{8} \sum_P \Delta_{Fp_1 p_2} \Delta_{Fp_3 p_4}.
\tag{5.80}
$$

There are 24 terms in the sum. Since $\Delta_{Fp_1 p_2} = \Delta_{Fp_2 p_1}$, $\Delta_{Fp_3 p_4} = \Delta_{Fp_4 p_3}$, and $\Delta_{Fp_1 p_2} \Delta_{Fp_3 p_4} = \Delta_{Fp_3 p_4} \Delta_{Fp_2 p_1}$, we obtain a $2 \times 2 \times 2 = 8$ factor from the sum, which just cancels with the factor $1/8$. Only $\frac{24}{8} = 3$ distinct terms are left, which can be written explicitly as

$$
\begin{aligned}
G_0(x_1, x_2, x_3, x_4) = &- \Delta_F(x_1 - x_2) \Delta_F(x_3 - x_4) \\
&- \Delta_F(x_1 - x_3) \Delta_F(x_2 - x_4) \\
&- \Delta_F(x_1 - x_4) \Delta_F(x_2 - x_3).
\end{aligned}
\tag{5.81}
$$

5.2.8   *Feynman rules*

We can express the expansion in a graphical way using Feymann diagrams. The Feynman rules set up the connection between the algebraic and graphical representation. For the free scalar field, they are given by

1) Each Feynman propagator $i\Delta_F(x - y)$ is represented by a line as shown in Fig.5.1.

x y

Fig. 5.1

2) Each source $iJ(x)$ is represented by a cross as shown in Fig.5.2.

Fig. 5.2

3) There is an integration over all the spacetime coordinates.

4) There is a combinatorial factor for each diagram that takes into account the symmetry of the diagram under exchange of the external lines.

Using the Feynman diagrams, we can express $G_0(x_1, x_2, x_3, x_4)$ in Eq.(5.81) as Fig.5.3.

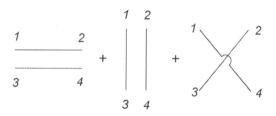

Fig. 5.3

Since $\mathcal{Z}_0[J] = e^{iW_0[J]}$, we have

$$iW_0[J] = -\frac{i}{2} \int d^4x d^4y J(x)\Delta_F(x-y)J(y), \qquad (5.82)$$

which can be expressed by a Feynman diagram shown in Fig.5.4. The $\frac{1}{2}$ factor in Eq.(5.82) comes from the symmetry of exchanging the endpoints in Fig.5.4, which corresponds to the invariance of the integral in Eq.(5.82) when the integration variables x and y are exchanged.

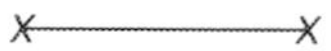

Fig. 5.4 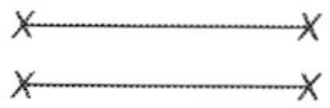   Fig. 5.5

For the expansion of $\mathcal{Z}_0[J]$, we have

$$\mathcal{Z}_0[J] = e^{iW_0[J]}$$

$$= 1 + iW_0[J] + \frac{1}{2}(iW_0[J])^2 + \frac{1}{3!}(iW_0[J])^3 + \cdots. \qquad (5.83)$$

There are two types of Feynman diagrams in the diagram representation of Eq.(5.83): the connected graphs that all parts are tied together, such as Fig.5.4 for $iW_0[J]$, and the unconnected graphs such as Fig.5.5 for $(iW_0[J])^2$. The connected Green's function defined by Eq.(5.57), such as

$$G_c(x_1, x_2) = \frac{1}{i} \left. \frac{\delta^2 W_0}{\delta J(x_1)\delta J(x_2)} \right|_{J=0} = i\Delta_F(x_1 - x_2) \tag{5.84}$$

can be represented as a Feynman diagram shown in Fig.5.1. Other G_c for the free field are zero.

5.3   Interacting scalar field

We can add any interaction term $V(\phi)$ in the Lagrangian density of matter to form the interacting scalar field without changing the total Lagrangian in Eq.(3.38). The form of the interaction term $V(\phi)$ should be determined in such a way that the ground state with the lowest energy can be achieved by adding the interacting term $V(\phi)$. Now we consider the Lagrangian density with an interaction term

$$\mathcal{L} = \mathcal{L}_0 - V(\phi), \tag{5.85}$$

where $\mathcal{L}_0 = \frac{1}{2}(\partial_\mu \phi \partial^\mu \phi - m^2 \phi^2)$ is the Lagrangian density for free scalar bosons. $V(\phi)$ is the self-interaction term, such as $\lambda\phi^4$ term in the Higgs mechanism. In the following, we will discuss the perturbation method to calculate the n-point functions defined by Eq.(5.51)

$$G(x_1, x_2, \cdots, x_n) = \frac{\int D\phi\, \phi(x_1)\phi(x_2)\cdots\phi(x_n)e^{iS[\phi]}}{\int D\phi\, e^{iS[\phi]}}. \tag{5.86}$$

We expand the action exponential in terms of the powers of the interaction

$$e^{iS[\phi]} = \sum_{N=0}^{\infty} \frac{1}{N!} \left(-i \int d^4x V \right)^N e^{iS_0[\phi]}. \tag{5.87}$$

Inserting Eq.(5.87) into Eq.(5.86), we have

$$G(x_1, x_2, \cdots, x_n)$$

$$= \frac{\int D\phi\, \phi(x_1)\phi(x_2)\cdots\phi(x_n) \sum_{N=0}^{\infty} \frac{1}{N!}(-i\int d^4x V)^N e^{iS_0[\phi]}}{\int D\phi \sum_{N=0}^{\infty} \frac{1}{N!}(-i\int d^4x V)^N e^{iS_0[\phi]}}. \tag{5.88}$$

5.3.1 *Perturbation expansion*

We can use Eq.(5.88) to do the perturbation calculations of the Green's functions up to any orders in V.

The generating functional is given by

$$\mathcal{Z}[J] = \mathcal{N}_0 \int D\phi e^{i \int d^4 x [\mathcal{L}_0 - V(\phi) + J\phi]}$$
$$= \mathcal{N}_0 \int D\phi e^{-i \int d^4 x V(\phi)} e^{i \int d^4 x (\mathcal{L}_0 + J\phi)}. \tag{5.89}$$

where $\mathcal{N}_0 = Z[0]^{-1}$. Using the relation

$$\frac{1}{i} \frac{\delta}{\delta J(y)} e^{i \int d^4 x (\mathcal{L}_0 + J\phi)} = \phi(y) e^{i \int d^4 x (\mathcal{L}_0 + J\phi)}, \tag{5.90}$$

we obtain

$$e^{-i \int d^4 y V(\phi(y))} e^{i \int d^4 x (\mathcal{L}_0 + J\phi)} = e^{-i \int d^4 y V(\frac{1}{i} \frac{\delta}{\delta J(y)})} e^{i \int d^4 x (\mathcal{L}_0 + J\phi)}. \tag{5.91}$$

Thus, $\mathcal{Z}[J]$ in Eq.(5.89) becomes

$$\mathcal{Z}[J] = \mathcal{N}_0 e^{-i \int d^4 y V(\frac{1}{i} \frac{\delta}{\delta J(y)})} \int D\phi e^{i \int d^4 x (\mathcal{L}_0 + J\phi)}. \tag{5.92}$$

Since the V dependent factor is now taken out of the functional integral, the functional integral is the one for the free field and can be expressed in terms of the free Feynman propagator. We have

$$\mathcal{Z}[J] = \mathcal{N}_0 e^{-i \int d^4 z V(\frac{1}{i} \frac{\delta}{\delta J(z)})} e^{-\frac{i}{2} \int d^4 x d^4 y J(x) \Delta_F(x-y) J(y)}$$
$$= \mathcal{N}_0 e^{-i \int d^4 x V(\frac{1}{i} \frac{\delta}{\delta J(x)})} \mathcal{Z}_0[J]. \tag{5.93}$$

Expanding the exponential factor $e^{-i \int d^4 y V(\frac{1}{i} \frac{\delta}{\delta J(y)})}$ in powers of V, we have

$$\mathcal{Z}[J] = \mathcal{N}_0 \sum_{N=0}^{\infty} \frac{1}{N!} \left[-i \int d^4 x V \left(\frac{1}{i} \frac{\delta}{\delta J(y)} \right) \right]^N \mathcal{Z}_0[J]. \tag{5.94}$$

Using the expansion Eq.(5.94), we can calculate the n-point Green's function perturbatively.

$$G(x_1, x_2, \cdots, x_n)$$
$$= \left(\frac{1}{i} \right)^n \frac{\delta^n \mathcal{Z}[J]}{\delta J(x_1) \delta J(x_2) \cdots \delta J(x_n)} \bigg|_{J=0}$$
$$= \frac{\int D\phi \phi(x_1) \phi(x_2) \cdots \phi(x_n) \sum_{N=0}^{\infty} \frac{1}{N!} \left[-i \int d^4 x V(\phi(x)) \right]^N e^{iS_0[\phi]}}{\int D\phi \sum_{N=0}^{\infty} \frac{1}{N!} \left[-i \int d^4 x V(\phi(x)) \right]^N e^{iS_0[\phi]}}. \tag{5.95}$$

We can also make the expansion for $iW[J] = \ln \mathscr{Z}[J]$ which is related to the connected diagrams and reads

$$
\begin{aligned}
iW[J] &= \ln \mathscr{Z}[J] \\
&= \ln \mathscr{N}_0 + \ln \left(e^{-i \int d^4x V(\frac{1}{i} \frac{\delta}{\delta J(x)})} e^{iW_0[J]} \right) \\
&= \ln \mathscr{N}_0 + iW_0 + \ln(e^{-iW_0} e^{-i \int d^4x V} e^{iW_0}) \\
&= \ln \mathscr{N}_0 + iW_0[J] + \ln \left(1 + e^{-iW_0[J]} (e^{-i \int d^4x V(\frac{1}{i} \frac{\delta}{\delta J(x)})} - 1) e^{iW_0[J]} \right).
\end{aligned}
\tag{5.96}
$$

We define a functional $\varepsilon[J]$ as

$$
\begin{aligned}
\varepsilon[J] &\equiv e^{-iW_0[J]} \left[e^{-i \int d^4x V(\frac{1}{i} \frac{\delta}{\delta J})} - 1 \right] e^{iW_0[J]} \\
&= e^{-iW_0[J]} \left\{ -i \int d^4x V \left(\frac{1}{i} \frac{\delta}{\delta J} \right) + \frac{1}{2!} \left[-i \int d^4x V \left(\frac{1}{i} \frac{\delta}{\delta J} \right) \right]^2 \right. \\
&\quad \left. + \cdots \right\} e^{iW_0[J]}.
\end{aligned}
\tag{5.97}
$$

Inserting the expansion formula Eq.(5.97) for $\varepsilon[J]$ into Eq.(5.96), we have

$$
\begin{aligned}
iW[J] &= \ln \mathscr{N}_0 + iW_0[J] + \left(\varepsilon[J] - \frac{1}{2} \varepsilon^2[J] + \cdots \right) \\
&= \ln \mathscr{N}_0 + iW_0[J] + e^{-iW_0[J]} \left[-i \int d^4x V \left(\frac{1}{i} \frac{\delta}{\delta J} \right) \right] e^{iW_0[J]} \\
&\quad + \frac{1}{2!} e^{-iW_0[J]} \left[-i \int d^4x V \left(\frac{1}{i} \frac{\delta}{\delta J} \right) \right]^2 e^{iW_0[J]} \\
&\quad - \frac{1}{2} \left\{ e^{-iW_0[J]} \left[-i \int d^4x V \left(\frac{1}{i} \frac{\delta}{\delta J} \right) \right] e^{iW_0[J]} \right\}^2 + \mathcal{O}(V^3) \\
&= \ln \mathscr{N}_0 + iW_0[J] + iW_1[J] + iW_2[J] - \frac{1}{2} (iW_1[J])^2 + \mathcal{O}(V^3).
\end{aligned}
\tag{5.98}
$$

with

$$
iW_0[J] = -\frac{i}{2} \int d^4x d^4y J(x) \Delta_F(x-y) J(y),
\tag{5.99a}
$$

$$
iW_1[J] = e^{-iW_0[J]} \left[-i \int d^4x V \left(\frac{1}{i} \frac{\delta}{\delta J} \right) \right] e^{iW_0[J]},
\tag{5.99b}
$$

$$
iW_2[J] = \frac{1}{2!} e^{-iW_0[J]} \left[-i \int d^4x V \left(\frac{1}{i} \frac{\delta}{\delta J} \right) \right]^2 e^{iW_0[J]}.
\tag{5.99c}
$$

Inserting the expansion Eq.(5.98) into Eq.(5.57), we can calculate the connected Green's functions of the interacting field.

5.3.2 Perturbation ϕ^4 theory

Now we consider the interaction term

$$V(\phi) = \frac{g}{4!}\phi^4, \tag{5.100}$$

where g is a constant, which is called the coupling constant. We will show that the ϕ^4 type interaction term is the only interaction term in four-dimensional spacetime leading to meaningful results for scalar bosons. Now we calculate the expansion of $iW[J]$ using Eq.(5.98).

$$iW[J] = \ln\mathcal{N}_0 + iW_0[J] + iW_1[J] + iW_2[J] - \frac{1}{2}(iW_1[J])^2 + \mathcal{O}(V^3) \tag{5.101}$$

with

$$iW_0[J] = -\frac{i}{2}\int d^4x d^4y J(x)\Delta_F(x-y)J(y), \tag{5.102a}$$

$$iW_1[J] = e^{-iW_0[J]}\left[-i\int d^4x \frac{g}{4!}\left(\frac{1}{i}\frac{\delta}{\delta J}\right)^4\right]e^{iW_0[J]}, \tag{5.102b}$$

$$iW_2[J] = \frac{1}{2!}e^{-iW_0[J]}\left[-i\int d^4x \frac{g}{4!}\left(\frac{1}{i}\frac{\delta}{\delta J}\right)^4\right]^2 e^{iW_0[J]}. \tag{5.102c}$$

5.3.2.1 Generating functional up to $\mathcal{O}(g)$

We first calculate the term linear in g

$$\begin{aligned}
iW_1[J] &= -\frac{ig}{4!}e^{-iW_0[J]}\int d^4x \frac{\delta^4}{\delta J^4(x)}e^{iW_0[J]} \\
&= -\frac{ig}{4!}e^{-iW_0[J]}\int d^4x \frac{\delta^4}{\delta J^4(x)}e^{-\frac{i}{2}\int d^4x d^4y J(x)\Delta_F(x-y)J(y)} \\
&= -\frac{ig}{4!}\Big\{-3\int \Delta_F(x-x)\Delta_F(x-x)d^4x \\
&\quad + 6i\int \Delta_F(y-x)\Delta_F(x-x)\Delta_F(x-z)J(y)J(z)d^4x d^4y d^4z \\
&\quad + \int \Delta_F(x-y)\Delta_F(x-z)\Delta_F(x-v)\Delta_F(x-w) \\
&\quad \times J(y)J(z)J(v)J(w)d^4x d^4y d^4z d^4v d^4w\Big\}.
\end{aligned} \tag{5.103}$$

The second term is quadratic in J and thus contributes to the two-point functions. The last term contains four powers of J and thus contributes to the four-point function. We can express the expansion Eq.(5.101) graphically using the following Feynman rules:

(1) Propagator: $i\Delta_F(x - y)$ is represented by a line as shown in Fig.5.1.
(2) Source: $iJ(x) =$ is represented by a cross as shown in Fig.5.2.
(3) There is an integration over the spacetime coordinates for each source.
(4) There is an a symmetry factor for each diagram.
(5) Each interaction factor $\frac{ig}{4!}$ is represented by a dot as shown in Fig.5.6.
(6) There is an integration $\int d^4x$ for each loop.

Fig. 5.6   Fig. 5.7

Generally the connected generating functional $iW[J]$ for interacting field is represented by a double line as shown in Fig.5.7. We can represent the expansion by Feynman diagrams

$$iW[J] = \ln\mathcal{N}_0 + iW_0[J] + iW_1[J] + \mathcal{O}(g^2)$$
$$= \ln\mathcal{N}_0 + \text{Fig.5.8} + (\text{Fig.5.9} + \text{Fig.5.10} + \text{Fig.5.11}) + \mathcal{O}(g^2).$$
$$(5.104)$$

$iW_0[J]$ is the zeroth order term, which is given by Eq.(5.82). The graphs in the parentheses are the first order terms given by Eq.(5.102b). The first one corresponds to the first integral in Eq.(5.103). There are no external lines in the graph, which describes the vacuum processes. The second graph corresponds to the second integral in Eq.(5.103). It has a single loop attached and is called the tadpole diagram, which gives a mass change due to the self-interaction. The third graph is the last integral in Eq.(5.103), which describes the interaction process.

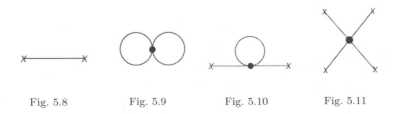

Fig. 5.8   Fig. 5.9   Fig. 5.10   Fig. 5.11

Now we show the vacuum contribution in Eq.(5.101) can be canceled out by the normalization term $\ln \mathcal{N}_0$. Since $\mathcal{N}_0 = Z[0]^{-1}$, we have

$$
\begin{aligned}
\ln \mathcal{N}_0 &= -\ln Z[0] \\
&= -\ln \int D\phi \, e^{i \int d^4 x (\mathcal{L}_0 - V)} \Big|_{J=0} \\
&= -\ln \left(e^{-i \int d^4 x V(\frac{1}{i} \frac{\delta}{\delta J(x)})} e^{-\frac{i}{2} \int d^4 x d^4 y J(x) \Delta_F(x-y) J(y)} \right) \Big|_{J=0}.
\end{aligned}
\tag{5.105}
$$

Expanding $\ln \mathcal{N}_0$ similarly as we did in Eq.(5.96) and using $W_0[0] = 0$, we have

$$
\begin{aligned}
\ln \mathcal{N}_0 &= -\ln \left[e^{-i \int d^4 x V(\frac{1}{i} \frac{\delta}{\delta J(x)})} e^{iW_0[J]} \right] \Big|_{J=0} \\
&= -\ln \left[1 + \left(e^{-i \int d^4 x V(\frac{1}{i} \frac{\delta}{\delta J(x)})} - 1 \right) e^{iW_0[J]} \right] \Big|_{J=0} \\
&= i \int d^4 x V \left(\frac{1}{i} \frac{\delta}{\delta J(x)} \right) e^{iW_0[J]} \Big|_{J=0} \\
&\quad - \frac{1}{2!} \left[-i \int d^4 x V \left(\frac{1}{i} \frac{\delta}{\delta J(x)} \right) \right]^2 e^{iW_0[J]} \Big|_{J=0} \\
&\quad + \frac{1}{2} \left[-i \int d^4 x V \left(\frac{1}{i} \frac{\delta}{\delta J(x)} \right) e^{iW_0[J]} \right]^2 \Big|_{J=0} + \cdots \\
&= -iW_0[0] - iW_1[0] - iW_2[0] + \frac{1}{2}(iW_1[0])^2 \cdots \\
&= -iW[0].
\end{aligned}
\tag{5.106}
$$

The vacuum terms are those without external lines. When we take $J = 0$ in $W[J]$, all the terms with external lines vanish and only vacuum terms remain. Thus $\ln \mathcal{N}_0 = -\ln Z[0]$ cancels with the vacuum contribution in the expansion of $iW[J]$ in Eq.(5.104). Then $iW[J]$ can be evaluated using the Feynman diagram shown in Fig.5.12.

Fig. 5.12

5.3.2.2   *Two-point function*

Using the expansion Eq.(5.96), we can calculate the connected n-point functions. We first consider the connected two-point function.

Terms up to $\mathcal{O}(g)$

Using Eq.(5.101), we can obtain the expansion up to terms linear in the coupling strength.

$$
\begin{aligned}
G_c(x_1, x_2) &= \frac{1}{i} \frac{\delta^2 W}{\delta J(x_1)\delta J(x_2)}\bigg|_{J=0} \\
&= -i \frac{\delta^2 W_0}{\delta J(x_1)\delta J(x_2)}\bigg|_{J=0} - i \frac{\delta^2 W_1}{\delta J(x_1)\delta J(x_2)}\bigg|_{J=0} + \mathcal{O}(g^2) \\
&= i\Delta_F(x_1 - x_2) + \frac{ig}{4!}12i \int d^4x \Delta_F(x - x)\Delta_F(x - x_1) \qquad (5.107)\\
&\quad \times \Delta_F(x - x_2) + \mathcal{O}(g^2) \\
&= i\Delta_F(x_1 - x_2) - \frac{g}{2} \int d^4x \Delta_F(x_2 - x) \\
&\quad \times \Delta_F(x - x)\Delta_F(x - x_1) + \mathcal{O}(g^2).
\end{aligned}
$$

$G_c(x_1, x_2)$ is also denoted as $G_c^{(2)}$.

The first term in Eq.(5.107) is the term of $\mathcal{O}(g^0)$, which is the Feynman propagator for a free field.

$$
\Delta_F(x_1 - x_2) = \int \frac{d^4q}{(2\pi)^4} e^{-iq\cdot(x_1 - x_2)} \frac{1}{q^2 - m^2 + i\epsilon}. \qquad (5.108)
$$

We can also evaluate the two-point function in the momentum space by the Fourier transformation

$$
\begin{aligned}
(2\pi)^4 &\delta^4(p_1 + p_2)G_c(p_1, p_2) \\
&= \int d^4x_1 \int d^4x_2 e^{-i(p_1\cdot x_1 + p_2\cdot x_2)} G_c(x_1, x_2).
\end{aligned} \qquad (5.109)
$$

For a free field, we have

$$
\begin{aligned}
(2\pi)^4 &\delta^4(p_1 + p_2)G_c(p_1, p_2) \\
&= \int d^4x_1 \int d^4x_2 e^{-i(p_1\cdot x_1 + p_2\cdot x_2)} \\
&\quad \times \left[\int \frac{d^4q}{(2\pi)^4} e^{-iq\cdot(x_1 - x_2)} \frac{i}{q^2 - m^2 + i\epsilon}\right] \\
&= (2\pi)^4 \delta^4(p_1 + p_2)\frac{i}{p_1^2 - m^2 + i\epsilon}.
\end{aligned} \qquad (5.110)
$$

Thus we have the momentum representation of the two-point function

$$G_{c0}(p,-p) = G_0(p,-p) = \frac{i}{p^2 - m^2 + i\epsilon} = i\Delta_F(p). \qquad (5.111)$$

The second term in Eq.(5.107) is the term of $\mathcal{O}(g)$. Its fourier transformation is given by

$$-\frac{g}{2} \int d^4x_1 \int d^4x_2 \Big\{ e^{-i(p_1 \cdot x_1 + p_2 \cdot x_2)} \Big[\int d^4x \int \frac{d^4q_1}{(2\pi)^4} \frac{d^4q_2}{(2\pi)^4} \frac{d^4q_3}{(2\pi)^4}$$

$$\frac{e^{-iq_2 \cdot (x_2 - x)} e^{-iq_1 \cdot (x - x_1)}}{(q_1^2 - m^2 + i\epsilon)(q_2^2 - m^2 + i\epsilon)(q_3^2 - m^2 + i\epsilon)} \Big] \Big\}$$

$$= -(2\pi)^4 \delta^4(p_1 + p_2) \frac{g}{2} \frac{1}{p_1^2 - m^2 + i\epsilon}$$

$$\times \int \frac{d^4q}{(2\pi)^4} \frac{1}{q^2 - m^2 + i\epsilon} \frac{1}{p_2^2 - m^2 + i\epsilon}. \qquad (5.112)$$

Thus we have the expansion in the momentum representation

$$G_c(p,-p) = \frac{i}{p^2 - m^2 + i\epsilon} + \frac{i}{p^2 - m^2 + i\epsilon}$$

$$\times \Big[\frac{-ig}{4!} 12 \int \frac{d^4q}{(2\pi)^4} \frac{i}{q^2 - m^2 + i\epsilon} \Big] \frac{i}{p^2 - m^2 + i\epsilon}. \qquad (5.113)$$

Similar to the coordinate space representation, we can use the Feynman rules in the momentum space to construct the expansion in the momentum representation. The Feynman rules in the momentum space are given by

(1) Each free propagator line corresponds to a factor $\frac{i}{q^2 - m^2 + i\epsilon}$.
(2) Each vertex is associated with a factor $-\frac{ig}{4!}$.
(3) The sum of all momenta flowing into a vertex should be zero.
(4) Each internal line is associated with an integration $\int \frac{d^4q}{(2\pi)^4}$.
(5) There is a symmetry factor for each diagram.

We use Σ to denote the integral in the second term of Eq.(5.113)

$$\Sigma = \frac{g}{2} \int \frac{d^4q}{(2\pi)^4} \frac{i}{q^2 - m^2 + i\epsilon}. \qquad (5.114)$$

Then the expansion for the two-point function can be rewritten as

$$G_c(p, -p) = G_0 + G_0 \frac{\Sigma}{i} G_0$$

$$= G_0(1 + \frac{\Sigma}{i} G_0)$$

$$\approx G_0(1 - \Sigma \frac{G_0}{i})^{-1} \qquad (5.115)$$

$$= \frac{i}{p^2 - m^2 + i\epsilon} \frac{1}{1 - \dfrac{\Sigma}{p^2 - m^2 + i\epsilon}}$$

$$= \frac{i}{p^2 - m^2 - \Sigma + i\epsilon}.$$

From this equation, it can be seen that Σ is the modification to the mass due to the self-interaction and thus is called the self-energy.

Terms up to $\mathcal{O}(g^2)$

From Eq.(5.98), we can see that there are two different terms in the $\mathcal{O}(g^2)$ contribution. One is $iW_2[J]$ which is the genuine term of the second order in V and the other is $-\frac{1}{2}(iW_1[J])^2$, which can be considered as the iteration of the first order term. We draw the corresponding Feynman diagrams in Fig.5.13. The first graph in Fig.5.13 is the Feynman diagram constructed by an iteration of the first order tadpole diagram. It is the square of the two first-order terms. As shown in Fig.5.13, this graph consists two parts of the first order in V connected by an internal line. Such a graph can be split into two unconnected parts when one internal line is cut. We call this kind of graphs *one-particle reducible*. Otherwise, it is called *one-particle-irreducible*(1PI).

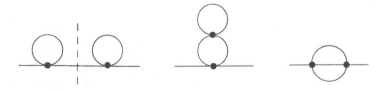

Fig. 5.13

In order to describe the 1PI graphs, we introduce a vertex function $\Gamma_{amp}^{(n)}(p_1, p_2, \cdots, p_n)$. It is also called *connected proper vertex function*. The n-point vertex function is defined as

$$\Gamma_{amp}^{(n)}(p_1, p_2, \cdots, p_n)$$
$$\equiv G_c^{-1}(p_1, -p_1)G_c^{-1}(p_2, -p_2)\cdots G_c^{-1}(p_n, -p_n)G_c(p_1, p_2, \cdots, p_n).$$

$$(5.116)$$

$\Gamma_{amp}^{(n)}(p_1, p_2, \cdots, p_n)$ is also called the *amputated Green's function* because it is the connected n-point function with external lines truncated, which is the reason we add subscript 'amp'. To simplify the notation, we usually omit the subscript 'amp'.

The free 1PI two-point function $\Gamma_0^{(2)}$ is given by

$$\Gamma_0^{(2)}(p, -p) = G_0^{-1}(p, -p) = \frac{1}{i}(p^2 - m^2). \qquad (5.117)$$

The 1PI part of the first order term is given by

$$\Gamma_1^{(2)}(p, -p) = G_0^{-1}(p, -p)G_0^{-1}(-p, p)G_{c1}(p, -p)$$
$$= 12\frac{-ig}{4!}\int \frac{d^4q}{(2\pi)^4}\frac{i}{q^2 - m^2 + i\epsilon}, \qquad (5.118)$$

where $G_{c1}(p, -p)$ is the second term in Eq.(5.113) contributed by the first order tadpole diagram in the first graph of Fig.5.13. The factor 12 in Eq.(5.118) comes from the symmetry factor.

For the 1PI graph of the second order in V shown as the second graph in Fig.5.13, we have

$$\Gamma_{2a}^{(2)}(p, -p)$$
$$= 12^2\left(\frac{-ig}{4!}\right)^2\int\frac{d^4q_1}{(2\pi)^4}\frac{d^4q_2}{(2\pi)^4}\frac{i}{q_1^2 - m^2 + i\epsilon}\frac{i(2\pi)^4\delta^4(q_1 - q_2)}{q_2^2 - m^2 + i\epsilon} \qquad (5.119)$$
$$\times\int\frac{d^4q_3}{(2\pi)^4}\frac{i}{q_3^2 - m^2 + i\epsilon}.$$

Another 1PI graph of the second order in V shown as the third graph in Fig.5.13 contributes to the vertex function a term

$$\Gamma_{2b}^{(2)}(p, -p)$$
$$= 4\cdot 4!\left(\frac{-ig}{4!}\right)^2\int\frac{d^4q_1}{(2\pi)^4}\frac{d^4q_2}{(2\pi)^4}\frac{d^4q_3}{(2\pi)^4}\delta^4(p - (q_1 + q_2 + q_3)) \qquad (5.120)$$
$$\times\frac{i}{q_1^2 - m^2 + i\epsilon}\frac{i}{q_2^2 - m^2 + i\epsilon}\frac{i}{q_3^2 - m^2 + i\epsilon}.$$

Using Eq.(5.101), we can calculate the expansion of $G_c^{(2)}$. The Feynman diagrams of $\mathcal{O}(g)$ and $\mathcal{O}(g^2)$ terms are shown in Fig.5.14.

The expansion of $G_c^{(2)}$ contains only the connected diagrams. Since $G_c^{(2)}$ is a summation of the contributions of all the connected diagrams, we call $G_c^{(2)}$ the complete or dressed propagator. The summation of the connected diagrams is denoted as the diagram shown in Fig.5.15.

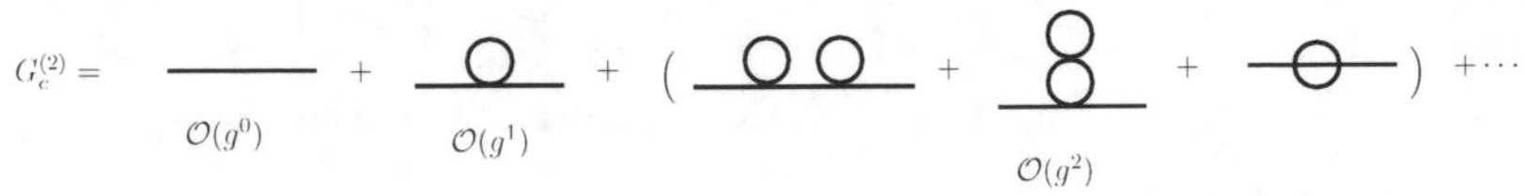

$G_c^{(2)} =$ 〔diagram〕

Fig. 5.14 $G_c^{(2)}$ as a summation of the connected diagrams.

Fig. 5.15   Dressed propagator.

Since the 1-particle-reducible graphs such as the first diagram shown in Fig.5.13 can be expressed as a product of the 1-particle-irreducible graphs, we can use the 1-particle-irreducible graphs to calculate $G_c^{(2)}$. We show the summation of 1PI graphs in Fig.5.16. This summation of 1PI graphs is

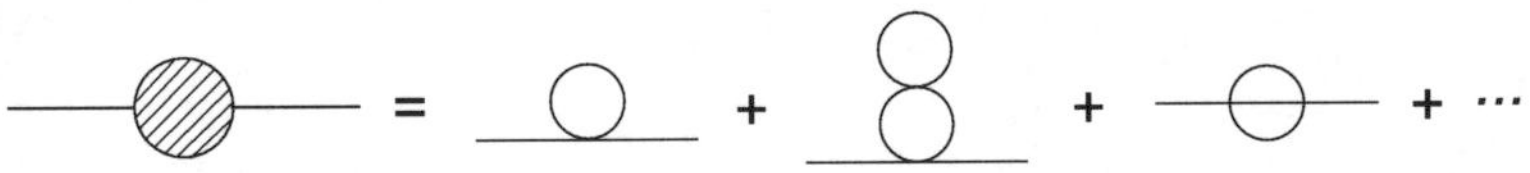

Fig. 5.16   Self-energy as a summation of 1PI graphs.

defined as $\frac{1}{i}\Sigma$, where Σ is called the self-energy. Then the expansion of the complete propagator $G_c^{(2)}$ can be written in terms of the self-energy as

$$G_c^{(2)}(p) = G_0(p) + G_0(p)\frac{1}{i}\Sigma(p)G_0(p)$$

$$+ G_0(p)\frac{1}{i}\Sigma(p)G_0(p)\frac{1}{i}\Sigma(p)G_0(p) + \cdots$$

$$= G_0 \left(1 + \frac{1}{i} \Sigma G_0 + \frac{1}{i} \Sigma G_0 \frac{1}{i} \Sigma G_0 + \cdots \right)$$

$$= G_0 \left(1 - \frac{1}{i} \Sigma G_0 \right)^{-1}$$

$$= \left(G_0^{-1} - \frac{1}{i} \Sigma \right)^{-1}$$

$$= \frac{1}{p^2 - m^2 - \Sigma(p)}.$$

$$(5.121)$$

Eq.(5.115), in which $G_c^{(2)}$ is expanded up to $\mathcal{O}(g)$ terms, is an approximated version of Eq.(5.121). The expansion in Eq.(5.121) can be expressed in terms of the diagrams shown in Fig.5.17. Eq.(5.121) can be rewritten as

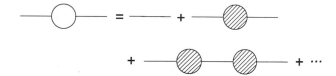

Fig. 5.17   Expansion of the complete propagator $G_c^{(2)}$ in terms of the self-energy.

$$[G_c^{(2)}(p)]^{-1} = G_0^{-1}(p) - \frac{1}{i} \Sigma(p). \qquad (5.122)$$

Using the connected proper vertex function, we have

$$\Gamma_{\text{amp}}^{(2)} = [G_c^{(2)}]^{-1}. \qquad (5.123)$$

Since the expansion of $\Gamma_{\text{amp}}^{(2)}$ is the inverse of the $G_c^{(2)}$ expansion, the self-energy $\Sigma(p)$ up to $\mathcal{O}(g^2)$ is then given by

$$\frac{1}{i} \Sigma = \Gamma_1^{(2)} + \Gamma_{2a}^{(2)} + \Gamma_{2b}^{(2)}. \qquad (5.124)$$

5.3.2.3   *Four-point function*

The three-point function is zero. Now we discuss the four-point function.

Terms up to $\mathcal{O}(g)$

The expansion of the four-point function up to terms of $\mathcal{O}(g)$ is given by

$$G_c(x_1, x_2, x_3, x_4) = \left(\frac{1}{i} \right)^3 \frac{\delta^4 W}{\delta J(x_1) \cdots \delta J(x_4)} \bigg|_{J=0}$$

$$= i \left. \frac{\delta^4 W_0}{\delta J(x_1) \cdots \delta J(x_4)} \right|_{J=0} + i \left. \frac{\delta^4 W_1}{\delta J(x_1) \cdots \delta J(x_4)} \right|_{J=0}$$
$$+ \mathcal{O}(g^2).$$

$$(5.125)$$

Since W_0 depends on J only quadratically, the first term in Eq.(5.125) vanishes. Only the second term contributes. Using Eq.(5.103), we have

$$G_c(x_1, x_2, x_3, x_4)$$
$$= -ig \int d^4 x \Delta_F(x - x_1) \Delta_F(x - x_2) \Delta_F(x - x_3) \Delta_F(x - x_4).$$

$$(5.126)$$

The momentum representation is given by

$$G_c(p_1, p_2, p_3, p_4) = -ig \prod_{k=1}^{4} \frac{i}{p_k^2 - m^2 + i\epsilon}. \tag{5.127}$$

Terms up to $\mathcal{O}(g^2)$

The Feynman diagrams for the terms of $\mathcal{O}(g^2)$ is shown in Fig.5.18. The four diagrams on the first line are the vertices with self-energy insertion on each of the external legs. The diagrams on the second line are the genuine contributions.

The contribution from the diagrams on the first line of Fig.5.18 is given by

$$G_{2a}(p_1, p_2, p_3, p_4) = 4! \prod_{k=1}^{4} \frac{i}{p_k^2 - m^2} 12 \left(\frac{-ig}{4!} \right)^2 \int \frac{d^4 q}{(2\pi)^4} \frac{i}{q^2 - m^2}$$
$$\times \sum_{l=1}^{4} \frac{i}{p_l^2 - m^2}. \tag{5.128}$$

From the diagrams on the second line of Fig.5.18, the contribution has the form

$$G_{2b}(p_1, p_2, p_3, p_4) = \frac{(4!)^2}{2} \prod_{k=1}^{4} \frac{i}{p_k^2 - m^2} \left(\frac{-ig}{4!} \right)^2 \int \frac{d^4 q_1}{(2\pi)^4} \frac{d^4 q_2}{(2\pi)^4} \frac{i}{q_1^2 - m^2}$$
$$\times \frac{i}{q_2^2 - m^2} (2\pi)^4 \sum_{kl} \delta^4(q_1 + q_2 - (p_k + p_l)),$$

$$(5.129)$$

where the sum over (kl) runs over the pairs of number (1,2),(1,3), and (1,4).

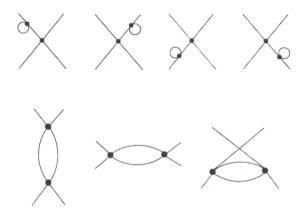

Fig. 5.18

5.4 Divergences in n-point functions

5.4.1 *Divergences in integrations*

One of the difficulties in the quantum field theory is the divergence problem. Many integrals in the two and four-point functions diverge. The divergence problem can be solved by the renormalization procedure. The renormalization procedure gives the effective field. The reason behind the renormalization procedure is that the quasi-particle states described by a momentum $\mathbf{p}$ are nonlocal ones. Although we can add any forms of the mass and interaction terms into the Lagrangian of matter without changing the total Lagrangian, the actual forms of the mass and interaction terms are those achieving the lowest energy for the ground state. The physical mass and other quantities of quasi-particles should be finite. Actually we have only a particular form of mass and interaction terms to give physical results.

Now we discuss the divergence problem. Let us consider the two-point function $G_c(p, -p)$ in Eq.(5.115).

$$G_c(p, -p) = \frac{i}{p^2 - m^2 - \Sigma + i\epsilon} \qquad (5.130)$$

with

$$\Sigma = \frac{g}{2} i \Delta_F(0) = \frac{g}{2} \int \frac{d^4 q}{(2\pi)^4} \frac{i}{q^2 - m^2 + i\epsilon}. \tag{5.131}$$

To perform the integration over q, we rewrite the expression of the self-energy Σ in Eq.(5.131) as

$$\begin{aligned}
\Sigma &= \frac{g}{2} \int \frac{d^4 q}{(2\pi)^4} \frac{i}{q^2 - m^2 + i\epsilon} \\
&= \frac{g}{2} \int \frac{d^3 q dq_0}{(2\pi)^4} \frac{i}{q_0^2 - \mathbf{q}^2 - m^2 + i\epsilon} \\
&= \frac{g}{2} \int \frac{d^3 q dq_0}{(2\pi)^4} \frac{i}{2\omega_{\mathbf{q}}} \left(\frac{1}{q_0 - \omega_{\mathbf{q}} + i\delta} - \frac{1}{q_0 + \omega_{\mathbf{q}} - i\delta} \right)
\end{aligned} \tag{5.132}$$

with $\omega_{\mathbf{q}} = \sqrt{\mathbf{q}^2 + m^2}$. There are two poles located at $\pm \omega_{\mathbf{q}} \mp i\delta$. According to Cauchy's theorem, the integral over q_0 can be evaluated by closing the integral route to enclose the poles and giving each pole a value of $(2\pi i)\times$ (the residue) $\times$ (the sign of the direction of the integral route).

The self-energy Σ becomes

$$\Sigma = \frac{g}{2} \int \frac{d^3 q}{(2\pi)^3} \frac{1}{\omega_{\mathbf{q}}} = \frac{g}{2} \int \frac{d^3 q}{(2\pi)^3} \frac{1}{\sqrt{\mathbf{q}^2 + m^2}}. \tag{5.133}$$

The integration is divergent. By introducing an upper bound Λ for the integral over $|\mathbf{q}|$, the integral diverges as Λ^2 when $\Lambda \to \infty$. It is called the quadratic divergence.

5.4.2 *Power counting*

We can use power counting to determine the degree of divergence of an integration. When an integration diverges as Λ^D, we say the degree of divergence is D. When $D > 0$, the integration diverges. $D = 0$ corresponds to a logarithmic divergence. $D < 0$ is the case of convergence. For example, the integration of Σ in Eq.(5.131) has $D = 2$ because $d^4 q$ gives 4 powers of q and the denominator in the integrand gives two powers of q.

For an interaction $\sim \phi^p$ in n dimensional spacetime, when a Feynman diagram has L loop and I internal lines, we have

$$D = nL - 2I \tag{5.134}$$

because each loop contributes an integral $\int d^n q$ and each internal propagator gives a power q^{-2}.

When a diagram has V interaction vertices, there are pV lines in total because each vertex contributes p lines. We denote the number of the external lines as E. One internal line originates and terminates at a vertex, consuming two legs of vertex. There are I internal lines. Thus we have

$$pV = E + 2I. \tag{5.135}$$

Each internal line carries an integral $\int d^n q$. In the mean time, each vertex contributes a delta-function associated with the conservation of momentum. Each delta-function, except for the one associated with the overall momentum conservation, decreases the actual integration number by one. Thus we have also the following relation between the number of loops L and the number of vertices V

$$L = I - (V - 1). \tag{5.136}$$

Using Eqs.(5.135) and (5.136) to eliminate L and I in Eq.(5.134), we have

$$D = n + \left(\frac{n(p-2)}{2} - p\right) V - \left(\frac{n}{2} - 1\right) E. \tag{5.137}$$

In a perturbation expansion, we increase the number of vertices V to get the high orders of perturbation expansion. When the factor before V in Eq.(5.137) is positive, the degree of divergence becomes larger with the increasing order of perturbation. We denote the factor before V in Eq.(5.137) by v

$$v = \frac{n(p-2)}{2} - p. \tag{5.138}$$

When $v > 0$, we have an infinite number of divergent terms. For four-dimensional spacetime ($n = 4$), if $p > 4$ (such as $\phi^5, \phi^6, \cdots$), the expansion terms are more and more divergent. $p = 4$ is the special case with $v = 0$. When $v = 0$, the degree of divergence $D = 4 - E$ is independent of V. All are divergent in the same manner, which makes the divergent parts cancel out possible. We will show that the renormalization procedure can remove the divergent part. Since $D = 4 - E$, there are only two types of divergent functions, the four-point functions with $E = 4$ and the two-point functions with $E = 2$.

When $n > 4$, there is no even integer p which gives $v \leq 0$. This poses a strong limitation on the dimension of spacetime in which the particles can have interactions that are renormalizable.

5.5 Dimensional regularization

In order to separate the divergent parts from the convergent parts, we need to introduce a parameter that could measure and remove the divergence.

This is called *regularization*. There are two important types of regularizations, one is the *Pauli-Villars regularization* and the other the *dimensional regularization*. The Pauli-Villars regularization uses the parameter Λ - the upper bound for the integrations over $|\mathbf{q}|$. The dimensional regularization uses the parameter $\varepsilon = 4 - n$. Both regularizations are equivalent. In the following, we will use the dimensional regularization.

Now we consider the action S in n-dimensional spacetime.

$$S = \int \mathcal{L} d^n x. \tag{5.139}$$

Since S is a dimensionless quantity. $\mathcal{L}$ must have the dimension l^{-n} with l as length dimension. Thus we have $[\phi] = l^{1-n/2}$, $[g] = l^{-(n+p(1-n/2))}$ and $[m] = l^{-1}$. In the renormalization procedure, we hope to keep the coupling constant g to be dimensionless. We add a mass factor to the ϕ^4 and rewrite the Lagrangian density with the interaction term as follows

$$\mathcal{L} = \mathcal{L}_0 - \frac{g}{4!} \mu^{4-n} \phi^4, \tag{5.140}$$

where μ is an arbitrary mass.

5.5.1   Two-point function

For the two point function, the divergent integration is contained in the self-energy term. We consider the lowest order term of the self-energy Σ, which is given by the tadpole diagram. Similar to Eq.(5.131), which gives the corresponding Σ in the four-dimensional spacetime, we have

$$\Sigma = \frac{g}{2} \mu^{4-n} \int \frac{d^n q}{(2\pi)^n} \frac{i}{q^2 - m^2 + i\epsilon}. \tag{5.141}$$

The integral in Eq.(5.141) can be evaluated using Eq.(F.35) in Appendix F. We have

$$\Sigma = \frac{g}{2} \mu^{4-n} \frac{1}{(2\pi)^n} m^{n-2} \pi^{\frac{n}{2}} \Gamma\left(1 - \frac{n}{2}\right). \tag{5.142}$$

The Γ-functions with negative integers as variable have poles at 0. Thus the expression of Σ is divergent at $n = 4$. In order to separate the divergent terms, we expand Γ around the pole. We have

$$\begin{aligned}
\Gamma\left(1 - \frac{n}{2}\right) &= \Gamma\left(-1 + \frac{\epsilon}{2}\right) \\
&= -\frac{2}{\epsilon} - 1 + \gamma + \mathcal{O}(\epsilon),
\end{aligned} \tag{5.143}$$

where $\gamma \approx 0.577$ is the Euler-Mascheroni constant. Inserting the expansion of Γ into Eq.(5.142), we have

$$
\begin{aligned}
\Sigma &= \frac{g}{2} \frac{\mu^\epsilon}{(2\pi)^{4-\epsilon}} m^{2-\epsilon} \pi^{\frac{4-\epsilon}{2}} \left(-\frac{2}{\epsilon} - 1 + \gamma + \mathcal{O}(\epsilon) \right) \\
&= \frac{gm^2}{32\pi^2} \left(\frac{4\pi\mu^2}{m^2} \right)^{\frac{\epsilon}{2}} \left(-\frac{2}{\epsilon} - 1 + \gamma + \mathcal{O}(\epsilon) \right).
\end{aligned}
\tag{5.144}
$$

Using

$$
x^\epsilon = e^{\epsilon \ln x} \cong 1 + \epsilon \ln x,
\tag{5.145}
$$

Eq.(5.144) becomes

$$
\begin{aligned}
\Sigma &= \frac{gm^2}{32\pi^2} \left[1 + \frac{\epsilon}{2} \ln \left(\frac{4\pi\mu^2}{m^2} \right) \right] \left(-\frac{2}{\epsilon} - 1 + \gamma + \mathcal{O}(\epsilon) \right) \\
&= \frac{gm^2}{32\pi^2} \left[-\frac{2}{\epsilon} - 1 + \gamma - \ln \left(\frac{4\pi\mu^2}{m^2} \right) \right] + \mathcal{O}(\epsilon) \\
&= -\frac{gm^2}{16\pi^2} \frac{1}{\epsilon} - \frac{gm^2}{32\pi^2} \left[1 - \gamma + \ln \left(\frac{4\pi\mu^2}{m^2} \right) \right] + \mathcal{O}(\epsilon).
\end{aligned}
\tag{5.146}
$$

The self-energy Σ diverges as $\frac{1}{\epsilon}$. This divergent term has been separated from the rest convergent terms.

5.5.2 *Four-point function*

Now we consider the regularization of the four-point function. The divergent term of $\mathcal{O}(g^2)$ comes from the 1PI vertex Feynman diagrams in the second line of Fig.5.18. The three contributions can be evaluated similarly. We take the middle graph on the second line of Fig.5.18 as an example and denote the contribution as $\Delta\Gamma^{(4)}$. We evaluate the connected vertex function using Eq.(5.129). We have

$$
\begin{aligned}
\Delta\Gamma^{(4)}(p_1, p_2, p_3, p_4) &= \left(-\frac{ig}{4!} \mu^{4-n} \right)^2 \frac{(4!)^2}{2} \\
&\quad \times \int \frac{d^n q}{(2\pi)^n} \frac{i}{q^2 - m^2} \frac{i}{(p-q)^2 - m^2}
\end{aligned}
\tag{5.147}
$$

with $p = p_1 + p_3 = p_2 + p_4$.

The integration can be evaluated using several mathematical tricks. Using the integration identity

$$
\frac{1}{ab} = \int_0^1 \frac{dz}{[az + b(1-z)]^2},
\tag{5.148}
$$

we transform the integrand in Eq.(5.147) into the following form

$$
\frac{1}{q^2 - m^2} \frac{1}{(p-q)^2 - m^2}
$$

$$
= \int_0^1 \frac{dz}{\{(q^2 - m^2)z + [(p-q)^2 - m^2](1-z)\}^2} \tag{5.149}
$$

$$
= \int_0^1 \frac{dz}{[q^2 - 2pq(1-z) + p^2(1-z) - m^2]^2}.
$$

Changing the variable q by $q' = q - p(1-z)$ in the integration of Eq.(5.147), we have

$$
\Delta\Gamma^{(4)}(p_1, p_2, p_3, p_4)
$$

$$
= \frac{1}{2}g^2\mu^{2(4-n)} \int \frac{d^n q'}{(2\pi)^n} \int_0^1 \frac{dz}{[q'^2 - m^2 + sz(1-z)]^2} \tag{5.150}
$$

with $s = p^2 = (p_1 + p_3)^2$.

We interchange the order of the integration over q' with that over z. The integration over q' can be evaluated in a similar way with that used for Eq.(5.141). We have

$$
\int \frac{d^n q'}{(2\pi)^n} \frac{1}{[q'^2 - m^2 + sz(1-z)]^2}
$$

$$
= \frac{i}{(2\pi)^n}[m^2 - sz(1-z)]^{\frac{n-4}{2}} \pi^{\frac{n}{2}} \frac{\Gamma(2 - \frac{n}{2})}{\Gamma(2)}. \tag{5.151}
$$

The four-point function becomes

$$
\Delta\Gamma^{(4)}(p_1, p_2, p_3, p_4)
$$

$$
= \frac{1}{2}g^2\mu^{2(4-n)}\frac{i\pi^{\frac{n}{2}}\Gamma(2 - \frac{n}{2})}{(2\pi)^n} \int_0^1 dz[m^2 - sz(1-z)]^{\frac{n-4}{2}} \tag{5.152}
$$

$$
= ig^2\mu^\epsilon \frac{1}{32\pi^2}\Gamma\left(\frac{\epsilon}{2}\right) \int_0^1 dz \left[\frac{m^2 - sz(1-z)}{4\pi\mu^2}\right]^{-\frac{\epsilon}{2}}.
$$

The integral in Eq.(5.152) is convergent. The divergent part is contained in $\Gamma(\frac{\epsilon}{2})$. Using Eq.(5.145) and $\Gamma(\frac{\epsilon}{2}) = \frac{2}{\epsilon} - \gamma + \mathcal{O}(\epsilon)$, we have

$$
\Delta\Gamma^{(4)}(p_1, p_2, p_3, p_4)
$$

$$
= g^2 \frac{i\mu^\epsilon}{32\pi^2}\left[\frac{2}{\epsilon} - \gamma + \mathcal{O}(\epsilon)\right]\left[1 - \frac{\epsilon}{2}\int_0^1 dz\left(\frac{m^2 - sz(1-z)}{4\pi\mu^2}\right)\right] \tag{5.153}
$$

$$
= g^2 \frac{i\mu^\epsilon}{16\pi^2}\frac{1}{\epsilon} - g^2 \frac{i\mu^\epsilon}{32\pi^2}\left[\gamma + \int_0^1 dz\left(\frac{m^2 - sz(1-z)}{4\pi\mu^2}\right)\right] + \mathcal{O}(\epsilon).
$$

In Eq.(5.153), we have separated the four-point function into a divergent part and a convergent part. The integral is a function of p^2, m^2 and μ^2, which will be denoted as $\Gamma(s, m, \mu)$,

$$\Gamma(s, m, \mu) = \int_0^1 dz \left(\frac{m^2 - sz(1 - z)}{4\pi\mu^2} \right). \qquad (5.154)$$

The results for the middle graph on the second line in Fig.5.18 can be used for the other two graphs on the second line in Fig.5.18 with appropriate replacement of the momentums at external vertices. We introduce three Lorentz invariant Mandelstam variables $s, t,$ and u

$$s = (p_1 + p_3)^2 = p^2, \qquad (5.155a)$$

$$t = (p_1 + p_2)^2, \qquad (5.155b)$$

$$u = (p_1 + p_4)^2. \qquad (5.155c)$$

These variables are responsible to the variable change in the last summation of Eq.(5.129). Thus the summation (denoted as $\Gamma_1^{(4)}$) of the contributions from all the three diagrams in the second line of Fig.5.18 is given by

$$\begin{aligned}
&\Gamma_1^{(4)}(p_1, p_2, p_3, p_4) \\
&= \frac{3ig^2\mu^\epsilon}{16\pi^2} \frac{1}{\epsilon} - \frac{ig^2\mu^\epsilon}{32\pi^2} \left[3\gamma + \Gamma(s, m, \mu) + \Gamma(t, m, \mu) + \Gamma(u, m, \mu) \right].
\end{aligned} \qquad (5.156)$$

In Eq.(5.156), the vertex correction $\Gamma_1^{(4)}$ has been split up into a divergent and a convergent part.

Summing with the zeroth order term, the total 1PI four-point function is given by

$$\begin{aligned}
\Gamma^{(4)}(p_1, p_2, p_3, p_4) &= -ig\mu^\epsilon + \Gamma_1^{(4)}(p_1, p_2, p_3, p_4) \\
&= -ig\mu^\epsilon + \frac{3ig^2\mu^\epsilon}{16\pi^2} \frac{1}{\epsilon} - \frac{ig^2\mu^\epsilon}{32\pi^2} [3\gamma \\
&\quad + \Gamma(s, m, \mu) + \Gamma(t, m, \mu) + \Gamma(u, m, \mu)] \qquad (5.157) \\
&= -ig\mu^\epsilon \left\{ 1 - g \left[\frac{3}{16\pi^2} \frac{1}{\epsilon} - \frac{1}{32\pi^2} [3\gamma + \Gamma(s, m, \mu) \right. \right. \\
&\quad \left. \left. + \Gamma(t, m, \mu) + \Gamma(u, m, \mu)] \right] \right\}.
\end{aligned}$$

We can define an effective coupling constant $\tilde{g}$ by

$$g \left(1 - \frac{\Gamma_1^{(4)}}{ig\mu^\epsilon} \right) \to \tilde{g}. \qquad (5.158)$$

In terms of $\tilde{g}$, we can express the vertex function as

$$\Gamma^{(4)}(p_1, p_2, p_3, p_4) = -i\tilde{g}\mu^\epsilon. \qquad (5.159)$$

5.6   Renormalization for scalar field

In the mass and interaction terms in Eq.(5.85), there are two parameters m and g, which can be changed without changing the total Lagrangian density in Eq.(3.38). We can change the parameters in the mass and interaction terms to obtain a physical convergent results. This process is called the renormalization procedure. In order to eliminate the divergence, we add the counter terms

$$\mathcal{L}_{counter} = -\frac{1}{2}\delta m^2 \phi^2 - \frac{g\mu^\epsilon}{4!}(Z_g Z^{-2} - 1)\phi^4. \tag{5.160}$$

We also make a transformation

$$\phi \to \sqrt{Z}\phi, \tag{5.161}$$

which is called the wave function renormalization. Eq.(5.161) is equivalent to an counter term $\frac{1}{2}(Z-1)[(\partial\phi)^2 - m^2\phi^2]$ for the kinetic and mass terms. Eqs.(5.160) and (5.161) together are equivalent to a counter Lagrangian density

$$\mathcal{L}'_{counter} = \frac{1}{2}(Z-1)[(\partial\phi)^2 - m^2\phi^2] - \frac{1}{2}\delta m^2 Z\phi^2 - \frac{g\mu^\epsilon}{4!}(Z_g - 1)\phi^4, \tag{5.162}$$

where Z, δm^2 and Z_g are the renormalization parameters determined in the following way.

The general form of the two-point function for the original Lagrangian is given by

$$G_c(p, -p) = \frac{1}{p^2 - m^2 - \Sigma + i\epsilon}. \tag{5.163}$$

The self-energy Σ is a function of the momentum p and can be expanded around the on-shell point $p^2 = m^2$. m is the physical observable mass.

$$\Sigma(p^2) = \Sigma(m^2) + (p^2 - m^2)\Sigma_1 + \Sigma_2(p^2), \tag{5.164}$$

where

$$\Sigma_1 = \left.\frac{\partial\Sigma}{\partial p^2}\right|_{p^2=m^2}. \tag{5.165}$$

In Eq.(5.164), we have written up the first two terms of the expansion around $p^2 = m^2$ explicitly. The remainder of the expansion is put into the $\Sigma_2(p^2)$ term with $\Sigma_2(m^2) = 0$.

We define

$$\delta m^2 \equiv \Sigma(m^2) \tag{5.166}$$

and

$$Z \equiv \frac{1}{1 - \Sigma_1}. \tag{5.167}$$

In terms of $\Gamma_1^{(4)}$ determined by Eq.(5.156), we define

$$Z_g \equiv \left(1 - \frac{\Gamma_1^{(4)}}{ig\mu^\epsilon}\right)_r^{-1} = (1 - g\delta\Gamma)_r^{-1} \tag{5.168}$$

with

$$\delta\Gamma \equiv \frac{\Gamma_1^{(4)}}{ig^2\mu^\epsilon}, \tag{5.169}$$

where r denotes the so-called symmetric point. At r,

$$p_i \cdot p_j = m^2 \left(\frac{4}{3}\delta_{ij} - \frac{1}{3}\right) \tag{5.170}$$

with $i, j = 1, \cdots, 4$ denoting the external lines. At the point r, all particles are on shell with $s = t = u = 4m^2/3$.

From the self-energy expansion up to the one-loop diagram,

$$\Sigma(p^2) = -g\frac{m^2}{16\pi^2}\frac{1}{\epsilon} - g\frac{m^2}{32\pi^2}\left[1 - \gamma + \ln\left(\frac{4\pi\mu^2}{m^2}\right)\right] + \mathcal{O}(\epsilon), \quad (5.171)$$

we have

$$\Sigma(m^2) = -g\frac{m^2}{16\pi^2}\frac{1}{\epsilon} - g\frac{m^2}{32\pi^2}\left[1 - \gamma + \ln\left(\frac{4\pi\mu^2}{m^2}\right)\right] + \mathcal{O}(\epsilon), \quad (5.172a)$$

$$\Sigma_1 = \Sigma_2 = 0. \tag{5.172b}$$

We can see that the divergent term is contained in the $\Sigma(m^2)$ term. In the one-loop expansion, we have $\Sigma_1 = 0$ and thus $Z = 1$. But in the two-loop approximation, it can be shown that $\Sigma_1 \neq 0$ because the graph for the two-loop approximation is p-dependent. Thus in general $Z \neq 1$.

Using the renormalized Lagrangian density

$$\mathcal{L}_r = \frac{1}{2}Z[(\partial\phi)^2 - (m^2 + \delta m^2)\phi^2] - \frac{g\mu^\epsilon}{4!}Z_g\phi^4, \tag{5.173}$$

we have the two-point function

$$G(p, -p) =$$

$$\frac{i}{Z(p^2 - m^2) - Z\Sigma(m^2) - Z(p^2 - m^2)\Sigma_1 - Z\Sigma_2(p^2) + \delta m^2 Z + i\epsilon} \tag{5.174}$$

$$= \frac{i}{p^2 - m^2 - Z\Sigma_2(p^2) + i\epsilon}.$$

Since $\Sigma_2(m^2) = 0$, $G(p, -p)$ has the pole at the physical mass m with residue i. $G(p, -p)$ has no those divergent quantities contained in $\Sigma(m^2)$ and Σ_1. The new effective coupling $\tilde{g}$ defined by Eq.(5.158) is given by

$$
\begin{aligned}
\tilde{g} &= gZ_g[1 - gZ_g\delta\Gamma(s,t,u)] \\
&= \frac{g}{1 - g\delta\Gamma(r)}\left[1 - \frac{g}{1 - g\delta\Gamma(r)}\delta\Gamma(s,t,u)\right] \\
&= g\{1 - g[\delta\Gamma(s,t,u) - \delta\Gamma(r)]\} + \mathcal{O}(g^2).
\end{aligned}
\tag{5.175}
$$

Since the divergent term $\sim \frac{1}{\epsilon}$ does not depend on the variables s, t, u, the substraction $\delta\Gamma(s,t,u) - \delta\Gamma(r)$ removes the divergent parts. $\tilde{g}$ does not contain the divergent term $\sim \frac{1}{\epsilon}$ and is thus finite. We obtain the renormalized 1PI four-point function as

$$
\begin{aligned}
\Gamma^{(4)}&(p_1, p_2, p_3, p_4) \\
&= -ig\left\{1 - \frac{g}{32\pi^2}\left[G(s, m^2) + G(t, m^2) + G(u, m^2)\right.\right. \\
&\qquad \left.\left. - 3G\left(\frac{3}{4}m^2, m^2\right)\right]\right\} + \mathcal{O}(g^2)
\end{aligned}
\tag{5.176}
$$

with

$$
G(s, m^2) = \int_0^1 \ln(m^2 - sz(1 - z))dz.
\tag{5.177}
$$

$\Gamma^{(4)}(p_1, p_2, p_3, p_4)$ does not contain the divergent term.

We can introduce the bare field ϕ_0, bare mass m_0 and bare coupling constant g_0 to simplify the expression of the renormalized Lagrangian. We define

$$
\phi_0 \equiv \sqrt{Z}\phi,
\tag{5.178a}
$$

$$
m_0^2 \equiv m^2 + \delta m^2,
\tag{5.178b}
$$

$$
g_0 \equiv g\mu^\epsilon\frac{Z_g}{Z^2}.
\tag{5.178c}
$$

Then we can express the complete Lagrangian density in terms of the bare quantities by

$$
\mathcal{L}_b = \frac{1}{2}[(\partial\phi_0)^2 - m_0^2\phi_0^2] - \frac{g_0}{4!}\phi_0^4.
\tag{5.179}
$$

This bare Lagrangian density has the same form as the original one and leads to the finite physical quantities.

5.7 Effective potential

Due to the renormalization, the emergent values or the measured values of physical quantities are different with the bare values in the original Lagrangian. Although the relation between the measured values and the bare values are complicated. We can use the effective potential to simplify the relation.

As an example, we consider the case of a scalar boson field. The underlining principle is the same and can be applied for all other fields. For a scalar boson field, the Lagrangian density is given by

$$\mathcal{L} = \frac{1}{2}\partial_\mu\phi\partial^\mu\phi - V(\phi) \tag{5.180}$$

with

$$V(\phi) = \frac{1}{2}m_0^2\phi^2 + \frac{g_0}{4!}\phi^4. \tag{5.181}$$

We consider the calculations of the corresponding classical field. Since the loop expansion is an expansion in $\hbar$, we consider the loop expansion. The physical classical quantities are related to the renormalized quantities. The renormalized mass m is given by

$$m^2 = -i\Gamma_{amp}^{(2)}(0) \tag{5.182}$$

and the renormalized coupling constant g is related to the vertex function $\Gamma^{(4)}$ by

$$g = i\Gamma_{amp}^{(4)}(p_i = 0). \tag{5.183}$$

The classical field ϕ_c is defined as expectation value $\langle\phi\rangle$. We will show that ϕ_c obeys the Euler-Lagrange equation with the Lagrangian whose parameters are the renormalized ones. The connected generating functional W is given by

$$e^{iW[J]} = \frac{\langle 0^+|0^-\rangle_J}{\langle 0^+|0^-\rangle_0}. \tag{5.184}$$

Thus the classic field ϕ_c is related to the connected generating functional by

$$\phi_c(x) = \frac{\langle 0^+|\phi(x)|0^-\rangle_J}{\langle 0^+|0^-\rangle_J}$$
$$= \frac{\delta W[J]}{\delta J(x)}. \tag{5.185}$$

ϕ_c depends on the source $J(x)$. The vacuum expectation value $\langle \phi \rangle_0$ is given by

$$\langle \phi \rangle_0 = \lim_{J \to 0} \phi_c. \tag{5.186}$$

We introduce a vertex function $\Gamma[\phi_c]$ which is related to the connected generating functional by

$$\Gamma[\phi_c] \equiv W[J] - \int d^4 x J(x) \phi_c(x). \tag{5.187}$$

Eq.(5.187) gives

$$\frac{\delta \Gamma[\phi_c]}{\delta \phi_c(x)} = -J(x). \tag{5.188}$$

When $J(x) \to 0$, ϕ_c is a constant. According to Eq.(5.187), ϕ_c becomes the solution of the equation

$$\left. \frac{d\Gamma[\phi]}{d\phi_c} \right|_{\phi_c} = 0. \tag{5.189}$$

For a classical system, the vacuum state $|0\rangle$ should be replaced by a state which has the expectation value of constant in microscopic scale and depend on position in macroscopic scale. The dependence of ϕ_c on the coordinates can be resulted from the boundary condition for a finite system. Then Eq.(5.188) has the form

$$\frac{\delta \Gamma[\phi_c]}{\delta \phi_c(x)} = 0. \tag{5.190}$$

$\Gamma[\phi_c]$ can be expanded in ϕ_c as

$$\Gamma[\phi_c] = \sum_{n=0}^{\infty} \frac{1}{n!} \int d^4 x_1 \cdots d^4 x_n \Gamma^{(n)}(x_1, \cdots, x_n) \phi_c(x_1) \cdots \phi_c(x_n). \tag{5.191}$$

After Fourier transformation, we have

$$\Gamma[\phi_c] = \sum_{n=0}^{\infty} \frac{1}{n!} \int d^4 p_1 \cdots d^4 p_n \delta^4(p_1 + \cdots + p_n)$$
$$\Gamma^{(n)}(p_1, \cdots, p_n) \tilde{\phi}_c(p_1) \cdots \tilde{\phi}_c(p_n). \tag{5.192}$$

We can also expand $\Gamma[\phi_c]$ in the Lagrangian form in terms of ϕ_c and its derivatives as follows:

$$\Gamma[\phi_c] = \int d^4 x \left[-U(\phi_c(x)) + \frac{1}{2}(\partial_\mu \phi_c)^2 + \cdots \right]$$
$$\equiv \int d^4 x \mathcal{L}_c, \tag{5.193}$$

where $U(\phi_c(x))$ is called the *effective potential* and $\Gamma[\phi_c]$ is thus also called the *effective action*. Inserting Eq.(5.193) into Eq.(5.190), we have

$$\frac{\partial \mathcal{L}_c}{\partial \phi_c} - \frac{\partial}{\partial x^\mu}\left(\frac{\partial \mathcal{L}_c}{\partial(\partial_\mu \phi_c)}\right) = 0, \tag{5.194}$$

which is the Euler-Lagrange equation.

Now we discuss the relation of $\Gamma^{(n)}$ in Eq.(5.192) with the amputated Green's function $\Gamma^{(n)}_{amp}$. According to Eq.(5.116), the amputated Green's function is defined by

$$\begin{aligned}
&\Gamma^{(n)}_{amp}(p_1, p_2, \cdots, p_n) \\
&= [G_c^{-1}(p_1, -p_1) G_c^{-1}(p_2, -p_2) \cdots G_c^{-1}(p_n, -p_n) G_c(p_1, p_2, \cdots, p_n).
\end{aligned} \tag{5.195}$$

In terms of the spacetime coordinates, we have

$$\begin{aligned}
&\Gamma^{(n)}_{amp}(x_1, x_2, \cdots, x_n) \\
&= \int d^4 y_1 \int d^4 y_2 \cdots \int d^4 y_n G_c^{(n)}(y_1, y_2, \cdots, y_n) \\
&\quad \times [G_c^{(2)}(y_1 - x_1)]^{-1}[G_c^{(2)}(y_2 - x_2)]^{-1} \cdots [G_c^{(2)}(y_n, x_n)]^{-1}.
\end{aligned} \tag{5.196}$$

To simplify the notation, we rewrite Eq.(5.196) in a compact form

$$\Gamma^{(n)}_{amp} = G_c^{(n)}(G_c^{(2)})^{-n}. \tag{5.197}$$

The generating functional $\Gamma_{amp}[J]$ of the amputated Green's function is defined by

$$\begin{aligned}
\Gamma_{amp}[J] &\equiv \sum_n \frac{1}{n!}\Gamma^{(n)}_{amp} J^n \\
&= \sum_n \frac{1}{n!} G_c^{(n)}(G_c^{(2)})^{-n} J^n \\
&= W[(G_c^{(2)})^{-1} J]
\end{aligned} \tag{5.198}$$

or

$$\Gamma_{amp}[G_c^{(2)} J] = W[J]. \tag{5.199}$$

Thus $W[J]$ is also the generating functional for the amputated Green's function.

Using Eq.(5.191), we have

$$\Gamma^{(n)}(x_1, x_2, \cdots, x_n) = \left.\frac{\delta^n \Gamma(\phi)}{\delta\phi(x_1) \cdots \delta\phi(x_n)}\right|_{\phi=\phi_c}. \tag{5.200}$$

According to Eq.(5.184), $W[0] = 0$. Then we have

$$\Gamma^{(0)} = 0. \tag{5.201}$$

Using Eq.(5.190), we have

$$\Gamma^{(1)} = \frac{\delta\Gamma}{\delta\phi_c} = 0. \tag{5.202}$$

Using the identity relation $\frac{\delta J}{\delta J} = 1$, we have

$$\frac{\delta J}{\delta\phi_c}\frac{\delta\phi_c}{\delta J} = -\frac{\delta^2\Gamma}{\delta\phi_c^2}\frac{\delta^2 W}{\delta J^2} = 1. \tag{5.203}$$

Taking $J = 0$, we have

$$-\Gamma^{(2)}(iG_c^{(2)}) = 1. \tag{5.204}$$

or

$$\Gamma^{(2)} = i(G_c^{(2)})^{-1} = i\Gamma_{amp}^{(2)}. \tag{5.205}$$

Taking the functional derivative $\frac{\delta}{\delta\phi_c}$ over Eq.(5.203), we have

$$\frac{\delta^3\Gamma}{\delta\phi_c^3}\frac{\delta^2 W}{\delta J^2} - \frac{\delta^2\Gamma}{\delta\phi_c^2}\frac{\delta^3 W}{\delta J^3}\frac{\delta^2\Gamma}{\delta\phi_c^2} = 0. \tag{5.206}$$

Multiplying Eq.(5.206) with $\frac{\delta^2\Gamma}{\delta\phi_c^2}$ and using Eq.(5.203), we have

$$\frac{\delta^3\Gamma}{\delta\phi_c^3} = -\frac{\delta^3 W}{\delta J^3}\left(\frac{\delta^2\Gamma}{\delta\phi_c^2}\right)^3. \tag{5.207}$$

Taking $J = 0$, we have

$$\Gamma^{(3)} = -iG_c^{(3)}(G_c^{(2)})^{-3} = -i\Gamma_{amp}^{(3)}. \tag{5.208}$$

Similarly, we can obtain the following relation by further taking the functional derivative $\frac{\delta}{\delta\phi_c}$.

$$\frac{\delta^4\Gamma}{\delta\phi_c^4} = \frac{\delta^4 W}{\delta J^4}\left(\frac{\delta^2\Gamma}{\delta\phi_c^2}\right)^4 - 3\left(\frac{\delta^3 W}{\delta J^3}\right)^2\left(\frac{\delta^2\Gamma}{\delta\phi_c^2}\right)^5. \tag{5.209}$$

Taking $J = 0$, we have

$$\Gamma^{(4)} = -i\Gamma_{amp}^{(4)} - i3\Gamma_{amp}^{(3)}\Gamma_{amp}^{(3)}. \tag{5.210}$$

For the case of ϕ^4 potential, we have

$$\Gamma^{(4)} = -i\Gamma_{amp}^{(4)}. \tag{5.211}$$

Since $\Gamma^{(n)}$ has only minor difference with $\Gamma_{amp}^{(n)}$, we often do not distinguish them and use the same name to call them.

When $\phi_c = a$ is a constant in the case of $J(x) = 0$, using Eq.(5.193), we have

$$\Gamma[a] = -\Omega U(a), \tag{5.212}$$

where Ω is the total volume of the spacetime. Comparing with Eq.(5.192), we have

$$U(a) = -\sum_{n=0}^{\infty} \frac{1}{n!} a^n \Gamma^{(n)}(p_i = 0).$$ (5.213)

The relations to the renormalized quantities now read

$$\left. \frac{d^2 U(\phi_c)}{d\phi_c^2} \right|_{\phi_c} = m^2,$$ (5.214a)

$$\left. \frac{d^4 U(\phi_c)}{d\phi_c^4} \right|_{\phi_c} = g.$$ (5.214b)

Eq.(5.189) for the vacuum expectation value becomes

$$\left. \frac{dU(\phi_c)}{d\phi_c} \right|_{\phi_c} = 0.$$ (5.215)

Inserting Eq.(5.214) into Eq.(5.193), we have

$$\Gamma[\phi_c] = \int d^4 x \left[\frac{1}{2} (\partial_\mu \phi_c)^2 - m^2 \phi_c^2 - \frac{g}{4!} \phi_c^4 \right].$$ (5.216)

Using the Euler-Lagrange equation Eq.(5.194) and neglecting the interaction term, we have the classical Klein-Gordon field equation with renormalized mass

$$(\Box + m^2)\phi_c(x) = 0.$$ (5.217)

We can also construct the classical Lagrangian density $\mathcal{L}$ directly using the renormalized quantities.

$$\mathcal{L}_c = \frac{1}{2} \partial_\mu \phi \partial^\mu \phi - V(\phi)$$ (5.218)

with

$$V(\phi) = \frac{1}{2} m^2 \phi^2 + \frac{g}{4!} \phi^4.$$ (5.219)

The classical action is given by

$$S_c[\phi] = \int d^4 x \mathcal{L}_c.$$ (5.220)

We choose a constant source function J to give a constant average field. Now we calculate $W[J]$ by the saddle-point approximation of path integral, which is also called the stationary phase approximation or the classical approximation.

$W[J]$ is calculated by

$$e^{\frac{i}{\hbar}W[J]} = \int D\phi \, e^{\frac{i}{\hbar}S_c[\phi,J]}, \tag{5.221}$$

where

$$S_c[\phi, J] = \int d^4x [\mathcal{L}_c + \phi(x)J(x)]. \tag{5.222}$$

We have used Planck constant $\hbar$ explicitly because we will use saddle-point approximation for the calculations of the classical case in which the action is much larger than $\hbar$ and $\hbar$ can be considered as a small quantity. The saddle-point position is determined by

$$\left. \frac{\delta S_c[\phi, J]}{\delta\phi(x)} \right|_{\phi_0} = -J(x). \tag{5.223}$$

Expanding the action around ϕ_0 gives

$$\begin{aligned}
S_c[\phi, J] =& S_c[\phi_0, J] + \int d^4x [\phi(x) - \phi_0] \left. \frac{\delta S_c}{\delta\phi(x)} \right|_{\phi_0} \\
&+ \frac{1}{2} \int d^4x d^4y [\phi(x) - \phi_0][\phi(y) - \phi_0] \left. \frac{\delta^2 S_c}{\delta\phi(x)\delta\phi(y)} \right|_{\phi_0} + \cdots \\
=& S_c[\phi_0] - \int d^4x \phi(x) J(x) \\
&+ \frac{1}{2} \int d^4x d^4y [\phi(x) - \phi_0] \left. \frac{\delta^2 S_c}{\delta\phi(x)\delta\phi(y)} \right|_{\phi_0} [\phi(y) - \phi_0] + \cdots .
\end{aligned} \tag{5.224}$$

Performing the functional differentiation of the action gives

$$\left. \frac{\delta^2 S_c}{\delta\phi(x)\delta\phi(y)} \right|_{\phi_0} = -[\Box + V''(\phi_0)]\delta(x - y). \tag{5.225}$$

Substituting Eq.(5.224) into Eq.(5.221) and performing the functional integration, we have

$$e^{\frac{i}{\hbar}W[J]} = e^{\frac{i}{\hbar}S_c[\phi_0,J]} \{ \det[\Box + V''(\phi_0)] \}^{-\frac{1}{2}}. \tag{5.226}$$

Using the relation

$$\det A = e^{\text{Tr} \ln A}, \tag{5.227}$$

we have

$$W[J] = S_c[\phi_0] + \int d^4x \phi_0(x) J(x) + \frac{i\hbar}{2} \text{Tr} \ln[\Box + V''(\phi_0)]. \tag{5.228}$$

We express ϕ_0 in terms of ϕ_c. Denoting $\phi_1 \equiv \phi_c - \phi_0$, we have

$$S_c[\phi_0] = S_c[\phi_c - \phi_1]$$

$$= S_c[\phi_c] - \int d^4x \phi_1(x) \left. \frac{\delta S_c}{\delta \phi(x)} \right|_{\phi_0} + \cdots \qquad (5.229)$$

$$= S_c[\phi_c] + \int d^4x \phi_1(x) J(x) + \cdots .$$

Using Eqs.(5.228) and (5.229), we calculate $\Gamma[\phi_c]$ in Eq.(5.187).

$$\begin{aligned}
\Gamma[\phi_c] =& S_c[\phi_0] + \int d^4x \phi_0(x) J(x) + \frac{i\hbar}{2} \mathrm{Tr} \ln[\Box + V''(\phi_0)] \\
& - \int d^4x J(x) \phi_c(x) \\
=& S_c[\phi_0] - \int d^4x \phi_1(x) J(x) + \frac{i\hbar}{2} \mathrm{Tr} \ln[\Box + V''(\phi_0)] \\
=& S_c[\phi_c] + \frac{i\hbar}{2} \mathrm{Tr} \ln[\Box + V''(\phi_0)].
\end{aligned} \qquad (5.230)$$

When the source field is a constant, which is valid in microscopic scale, we have

$$\phi_c(x) = a. \qquad (5.231)$$

Thus we have

$$S_c[a] = -\Omega V(a), \qquad (5.232)$$

which gives

$$U(a) = V(a) - \frac{i\hbar}{2} \Omega^{-1} \mathrm{Tr} \ln[\Box + V''(a)]. \qquad (5.233)$$

In the classical limit, $\hbar \to 0$, which corresponds to the tree approximation. Eqs.(5.230) and (5.233) show that the effective action $\Gamma[a]$ becomes the same as the classical action and the effective potential $U(a)$ becomes the same as the classical potential. Eq.(5.190) with the $\Gamma[\phi_c]$ expanded in the Lagrangian form in Eq.(5.193) is equivalent to the Euler-Lagrange equation. Since the effective action is the same as the classical action, we can use the classical Lagrangian density in the Euler-Lagrange equation.

Since the divergence comes from the large k, the renomalization procedure can remove the integration over large k, which gives an effective field and effective Lagrangian with effective potential in low energy. The effective Lagrangian in low energy is applicable in quantum mechanics. In the next section, we will further explain the meaning of the renormalization procedure.

For the non-vacuum case, we shall use the Riemann spacetime. We can first carry out the renormalization procedure in the local flat metric approximately, which is feasible because the divergence comes from the large **k** which is effective locally. The renomalization procedure gives the effective field for the effective potential. Then we use the effective field and effective potential in the action in the Riemann spacetime. Thus we have the effective total action for the effective field in the Riemann spacetime, which is invariant under an infinitesimal spacetime translation. Similar to the procedure leading to Eq.(3.23), we obtain the classical Einstein equations. Since the gravitational constant G is renormalized by the renormalization constants, the renormalized gravitational constant is correlated with the parameters in the ground state of the total Lagrangian.

5.8   Reduction theory

5.8.1   *Quasi-particles*

A stable state should have a definite energy. A point-like particle does not have a definite energy. Thus a stable state can not be a pure point-like particle state. For a free field, a quanta generated by $\hat{a}^\dagger(p)$ has a definite energy. Its state is a plane wave. Such a state(quanta) is usually called quasi-particle state which is non-local. Usually, we refer particles to both the basic point-like particles and quasi-particle states. However, it is conceptually important to distinguish the pure point-like particles and quasi-particles. For a free field, it is not always necessary to distinguish the point-like particle state and quasi-particle state. They can be considered as particle states in real space and momentum space, respectively. For an interacting field, it is important not only conceptually but also technically to distinguish the pure point-like particles and quasi-particles. Especially, the experimentally measured properties are those of quasi-particles, which are the renormalized quantities. That is why we need the techniques of renormalization and the S matrix to deal with the quantum fields with interactions. Since we will deal with both the quasi-particles of free field and the quasi-particles of interacting field in the following sections, for the simplicity of language, the quasi-particle state of free field is simply called particle state and quasi-particle state is used only for the interacting field.

5.8.2 *Scattering matrix*

In the quantum field theory with interaction, one particle state is not the energy eigenstate and has not a definite single-particle energy in general. However, the concept of single-particle state is useful and widely used in particle physics and quantum mechanics. In particular, experimentally, the measured properties are mostly those of single quasi-particle states. In order to use single-particle states to describe the physics of systems, we need an equivalent formulation in terms of single-particle states for the quantum fields with interaction.

For a scattering process due to interaction, the incoming state $|\alpha; \text{in}\rangle$ at $t \to -\infty$ transforms into an outgoing state $|\beta; \text{out}\rangle$ at $t \to \infty$. The incoming state $|\alpha; \text{in}\rangle$ consists of several single quasi-particle states which are separated and have small interaction. Strictly, they are quasi-particle states. The interaction can not be switched off. The system with interaction can only be approximated as a system with weakly interacted quasi-particles with renormalized particle parameters. When these single quasi-particles approach each other, they interact and are scattered into the outgoing state with the quasi-particles separated and having small interaction. The incoming state $|\alpha; \text{in}\rangle$ and the outgoing state $|\beta; \text{out}\rangle$ are called the asymptotic states. The quasi-particles are free quasi-particles only in these asymptotic states outside of the interaction range. They are described by the similar equations of motion for free particles. These particles should be stable. Thus they are quasi-particle states with α (or β) characterized by the single-particle energies and momentums. We introduce the S matrix defined by

$$S_{\beta\alpha} \equiv \langle \beta; \text{out}|\alpha; \text{in}\rangle. \tag{5.234}$$

to describe the scattering process. The S matrix is the transition amplitude for an incoming state $|\alpha; \text{in}\rangle$ to be scattered into the outgoing state $|\beta; \text{out}\rangle$. Correspondingly, we can also define an operator $\hat{S}$ related to the S matrix by

$$\langle \beta; \text{out}| = \langle \beta; \text{in}|\hat{S}, \tag{5.235}$$

which transforms the incoming state $\langle \beta; \text{in}|$ into the outgoing state $\langle \beta; \text{out}|$. Then we have

$$S_{\beta\alpha} = \langle \beta; \text{in}|\hat{S}|\alpha; \text{in}\rangle. \tag{5.236}$$

Since

$$\langle \beta; \text{in}|\hat{S}\hat{S}^\dagger|\alpha; \text{in}\rangle = \langle \beta; \text{out}|\alpha; \text{out}\rangle = \delta_{\alpha\beta}, \tag{5.237}$$

we have

$$\hat{S}\hat{S}^{\dagger} = 1. \tag{5.238}$$

Eq.(5.238) shows that $\hat{S}$ is a unitary operator. We can also express $S_{\beta\alpha}$ in terms of the time translation operator $\hat{O}$ defined by Eq.(2.85). Using Eq.(2.94), we can express the transition amplitude as

$$\langle \beta; \text{out}, t | \alpha; \text{in}, t' \rangle = \langle \beta; \text{out} | e^{-i(t'-t)\hat{G}_t} | \alpha; \text{in} \rangle. \tag{5.239}$$

We introduce

$$U(t, t') = e^{-i(t'-t)\hat{G}_t}. \tag{5.240}$$

Then Eq.(5.239) becomes

$$\langle \beta; \text{out}, t | \alpha; \text{in}, t' \rangle = \langle \beta; \text{out} | U(t, t') | \alpha; \text{in} \rangle. \tag{5.241}$$

In particular, when $t' \to -\infty$ and $t \to +\infty$, we have

$$\langle \beta; \text{out}, +\infty | \alpha; \text{in}, -\infty \rangle = \langle \beta; \text{out} | U(+\infty, -\infty) | \alpha; \text{in} \rangle. \tag{5.242}$$

5.8.3   *Quasi-particle operators for scalar fields*

Now we use $\hat{a}_{\mathbf{p},\text{in}}$ and $\hat{a}^{\dagger}_{\mathbf{p},\text{in}}$ to describe the incoming quasi-particles with momentum $\mathbf{p}$. $\hat{a}_{\mathbf{p},\text{in}}$ and $\hat{a}^{\dagger}_{\mathbf{p},\text{in}}$ satisfy the commutation relations Eqs.(2.156) and (2.157) when the quasi-particles are bosons. For fermions, the commutation relations Eqs.(2.405), (2.406) and (2.407) should be used.

First we consider the complex scalar bosons. We have

$$[\hat{a}_{\mathbf{p},\text{in}}, \hat{a}^{\dagger}_{\mathbf{p}',\text{in}}] = \delta^3(\mathbf{p} - \mathbf{p}'), \tag{5.243a}$$

$$[\hat{a}_{\mathbf{p},\text{in}}, \hat{a}_{\mathbf{p}',\text{in}}] = [\hat{a}^{\dagger}_{\mathbf{p},\text{in}}, \hat{a}^{\dagger}_{\mathbf{p}',\text{in}}] = 0, \tag{5.243b}$$

and

$$[\hat{b}_{\mathbf{p},\text{in}}, \hat{b}^{\dagger}_{\mathbf{p}',\text{in}}] = \delta^3(\mathbf{p} - \mathbf{p}'), \tag{5.244a}$$

$$[\hat{b}_{\mathbf{p},\text{in}}, \hat{b}_{\mathbf{p}',\text{in}}] = [\hat{b}^{\dagger}_{\mathbf{p},\text{in}}, \hat{b}^{\dagger}_{\mathbf{p}',\text{in}}] = 0, \tag{5.244b}$$

where $\hat{b}_{\mathbf{p},\text{in}}$ and $\hat{b}^{\dagger}_{\mathbf{p},\text{in}}$ are the annihilation and creation operators for quasi-antiparticles. The corresponding field operators are expressed as an expansion with respect to the plane-wave basis $u_{\mathbf{p}}(x)$ given by Eq.(2.158). We have

$$\hat{\phi}_{\text{in}}(\mathbf{x}, t) = \int d^3p [\hat{a}_{\mathbf{p},\text{in}} u_{\mathbf{p}}(\mathbf{x}, t) + \hat{b}^{\dagger}_{\mathbf{p},\text{in}} u^*_{\mathbf{p}}(\mathbf{x}, t)], \tag{5.245a}$$

$$\hat{\phi}^{\dagger}_{\text{in}}(\mathbf{x}, t) = \int d^3p [\hat{a}^{\dagger}_{\mathbf{p},\text{in}} u^*_{\mathbf{p}}(\mathbf{x}, t) + \hat{b}_{\mathbf{p},\text{in}} u_{\mathbf{p}}(\mathbf{x}, t)]. \tag{5.245b}$$

Since

$$(\Box + m^2)u_{\mathbf{p}}(\mathbf{x}, t) = 0, \tag{5.246}$$

the field operators satisfy the Klein-Gordon equation for noninteracting fields

$$(\Box + m^2)\hat{\phi}_{\text{in}}(x) = 0. \tag{5.247}$$

It is noted that m is the mass of quasi-particle, which is the renormalized mass. Similarly, we can introduce the operators for the outgoing quasi-particles by replacing subscript 'in' with 'out'. We have

$$[\hat{a}_{\mathbf{p},\text{out}}, \hat{a}^\dagger_{\mathbf{p}',\text{out}}] = \delta^3(\mathbf{p} - \mathbf{p}'), \tag{5.248a}$$

$$[\hat{a}_{\mathbf{p},\text{out}}, \hat{a}_{\mathbf{p}',\text{out}}] = [\hat{a}^\dagger_{\mathbf{p},\text{out}}, \hat{a}^\dagger_{\mathbf{p}',\text{out}}] = 0, \tag{5.248b}$$

$$[\hat{b}_{\mathbf{p},\text{out}}, \hat{b}^\dagger_{\mathbf{p}',\text{out}}] = \delta^3(\mathbf{p} - \mathbf{p}'), \tag{5.248c}$$

$$[\hat{b}_{\mathbf{p},\text{out}}, \hat{b}_{\mathbf{p}',\text{out}}] = [\hat{b}^\dagger_{\mathbf{p},\text{out}}, \hat{b}^\dagger_{\mathbf{p}',\text{out}}] = 0 \tag{5.248d}$$

and

$$\hat{\phi}_{\text{out}}(\mathbf{x}, t) = \int d^3p[\hat{a}_{\mathbf{p},\text{out}}u_{\mathbf{p}}(\mathbf{x}, t) + \hat{b}^\dagger_{\mathbf{p},\text{out}}u^*_{\mathbf{p}}(\mathbf{x}, t)], \tag{5.249a}$$

$$\hat{\phi}^\dagger_{\text{out}}(\mathbf{x}, t) = \int d^3p[\hat{a}^\dagger_{\mathbf{p},\text{out}}u^*_{\mathbf{p}}(\mathbf{x}, t) + \hat{b}_{\mathbf{p},\text{out}}u_{\mathbf{p}}(\mathbf{x}, t)]. \tag{5.249b}$$

The field operators $\hat{\phi}_{\text{out}}(\mathbf{x}, t)$ and $\hat{\phi}^\dagger_{\text{out}}(\mathbf{x}, t)$ satisfy the Klein-Gordon equation

$$(\Box + m^2)\hat{\phi}_{\text{out}}(x) = 0, \tag{5.250a}$$

$$(\Box + m^2)\hat{\phi}^\dagger_{\text{out}}(x) = 0. \tag{5.250b}$$

Strictly speaking, the quasi-particle states should be described by wave packets. The plane wave $u_{\mathbf{p}}(\mathbf{x}, t)$ in Eq.(5.246) should be replaced by a spatially localized wave packet $u_\alpha(\mathbf{x}, t)$. $u_\alpha(\mathbf{x}, t)$ is a complete set of the localized solutions of the Klein-Gordon equation

$$(\Box + m^2)u_\alpha(\mathbf{x}, t) = 0. \tag{5.251}$$

Then δ-function in the commutation relations for quasi-particles is replaced by a peak function which can be considered as a δ-function approximately. Since $\hat{\phi}_{\text{in(out)}}$ is the field operator for quasi-particles, $\langle\alpha|\hat{\phi}_{\text{in(out)}}(x)|\beta\rangle$ would be different with $\lim_{x_0 \to -\infty(+\infty)}\langle\alpha|\hat{\phi}(x)|\beta\rangle$. We have

$$\lim_{x_0 \to -\infty} \langle\alpha|\hat{\phi}(x)|\beta\rangle = \sqrt{Z}\langle\alpha|\hat{\phi}_{\text{in}}(x)|\beta\rangle, \tag{5.252a}$$

$$\lim_{x_0 \to +\infty} \langle\alpha|\hat{\phi}(x)|\beta\rangle = \sqrt{Z}\langle\alpha|\hat{\phi}_{\text{out}}(x)|\beta\rangle, \tag{5.252b}$$

where Z is a renormalization constant. For the free field without interaction, $Z = 1$. Eq.(5.252) is called the asymptotic conditions for the field operators. These asymptotic conditions are different with the asymptotic conditions

$$\lim_{x_0 \to -\infty} \hat{\phi}(x) = \hat{\phi}_{\text{in}}(x), \tag{5.253a}$$

$$\lim_{x_0 \to +\infty} \hat{\phi}(x) = \hat{\phi}_{\text{out}}(x) \tag{5.253b}$$

for free fields. The asymptotic conditions Eq.(5.252) for an interacting field gives weak limit conditions (also called weak operator convergence) in contrast with the ordinary limit conditions for the operators

$$\lim_{x_0 \to -\infty} \hat{\phi}(x) = \sqrt{Z}\hat{\phi}_{\text{in}}(x), \tag{5.254a}$$

$$\lim_{x_0 \to +\infty} \hat{\phi}(x) = \sqrt{Z}\hat{\phi}_{\text{out}}(x). \tag{5.254b}$$

5.8.4   *Lehmann-Källen spectral representation*

In order to understand the meaning of renormalization constant Z, we discuss the invariant function of the interacting field. We introduce the vacuum expectation of a product of field operators defined by

$$W(x - y) = \langle 0|\hat{\phi}(x)\hat{\phi}^\dagger(y)|0\rangle. \tag{5.255}$$

$W(x - y)$ is called the Wightman function. We denote $|\alpha\rangle$ as a complete set of quasi-particle states. The completeness relation is given by

$$\sum_\alpha |\alpha\rangle\langle\alpha| = 1. \tag{5.256}$$

The set contains the single-particle states $|\mathbf{p}\rangle$ and also many-particle states. Since $|\alpha\rangle$ are the quasi-particle states, they are the eigenstates of the momentum operator

$$\hat{p}_\mu|\alpha\rangle = p_\mu^{(\alpha)}|\alpha\rangle. \tag{5.257}$$

Since the energies and masses of particles are positive, we have

$$p_0^{(\alpha)} \geq 0 \quad \text{and} \quad (p_\mu^{(\alpha)})^2 \geq 0, \tag{5.258}$$

which means that these states have a space-like four-momentum.

The four-momentum operator is the generator of infinitesimal spacetime translation transformation. We have

$$\partial^\mu \hat{\phi}(x) = i[\hat{p}^\mu, \hat{\phi}(x)]. \tag{5.259}$$

Eq.(5.259) is equivalent to translation transformation

$$\hat{\phi}(x) = e^{i\hat{p}\cdot x}\hat{\phi}(0)e^{-i\hat{p}\cdot x}. \tag{5.260}$$

Using Eq.(5.260), Eq.(5.255) becomes

$$W(x-y) = \langle 0|e^{i\hat{p}\cdot x}\hat{\phi}(0)e^{-i\hat{p}\cdot x}e^{i\hat{p}\cdot y}\hat{\phi}^\dagger(0)e^{-i\hat{p}\cdot y}|0\rangle. \tag{5.261}$$

Inserting the completeness relation Eq.(5.256) into Eq.(5.261), we have

$$W(x-y) = \sum_\alpha \langle 0|e^{i\hat{p}\cdot x}\hat{\phi}(0)e^{-i\hat{p}\cdot x}|\alpha\rangle\langle\alpha|e^{i\hat{p}\cdot y}\hat{\phi}^\dagger(0)e^{-i\hat{p}\cdot y}|0\rangle$$
$$= \sum_\alpha e^{-ip^{(\alpha)}\cdot(x-y)}|\langle 0|\hat{\phi}(0)|\alpha\rangle|^2. \tag{5.262}$$

We introduce the spectral density $\rho(p^2)$ defined as

$$\Theta(p_0)\rho(p^2) \equiv (2\pi)^3 \sum_\alpha \delta^4(p-p^{(\alpha)})|\langle 0|\hat{\phi}(0)|\alpha\rangle|^2. \tag{5.263}$$

Since $p_0^{(\alpha)} \geq 0$, the right hand side of Eq.(5.263) is zero when $p_0 < 0$. Thus we add a factor $\Theta(p_0)$ on the left hand side of Eq.(5.263). $\rho(p^2)$ is positive definite. In terms of $\rho(p^2)$, Eq.(5.262) is expressed as

$$W(x-y) = \int \frac{d^4p}{(2\pi)^3}\Theta(p_0)\rho(p^2)e^{-ip\cdot(x-y)}. \tag{5.264}$$

Using mathematical formula

$$\rho(p^2) = \int_0^\infty ds\delta(p^2 - s)\rho(s), \tag{5.265}$$

Eq.(5.264) becomes

$$W(x-y) = \int_0^\infty ds\rho(s)\int \frac{d^4p}{(2\pi)^3}\Theta(p_0)\delta(p^2-s)e^{-ip\cdot(x-y)}. \tag{5.266}$$

In Eq.(5.266), the spectral density is outside the momentum integral. Similar to Eq.(5.10), we extend the three-dimensional integration in Eq.(5.8a) to a four-dimensional one.

$$i\triangle^{(+)}(x-y) = \int \frac{d^3p}{2\omega_\mathbf{p}(2\pi)^3}e^{-ip\cdot(x-y)}$$
$$= \int \frac{d^4p}{(2\pi)^3}\frac{1}{2\omega_\mathbf{p}}\delta(p_0 - \omega_\mathbf{p})e^{-i[p_0\cdot(x_0-y_0)-\mathbf{p}\cdot(\mathbf{x}-\mathbf{y})]}. \tag{5.267}$$

Using the identity

$$\frac{1}{2\omega_\mathbf{p}}\delta(p_0 - \omega_\mathbf{p}) = \Theta(p_0)\delta(p_0^2 - \omega_\mathbf{p}^2), \tag{5.268}$$

we have

$$
\begin{aligned}
i\triangle^{(+)}(x-y;m^2) &= \int \frac{d^4p}{(2\pi)^3}\Theta(p_0)\delta(p_0^2-\omega_{\mathbf{p}}^2)e^{-ip\cdot(x-y)}\\
&= \int \frac{d^4p}{(2\pi)^3}\Theta(p_0)\delta(p_0^2-\mathbf{p}^2-m^2)e^{-ip\cdot(x-y)} \qquad (5.269)\\
&= \int \frac{d^4p}{(2\pi)^3}\Theta(p_0)\delta(p^2-m^2)e^{-ip\cdot(x-y)}.
\end{aligned}
$$

Using Eq.(5.269), Eq.(5.266) becomes

$$
W(x-y) = \int_0^\infty ds\rho(s)i\triangle^{(+)}(x-y;s). \qquad (5.270)
$$

Eq.(5.270) is called the spectral representation or Lehmann-Källen representation of Wightman function. Other invariant functions of interacting fields can also be written in a way of the spectral representation. For example, the spectral representation of the Feynman propagator for an interacting scalar field is given by

$$
\begin{aligned}
i\triangle_F'(x-y) &= \langle 0|T(\hat{\phi}(x)\hat{\phi}^\dagger(y))|0\rangle\\
&= \Theta(x_0-y_0)W(x-y) + \Theta(y_0-x_0)W(y-x)\\
&= \int_0^\infty ds\rho(s)[\Theta(x_0-y_0)i\triangle^{(+)}(x-y;s)\\
&\quad + \Theta(y_0-x_0)i\triangle^{(+)}(y-x;s)] \qquad (5.271)\\
&= \int_0^\infty ds\rho(s)i\triangle_F(x-y;s).
\end{aligned}
$$

In the momentum space, Eq.(5.271) becomes

$$
i\triangle_F'(p) = \int_0^\infty ds\rho(s)\frac{1}{p^2-s+i\epsilon}. \qquad (5.272)
$$

Similarly, we can express the Pauli-Jordan function for an interacting scalar field as

$$
\begin{aligned}
i\triangle'(x-y) &= \langle 0|[\hat{\phi}(x),\hat{\phi}^\dagger(y)]|0\rangle\\
&= W(x-y) - W(y-x). \qquad (5.273)
\end{aligned}
$$

Thus the spectral representation of the Pauli-Jordan function for an interacting scalar field has the form

$$
i\triangle'(x-y) = \int_0^\infty ds\rho(s)i\triangle(x-y;s), \qquad (5.274)
$$

where $\triangle(x - y; s)$ is the Pauli-Jordan function for the free field given by Eq.(5.13). Using the equal-time commutation relation Eq.(2.60), we have

$$
\begin{aligned}
i\delta^3(\mathbf{x} - \mathbf{y}) &= [\hat{\phi}(x), \hat{\pi}(y)]_{x_0 \to y_0} \\
&= [\hat{\phi}(x), \dot{\hat{\phi}}^\dagger(y)]_{x_0 \to y_0} \\
&= \partial_{y_0} \langle 0|[\hat{\phi}(x), \hat{\phi}^\dagger(y)]|0\rangle_{x_0 = y_0} \\
&= i\partial_{y_0} \Delta'(x - y)\Big|_{x_0 = y_0} \\
&= \int_0^\infty ds\rho(s) i\partial_{y_0} \triangle(x - y; s)\Big|_{x_0 = y_0} \\
&= \int_0^\infty ds\rho(s) i\delta^3(\mathbf{x} - \mathbf{y}).
\end{aligned}
\tag{5.275}
$$

In the derivation of the last line, we have used Eq.(5.19). Thus we obtain the sum rule for the spectral density

$$
\int_0^\infty ds\rho(s) = 1.
\tag{5.276}
$$

Both single quasi-particle and many quasi-particle states contribute to the spectral density. In order to isolate the contribution of single quasi-particle states, we split the summation in Eq.(5.263) into the contribution of single quasi-particle states $|p'\rangle$ and that of many quasi-particle states $|\alpha(n > 1)\rangle$. We have

$$
\begin{aligned}
\Theta(p_0)\rho(p^2) &= (2\pi)^3 \int d^3p' \delta^4(p - p')|\langle 0|\hat{\phi}(0)|p'\rangle|^2 \\
&+ (2\pi)^3 \sum_{\alpha(n>1)} \delta^4(p - p^{(\alpha)})|\langle 0|\hat{\phi}(0)|\alpha\rangle|^2.
\end{aligned}
\tag{5.277}
$$

Using Eq.(5.252a), we obtain

$$
\begin{aligned}
\langle 0|\hat{\phi}(0)|p'\rangle &= \sqrt{Z}\langle 0|\hat{\phi}_{\text{in}}(0)|p'\rangle \\
&= \sqrt{Z}u_{p'}(0) \\
&= \sqrt{Z}\frac{1}{\sqrt{(2\pi)^3 2\omega_{\mathbf{p}'}}}.
\end{aligned}
\tag{5.278}
$$

Using mathematical identity

$$
\delta(f(x)) = \sum_i \frac{\delta(x - x_i)}{|f'(x_i)|},
\tag{5.279}
$$

where x_i are the zeroes of $f(x)$, we have

$$
\frac{1}{2\omega_{\mathbf{p}'}} = \int_{-\infty}^\infty dp'_0 \Theta(p'_0)\delta(p'^2 - m^2).
\tag{5.280}
$$

Then Eq.(5.277) becomes

$$
\begin{aligned}
\Theta(p_0)\rho(p^2) &= (2\pi)^3 \int d^3p'\delta^4(p-p')Z\frac{1}{(2\pi)^3 2\omega_{\mathbf{p}'}} \\
&\quad + (2\pi)^3 \sum_{\alpha(n>1)} \delta^4(p-p^{(\alpha)})|\langle 0|\hat{\phi}(0)|\alpha\rangle|^2 \\
&= Z\int d^4p'\Theta(p'_0)\delta^4(p-p')\delta(p^2-m^2) \\
&\quad + (2\pi)^3 \sum_{\alpha(n>1)} \delta^4(p-p^{(\alpha)})|\langle 0|\hat{\phi}(0)|\alpha\rangle|^2 \\
&= Z\Theta(p'_0)\delta(p^2-m^2) + (2\pi)^3 \sum_{\alpha(n>1)} \delta^4(p-p^{(\alpha)})|\langle 0|\hat{\phi}(0)|\alpha\rangle|^2.
\end{aligned}
\tag{5.281}
$$

Thus the sum rule Eq.(5.276) can be written as

$$
Z + \int_{m_2^2}^{\infty} ds\rho(s) = 1,
\tag{5.282}
$$

where $m_2 = 2m$ is the mass of two quasi-particle states. Since $\rho(s) > 0$, we have

$$
0 < Z < 1,
\tag{5.283}
$$

For an interacting field, single quasi-particle states come not only from single-particle states of the free field but also from many particle states of the free field, and *vise versa*. The matrix element $\langle 0|\hat{\phi}(0)|\alpha\rangle$ contains the contribution of many particle states. $\sqrt{Z}$ in Eq.(5.252) accounts for the contribution of many particle states. Inserting the decomposition of the spectral density Eq.(5.281) into Eq.(5.271), the Feynman propagator of an interacting field becomes

$$
\begin{aligned}
i\triangle'_F(p) &= \int_0^{\infty} ds\rho(s)i\triangle_F(p;s) \\
&= Zi\triangle_F(p;m^2) + \int_{m_2^2}^{\infty} ds\rho(s)i\triangle_F(p;s),
\end{aligned}
\tag{5.284}
$$

which splits the contribution of single-particle states and that of many-particle states. Multiplying $p^2 - m^2$ on both sides of Eq.(5.284), we have

$$
Z = \lim_{p^2 \to m^2} (p^2 - m^2)\triangle'_F(p).
\tag{5.285}
$$

Thus the renormalization constant Z can be calculated by extracting the residue of the single-particle pole of the Feynman propagator for an interacting field. Eq.(5.285) shows that the Z used here is the same with the Z defined by Eq.(5.167).

5.8.5 Yang-Feldman equation

For an interacting scalar field, the equation of motion is given by

$$(\Box^2 + m^2)\hat{\phi}(x) = \hat{j}(x). \tag{5.286}$$

$\hat{j}(x)$ is called the source and is a function of the field operator $\hat{\phi}(x)$. We usually split the Lagrangian density of the interacting field into the Lagrangian density of the free field $\mathcal{L}_0$ and the interaction term $\mathcal{L}_1$

$$\mathcal{L} = \mathcal{L}_0 + \mathcal{L}_1. \tag{5.287}$$

Since the interaction term does not contain the derivative of field, we have

$$\hat{j}(x) = \frac{\partial \hat{\mathcal{L}}_1}{\partial \hat{\phi}(x)}. \tag{5.288}$$

In Eq.(5.286), we have used the renormalized mass m. Thus $\hat{j}(x)$ contains a mass correction term

$$\hat{j}_{\delta m}(x) = (m^2 - m_0^2)\hat{\phi}(x) \equiv \delta m^2 \hat{\phi}(x), \tag{5.289}$$

where m_0 is the bare mass.

We introduce the annihilation operators $\hat{a}_{\mathbf{p}}$ and $\hat{b}_{\mathbf{p}}$ for the interacting field by

$$\hat{a}_{\mathbf{p}}(x_0) = i \int d^3x u_{\mathbf{p}}^*(x) \overleftrightarrow{\partial_0} \hat{\phi}(x), \tag{5.290a}$$

$$\hat{a}_{\mathbf{p}}^\dagger(x_0) = -i \int d^3x u_{\mathbf{p}}(x) \overleftrightarrow{\partial_0} \hat{\phi}^\dagger(x), \tag{5.290b}$$

$$\hat{b}_{\mathbf{p}}(x_0) = i \int d^3x u_{\mathbf{p}}^*(x) \overleftrightarrow{\partial_0} \hat{\phi}^\dagger(x) \tag{5.290c}$$

$$\hat{b}_{\mathbf{p}}^\dagger(x_0) = -i \int d^3x u_{\mathbf{p}}(x) \overleftrightarrow{\partial_0} \hat{\phi}(x). \tag{5.290d}$$

The field operators of the interacting complex scalar field take the form

$$\hat{\phi}(x) = \int d^3p [\hat{a}_{\mathbf{p}}(x_0)u_{\mathbf{p}}(x) + \hat{b}_{\mathbf{p}}^\dagger(x_0)u_{\mathbf{p}}^*(x)], \tag{5.291a}$$

$$\hat{\phi}^\dagger(x) = \int d^3p [\hat{a}_{\mathbf{p}}^\dagger(x_0)u_{\mathbf{p}}^*(x) + \hat{b}_{\mathbf{p}}(x_0)u_{\mathbf{p}}(x)]. \tag{5.291b}$$

Using Eq.(5.290), we have

$$\sqrt{Z}\langle\alpha|\hat{a}_{\mathbf{p},\text{in}}|\beta\rangle = \lim_{x_0 \to -\infty} \langle\alpha|\hat{a}_{\mathbf{p}}|\beta\rangle$$
$$= \lim_{x_0 \to -\infty} i \int d^3x u_{\mathbf{p}}^*(x) \overleftrightarrow{\partial_0} \langle\alpha|\hat{\phi}(x)|\beta\rangle. \tag{5.292}$$

Using the mathematical identity

$$\int d^3x' F(\mathbf{x}', -\infty) = \int d^3x' F(\mathbf{x}', x_0) - \int d^3x' \int_{-\infty}^{x_0} dx_0' \partial_0' F(\mathbf{x}', x_0'), \tag{5.293}$$

Eq.(5.292) becomes

$$\sqrt{Z}\langle\alpha|\hat{a}_{\mathbf{p},\text{in}}|\beta\rangle = i\int d^3x' u_{\mathbf{p}}^*(x')\overleftrightarrow{\partial_0'}\langle\alpha|\hat{\phi}(x')|\beta\rangle\Big|_{x_0'=x_0}$$
$$-i\int d^3x'\int_{-\infty}^{x_0}dx_0'\partial_0'[u_{\mathbf{p}}^*(x')\overleftrightarrow{\partial_0'}\langle\alpha|\hat{\phi}(x')|\beta\rangle]$$
$$= \langle\alpha|\hat{a}_{\mathbf{p}}(x_0)|\beta\rangle \qquad (5.294)$$
$$-i\int d^3x'\int_{-\infty}^{x_0}dx_0'[u_{\mathbf{p}}^*(x')\partial_0'^2\langle\alpha|\hat{\phi}(x')|\beta\rangle$$
$$-(\partial_0'^2 u_{\mathbf{p}}^*(x'))\langle\alpha|\hat{\phi}(x')|\beta\rangle].$$

Since $u_{\mathbf{p}}(x)$ satisfies the Klein-Gordon equation, we have

$$\partial_0^2 u_{\mathbf{p}}(x) = (\nabla^2 - m^2)u_{\mathbf{p}}(x). \qquad (5.295)$$

Employing Eq.(5.295) and integrating by parts, Eq.(5.294) becomes

$$\sqrt{Z}\langle\alpha|\hat{a}_{\mathbf{p},\text{in}}|\beta\rangle = \langle\alpha|\hat{a}_{\mathbf{p}}(x_0)|\beta\rangle$$
$$-i\int d^3x'\int_{-\infty}^{x_0}dx_0' u_{\mathbf{p}}^*(x')(\Box'^2 + m^2)\langle\alpha|\hat{\phi}(x')|\beta\rangle.$$
$$(5.296)$$

Using Eq.(5.286), we obtain

$$\sqrt{Z}\langle\alpha|\hat{a}_{\mathbf{p},\text{in}}|\beta\rangle = \langle\alpha|\hat{a}_{\mathbf{p}}(x_0)|\beta\rangle - i\int d^3x'\int_{-\infty}^{x_0}dx_0' u_{\mathbf{p}}^*(x')\langle\alpha|\hat{j}(x')|\beta\rangle.$$
$$(5.297)$$

Similarly, we have

$$\sqrt{Z}\langle\alpha|\hat{b}_{\mathbf{p},\text{in}}^\dagger|\beta\rangle = \langle\alpha|\hat{b}_{\mathbf{p}}^\dagger(x_0)|\beta\rangle + i\int d^3x'\int_{-\infty}^{x_0}dx_0' u_{\mathbf{p}}(x')\langle\alpha|\hat{j}(x')|\beta\rangle.$$
$$(5.298)$$

Multiplying Eq.(5.297) by $u_{\mathbf{p}}(x)$ and Eq.(5.298) by $u^*(x)$, then adding them and integrating over $\mathbf{p}$, we have

$$\sqrt{Z}\langle\alpha|\hat{\phi}_{\mathbf{p},\text{in}}(x)|\beta\rangle = \langle\alpha|\hat{\phi}_{\mathbf{p}}(x)|\beta\rangle - i\int d^3x'\int_{-\infty}^{x_0}dx_0'\int d^3p \qquad (5.299)$$
$$[u_{\mathbf{p}}(x)u_{\mathbf{p}}^*(x') - u_{\mathbf{p}}^*(x)u_{\mathbf{p}}(x')]\langle\alpha|\hat{j}(x')|\beta\rangle.$$

The momentum integral in the second term on the right hand side of Eq.(5.299) gives the Pauli-Jordan function.

$$\int d^3p[u_{\mathbf{p}}(x)u_{\mathbf{p}}^*(x') - u_{\mathbf{p}}^*(x)u_{\mathbf{p}}(x')]$$
$$= \int d^3p\frac{1}{2\omega_{\mathbf{p}}(2\pi)^3}[e^{-ip\cdot(x-x')} - e^{ip\cdot(x-x')}] \qquad (5.300)$$
$$= i\triangle(x - x').$$

In the derivation of Eq.(5.300), we have used Eq.(5.7). Thus Eq.(5.299) becomes

$$\sqrt{Z}\langle\alpha|\hat{\phi}_{\mathbf{p},\text{in}}(x)|\beta\rangle = \langle\alpha|\hat{\phi}_{\mathbf{p}}(x)|\beta\rangle$$

$$+ \int d^4x' \Theta(x_0 - x_0') \triangle(x - x')\langle\alpha|\hat{j}(x')|\beta\rangle. \tag{5.301}$$

Using Eq.(5.36a), we have

$$\langle\alpha|\hat{\phi}_{\mathbf{p}}(x)|\beta\rangle = \sqrt{Z}\langle\alpha|\hat{\phi}_{\mathbf{p},\text{in}}(x)|\beta\rangle - \int d^4x' \triangle_R(x - x')\langle\alpha|\hat{j}(x')|\beta\rangle. \tag{5.302}$$

Similar calculations can be applied to the case of limit $x_0 \to +\infty$, which gives

$$\langle\alpha|\hat{\phi}_{\mathbf{p}}(x)|\beta\rangle = \sqrt{Z}\langle\alpha|\hat{\phi}_{\mathbf{p},\text{out}}(x)|\beta\rangle - \int d^4x' \triangle_A(x - x')\langle\alpha|\hat{j}(x')|\beta\rangle. \tag{5.303}$$

Eq.(5.302) and Eq.(5.303) are called the Yang-Feldman equations.

5.8.6 *LSZ reduction formula*

We consider the S matrix element for a scattering process of $n - m$ quasi-particles.

$$S_{\text{fi}} = \langle q_1, \cdots, q_m; \text{out}|p_1, \cdots, p_n; \text{in}\rangle. \tag{5.304}$$

The scattering process containing quasi-antiparticles can be calculated similarly. For simplicity, we consider first the real scalar field, for which $\hat{\phi} = \hat{\phi}^\dagger$. We want to express S_{fi} in terms of n-particle Green's functions defined as

$$G^n(x_1, \cdots, x_n) = \langle 0|T\hat{\phi}(x_1) \cdots \hat{\phi}(x_n)|0\rangle. \tag{5.305}$$

First, we extract the creation operator $\hat{a}_{\mathbf{p},\text{in}}^\dagger$ from the initial state $|p_1, p_2, \cdots, p_n; \text{in}\rangle$ in Eq.(5.304). We have

$$S_{\text{fi}} = \langle q_1, \cdots, q_m; \text{out}|\hat{a}_{\mathbf{p}_1,in}^\dagger|p_2, \cdots, p_n; \text{in}\rangle$$

$$= \langle q_1, \cdots, q_m; \text{out}|\hat{a}_{\mathbf{p}_1,out}^\dagger|p_2, \cdots, p_n; \text{in}\rangle$$

$$+ \langle q_1, \cdots, q_m; \text{out}|(\hat{a}_{\mathbf{p}_1,\text{in}}^\dagger - \hat{a}_{\mathbf{p}_1,\text{out}}^\dagger)|p_2, \cdots, p_n; \text{in}\rangle$$

$$= \sum_{k=1}^{m} \langle q_1, \cdots, q_m; \text{out}|p_2, \cdots, p_n; \text{in}\rangle \delta^3(\mathbf{q}_k - \mathbf{p}_1)$$

$$+ \langle q_1, \cdots, q_m; \text{out}|(\hat{a}_{\mathbf{p}_1,\text{in}}^\dagger - \hat{a}_{\mathbf{p}_1,\text{out}}^\dagger)|p_2, \cdots, p_n; \text{in}\rangle. \tag{5.306}$$

The first term in Eq.(5.306) contributes only when one of momenta $\mathbf{q}_1, \cdots, \mathbf{q}_m$ is equal to $\mathbf{p}_1$. This is the case that one of the incoming quasi-particle run through without scattering. Since we calculate the scattering process, we consider the case that all momenta of the incoming and outgoing quasi-particles are different. Thus we neglect the first term in Eq.(5.306).

The field operators for real scalar bosons are given by

$$\hat{\phi}_{\text{in}}(\mathbf{x}, x_0) = \int d^3 p [\hat{a}_{\mathbf{p},\text{in}} u_{\mathbf{p}}(\mathbf{x}, x_0) + \hat{a}^\dagger_{\mathbf{p},\text{in}} u^*_{\mathbf{p}}(\mathbf{x}, x_0)], \tag{5.307a}$$

$$\hat{\phi}_{\text{out}}(\mathbf{x}, x_0) = \int d^3 p [\hat{a}_{\mathbf{p},\text{out}} u_{\mathbf{p}}(\mathbf{x}, x_0) + \hat{a}^\dagger_{\mathbf{p},\text{out}} u^*_{\mathbf{p}}(\mathbf{x}, x_0)]. \tag{5.307b}$$

The inverse relations of Eq.(5.307) are

$$\hat{a}_{\mathbf{p},\text{in}} = i \int d^3 x u^*_{\mathbf{p}}(\mathbf{x}, x_0) \overleftrightarrow{\partial_0} \hat{\phi}_{\text{in}}(\mathbf{x}, x_0), \tag{5.308a}$$

$$\hat{a}^\dagger_{\mathbf{p},\text{in}} = -i \int d^3 x u_{\mathbf{p}}(\mathbf{x}, x_0) \overleftrightarrow{\partial_0} \hat{\phi}_{\text{in}}(\mathbf{x}, x_0), \tag{5.308b}$$

$$\hat{a}_{\mathbf{p},\text{out}} = i \int d^3 x u^*_{\mathbf{p}}(\mathbf{x}, x_0) \overleftrightarrow{\partial_0} \hat{\phi}_{\text{out}}(\mathbf{x}, x_0), \tag{5.308c}$$

$$\hat{a}^\dagger_{\mathbf{p},\text{out}} = -i \int d^3 x u_{\mathbf{p}}(\mathbf{x}, x_0) \overleftrightarrow{\partial_0} \hat{\phi}_{\text{out}}(\mathbf{x}, x_0) \tag{5.308d}$$

Using Eq.(5.308), we have

$$\hat{a}^\dagger_{\mathbf{p}_1,\text{in}} - \hat{a}^\dagger_{\mathbf{p}_1,\text{out}} = i \int d^3 x [\hat{\phi}_{\text{in}}(x) - \hat{\phi}_{\text{out}}(x)] \overleftrightarrow{\partial_0} u_{\mathbf{p}_1}(x). \tag{5.309}$$

Since x_0 can be any point of time, we can take the limit $x_0 \to \infty$ or $x_0 \to -\infty$. Eq.(5.309) can be rewritten as

$$\hat{a}^\dagger_{\mathbf{p}_1,\text{in}} - \hat{a}^\dagger_{\mathbf{p}_1,\text{out}} = i \int d^3 x [\lim_{x_0 \to -\infty} \hat{\phi}_{\text{in}}(x) \overleftrightarrow{\partial_{x_0}} u_{\mathbf{p}_1}(x)$$

$$- \lim_{x_0 \to \infty} \hat{\phi}_{\text{out}}(x) \overleftrightarrow{\partial_{x_0}} u_{\mathbf{p}_1}(x)] \tag{5.310}$$

$$\Rightarrow -\frac{i}{\sqrt{Z}} (\lim_{x_0 \to \infty} - \lim_{x_0 \to -\infty}) \int d^3 x \hat{\phi}(x) \overleftrightarrow{\partial_{x_0}} u_{\mathbf{p}_1}(x),$$

where $\Rightarrow$ is the equal mark in the meaning of weak convergence Eq.(5.252).

Using the mathematical identity

$$(\lim_{x_0 \to \infty} - \lim_{x_0 \to -\infty}) F(x) = \int_{-\infty}^{\infty} dx_0 \partial_{x_0} F(x), \tag{5.311}$$

Eq.(5.310) becomes

$$\hat{a}^\dagger_{\mathbf{p}_1,\text{in}} - \hat{a}^\dagger_{\mathbf{p}_1,\text{out}} \Rightarrow -\frac{i}{\sqrt{Z}} \int d^4 x \partial_{x_0} [\hat{\phi}(x) \overleftrightarrow{\partial_{x_0}} u_{\mathbf{p}_1}(x)]$$

$$= -\frac{i}{\sqrt{Z}} \int d^4 x [\hat{\phi}(x) \overrightarrow{\partial^2_{x_0}} u_{\mathbf{p}_1}(x) - \hat{\phi}(x) \overleftarrow{\partial^2_{x_0}} u_{\mathbf{p}_1}(x)]$$

$$= -\frac{i}{\sqrt{Z}} \int d^4 x [\hat{\phi}(x) (\nabla^2 - m^2) u_{\mathbf{p}_1}(x) - \hat{\phi}(x) \overleftarrow{\partial^2_{x_0}} u_{\mathbf{p}_1}(x)] \tag{5.312}$$

$$= \frac{i}{\sqrt{Z}} \int d^4 x \hat{\phi}(x) (\overleftarrow{\Box^2} + m^2) u_{\mathbf{p}_1}(x).$$

In the derivation of the third line of Eq.(5.312), we have used Eq.(5.251). Thus Eq.(5.306) becomes

$$S_\mathrm{fi} = \frac{i}{\sqrt{Z}} \int d^4x \langle q_1, \cdots, q_m; \mathrm{out}| \hat{\phi}(x) |p_2, \cdots, p_n; \mathrm{in}\rangle$$
$$(\overleftarrow{\Box^2 + m^2}) u_{\mathbf{p}_1}(x). \tag{5.313}$$

The number of quasi-particles in the initial state has been reduced from n to $n-1$ in the expression of S_fi. We can continue the similar procedure to further reduce the number of quasi-particles in the expression of S_fi. We extract the second quasi-particle from the initial state $|p_2, \cdots, p_n; \mathrm{in}\rangle$. Now we consider the factor $\langle q_1, \cdots, q_m; \mathrm{out}|\hat{\phi}(x)|p_2, \cdots, p_n; \mathrm{in}\rangle$ in Eq.(5.313).

$$\langle q_1, \cdots, q_m; \mathrm{out}|\hat{\phi}(x)|p_2, \cdots, p_n; \mathrm{in}\rangle$$
$$= \langle q_1, \cdots, q_m; \mathrm{out}|\hat{\phi}(x)\hat{a}^\dagger_{\mathbf{p}_2, \mathrm{in}}|p_3, \cdots, p_n; \mathrm{in}\rangle$$
$$= i \lim_{y_0 \to -\infty} \langle q_1, \cdots, q_m; \mathrm{out}|$$
$$\int d^3y \hat{\phi}(x)\hat{\phi}_\mathrm{in}(y) \overleftrightarrow{\partial}_{y_0} u_{\mathbf{p}_2}(y)|p_3, \cdots, p_n; \mathrm{in}\rangle \tag{5.314}$$
$$= \frac{i}{\sqrt{Z}} \lim_{y_0 \to -\infty} \langle q_1, \cdots, q_m; \mathrm{out}|$$
$$\int d^3y \hat{\phi}(x)\hat{\phi}(y) \overleftrightarrow{\partial}_{y_0} u_{\mathbf{p}_2}(y)|p_3, \cdots, p_n; \mathrm{in}\rangle.$$

Because of the limit $y_0 \to -\infty$, we can introduce a time-ordered product in Eq.(5.314). Then Eq.(5.314) can be rewritten as

$$\langle q_1, \cdots, q_m; \mathrm{out}|\hat{\phi}(x)|p_2, \cdots, p_n, \mathrm{in}\rangle$$
$$= \frac{i}{\sqrt{Z}} \lim_{y_0 \to -\infty} \langle q_1, \cdots, q_m; \mathrm{out}|$$
$$\int d^3y T(\hat{\phi}(x)\hat{\phi}(y)) \overleftrightarrow{\partial}_{y_0} u_{\mathbf{p}_2}(y)|p_3, \cdots, p_n; \mathrm{in}\rangle. \tag{5.315}$$

Using the mathematical identity Eq.(5.311), Eq.(5.315) becomes

$$\langle q_1, \cdots, q_m; \mathrm{out}|\hat{\phi}(x)|p_2, \cdots, p_n, \mathrm{in}\rangle$$
$$= \frac{i}{\sqrt{Z}} \lim_{y_0 \to \infty} \langle q_1, \cdots, q_m; \mathrm{out}|$$
$$\int d^3y T(\hat{\phi}(x)\hat{\phi}(y)) \overleftrightarrow{\partial}_{y_0} u_{\mathbf{p}_2}(y)|p_3, \cdots, p_n; \mathrm{in}\rangle \tag{5.316}$$
$$- \frac{i}{\sqrt{Z}} \langle q_1, \cdots, q_m; \mathrm{out}|$$
$$\int d^4y \partial_{y_0}(T(\hat{\phi}(x)\hat{\phi}(y)) \overleftrightarrow{\partial}_{y_0} u_{\mathbf{p}_2}(y))|p_3, \cdots, p_n; \mathrm{in}\rangle.$$

The first term in Eq.(5.316) can be rewritten as

$$\frac{i}{\sqrt{Z}} \lim_{y_0 \to \infty} \langle q_1, \cdots, q_m; \text{out}| \int d^3 y T(\hat{\phi}(x)\hat{\phi}(y)) \overleftrightarrow{\partial_{y_0}} u_{\mathbf{p}_2}(y) | p_3, \cdots, p_n; \text{in} \rangle$$

$$= \frac{i}{\sqrt{Z}} \lim_{y_0 \to \infty} \langle q_1, \cdots, q_m; \text{out}| \int d^3 y \hat{\phi}(y)\hat{\phi}(x) \overleftrightarrow{\partial_{y_0}} u_{\mathbf{p}_2}(y) | p_3, \cdots, p_n; \text{in} \rangle$$

$$= i \lim_{y_0 \to \infty} \langle q_1, \cdots, q_m; \text{out}| \int d^3 y \hat{\phi}_{out}(y)\hat{\phi}(x) \overleftrightarrow{\partial_{y_0}} u_{\mathbf{p}_2}(y) | p_3, \cdots, p_n; \text{in} \rangle$$

$$= \langle q_1, \cdots, q_m; \text{out}| \hat{a}^\dagger_{\mathbf{p}_2, out} \hat{\phi}(x) | p_3, \cdots, p_n; \text{in} \rangle$$

$$= 0$$

$$(5.317)$$

because we consider only the case that all momenta of the incoming and outgoing quasi-particles are different. The second term of Eq.(5.316) can be transformed in a similar way as was made in Eq.(5.312). We have

$$\langle q_1, \cdots, q_m; \text{out}| \hat{\phi}(x) | p_2, \cdots, p_n; \text{in} \rangle$$

$$= \frac{i}{\sqrt{Z}} \langle q_1, \cdots, q_m; \text{out}| \qquad (5.318)$$

$$\int d^4 y T(\hat{\phi}(x)\hat{\phi}(y)) (\overleftarrow{\square^2} + m^2) u_{\mathbf{p}_2}(y) | p_3, \cdots, p_n; \text{in} \rangle.$$

Thus we obtain the S matrix element

$$S_{\text{fi}} = \left(\frac{i}{\sqrt{Z}} \right)^2 \int d^4 x d^4 y \langle q_1, \cdots, q_m; \text{out}| T(\hat{\phi}(x)\hat{\phi}(y))$$

$$| p_3, \cdots, p_n; \text{in} \rangle (\overleftarrow{\square_x^2} + m^2)(\overleftarrow{\square_y^2} + m^2) u_{\mathbf{p}_1}(x) u_{\mathbf{p}_2}(y). \qquad (5.319)$$

Continuing the procedure, we can extract all the quasi-particles from the initial state. We have

$$S_{\text{fi}} = \left(\frac{i}{\sqrt{Z}} \right)^n \int d^4 x_1 \cdots d^4 x_n \langle q_1, \cdots, q_m; \text{out}| T(\hat{\phi}(x_1) \cdots \hat{\phi}(x_n)) | 0 \rangle$$

$$(\overleftarrow{\square_{x_1}^2} + m^2) \cdots (\overleftarrow{\square_{x_n}^2} + m^2) u_{\mathbf{p}_1}(x_1) \cdots u_{\mathbf{p}_n}(x_n).$$

$$(5.320)$$

Similarly we can extract the quasi-particles from the final state. As an example, we extract the first quasi-particle from the final state. We need to treat $\langle q_1, \cdots, q_m; \text{out}|0\rangle$. Extracting the first quasi-particle, we obtain

$$
\langle q_1, \cdots, q_m; \text{out}|T(\hat{\phi}(x_1)\cdots\hat{\phi}(x_n))|0\rangle
$$

$$
= \langle q_2, \cdots, q_m; \text{out}|\hat{a}_{\mathbf{q}_1,\text{out}}T(\hat{\phi}(x_1)\cdots\hat{\phi}(x_n))|0\rangle
$$

$$
= \frac{i}{\sqrt{Z}} \lim_{y_0\to\infty} \langle q_2, \cdots, q_m; \text{out}|
$$

$$
\int d^3 y\, u^*_{\mathbf{q}_1}(y)\overleftrightarrow{\partial_{y_0}}\hat{\phi}(y)T(\hat{\phi}(x_1)\cdots\hat{\phi}(x_n))|0\rangle
$$

$$
= \frac{i}{\sqrt{Z}} \lim_{y_0\to\infty} \langle q_2, \cdots, q_m; \text{out}|
$$

$$
\int d^3 y\, u^*_{\mathbf{q}_1}(y)\overleftrightarrow{\partial_{y_0}}T(\hat{\phi}(y)\hat{\phi}(x_1)\cdots\hat{\phi}(x_n))|0\rangle \tag{5.321}
$$

$$
= \langle q_2, \cdots, q_m; \text{out}|T(\hat{\phi}(x_1)\cdots\hat{\phi}(x_n))\hat{a}_{\mathbf{q}_1,\text{in}}|0\rangle
$$

$$
+ \frac{i}{\sqrt{Z}}\langle q_2, \cdots, q_m; \text{out}|
$$

$$
\int d^4 y\, \partial_{y_0}[u^*_{\mathbf{q}_1}(y)\overleftrightarrow{\partial_{y_0}}T(\hat{\phi}(y)\hat{\phi}(x_1)\cdots\hat{\phi}(x_n))]|0\rangle
$$

$$
= \frac{i}{\sqrt{Z}}\langle q_2, \cdots, q_m; \text{out}|
$$

$$
\int d^4 y\, u^*_{\mathbf{q}_1}(y)(\overrightarrow{\Box_y^2 + m^2})T(\hat{\phi}(y)\hat{\phi}(x_1)\cdots\hat{\phi}(x_n))|0\rangle.
$$

Continuing the procedure, we have

$$
S_{\text{fi}} = \langle q_1, \cdots, q_m; \text{out}|p_1, \cdots, p_n; \text{in}\rangle
$$

$$
= \langle q_1, \cdots, q_m; \text{in}|\hat{S}|p_1, \cdots, p_n; \text{in}\rangle
$$

$$
= \left(\frac{i}{\sqrt{Z}}\right)^{n+m} \int d^4 y_1 \cdots d^4 y_m d^4 x_1 \cdots d^4 x_n
$$

$$
\times\, u^*_{\mathbf{q}_1}(y_1)\cdots u^*_{\mathbf{q}_m}(y_m)(\overrightarrow{\Box_{y_1}^2 + m^2})\cdots(\overrightarrow{\Box_{y_m}^2 + m^2}) \tag{5.322}
$$

$$
\times\, \langle 0|T(\hat{\phi}(x_1)\cdots\hat{\phi}(x_n)\hat{\phi}(y_1)\cdots\hat{\phi}(y_m))|0\rangle
$$

$$
\times\, (\overleftarrow{\Box_{x_1}^2 + m^2})\cdots(\overleftarrow{\Box_{x_n}^2 + m^2})u_{\mathbf{p}_1}(x_1)\cdots u_{\mathbf{p}_n}(x_n).
$$

Eq.(5.322) is called the LSZ (Lehman-Symanzik-Zimmermann) reduction formula for scalar particles. Eq.(5.322) relates the S matrix to the $n + m$ particle Green's functions. Since the operator $\Box^2 + m^2$ is the inverse of the boson propagator, the S matrix is determined by the amputated Green's function. Eq.(5.322) can also be expressed in the momentum space. Using

Eq.(5.58), we have

$$
\begin{aligned}
S_{\text{fi}} = {}&(4\pi)^4 \left(\frac{1}{\sqrt{Z}}\right)^{n+m} N_{q_1} \cdots N_{q_m} N_{p_1} \cdots N_{p_n} \\
&\times (q_1^2 - m^2) \cdots (q_m^2 - m^2)(p_1^2 - m^2) \cdots (p_n^2 - m^2) \quad (5.323) \\
&\times G^{(n+m)}(q_1, \cdots, q_m, -p_1, \cdots, -p_n) \\
&\times \delta^4(q_1 + \cdots + q_m - p_1 \cdots - p_n),
\end{aligned}
$$

where $N_p = [(2\pi)^3 2\omega_{\mathbf{p}}]^{-\frac{1}{2}}$ is the normalization constant.

For the complex scalar bosons, there are anti-particles. We can treat the anti-particles in a similar way. The LSZ reduction formula for the complex scalar bosons reads

$$
\begin{aligned}
S_{\text{fi}} = {}&\langle q_1, \cdots, q_m, q_1', \cdots, q_{m'}'; \text{out}|p_1, \cdots, p_n, p_1', \cdots, p_{n'}'; \text{in}\rangle \\
= {}&\left(\frac{i}{\sqrt{Z}}\right)^{n+n'+m+m'} \\
&\times \int d^4 y_1 \cdots d^4 y_m d^4 y_1' \cdots d^4 y_{m'}' d^4 x_1 \cdots d^4 x_n d^4 x_1' \cdots d^4 x_{n'}' \\
&\times u_{\mathbf{q}_1}^*(y_1) \cdots u_{\mathbf{q}_m}^*(y_m) u_{\mathbf{q}_1'}^*(y_1') \cdots u_{\mathbf{q}_{m'}'}^*(y_{m'}') \\
&\times (\overrightarrow{\Box_{y_1}^2 + m^2}) \cdots (\overrightarrow{\Box_{y_m}^2 + m^2})(\overrightarrow{\Box_{y_1'}^2 + m^2}) \cdots (\overrightarrow{\Box_{y_{m'}'}^2 + m^2}) \quad (5.324) \\
&\times \langle 0|T(\hat{\phi}(x_1') \cdots \hat{\phi}(x_{n'}')\hat{\phi}^\dagger(x_1) \cdots \hat{\phi}^\dagger(x_n) \\
&\times \hat{\phi}^\dagger(y_1') \cdots \hat{\phi}^\dagger(y_{m'}')\hat{\phi}(y_1) \cdots \hat{\phi}(y_m))|0\rangle \\
&\times (\overleftarrow{\Box_{x_1}^2 + m^2}) \cdots (\overleftarrow{\Box_{x_n}^2 + m^2})(\overleftarrow{\Box_{x_1'}^2 + m^2}) \cdots (\overleftarrow{\Box_{x_{n'}'}^2 + m^2}) \\
&\times u_{\mathbf{p}_1}(x_1) \cdots u_{\mathbf{p}_n}(x_n) u_{\mathbf{p}_1'}(x_1') \cdots u_{\mathbf{p}_{n'}'}(x_{n'}').
\end{aligned}
$$

In Eq.(5.324), the quantities of quasi-antiparticles are distinguished by superscript $'$.

5.8.7   *LSZ reduction formula for spin-$\frac{1}{2}$ particles*

Similar to the treatment of the LSZ reduction formula for the scalar (spin-0) bosons. We can derive the LSZ reduction formula for the spinor (spin-$\frac{1}{2}$) fermions.

We introduce the creation and annihilation operators $\hat{b}_{\text{in/out}}^\dagger$, $\hat{d}_{\text{in/out}}^\dagger$, $\hat{b}_{\text{in/out}}$ and $\hat{d}_{\text{in/out}}$ for the interacting spinor fermion field. They satisfy the anti-commutation relations.

$$\{\hat{b}_{\text{in/out}}(\mathbf{p}, s), \hat{b}^\dagger_{\text{in/out}}(\mathbf{p}', s')\} = \delta_{ss'}\delta^3(\mathbf{p} - \mathbf{p}'), \tag{5.325a}$$

$$\{\hat{d}_{\text{in/out}}(\mathbf{p}, s), \hat{d}^\dagger_{\text{in/out}}(\mathbf{p}', s')\} = \delta_{ss'}\delta^3(\mathbf{p} - \mathbf{p}'), \tag{5.325b}$$

$$\{\hat{b}_{\text{in/out}}(\mathbf{p}, s), \hat{b}_{\text{in/out}}(\mathbf{p}', s')\} = 0, \tag{5.325c}$$

$$\{\hat{b}^\dagger_{\text{in/out}}(\mathbf{p}, s), \hat{b}^\dagger_{\text{in/out}}(\mathbf{p}', s')\} = 0, \tag{5.325d}$$

$$\{\hat{d}_{\text{in/out}}(\mathbf{p}, s), \hat{d}_{\text{in/out}}(\mathbf{p}', s')\} = 0, \tag{5.325e}$$

$$\{\hat{d}^\dagger_{\text{in/out}}(\mathbf{p}, s), \hat{d}^\dagger_{\text{in/out}}(\mathbf{p}', s')\} = 0. \tag{5.325f}$$

The quasi-particle field operators are defined by the expansion formula

$$\hat{\psi}_{\text{in/out}}(\mathbf{x}, t)$$
$$= \int d^3p \sum_s [\hat{b}_{\text{in/out}}(\mathbf{p}, s)u_{\mathbf{p}s}(x) + \hat{d}^\dagger_{\text{in/out}}(\mathbf{p}, s)v_{\mathbf{p}s}(x)], \tag{5.326a}$$

$$\hat{\psi}^\dagger_{\text{in/out}}(\mathbf{x}, t)$$
$$= \int d^3p \sum_s [\hat{b}^\dagger_{\text{in/out}}(\mathbf{p}, s)u^\dagger_{\mathbf{p}s}(x) + \hat{d}_{\text{in/out}}(\mathbf{p}, s)v^\dagger_{\mathbf{p}s}(x)], \tag{5.326b}$$

where $u_{\mathbf{p}s}(x)$ and $v_{\mathbf{p}s}(x)$ are the normalized Dirac plane waves of particles and anti-particles.

$$u_{\mathbf{p}s}(x) = \frac{1}{(2\pi)^{\frac{3}{2}}}\sqrt{\frac{m}{\omega_{\mathbf{p}}}}u(\mathbf{p}, s)e^{-ip\cdot x}, \tag{5.327a}$$

$$v_{\mathbf{p}s}(x) = \frac{1}{(2\pi)^{\frac{3}{2}}}\sqrt{\frac{m}{\omega_{\mathbf{p}}}}v(\mathbf{p}, s)e^{ip\cdot x}. \tag{5.327b}$$

$u_{\mathbf{p}s}(x)$ and $v_{\mathbf{p}s}(x)$ satisfy the free Dirac equation.

$$(i\slashed{\partial} - m)u_{\mathbf{p}s}(x) = (\slashed{p} - m)u_{\mathbf{p}s}(x) = 0, \tag{5.328a}$$

$$(i\slashed{\partial} - m)v_{\mathbf{p}s}(x) = (-\slashed{p} - m)v_{\mathbf{p}s}(x) = 0. \tag{5.328b}$$

The inverse formula of Eq.(5.326) has the form

$$\hat{b}^\dagger_{\text{in/out}}(\mathbf{p}, s) = \int d^3x \hat{\psi}^\dagger_{\text{in/out}}(x)u_{\mathbf{p}s}(x), \tag{5.329a}$$

$$\hat{d}^\dagger_{\text{in/out}}(\mathbf{p}, s) = \int d^3x v^\dagger_{\mathbf{p}s}(x)\hat{\psi}_{\text{in/out}}(x), \tag{5.329b}$$

$$\hat{b}_{\text{in/out}}(\mathbf{p}, s) = \int d^3x u^\dagger_{\mathbf{p}s}(x)\hat{\psi}_{\text{in/out}}(x), \tag{5.329c}$$

$$\hat{d}_{\text{in/out}}(\mathbf{p}, s) = \int d^3x \hat{\psi}^\dagger_{\text{in/out}}(x)v_{\mathbf{p}s}(x). \tag{5.329d}$$

The corresponding weak limit conditions for the operators of the spinor fermions are given by

$$\lim_{x_0 \to -\infty} \langle \alpha | \hat{\psi}(x) | \beta \rangle = \sqrt{Z_2} \langle \alpha | \hat{\psi}_{\text{in}}(x) | \beta \rangle, \tag{5.330a}$$

$$\lim_{x_0 \to +\infty} \langle \alpha | \hat{\psi}(x) | \beta \rangle = \sqrt{Z_2} \langle \alpha | \hat{\psi}_{\text{out}}(x) | \beta \rangle, \tag{5.330b}$$

where Z_2 is the renormalization constant for fermions.

Now we calculate the S matrix for a scattering process involving n quasi-particles and $\bar{n}$ quasi-antiparticles in the initial state, and m quasi-particles and $\bar{m}$ quasi-antiparticles in the final state.

$$\begin{aligned} S_{\text{fi}} &= \langle f; \text{out} | i; \text{in} \rangle \\ &= \langle q_1 r_1, \cdots, q_m r_m, \bar{q}_1 \bar{r}_1, \cdots, \bar{q}_{\bar{m}} \bar{r}_{\bar{m}}; \text{out} \\ &\quad | p_1 s_1, \cdots, p_n s_n, \bar{p}_1 \bar{s}_1, \cdots, \bar{p}_{\bar{n}} \bar{s}_{\bar{n}}; \text{in} \rangle. \end{aligned} \tag{5.331}$$

We extract the first quasi-particle characterized by $p_1 s_1$ in the initial state as an example. To simplify the notation, we use the abbreviation $|i; \text{in}\rangle = \hat{b}_{\text{in}}^\dagger(\mathbf{p}_1, s_1) | i - (p_1 s_1); \text{in} \rangle$, we have

$$\begin{aligned} S_{\text{fi}} &= \langle f; \text{out} | \hat{b}_{\text{in}}^\dagger(\mathbf{p}_1, s_1) | i - (p_1 s_1); \text{in} \rangle \\ &= \lim_{x_0 \to -\infty} \int d^3 x \langle f; \text{out} | \hat{\psi}_{\text{in}}^\dagger(x) | i - (p_1 s_1); \text{in} \rangle u_{\mathbf{p}_1 s_1}(x) \\ &= \frac{1}{\sqrt{Z_2}} \lim_{x_0 \to -\infty} \int d^3 x \langle f; \text{out} | \hat{\psi}^\dagger(x) | i - (p_1 s_1); \text{in} \rangle u_{\mathbf{p}_1 s_1}(x) \\ &= \frac{1}{\sqrt{Z_2}} \lim_{x_0 \to \infty} \int d^3 x \langle f; \text{out} | \hat{\psi}^\dagger(x) | i - (p_1 s_1); \text{in} \rangle u_{\mathbf{p}_1 s_1}(x) \\ &\quad - \frac{1}{\sqrt{Z_2}} (\lim_{x_0 \to \infty} - \lim_{x_0 \to -\infty}) \int d^3 x \langle f; \text{out} | \hat{\psi}^\dagger(x) | i - (p_1 s_1); \text{in} \rangle u_{\mathbf{p}_1 s_1}(x) \\ &= \langle f; \text{out} | \hat{b}_{\text{out}}^\dagger(\mathbf{p}_1, s_1) | i - (p_1 s_1); \text{in} \rangle \\ &\quad - \frac{1}{\sqrt{Z_2}} (\lim_{x_0 \to \infty} - \lim_{x_0 \to -\infty}) \int d^3 x \langle f; \text{out} | \hat{\psi}^\dagger(x) | i - (p_1 s_1); \text{in} \rangle u_{\mathbf{p}_1 s_1}(x). \end{aligned} \tag{5.332}$$

We consider the case that all momenta of the initial and final quasi-particles are different. Thus the first term vanishes. The second term can be rewritten as a four-dimensional integral

$$\begin{aligned} &- \frac{1}{\sqrt{Z_2}} (\lim_{x_0 \to \infty} - \lim_{x_0 \to -\infty}) \int d^3 x \langle f; \text{out} | \hat{\psi}^\dagger(x) | i - (p_1 s_1); \text{in} \rangle u_{\mathbf{p}_1 s_1}(x) \\ &= -\frac{1}{\sqrt{Z_2}} \int d^4 x \langle f; \text{out} | \partial_0 [\hat{\psi}^\dagger(x) u_{\mathbf{p}_1 s_1}(x)] | i - (p_1 s_1); \text{in} \rangle \\ &= -\frac{1}{\sqrt{Z_2}} \int d^4 x \langle f; \text{out} | \hat{\psi}^\dagger(x) (\overleftarrow{\partial_0} + \overrightarrow{\partial_0}) u_{\mathbf{p}_1 s_1}(x) | i - (p_1 s_1); \text{in} \rangle. \end{aligned} \tag{5.333}$$

Since $u_{\mathbf{p}_1 s_1}(x)$ satisfies the free Dirac equation, we have

$$\partial_0 u_{\mathbf{p}_1 s_1}(x) = i\gamma_0(i\gamma_k\partial^k - m)u_{\mathbf{p}_1 s_1}(x). \tag{5.334}$$

Inserting Eq.(5.334) into Eq.(5.333) gives

$$
\begin{aligned}
S_{\mathrm{fi}} &= -\frac{1}{\sqrt{Z_2}} \int d^4x \langle f; \text{out}|\hat{\psi}^\dagger(x)(\overleftarrow{\partial_0} + \overrightarrow{\partial_0})u_{\mathbf{p}_1 s_1}(x)|i - (p_1 s_1); \text{in}\rangle \\
&= -\frac{1}{\sqrt{Z_2}} \int d^4x \langle f; \text{out}|\hat{\psi}^\dagger(x)[\overleftarrow{\partial_0} + i\gamma_0(i\gamma_k\partial^k - m)] \\
&\quad \times u_{\mathbf{p}_1 s_1}(x)|i - (p_1 s_1); \text{in}\rangle \\
&= -\frac{\imath}{\sqrt{Z_2}} \int d^4x \langle f; \text{out}|\hat{\psi}^\dagger(x)\gamma^0[-i\gamma_0\overleftarrow{\partial_0} + i\gamma_k\partial^k - m)] \\
&\quad \times u_{\mathbf{p}_1 s_1}(x)|i - (p_1 s_1); \text{in}\rangle \\
&= -\frac{i}{\sqrt{Z_2}} \int d^4x \langle f; \text{out}|\hat{\bar{\psi}}(x)|i - (p_1 s_1); \text{in}\rangle(\overleftarrow{-i\partial\!\!\!/_x} - m)u_{\mathbf{p}_1 s_1}(x).
\end{aligned}
\tag{5.335}
$$

In the derivation of the last line, we have performed an integration by parts. We can similarly extract an quasi-antiparticle $(\bar{p}_1 \bar{s}_1)$, which gives

$$S_{\mathrm{fi}} = -\frac{i}{\sqrt{Z_2}} \int d^4x \, \bar{v}_{\bar{\mathbf{p}}_1 \bar{s}_1}(x)(\overrightarrow{i\partial\!\!\!/_x - m})\langle f; \text{out}|\hat{\bar{\psi}}(x)|i - (\bar{p}_1 \bar{s}_1); \text{in}\rangle. \tag{5.336}$$

Extracting an quasi-particle $(q_1 r_1)$ in the final state gives

$$S_{\mathrm{fi}} = -\frac{i}{\sqrt{Z_2}} \int d^4x \, \bar{u}_{\mathbf{q}_1 r_1}(x)(\overrightarrow{i\partial\!\!\!/_x - m})\langle f - (q_1 r_1); \text{out}|\hat{\psi}(x)|i; \text{in}\rangle \tag{5.337}$$

and extracting an quasi-antiparticle $(\bar{q}_1 \bar{r}_1)$ in the final state gives

$$S_{\mathrm{fi}} = \frac{i}{\sqrt{Z_2}} \int d^4x \langle f - (\bar{q}_1 \bar{r}_1); \text{out}|\hat{\psi}(x)|i; \text{in}\rangle(\overleftarrow{-i\partial\!\!\!/_x - m})v_{\bar{\mathbf{q}}_1 \bar{r}_1}(x). \tag{5.338}$$

Continuing the extracting procedure leads to the general LSZ reduction formula for spinor fermions

$$
\begin{aligned}
S_{\mathrm{fi}} &= \left(-\frac{i}{\sqrt{Z_2}}\right)^{n+m} \left(\frac{i}{\sqrt{Z_2}}\right)^{\bar{n}+\bar{m}} \\
&\quad \times \int d^4x_1 \cdots d^4x_n d^4\bar{x}_1 \cdots d^4\bar{x}_{\bar{n}} d^4y_1 \cdots d^4y_m d^4\bar{y}_1 \cdots d^4\bar{y}_{\bar{m}} \\
&\quad \times \bar{u}_{\mathbf{q}_m r_m}(y_m)(\overrightarrow{i\partial\!\!\!/_{y_m} - m}) \cdots \bar{u}_{\mathbf{q}_1 r_1}(y_1)(\overrightarrow{i\partial\!\!\!/_{y_1} - m}) \\
&\quad \times \bar{v}_{\bar{\mathbf{p}}_{\bar{n}} \bar{s}_{\bar{n}}}(\bar{x}_{\bar{n}})(\overrightarrow{i\partial\!\!\!/_{\bar{x}_{\bar{n}}} - m}) \cdots \bar{v}_{\bar{\mathbf{p}}_1 \bar{s}_1}(\bar{x}_1)(\overrightarrow{i\partial\!\!\!/_{\bar{x}_1} - m}) \\
&\quad \times \langle 0|T(\hat{\bar{\psi}}(\bar{y}_{\bar{m}}) \cdots \hat{\bar{\psi}}(\bar{y}_1)\hat{\psi}(y_m) \cdots \hat{\psi}(y_1) \\
&\quad \times \hat{\bar{\psi}}(x_1) \cdots \hat{\bar{\psi}}(x_n)\hat{\psi}(\bar{x}_1) \cdots \hat{\psi}(\bar{x}_{\bar{n}}))|0\rangle \\
&\quad \times (\overleftarrow{-i\partial\!\!\!/_{x_1} - m})u_{\mathbf{p}_1 s_1}(x_1) \cdots (\overleftarrow{-i\partial\!\!\!/_{x_n} - m})u_{\mathbf{p}_n s_n}(x_n) \\
&\quad \times (\overleftarrow{-i\partial\!\!\!/_{\bar{y}_1} - m})v_{\bar{\mathbf{q}}_1 \bar{r}_1}(\bar{y}_1)(\overleftarrow{-i\partial\!\!\!/_{\bar{y}_{\bar{m}}} - m}) \cdots v_{\bar{\mathbf{q}}_{\bar{m}} \bar{r}_{\bar{m}}}(\bar{y}_{\bar{m}}),
\end{aligned}
\tag{5.339}
$$

which shows that the S matrix is related to the many-particle Green's functions.

From the LSZ reduction formula, we can see that the quasi-particles with interaction can be described by the similar forms of the Lagrangians and the equations of motion of the free particles, adding with the interaction modification, such as replacing mass with renormalized mass. It should be noted that the renormalized parameters are used in the Lagrangians and the equations of motion for the quasi-particles with interaction.

5.8.8   *Functional form of S matrix*

The S matrix can be expressed in path integral formulation, which can be calculated more easily. Now we show that the S matrix in Eq.(5.322) for scalar bosons can be expressed in the following functional form

$$
\hat{S} =: \exp\left(\frac{1}{\sqrt{Z}} \int \hat{\phi}_{\text{in}}(x)(\Box^2 + m^2)\frac{\delta}{\delta J(x)} d^4x\right) : Z[J]\big|_{J=0}
$$

$$
= \sum_{k=0}^{\infty}\left[\frac{1}{Z^{\frac{k}{2}}k!} \int d^4x_1 \cdots d^4x_k : \hat{\phi}_{\text{in}}(x_1)\cdots\hat{\phi}_{\text{in}}(x_k):\right. \tag{5.340}
$$

$$
\times (\Box_1^2 + m^2)\cdots(\Box_k^2 + m^2)\frac{\delta^k}{\delta J(x_1)\cdots\delta J(x_k)}Z[J]\Big]\Big|_{J=0},
$$

where :: denotes the normal ordered product of field operators. To show Eq.(5.340) is equivalent to Eq.(5.322), we calculate the matrix element $S_{\text{fi}} = \langle f; \text{out}|i; \text{in}\rangle$ using Eq.(5.340). We have

$$
S_{\text{fi}} = \langle q_1, \cdots, q_m; \text{in}|\hat{S}|p_{m+1}, \cdots, p_{m+n}; \text{in}\rangle
$$

$$
= \int \prod_{i=1}^{m} d^4x_i \prod_{j=m+1}^{m+n} d^4x_j \frac{1}{(\sqrt{Z})^{m+n}(m+n)!} \tag{5.341}
$$

$$
\times \langle q_1, \cdots, q_m; \text{in}| : \hat{\phi}_{\text{in}}(x_1)\cdots\hat{\phi}_{\text{in}}(x_{m+n}) : |p_{m+1}, \cdots, p_{m+n}; \text{in}\rangle
$$

$$
\times (\Box_1^2 + m^2)\cdots(\Box_{m+n}^2 + m^2)i^{m+n}G(x_1, \cdots, x_{m+n}).
$$

The nonzero terms contain m creation operators and n annihilation operators. Since each field operator has two terms added: one contains creation operator and the other annihilation operator, we have in total 2^{m+n} nonzero product terms. Using Eq.(5.307), we have

$$
\langle q_1, \cdots, q_m; \text{in}| : \hat{\phi}_{\text{in}}(x_1) \cdots \hat{\phi}_{\text{in}}(x_{m+n}) : |p_{m+1}, \cdots, p_{m+n}; \text{in}\rangle
$$

$$
= \int d^3 q_1' \cdots d^3 q_m' d^3 p_{m+1}' \cdots d^3 p_{m+n}' \frac{(m+n)!}{m!n!} \langle q_1, \cdots, q_m; \text{in}|
$$

$$
\times \prod_{k=1}^{m} u_{\mathbf{q}_k'}^*(x_k) \hat{a}_{\mathbf{q}_k'}^\dagger \prod_{l=m+1}^{m+n} u_{\mathbf{p}_l'}(x_l) \hat{a}_{\mathbf{p}_l'} |p_{m+1}, \cdots, p_{m+n}; \text{in}\rangle \qquad (5.342)
$$

$$
= (m+n)! \prod_{k=1}^{m} u_{\mathbf{q}_k}^*(x_k) \prod_{l=m+1}^{m+n} u_{\mathbf{p}_l}(x_l).
$$

The degeneracy term $\frac{(m+n)!}{m!n!}$ comes from the binomial expansion in the normal mode expression $(u\hat{a} + u^*\hat{a}^\dagger)^{m+n}$. The integration over $\mathbf{q}'$ and $\mathbf{p}'$ gives an extra degeneracy factor $m!n!$. Thus the final degeneracy factor in Eq.(5.342) is $(m+n)!$. Inserting Eq.(5.342) into Eq.(5.341), we obtain

$$
S_{fi} = \left(\frac{i}{\sqrt{Z}}\right)^{m+n} \int \prod_{i=1}^{m} d^4 x_i \prod_{j=m+1}^{m+n} d^4 x_j
$$

$$
\times \prod_{i=1}^{m} u_{\mathbf{q}_i}^*(x_i) \prod_{j=m+1}^{m+n} u_{\mathbf{p}_j}(x_j) \qquad (5.343)
$$

$$
\times (\Box_1^2 + m^2) \cdots (\Box_{m+n}^2 + m^2) G(x_1, \cdots, x_{m+n}),
$$

which is same as Eq.(5.322).

5.9 Generating functional for fermion field

Now we consider the generating functional for a fermion field. The Lagrangian density for a free fermion field is given by Eq.(2.233), which reads

$$
\mathcal{L}_0 = i\bar{\psi}\gamma^\mu \partial_\mu \psi - m\bar{\psi}\psi. \qquad (5.344)
$$

The generating functional for the free fermion field is defined by

$$
\mathcal{Z}_0[\eta, \bar{\eta}] \equiv \frac{1}{Z_0[0]} \int D\bar{\psi}D\psi \exp\Big\{ i \int [\bar{\psi}(x)(i\gamma \cdot \partial - m)\psi(x)
$$

$$
+ \bar{\eta}(x)\psi(x) + \bar{\psi}(x)\eta(x)]d^4 x \Big\}, \qquad (5.345)
$$

where

$$
Z_0[0] = \int D\bar{\psi}D\psi \exp\Big[i \int \bar{\psi}(x)(i\gamma \cdot \partial - m)\psi(x)d^4 x \Big]. \qquad (5.346)
$$

$\bar{\eta}(x)$ in Eq.(5.345) is the source for $\psi(x)$ and $\eta(x)$ is the source for $\bar{\psi}(x)$. To simplify the notation, we introduce the operator S_F defined by

$$S_F^{-1} \equiv i\gamma^\mu \partial_\mu - m. \tag{5.347}$$

S_F is called the Feynman propagator for spinor fermions. Using Eq.(5.40), we have

$$\begin{aligned} S_F^{-1}(i\gamma \cdot \partial + m)\Delta_F(x) &= (-\Box - m^2)\Delta_F(x) \\ &= \delta^4(x), \end{aligned} \tag{5.348}$$

where $\Delta_F(x)$ is the Feynman propagator for scalar bosons. Thus the Feynman propagator for spinor fermions can be expressed as

$$S_F(x) = (i\gamma^\mu \partial_\mu + m)\Delta_F(x). \tag{5.349}$$

In the momentum space, we have

$$S_F(p) = \frac{\not{p} + m}{p^2 - m^2 + i\epsilon} \tag{5.350}$$

or

$$S_F^{-1}(p) = \gamma_\mu p^\mu - m. \tag{5.351}$$

The Feynman propagator for spinor fermions can also be defined as

$$iS_{\alpha\beta}(x - y) \equiv \langle 0|T(\hat{\psi}_\alpha(x)\hat{\bar{\psi}}_\beta(y))|0\rangle. \tag{5.352}$$

In the following, we show that Eq.(5.352) is equivalent to Eq.(5.347). Eq.(5.352) can be rewritten as

$$\begin{aligned} iS_{\alpha\beta}(x - y) &= \Theta(x_0 - y_0)\langle 0|\hat{\psi}_\alpha(x)\hat{\bar{\psi}}_\beta(y)|0\rangle \\ &\quad - \Theta(y_0 - x_0)\langle 0|\hat{\bar{\psi}}_\beta(y)\hat{\psi}_\alpha(x)|0\rangle. \end{aligned} \tag{5.353}$$

Inserting the plane wave expansion Eq.(2.427) of the field operator $\hat{\psi}(x)$ into Eq.(5.353) and using the anti-commutation relations Eq.(2.426), we have

$$\begin{aligned} iS_{\alpha\beta}(x - y) &= \int \frac{d^3p}{(2\pi)^3} \frac{m}{\omega_\mathbf{p}} \Big[\Theta(x_0 - y_0)e^{-ip\cdot(x-y)}\sum_s u_\alpha(\mathbf{p}, s)\bar{u}_\beta(\mathbf{p}, s) \\ &\quad - \Theta(y_0 - x_0)e^{ip\cdot(x-y)}\sum_s v_\alpha(\mathbf{p}, s)\bar{v}_\beta(\mathbf{p}, s)\Big]. \end{aligned} \tag{5.354}$$

Using the free Dirac equation Eq.(2.429), we have

$$(H_\mathbf{p} - \omega_\mathbf{p})u(\mathbf{p}, s) = 0, \tag{5.355a}$$

$$(H_\mathbf{p} + \omega_\mathbf{p})v(-\mathbf{p}, s) = 0, \tag{5.355b}$$

where

$$H_{\mathbf{p}} \equiv \gamma^0 (\boldsymbol{\gamma} \cdot \mathbf{p} + m) \tag{5.356}$$

is the Dirac Hamiltonian. Eq.(5.355) can also be written as

$$(H_{\mathbf{p}} + \omega_{\mathbf{p}})u(\mathbf{p}, s) = 2\omega_{\mathbf{p}} u(\mathbf{p}, s), \tag{5.357a}$$

$$(H_{\mathbf{p}} - \omega_{\mathbf{p}})v(\text{-}\mathbf{p}, s) = -2\omega_{\mathbf{p}} v(\text{-}\mathbf{p}, s). \tag{5.357b}$$

The completeness relation Eq.(2.391c) can be expressed in terms of $u(\mathbf{p}, s)$ and $v(\text{-}\mathbf{p}, s)$

$$\sum_s [u_\alpha(\mathbf{p}, s)u_\beta^\dagger(\mathbf{p}, s) + v_\alpha(-\mathbf{p}, s)v_\beta^\dagger(-\mathbf{p}, s)] = \frac{\omega_{\mathbf{p}}}{m}\delta_{\alpha\beta}. \tag{5.358}$$

Multiplying both sides of Eq.(5.358) by $H_{\mathbf{p}} + \omega_{\mathbf{p}}$, and using Eq.(5.355) and Eq.(5.357), we have

$$2\omega_{\mathbf{p}} \sum_s u_\alpha(\mathbf{p}, s)u_\beta^\dagger(\mathbf{p}, s) = \frac{\omega_{\mathbf{p}}}{m}[\gamma^0(\boldsymbol{\gamma} \cdot \mathbf{p} + m) + \omega_{\mathbf{p}}]_{\alpha\beta}. \tag{5.359}$$

Multiplying both sides of Eq.(5.359) by γ^0 from the right, we obtain

$$\begin{aligned}
\sum_s u_\alpha(\mathbf{p}, s)\bar{u}_\beta^\dagger(\mathbf{p}, s) &= \frac{1}{2m}(\gamma^0 \omega_{\mathbf{p}} - \boldsymbol{\gamma} \cdot \mathbf{p} + m)_{\alpha\beta} \\
&= \left(\frac{\gamma_\mu p^\mu + m}{2m}\right)_{\alpha\beta}.
\end{aligned} \tag{5.360}$$

Similarly, we have

$$\sum_s v_\alpha(\mathbf{p}, s)v_\beta^\dagger(\mathbf{p}, s) = \left(\frac{\gamma_\mu p^\mu - m}{2m}\right)_{\alpha\beta}. \tag{5.361}$$

Inserting Eq.(5.360) and Eq.(5.361) into Eq.(5.354), we have

$$\begin{aligned}
iS_{\alpha\beta}(x - y) &= \int \frac{d^3p}{(2\pi)^3} \frac{1}{2\omega_{\mathbf{p}}} \big[\Theta(x_0 - y_0)e^{-ip\cdot(x-y)}(\not{p} + m)_{\alpha\beta} \\
&\quad - \Theta(y_0 - x_0)e^{ip\cdot(x-y)}(\not{p} - m)_{\alpha\beta}\big] \\
&= \Theta(x_0 - y_0)(i\not{\partial} + m)_{\alpha\beta}i\Delta^{(+)}(x - y) \\
&\quad - \Theta(y_0 - x_0)(i\not{\partial} + m)_{\alpha\beta}i\Delta^{(-)}(x - y).
\end{aligned} \tag{5.362}$$

In the derivation of the last line of Eq.(5.362), Eq.(5.8) has been used. Using Eq.(5.29), we obtain

$$\begin{aligned}
S_{\alpha\beta}(x - y) &= (i\not{\partial} + m)_{\alpha\beta}[\Theta(x_0 - y_0)\Delta^{(+)}(x - y) \\
&\quad - \Theta(y_0 - x_0)\Delta^{(-)}(x - y)] \\
&= (i\not{\partial} + m)_{\alpha\beta}\Delta_{\mathrm{F}}(x - y),
\end{aligned} \tag{5.363}$$

which gives Eq.(5.349). Thus Eq.(5.352) is equivalent to Eq.(5.347).

Using the Feynman propagator S_F, the generating functional can be expressed as

$$\mathcal{Z}_0[\eta, \bar{\eta}] = \frac{1}{Z_0[0]} \int D\bar{\psi} D\psi \exp\left[i \int (\bar{\psi} S_F^{-1} \psi + \bar{\eta}\psi + \bar{\psi}\eta) d^4 x\right]. \quad (5.364)$$

We define

$$Q(\psi, \bar{\psi}) \equiv \bar{\psi} S_F^{-1} \psi + \bar{\eta}\psi + \bar{\psi}\eta. \quad (5.365)$$

In order to calculate the generating functional, we first determine the values of ψ_m and $\bar{\psi}_m$, which minimize Q. ψ_m and $\bar{\psi}_m$ satisfy the equations $\frac{\delta Q}{\delta \bar{\psi}_m} = 0$ and $\frac{\delta Q}{\delta \psi_m} = 0$, respectively. We have

$$\psi_m = -S_F \eta, \quad (5.366a)$$

$$\bar{\psi}_m = -\bar{\eta} S_F. \quad (5.366b)$$

Thus the minimum value of Q is given by

$$Q_m = Q(\psi_m, \bar{\psi}_m) = -\bar{\eta} S_F \eta. \quad (5.367)$$

Then Q can be rewritten as

$$Q(\psi, \bar{\psi}) = Q_m + (\bar{\psi} - \bar{\psi}_m) S_F^{-1} (\psi - \psi_m). \quad (5.368)$$

Inserting Eq.(5.368) into Eq.(5.364), we have

$$\mathcal{Z}_0[\eta, \bar{\eta}] = \frac{1}{Z_0[0]} \int D\bar{\psi} D\psi \exp\left\{i \int [Q_m + (\bar{\psi} - \bar{\psi}_m) S_F^{-1} (\psi - \psi_m)] d^4 x\right\}$$

$$= \frac{1}{Z_0[0]} \exp\left[-i \int \bar{\eta}(x) S_F \eta(y) d^4 x d^4 y\right] \det(-i S_F^{-1}).$$

$$(5.369)$$

Since

$$Z_0[0] = \det(-i S_F^{-1}), \quad (5.370)$$

we have

$$\mathcal{Z}_0[\eta, \bar{\eta}] = \exp\left[-i \int \bar{\eta}(x) S_F(x - y) \eta(y) d^4 x d^4 y\right]. \quad (5.371)$$

Similar to the scalar field, we define the free n-point functions for the fermion field.

$$G_0^{(2n)}(y_1, \cdots, y_n; x_1, \cdots, x_n)$$

$$= \left(\frac{1}{i}\right)^{2n} \frac{\delta^{2n} \mathcal{Z}_0[\eta, \bar{\eta}]}{\delta\eta(x_n) \cdots \delta\eta(x_1) \delta\bar{\eta}(y_1) \cdots \delta\bar{\eta}(y_n)}\Bigg|_{\eta=\bar{\eta}=0}. \quad (5.372)$$

Using Eq.(5.371), we have

$$G_0^{(2)}(y;x) = \left(\frac{1}{i}\right)^2 \frac{\delta^2 \mathcal{Z}_0[\eta, \bar{\eta}]}{\delta\eta(x)\delta\bar{\eta}(y)}\Big|_{\eta=\bar{\eta}=0}$$

$$= -\frac{\delta}{\delta\eta(x)} \frac{\delta}{\delta\bar{\eta}(y)} \exp\left[-i \int \bar{\eta}(x) S_F(x-y)\eta(y) d^4x d^4y\right]\Big|_{\eta=\bar{\eta}=0}$$

$$= iS_F(x-y).$$

$$(5.373)$$

When interaction is considered, the Lagrangian density $\mathcal{L}$ becomes

$$\mathcal{L} = \mathcal{L}_0 + \mathcal{L}_{int}(\bar{\psi}, \psi). \qquad (5.374)$$

It is east to show that

$$\mathcal{Z}[\eta, \bar{\eta}] = \exp\left[i \int \mathcal{L}_{int}\left(\frac{1}{i}\frac{\delta}{\delta\eta}, \frac{1}{i}\frac{\delta}{\delta\bar{\eta}}\right) d^4x\right] \mathcal{Z}_0[\eta, \bar{\eta}], \qquad (5.375)$$

where $\mathcal{Z}_0[\eta, \bar{\eta}]$ is given by Eq.(5.371).

5.10 Feynman propagator for photon field

Now we consider the Feynman propagator of a photon field. The Lagrangian density $\mathcal{L}$ of a photon field is given by Eq.(2.550). Thus the generating functional for a photon field has the form

$$Z[J] = \int DA_\mu \exp\left[i \int (\mathcal{L} - j_e^\mu A_\mu) d^4x\right] \qquad (5.376)$$

with

$$\mathcal{L} = -\frac{1}{4} F_{\mu\nu} F^{\mu\nu} - \frac{1}{2\xi}(\partial_\mu A^\mu)^2 \qquad (5.377)$$

and

$$j_e^\mu = e\bar{\psi}\gamma^\mu\psi. \qquad (5.378)$$

The Lagrangian density $\mathcal{L}$ can be expressed in the following form

$$\mathcal{L} = -\frac{1}{4} F_{\mu\nu} F^{\mu\nu} - \frac{1}{2\xi}(\partial_\mu A^\mu)^2$$

$$= \frac{1}{2} A^\mu \left[\eta_{\mu\nu}\Box + \left(\frac{1}{\xi} - 1\right)\partial_\mu\partial_\nu\right] A^\nu$$

$$+ \frac{1}{2}\partial_\mu [A_\nu(\partial^\nu A^\mu) - (\partial_\nu A^\nu)A^\mu] \qquad (5.379)$$

$$- \frac{1}{2}\partial_\mu (A_\nu \partial^\mu A^\nu) - \frac{1}{2}\partial_\mu \left[\left(\frac{1}{\xi} - 1\right) A^\mu \partial_\nu A^\nu\right]$$

$$= \frac{1}{2} A^\mu \left[\eta_{\mu\nu}\Box + \left(\frac{1}{\xi} - 1\right)\partial_\mu\partial_\nu\right] A^\nu.$$

In the derivation of the last line of Eq.(5.379), we have discarded the divergence terms because they yield only surface contribution. In the momentum space, the operator $\eta_{\mu\nu}\Box + \left(\frac{1}{\xi} - 1\right)\partial_\mu\partial_\nu$ becomes $-\eta_{\mu\nu}k^2 + \left(1 - \frac{1}{\xi}\right)k_\mu k_\nu$. The inverse of $-\eta_{\mu\nu}k^2 + \left(1 - \frac{1}{\xi}\right)k_\mu k_\nu$ is $-\frac{1}{k^2}[\eta^{\mu\nu} + (\xi - 1)\frac{k^\mu k^\nu}{k^2}]$, which gives the Feynman propagator

$$D_{\mathrm{F}}^{\mu\nu}(k) = -\frac{1}{k^2}\left[\eta^{\mu\nu} + (\xi - 1)\frac{k^\mu k^\nu}{k^2}\right]. \tag{5.380}$$

The second term on the right hand side of Eq.(5.380) depend on the value of ξ, which is related to the choice of the gauge. Since ξ is a parameter which can be chosen to be any value, the physics is not affected by the choice of the gauge. The choice of $\xi = 1$ is called the Feynman gauge and the choice of $\xi = 0$ is called the Landau gauge.

When $\xi = 1$, the Lagrangian density takes the form

$$\mathcal{L} = -\frac{1}{4}F_{\mu\nu}F^{\mu\nu} - \frac{1}{2}(\partial_\mu A^\mu)^2. \tag{5.381}$$

After integration by parts and dropping out the surface terms, the Lagrangian density in Eq.(5.381) becomes

$$\mathcal{L} = \frac{1}{2}A^\mu\eta_{\mu\nu}\Box A^\nu. \tag{5.382}$$

The inverse of the operator $\eta_{\mu\nu}\Box$ is the Feynman propagator of the photon field $D_{\mathrm{F}}^{\mu\nu}(x - y)$. Thus we have

$$\begin{aligned}
iD_{\mathrm{F}}^{\mu\nu}(x - y) &= \langle 0|T(A^\mu(x)A^\nu(y))|0\rangle \\
&= -i\eta^{\mu\nu}\Delta_{\mathrm{F}}(x - y; m = 0) \\
&= -i\eta^{\mu\nu}\int \frac{d^4k}{(2\pi)^4}\frac{e^{-ik\cdot(x-y)}}{k^2 + i\epsilon}.
\end{aligned} \tag{5.383}$$

In the momentum space, the Feynman propagator takes the form

$$D_{\mathrm{F}}^{\mu\nu}(k) = \frac{-\eta^{\mu\nu}}{k^2 + i\epsilon}. \tag{5.384}$$

5.11   Coulomb interaction

We divide A_μ into two parts $A_\mu^\perp$ and $A_\mu^\parallel$. $A_\mu^\perp$ is perpendicular to j_e^μ and $A_\mu^\parallel$ is parallel to j_e^μ. Since $A_\mu^\perp$ is the radiation field without source, we can use the radiation gauge $\boldsymbol{\nabla}\cdot\mathbf{A}^\perp = 0$ and $A_0^\perp = 0$. For the two transverse components $\hat{A}_\mu^\perp(\mu = 1, 2)$, using Eq.(2.562), we have

$$\frac{\partial\hat{F}^{\perp\mu\nu}}{\partial x^\nu} = 0, \tag{5.385}$$

where

$$\hat{F}^{\perp\mu\nu} = \frac{\partial \hat{A}^{\perp\nu}}{\partial x_\mu} - \frac{\partial \hat{A}^{\perp\mu}}{\partial x_\nu}. \tag{5.386}$$

Since

$$\hat{\mathcal{L}}^\perp = -\frac{1}{4}\hat{F}^\perp_{\mu\nu}\hat{F}^{\perp\mu\nu}$$
$$= -\frac{1}{2}(\partial_\mu \hat{A}^\perp_\nu)(\partial^\mu \hat{A}^{\perp\nu}) + \frac{1}{2}(\partial_\mu \hat{A}^\perp_\nu)(\partial^\nu \hat{A}^{\perp\mu}), \tag{5.387}$$

we have

$$\frac{\partial \hat{\mathcal{L}}^\perp}{\partial \hat{A}^\perp_\mu} - \frac{\partial}{\partial x^\nu}\frac{\partial \hat{\mathcal{L}}^\perp}{\partial(\partial_\nu \hat{A}^\perp_\mu)} = \frac{\partial \hat{F}^{\perp\mu\nu}}{\partial x^\nu} = 0. \tag{5.388}$$

Thus the two transverse components $\hat{A}^\perp_\mu(\mu = 1, 2)$ satisfy the Euler-Lagrange equation in operator form. In the frame where $A^\perp_3 = 0$, Eq.(5.385) is valid for all the components. Since Eq.(5.385) is a tensor equation, It is valid for all the frames. Thus we have the Euler-Lagrange equation in operator form for $\hat{A}^\perp_\mu$.

$$\frac{\partial \hat{\mathcal{L}}^\perp}{\partial \hat{A}^\perp_\mu} - \frac{\partial}{\partial x^\mu}\frac{\partial \hat{\mathcal{L}}^\perp}{\partial(\partial_\mu \hat{A}^\perp_\mu)} = 0. \tag{5.389}$$

$A^\parallel_\mu$ is coupled to the source j^μ_e. Since we have the gauge transformation $A^\parallel_\mu \to A'^\parallel_\mu = A^\parallel_\mu - \partial_\mu\Lambda$ as given in Eq.(2.551), we can take $\Box\hat{A}^0 = -\hat{j}^0_e$. We also have the gauge transformation $\partial_\mu A^\parallel_\mu \to \partial_\mu A'^\parallel_\mu = \partial_\mu A^\parallel_\mu - \sigma$ as given in Eq.(2.552). Thus we can take $\partial_\mu \hat{A}^\parallel_\mu = 0$ (Lorentz gauge) with proper choice of σ. Then we can similarly prove that

$$\frac{\partial \hat{F}^{\parallel\mu\nu}}{\partial x^\nu} = -\hat{j}^\mu_e \tag{5.390}$$

in the frame $A^\parallel_3 = 0$. Eq.(5.390) is a tensor equation. It should be valid for any frame.

Thus we have the Euler-Lagrange equation in operator form for the massless vector bosons.

$$\frac{\partial \hat{\mathcal{L}}}{\partial \hat{A}_\mu} - \frac{\partial}{\partial x^\mu}\frac{\partial \hat{\mathcal{L}}}{\partial(\partial_\mu \hat{A}_\mu)} = 0 \tag{5.391}$$

with

$$\hat{A}_\mu = \hat{A}^\perp_\mu + \hat{A}^\parallel_\mu. \tag{5.392}$$

The Euler-Lagrange equation in operator form for the massless vector bosons leads to the following Maxwell equations in operator form.

$$\frac{\partial \hat{F}^{\mu\nu}}{\partial x^\nu} = -\hat{j}_e^\mu \qquad (5.393)$$

with

$$\hat{j}_e^\mu = e\hat{\bar{\psi}}\gamma^\mu\hat{\psi}. \qquad (5.394)$$

Since

$$\hat{F}^{\mu\nu} = \frac{\partial \hat{A}^\nu}{\partial x_\mu} - \frac{\partial \hat{A}^\mu}{\partial x_\nu}, \qquad (5.395)$$

$\hat{j}_e^\mu$ satisfies the equation of continuity for the electric current vector

$$\partial_\mu \hat{j}_e^\mu = 0. \qquad (5.396)$$

Inserting Eq.(5.395), Eq.(5.393) becomes

$$\Box \hat{A}^\mu - \partial^\mu(\partial_\nu \hat{A}^\nu) = \hat{j}_e^\mu. \qquad (5.397)$$

or in the spatial and time components separately

$$\Box \hat{\mathbf{A}} - \boldsymbol{\nabla}\left(\frac{\partial}{\partial t}\hat{A}_0 + \boldsymbol{\nabla}\cdot\hat{\mathbf{A}}\right) = \hat{\mathbf{j}}_e, \qquad (5.398a)$$

$$-\Delta\hat{A}_0 - \frac{\partial}{\partial t}(\boldsymbol{\nabla}\cdot\hat{\mathbf{A}}) = \hat{\rho} \qquad (5.398b)$$

with

$$\hat{\rho} = \hat{j}_e^0. \qquad (5.399)$$

As we have shown, $\hat{A}^\mu$ is not uniquely determined because there is a gauge transformation

$$\hat{A}'^\mu(x) = \hat{A}^\mu(x) - \partial^\mu\hat{\Lambda}(x). \qquad (5.400)$$

In section 2.6.2.2, we have used the Coulomb gauge to derive the Lagrangian density of the gauge bosons. The formulas in the Coulomb gauge are not in the covariance form. They are expressed in a special frame. Thus the calculations using the Coulomb gauge should be restricted in this frame. Through the gauge transformation Eq.(5.400), we could use other gauge in which the formulations can be expressed in a covariance form.

Suppose we have used other gauge and then we want to go back the Coulomb gauge. The Coulomb gauge can be satisfied if we choose the gauge transformation Eq.(5.400) in such a way that

$$\boldsymbol{\nabla}\cdot\hat{\mathbf{A}}' = 0, \qquad (5.401)$$

which gives

$$\nabla \cdot \hat{\mathbf{A}} - \Delta \hat{\Lambda} = 0 \tag{5.402}$$

or

$$\Delta \hat{\Lambda} = \nabla \cdot \hat{\mathbf{A}}. \tag{5.403}$$

Thus Λ is a solution of the Poisson equation Eq.(5.403). The Poisson equation Eq.(5.403) can be calculated using Green's function of the Laplacian defined by

$$\Delta G(\mathbf{x} - \mathbf{x}') = \delta^3(\mathbf{x} - \mathbf{x}'). \tag{5.404}$$

One can easily verify that $G(\mathbf{x} - \mathbf{x}')$ satisfying Eq.(5.404) has the following form

$$G(\mathbf{x} - \mathbf{x}') = -\frac{1}{4\pi} \frac{1}{|\mathbf{x} - \mathbf{x}'|}. \tag{5.405}$$

Thus the transformed vector potential operator has the form

$$\hat{\mathbf{A}}' = \hat{\mathbf{A}} - \nabla\hat{\Lambda} = \hat{\mathbf{A}} - \nabla \int d^3x\, G(\mathbf{x} - \mathbf{x}')\nabla' \cdot \hat{\mathbf{A}}(\mathbf{x}', t). \tag{5.406}$$

Eq.(5.406) can be understood as a projection of $\hat{\mathbf{A}}$ into the transverse parts which are the only components existed in the Coulomb gauge.

We split the vector potential operator $\hat{\mathbf{A}}$ into the transverse part $\hat{\mathbf{A}}_\perp$ and longitudinal part $\hat{\mathbf{A}}_\|$.

$$\hat{\mathbf{A}} = \hat{\mathbf{A}}_\perp + \hat{\mathbf{A}}_\|. \tag{5.407}$$

$\hat{\mathbf{A}}_\perp$ and $\hat{\mathbf{A}}_\|$ satisfy the constrained conditions

$$\nabla \cdot \hat{\mathbf{A}}_\perp = 0 \tag{5.408}$$

and

$$\nabla \times \hat{\mathbf{A}}_\| = 0. \tag{5.409}$$

The splitting can be realized by using the projection operators given by Eq.(2.554). Using the projection operators

$$(\hat{P}_\perp)_{ij} = \delta_{ij} - \partial_i \frac{1}{\Delta}\partial_j, \tag{5.410a}$$

$$(\hat{P}_\|)_{ij} = \partial_i \frac{1}{\Delta}\partial_j, \tag{5.410b}$$

we have

$$\hat{\mathbf{A}}_\perp = \hat{P}_\perp\hat{\mathbf{A}}, \tag{5.411a}$$

$$\hat{\mathbf{A}}_\| = \hat{P}_\|\hat{\mathbf{A}}. \tag{5.411b}$$

Thus Eq.(5.406) can be written as

$$\hat{\mathbf{A}}' = \hat{\mathbf{A}}_{\perp} = \hat{\mathbf{A}} - \boldsymbol{\nabla} \int d^3x\, G(\mathbf{x} - \mathbf{x}')\boldsymbol{\nabla}' \cdot \hat{\mathbf{A}}(\mathbf{x}', t). \qquad (5.412)$$

In the Coulomb gauge, Eq.(5.398) becomes

$$\Box\hat{\mathbf{A}} - \frac{\partial}{\partial t}\boldsymbol{\nabla}\hat{A}_0 = \hat{\mathbf{j}}_e, \qquad (5.413a)$$

$$\Delta\hat{A}_0 = -\hat{\rho}. \qquad (5.413b)$$

$\hat{A}_0$ satisfies the Poisson equation. Its solution takes the form

$$\hat{A}_0(\mathbf{x}, t) = \int d^3x' \frac{1}{4\pi} \frac{\hat{\rho}(\mathbf{x}', t)}{|\mathbf{x} - \mathbf{x}'|}, \qquad (5.414)$$

which is the instantaneous Coulomb potential. This equation does not violate the causality principle because $\hat{A}_0$ is an artificial component of the field operator.

Now we consider the free electromagnetic field or pure radiation field. In this case, $\hat{j}_e^{\mu} = 0$. Then $\hat{A}_0 = 0$. $\boldsymbol{\nabla} \cdot \hat{\mathbf{A}} = 0$ and $\hat{A}_0 = 0$ are the radiation gauge used in the sections 2.6.2.3 and 2.6.2.4 to derive the generator of time translation transformation $\hat{G}_t$ for massless vector bosons. Eq.(2.563) for $\hat{\mathbf{A}}$ can be generalized to express four vector $\hat{A}$ by introducing the transverse polarization vectors ϵ^{μ} in four dimensional spacetime with the following properties.

$$\epsilon^0(\mathbf{k}, l) = 0, \qquad (5.415a)$$

$$\mathbf{k} \cdot \boldsymbol{\epsilon}(\mathbf{k}, l) = 0, \qquad (5.415b)$$

$$\boldsymbol{\epsilon}(\mathbf{k}, l) \cdot \boldsymbol{\epsilon}(\mathbf{k}, l') = \delta_{ll'}. \qquad (5.415c)$$

Thus we have

$$\hat{A}^{\mu}(x) = \int \frac{d^3k}{\sqrt{2\omega_{\mathbf{k}}(2\pi)^3}} \sum_{l=1}^{2} \epsilon^{\mu}(\mathbf{k}, l)(\hat{a}_{\mathbf{k}l}e^{-ik\cdot x} + \hat{a}_{\mathbf{k}l}^{\dagger}e^{ik\cdot x}). \qquad (5.416)$$

Using Eq.(5.416), we can calculate the Feynman propagator $D_{\text{tr}}^{\mu\nu}$ for the transverse radiation field.

$$iD_{\text{tr}}^{\mu\nu}(x - y) = \langle 0|T(A^{\mu}(x)A^{\nu}(y))|0\rangle$$

$$
= \int \frac{d^3 k'}{\sqrt{2\omega_{\mathbf{k}}'(2\pi)^3}} \int \frac{d^3 k}{\sqrt{2\omega_{\mathbf{k}}(2\pi)^3}} \sum_{ll'=1}^{2} \epsilon^\mu(\mathbf{k}', l') \epsilon^\nu(\mathbf{k}, l)
$$

$$
\times \Big(\Theta(x_0 - y_0) e^{-i(k' \cdot x - k \cdot y)} \langle 0 | \hat{a}_{\mathbf{k}'l'} \hat{a}_{\mathbf{k}l}^\dagger | 0 \rangle
$$

$$
+ \Theta(y_0 - x_0) e^{i(k' \cdot x - k \cdot y)} \langle 0 | \hat{a}_{\mathbf{k}l} \hat{a}_{\mathbf{k}'l'}^\dagger | 0 \rangle \Big)
$$

$$
= \int \frac{d^3 k}{2\omega_{\mathbf{k}}(2\pi)^3} \sum_{l=1}^{2} \epsilon^\mu(\mathbf{k}, l) \epsilon^\nu(\mathbf{k}, l)
$$

$$
\times \Big[\Theta(x_0 - y_0) e^{-ik \cdot (x-y)} + \Theta(y_0 - x_0) e^{ik \cdot (x-y)} \Big]
$$

$$
= \int \frac{d^3 k}{2\omega_{\mathbf{k}}(2\pi)^3} \Lambda^{\mu\nu}(k) \Big[\Theta(x_0 - y_0) e^{-ik \cdot (x-y)}
$$

$$
+ \Theta(y_0 - x_0) e^{ik \cdot (x-y)} \Big]
$$

$$
\tag{5.417}
$$

with

$$
\Lambda^{\mu\nu}(k) = \sum_{l=1}^{2} \epsilon^\mu(\mathbf{k}, l) \epsilon^\nu(\mathbf{k}, l). \tag{5.418}
$$

Using Eqs.(5.415) and (2.571), we have

$$
\Lambda^{00} = 0, \quad \Lambda^{0i} = \Lambda^{i0} = 0, \tag{5.419a}
$$

$$
\Lambda^{ij} = \eta^{ij} - \frac{k^i k^j}{\mathbf{k}^2}. \tag{5.419b}
$$

Using Eq.(5.30), Eq.(5.417) becomes

$$
iD_{\mathrm{tr}}^{\mu\nu}(x - y) = \int \frac{d^3 k}{2\omega_{\mathbf{k}}(2\pi)^3} \Lambda^{\mu\nu}(k) \Big[\Theta(x_0 - y_0) e^{-ik \cdot (x-y)}
$$

$$
+ \Theta(y_0 - x_0) e^{ik \cdot (x-y)} \Big] \tag{5.420}
$$

$$
= i \int \frac{d^4 k}{(2\pi)^4} \Lambda^{\mu\nu}(k) \frac{e^{-ik \cdot (x-y)}}{k^2 + i\epsilon}.
$$

In addition to the two transverse polarization vectors, we introduce two other polarization vectors

$$
\epsilon_0^\mu = n^\mu \equiv (1, 0, 0, 0), \tag{5.421a}
$$

$$
\epsilon_3^\mu = \left(0, \frac{\mathbf{k}}{|\mathbf{k}|} \right) = \frac{k^\mu - (k \cdot n) n^\mu}{[(k \cdot n)^2 - k^2]^{\frac{1}{2}}}. \tag{5.421b}
$$

The four polarization vectors satisfy the completeness relation

$$\sum_{\lambda=0}^{3} \eta_{\lambda\lambda}\epsilon^{\mu}(\mathbf{k},\lambda)\epsilon^{\nu}(\mathbf{k},\lambda) = \eta^{\mu\nu}, \qquad (5.422)$$

which can be proved in a similar way as we did for Eq.(2.473). Using Eq.(5.422), we have

$$\Lambda^{\mu\nu}(k) = \sum_{\lambda=1}^{2} \epsilon^{\mu}(\mathbf{k},\lambda)\epsilon^{\nu}(\mathbf{k},\lambda)$$

$$= -\eta^{\mu\nu} - \frac{[k^{\mu} - (k \cdot n)n^{\mu}][k^{\nu} - (k \cdot n)n^{\nu}]}{(k \cdot n)^2 - k^2} + n^{\mu}n^{\nu} \qquad (5.423)$$

$$= -\eta^{\mu\nu} - \frac{k^{\mu}k^{\nu} - (k \cdot n)(n^{\mu}k^{\nu} + n^{\nu}k^{\mu})}{(k \cdot n)^2 - k^2} - \frac{k^2 n^{\mu}n^{\nu}}{(k \cdot n)^2 - k^2}.$$

The contribution from the first term in Eq.(5.423) is the same with the ordinary Feynman propagator given by Eq.(5.383). Since $k^{\mu}j_{\mu} = 0$, the second term in Eq.(5.423) makes no contribution in the actual calculations and can be omitted. The third term in Eq.(5.423) contributes to the Feynman propagator $iD_{\text{tr}}^{\mu\nu}(x - y)$ a term of form

$$-iD_{\text{coul}}^{\mu\nu}(x - y) = -i \int \frac{d^4k}{(2\pi)^4} e^{-ik \cdot (x-y)} \frac{n^{\mu}n^{\nu}}{k^2 + i\epsilon} \frac{k^2}{\mathbf{k}^2}$$

$$= -in^{\mu}n^{\nu} \int \frac{d^3k}{(2\pi)^3} \frac{1}{\mathbf{k}^2} e^{-ik \cdot (x-y)} \delta(x^0 - y^0) \qquad (5.424)$$

$$= -i\delta(x^0 - y^0) \frac{\eta^{\mu 0}\eta^{\nu 0}}{4\pi|\mathbf{x} - \mathbf{y}|}.$$

In the perturbation calculations, this term is cancelled by the Coulomb interaction in the Hamiltonian expressed in the Coulomb gauge. To show that there is a Coulomb interaction term in the Hamiltonian with the Coulomb gauge, we split the electric field $\mathbf{E}$ into the transverse and longitudinal components.

$$\mathbf{E} = \mathbf{E}_{\perp} + \mathbf{E}_{\parallel} \qquad (5.425)$$

with

$$\mathbf{E}_{\perp} = -\dot{\mathbf{A}} \qquad (5.426)$$

and

$$\mathbf{E}_{\parallel} = -\nabla A^0. \qquad (5.427)$$

The Lagrangian density can be expressed in terms of $\mathbf{E}$ and $\mathbf{B}$ as

$$\begin{aligned}
\mathcal{L} &= -\frac{1}{4}F_{\mu\nu}F^{\mu\nu} - j_e^\mu A_\mu \\
&= \frac{1}{2}\left(\mathbf{E}^2 - \mathbf{B}^2\right) - j_e^\mu A_\mu.
\end{aligned} \tag{5.428}$$

Inserting Eq.(5.425) into Eq.(5.428), we have

$$\mathcal{L} = \frac{1}{2}\left[\mathbf{E}_\perp^2 + \mathbf{E}_\parallel^2 + 2\mathbf{E}_\perp \cdot \mathbf{E}_\parallel - (\mathbf{\nabla} \times A)^2\right] - j_e^\mu A_\mu. \tag{5.429}$$

The mixed term $\mathbf{E}_\perp \cdot \mathbf{E}_\parallel$ makes a contribution to the Lagrangian of the form

$$\begin{aligned}
\int d^3x\, \mathbf{E}_\perp \cdot \mathbf{E}_\parallel &= \int d^3x\, \dot{\mathbf{A}} \cdot \nabla A^0 \\
&= -\int d^3x\, A^0 \mathbf{\nabla} \cdot \dot{\mathbf{A}} \\
&= 0.
\end{aligned} \tag{5.430}$$

Thus Eq.(5.429) becomes

$$\mathcal{L} = \frac{1}{2}\left[\dot{\mathbf{A}}^2 + \mathbf{E}_\parallel^2 - (\mathbf{\nabla} \times A)^2\right] - j_e^\mu A_\mu. \tag{5.431}$$

Now we calculate the corresponding Hamiltonian. The canonical energy-momentum tensor $\Theta^{\mu\nu}$ reads

$$\Theta^{\mu\nu} = \frac{\partial \mathcal{L}}{\partial(\partial_\mu A_\sigma)}\partial^\nu A_\sigma - \eta^{\mu\nu}\mathcal{L}. \tag{5.432}$$

Using the relation

$$\frac{\partial F_{\alpha\beta}F^{\alpha\beta}}{\partial(\partial_\mu A_\sigma)} = 4F^{\mu\sigma}, \tag{5.433}$$

we have

$$\Theta^{\mu\nu} = \frac{1}{4}\eta^{\mu\nu}F_{\alpha\beta}F^{\alpha\beta} - F^{\mu\sigma}\partial^\nu A_\sigma + \eta^{\mu\nu}j_e^\sigma A_\sigma. \tag{5.434}$$

We introduce the symmetric energy-momentum tensor

$$T^{\mu\nu} = \Theta^{\mu\nu} + \partial_\kappa \chi^{\kappa\mu\nu} \tag{5.435}$$

with

$$\chi^{\kappa\mu\nu} \equiv -F^{\kappa\mu}A^\nu = F^{\mu\kappa}A^\nu. \tag{5.436}$$

Using $\frac{\partial F^{\mu\nu}}{\partial x^\nu} = -j_e^\mu$, we find

$$\begin{aligned}
T^{\mu\nu} = &\frac{1}{4}\eta^{\mu\nu}F_{\alpha\beta}F^{\alpha\beta} + F^{\mu\sigma}F_\sigma{}^\nu \\
&+ \eta^{\mu\nu}j_e^\sigma A_\sigma - j_e^\mu A^\nu.
\end{aligned} \tag{5.437}$$

Then the Hamiltonian is given by

$$
\begin{aligned}
H &= \int d^3x\, T_0^0 \\
&= \int d^3x \left(\frac{1}{4} F_{\alpha\beta} F^{\alpha\beta} + F^{0\sigma} F_{\sigma 0} + j_e^\sigma A_\sigma - j_e^0 A_0 \right) \\
&= \int d^3x \left[\frac{1}{2} (\mathbf{E}^2 + \mathbf{B}^2) - \mathbf{j}_e \cdot \mathbf{A} \right] \\
&= \int d^3x \left[\frac{1}{2} (\mathbf{E}_\perp^2 + \mathbf{B}^2 + \mathbf{E}_\parallel^2) - \mathbf{j}_e \cdot \mathbf{A} \right].
\end{aligned}
\tag{5.438}
$$

In Eq.(5.438), the first two terms are the contributions from the radiation field. The third term is the Coulomb interaction of the charge density.

$$
\begin{aligned}
H_{\text{coul}} &= \int d^3x\, \frac{1}{2} \mathbf{E}_\parallel^2 \\
&= \int d^3x\, \frac{1}{2} (\nabla A_0)^2 \\
&= -\int d^3x\, \frac{1}{2} A_0 \nabla^2 A_0 \\
&= \frac{1}{2} \int d^3x \int d^3x'\, \frac{\rho(\mathbf{x}',t)\rho(\mathbf{x},t)}{4\pi|\mathbf{x}-\mathbf{x}'|} \\
&= \frac{1}{2} \int d^3x \int d^3x'\, \frac{j_{e0}(\mathbf{x}',t)j_{e0}(\mathbf{x},t)}{4\pi|\mathbf{x}-\mathbf{x}'|}.
\end{aligned}
\tag{5.439}
$$

In the derivation of Eq.(5.439), we have used Eqs.(5.413) and (5.414). Since $\mathcal{L}_{\text{int}} = -\mathcal{H}_{\text{int}}$, this Coulomb interaction between the charge densities $j_{e0}(\mathbf{x},t)$ contributes a term

$$
\left(\frac{1}{i} \right)^2 \frac{\delta^2 Z[j]}{\delta j_{e\mu}(x)\delta j_{e\nu}(y)} = i\delta(x_0 - y_0) \frac{\eta^{\mu 0}\eta^{\nu 0}}{4\pi|\mathbf{x}-\mathbf{y}|}
\tag{5.440}
$$

in propagator. From Eq.(5.423), we find

$$
\begin{aligned}
D_{\text{tr}}^{\mu\nu}(k) &= -\frac{\eta^{\mu\nu}}{k^2} - \frac{n^\mu n^\nu}{(k\cdot n)^2 - k^2} \\
&= D_{\text{F}}^{\mu\nu}(k) - D_{\text{coul}}^{\mu\nu}(k)
\end{aligned}
\tag{5.441}
$$

with

$$
D_{\text{F}}^{\mu\nu}(k) = -\frac{\eta^{\mu\nu}}{k^2}
\tag{5.442}
$$

and

$$
D_{\text{coul}}^{\mu\nu}(k) = \frac{n^\mu n^\nu}{(k\cdot n)^2 - k^2}.
\tag{5.443}
$$

Then

$$D_{\mathrm{F}}^{\mu\nu}(k) = D_{\mathrm{tr}}^{\mu\nu}(k) + D_{\mathrm{coul}}^{\mu\nu}(k). \tag{5.444}$$

If $j_{e0} = 0$, $A_0 = 0$ and there would be no Coulomb interaction. Therefore, the covariant Feynman propagator $D_{\mathrm{F}}^{\mu\nu}(k)$ adds up the contributions from the radiation field and the Coulomb interaction between the charge densities.

5.12 Ward-Takahashi identity

Now we consider the renormalization of the fermion and photon fields. There is a key formula in the proof of the renormalibility for these fields. This formula is the so-called Ward-Takahashi identity.

We start from the generating functional Z for a system of fermions (such as electrons) and photons. The theory about such systems are called quantum electrodynamics(QED). The generating functional Z is given by

$$Z = N \int DA_\mu D\bar{\psi} D\psi \exp\left(i \int \mathcal{L}_{\mathrm{qed}} d^4 x\right) \tag{5.445}$$

with

$$\begin{aligned} \mathcal{L}_{\mathrm{qed}} = &-\frac{1}{4} F_{\mu\nu} F^{\mu\nu} + i\bar{\psi}\gamma^\mu(\partial_\mu + ieA_\mu)\psi - m\bar{\psi}\psi \\ &- \frac{1}{2\xi}(\partial^\mu A_\mu)^2 + J^\mu A_\mu + \bar{\eta}\psi + \bar{\psi}\eta. \end{aligned} \tag{5.446}$$

The Lagrangian density $\mathcal{L}_{\mathrm{qed}}$ contains the term for the free photon field $-\frac{1}{4} F_{\mu\nu} F^{\mu\nu}$, the terms for the free fermion field $\bar{\psi}\gamma^\mu(i\partial_\mu - m)\psi$, and the interacting term $-e\bar{\psi}\gamma^\mu A_\mu\psi$. $-\frac{1}{2\xi}(\partial^\mu A_\mu)^2$ is the gauge fixing term. $J^\mu A_\mu$, $\bar{\eta}\psi$ and $\bar{\psi}\eta$ are the source terms for A_μ, ψ and $\bar{\psi}$, respectively. We have shown that Z should be gauge invariant. This condition for Z leads to a differential equation for Z.

The infinitesimal gauge transformations are given by

$$A_\mu \rightarrow A_\mu + \partial_\mu \Lambda, \tag{5.447a}$$

$$\psi \rightarrow \psi - ie\Lambda\psi, \tag{5.447b}$$

$$\bar{\psi} \rightarrow \bar{\psi} + ie\Lambda\bar{\psi}. \tag{5.447c}$$

The Lagrangian density consisting of the first three terms in Eq.(5.446) is gauge invariant. The remaining terms contribute Z by a factor after the gauge transformation

$$Q = \exp\left(i \int d^4 x \left[-\frac{1}{\xi}(\partial^\mu A_\mu)\Box\Lambda + J^\mu \partial_\mu \Lambda - ie\Lambda(\bar{\eta}\psi - \bar{\psi}\eta)\right]\right). \tag{5.448}$$

Since Λ is infinitesimal, Eq.(5.448) can be expanded into following expression:

$$
\begin{aligned}
Q &= 1 + i \int d^4x \left[-\frac{1}{\xi}(\partial^\mu A_\mu)\Box\Lambda + J^\mu \partial_\mu \Lambda - ie\Lambda(\bar\eta\psi - \bar\psi\eta) \right] \\
&= 1 + i \int d^4x \left[-\frac{1}{\xi}\Box(\partial^\mu A_\mu) - \partial_\mu J^\mu - ie(\bar\eta\psi - \bar\psi\eta) \right]\Lambda.
\end{aligned}
\tag{5.449}
$$

In the derivation of Eq.(5.449), we have integrated by parts to remove the derivative from Λ. The factor Q can be generated by the substitutions

$$
\psi \to \frac{1}{i}\frac{\delta}{\delta\bar\eta}, \quad \bar\psi \to \frac{1}{i}\frac{\delta}{\delta\eta}, \quad A_\mu \to \frac{1}{i}\frac{\delta}{\delta J^\mu}
\tag{5.450}
$$

when acting on Z. Since Z is gauge invariant, the factor Q given by Eq.(5.449), when acting on Z, is equal to 1. Since $\Lambda(x)$ is an arbitrary function, we have

$$
\begin{aligned}
&i\left[-\frac{1}{\xi}\Box(\partial^\mu A_\mu) - \partial_\mu J^\mu - ie(\bar\eta\psi - \bar\psi\eta) \right] Z[\eta, \bar\eta, J] \\
&= \left[-\frac{1}{\xi}\Box\partial^\mu \frac{\delta}{\delta J^\mu} - i\partial_\mu J^\mu - ie\left(\bar\eta\frac{\delta}{\delta\bar\eta} - \eta\frac{\delta}{\delta\eta} \right) \right] Z[\eta, \bar\eta, J_\mu] \\
&= 0.
\end{aligned}
\tag{5.451}
$$

Inserting $Z = e^{iW}$, Eq.(5.451) can be written as a differential equation for W.

$$
-\frac{1}{\xi}\Box\partial^\mu \frac{\delta W}{\delta J^\mu} - \partial_\mu J^\mu - ie\left(\bar\eta\frac{\delta W}{\delta\bar\eta} - \eta\frac{\delta W}{\delta\eta} \right) = 0.
\tag{5.452}
$$

We can also transform Eq.(5.452) into an equation for the vertex function Γ, where Γ is defined by

$$
\Gamma[\psi, \bar\psi, A_\mu] \equiv W[\eta, \bar\eta, J_\mu] - \int d^3x (\bar\eta\psi + \bar\psi\eta + J^\mu A_\mu).
\tag{5.453}
$$

Thus we have

$$
\frac{\delta\Gamma}{\delta A_\mu(x)} = -J^\mu(x),
\tag{5.454a}
$$

$$
\frac{\delta\Gamma}{\delta\psi(x)} = -\bar\eta(x),
\tag{5.454b}
$$

$$
\frac{\delta\Gamma}{\delta\bar\psi(x)} = -\eta(x),
\tag{5.454c}
$$

and

$$\frac{\delta W}{\delta J_\mu(x)} = A^\mu(x), \tag{5.455a}$$

$$\frac{\delta W}{\delta \bar\eta(x)} = \psi(x), \tag{5.455b}$$

$$\frac{\delta W}{\delta \eta(x)} = \bar\psi(x). \tag{5.455c}$$

Using Eqs.(5.454) and (5.455), we have

$$-\frac{1}{\xi}\Box\partial^\mu A_\mu(x) + \partial_\mu\frac{\delta\Gamma}{\delta A_\mu(x)} + ie\psi\frac{\delta\Gamma}{\delta\psi(x)} - ie\bar\psi\frac{\delta\Gamma}{\delta\bar\psi(x)} = 0. \tag{5.456}$$

Differentiating Eq.(5.456) with respect to $\bar\psi(x_1)$ and $\psi(y_1)$ at $\bar\psi = \psi = A_\mu = 0$, the first term vanishes and we have

$$\begin{aligned}
\frac{\partial}{\partial x^\mu}\frac{\delta^3\Gamma}{\delta\bar\psi(x_1)\delta\psi(y_1)\delta A_\mu(x)} &= ie\delta^4(x - x_1)\frac{\delta^2\Gamma}{\delta\bar\psi(x_1)\delta\psi(y_1)} \\
&\quad - ie\delta^4(x - y_1)\frac{\delta^2\Gamma}{\delta\bar\psi(x_1)\delta\psi(y_1)}.
\end{aligned} \tag{5.457}$$

Eq.(5.457) is an equation relating the derivative of the (1PI) vertex functions with the propagators of the interacting fields. This equation is often used in its form in momentum space. We define the vertex function $\Gamma_\mu(p, q, p')$ in momentum space by the transformation

$$\begin{aligned}
\int d^4x \int d^4x_1 \int d^4y_1 e^{i(p'\cdot x_1 - p\cdot y_1 - q\cdot x)}&\frac{\delta^3\Gamma}{\delta\bar\psi(x_1)\delta\psi(y_1)\delta A^\mu(x)} \\
&= e(2\pi)^4\delta^4(p' - p - q)\Gamma_\mu(p, q, p').
\end{aligned} \tag{5.458}$$

$\delta^2\Gamma/\delta\bar\psi\delta\psi$ is the inverse propagator $S_F'^{-1}$(here the prime is used to distinguish it from the free propagator), which can be derived in a similar way as for Eq.(5.205). We define $S_F'(p)$ by

$$\int d^4x_1 \int d^4y_1 e^{i(p'\cdot x_1 - p\cdot y_1)}\frac{\delta^2\Gamma}{\delta\bar\psi(x_1)\delta\psi(y_1)} = (2\pi)^4\delta^4(p' - p)S_F'^{-1}(p). \tag{5.459}$$

Multiplying Eq.(5.457) by $e^{i(p'\cdot x_1 - p\cdot y_1 - q\cdot x)}$ and integrating over x, x_1 and y_1, then integrating by parts for the integration on the left hand side of the equation, we obtain

$$q^\mu\Gamma_\mu(p, q, p + q) = S_F'^{-1}(p + q) - S_F'^{-1}(p). \tag{5.460}$$

Eq.(5.460) is called the *Ward-Takahashi identity*. When we take the limit $q_\mu \to 0$, Eq.(5.460) becomes

$$\frac{\partial S_F'^{-1}}{\partial p^\mu} = \Gamma_\mu(p, 0, p), \tag{5.461}$$

which is called the *Ward identity*. Since this identity relates the propagator with the vertex function and keep the gauge invariance to all orders in the perturbation expansion, it is a useful relation in the renormalization calculations. One can easily check that the Ward identity Eq.(5.461) holds in the lowest order.

$S_F'^{-1}$ becomes the free Feynman propagator to the lowest order. Using the free Feynman propagator $S_F^{-1}(p) = \gamma_\mu p^\mu - m$ in Eq.(5.351), we have

$$\frac{\partial S_F'^{-1}}{\partial p^\mu} = \gamma_\mu. \tag{5.462}$$

$\Gamma_\mu(p, 0, p)$ can be calculated using Eq.(5.458). The analogous derivation with that for Eq.(5.207) gives

$$\frac{\delta^3 \Gamma}{\delta \bar{\psi}(x_1) \delta \psi(y_1) \delta A_\mu(x)} = -\int d^4 u_1 d^4 v_1 d^4 u \Big[S_F^{-1}(u_1 - x_1) S_F^{-1}(v_1 - y_1)$$
$$(D_F^{\mu\nu}(u - x))^{-1} \frac{\delta^3 W[0]}{\delta \eta(u_1) \delta \bar{\eta}(v_1) \delta J^\nu(u)} \Big]. \tag{5.463}$$

The interaction term is $-e\bar{\psi}\gamma_\mu\psi A^\mu$. Thus we have

$$Z[\eta, \bar{\eta}, J_\mu] = N \exp\left(-ie \int d^4 z \frac{1}{i} \frac{\delta}{\delta \eta(z)} \gamma^\mu \frac{1}{i} \frac{\delta}{\delta \bar{\eta}(z)} \frac{1}{i} \frac{\delta}{\delta J^\mu(z)} \right) Z_0, \tag{5.464}$$

where Z_0 is the generating functional for the free electron and photon fields.

$$Z_0 = \exp\left[-i \int d^4 x d^4 y \bar{\eta}(x) S_F(x - y) \eta(y) \right]$$
$$\times \exp\left[-\frac{i}{2} \int d^4 x d^4 y J^\mu(x) D_{F\mu\nu}(x - y) J^\nu(y) \right]. \tag{5.465}$$

Using Eq.(5.465), we obtain, to the lowest order,

$$\frac{\delta^3 \mathcal{Z}[0]}{\delta \eta(u_1) \delta \bar{\eta}(v_1) \delta J^\mu(u)}$$
$$= ie \int d^4 z S_F(u_1 - z) S_F(v_1 - z) D_{F\mu\nu}(u - z) \gamma^\nu, \tag{5.466}$$

which gives

$$\frac{\delta^3 W[0]}{\delta \eta(u_1) \delta \bar{\eta}(v_1) \delta J^\mu(u)}$$
$$= -e \int d^4 z S_F(u_1 - z) S_F(v_1 - z) D_{F\mu\nu}(u - z) \gamma^\nu, \tag{5.467}$$

Expressing Eqs.(5.463) and (5.467) in momentum space, we obtain the vertex function $\Gamma_\mu(p, q, p+q)$ defined in Eq.(5.458).

$$\Gamma_\mu(p, q, p+q) = \gamma_\mu = \frac{\partial S_{\mathrm{F}}^{'-1}}{\partial p^\mu}, \qquad (5.468)$$

which shows that the Ward identity Eq.(5.461) is satisfied to the lowest order. The significance of the Ward identity lies in that it is an exact identity and holds to all orders in perturbation expansion.

5.13 Feynman rules in QED

In the QED calculations, we use the generating functional defined in Eq.(5.445). In the perturbation calculations, we have the following Feynman rules

(1) The free fermion propagator $iS_{\mathrm{F}} = \frac{i(\not{p}+m)}{p^2 - m^2 + i\epsilon}$ (in the Feynman gauge) is represented by an arrow line as shown in Fig.5.19. The arrow denotes the flow of charge. The momentum p is in the direction of arrow.
(2) The free photon propagator $iD_{\mathrm{F}}^{\mu\nu} = -\frac{i\eta^{\mu\nu}}{k^2 + i\epsilon}$ is represented by a wiggly line as shown in Fig.5.20.
(3) The fermion-photon vertex $-ie\gamma_\mu$ is represented by a diagram shown in Fig.5.21. The four-momentum is conserved at the vertex.
(4) Each internal line is associated with an integration $\int \frac{d^4k}{(2\pi)^4}$.
(5) There is a factor (-1) for each closed fermion loop due to the Grassmann algebra for fermions.

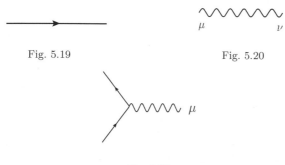

Fig. 5.19   Fig. 5.20

Fig. 5.21

5.14 Perturbation calculations of QED

Now we consider the perturbation calculations of QED. There are two kinds of particles in QED. One is the spinor fermions (electrons and positrons). The other is the massless vector bosons (photons). First, let us consider the self-energy. The electron self-energy diagram of the lowest order is shown in Fig.5.22. Using the Feynman rules, we have

$$\frac{1}{i}\Sigma(p) = (-ie)^2 \int \frac{d^4k}{(2\pi)^4} \gamma^\mu \frac{i}{\not{p} - \not{k} - m} \frac{-i\eta_{\mu\nu}}{k^2} \gamma^\nu, \tag{5.469}$$

where Σ is the self-energy of electron.

Similar to Eq.(5.121), the complete electron propagator iS'_F can be obtained using the Feynman diagrams in Fig.5.23. Using the Feynman rules, we have

$$\begin{aligned} iS'_F &= iS_F + iS_F \frac{\Sigma}{i} iS_F + iS_F \frac{\Sigma}{i} iS_F \frac{\Sigma}{i} iS_F + \cdots \\ &= iS_F \left(1 + \frac{\Sigma}{i} iS'_F \right). \end{aligned} \tag{5.470}$$

Solving iS'_F gives

$$S'^{-1}_F = S^{-1}_F - \Sigma \tag{5.471}$$

or

$$S'_F = \frac{S_F}{1 - \Sigma S_F} \tag{5.472}$$

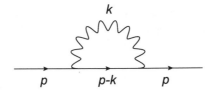

Fig. 5.22   Electron self-energy.

Fig. 5.23   Complete electron propagator as a sum of the Feynman diagrams.

The photon self-energy diagram of the lowest order is shown in Fig.5.24. The photon self-energy is also called the vacuum polarization. Using the Feynman rules, we obtain

$$i\Pi_{\mu\nu} = -(-ie)^2 \int \frac{d^4p}{(2\pi)^4} \text{Tr} \left(\gamma_\mu \frac{i}{\not{p} - m} \gamma_\nu \frac{i}{\not{p} - \not{k} - m} \right), \tag{5.473}$$

where $\Pi^{\alpha\beta}(k)$ is the photon self-energy.

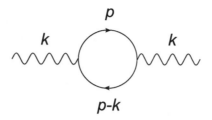

Fig. 5.24 Photon self-energy.

Fig. 5.25 Complete photon propagator as a sum of the Feynman diagrams.

The complete photon propagator $iD_F'^{\mu\nu}(k)$ can be obtained using the Feynman diagrams in Fig.5.25. The Feynman rules gives

$$iD_F'^{\mu\nu}(k) = iD_F^{\mu\nu}(k) + iD_F^{\mu\alpha}(k)i\Pi^{\alpha\beta}(k)iD_F^{\beta\nu}(k) + \cdots . \qquad (5.474)$$

The integrals in Eqs.(5.469) and (5.473) are divergent. The divergent part can be eliminated through renormalization procedure.

With the lowest corrections, the vertex function in Eq.(5.458) can be calculated using the Feynman diagrams in Fig.5.26. Using the Feynman rules, we have

$$-ie\Gamma_\mu(p,q,p+q) = -ie\gamma_\mu - ie\Lambda_\mu(p,q,p+q) \qquad (5.475)$$

with

$$-ie\Lambda_\mu(p,q,p+q) = (-ie)^3 \int \frac{d^4k}{(2\pi)^4} \frac{-i\eta_{\kappa\lambda}}{k^2}\gamma^\kappa \frac{i}{\gamma\cdot(p-k)-m}\gamma_\mu$$

$$\times \frac{i}{\gamma\cdot(p-k+q)-m}\gamma^\lambda \qquad (5.476)$$

$$= e^3 \int \frac{d^4k}{(2\pi)^4} \frac{1}{k^2}\gamma^\lambda S_F(p-k)\gamma_\mu S_F(p-k+q)\gamma_\lambda.$$

5.15 Divergences of QED

Similar to Section 5.5, we use the dimensional regularization to separate the divergent parts from the convergent parts. We first generalize the Lagrangian density Eq.(5.446) to $n = 4 - \varepsilon$ dimensional spacetime. $\mathcal{L}$

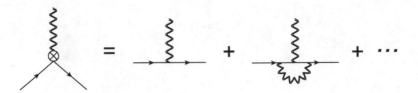

Fig. 5.26   Vertex function as a sum of the Feynman diagrams.

has the scale dimension l^{-n} with l as the length dimension, which is denoted as $[\mathcal{L}] = -n$. Thus we have $[\psi] = \frac{1-n}{2}$ and $[A_\mu] = 1 - \frac{n}{2}$. In order to make the interaction term to have the dimension l^{-n}, we need multiply $eA^\nu\bar{\psi}\gamma_\nu\psi$ by $\mu^{2-\frac{n}{2}}$, where μ is an arbitrary mass. Then we have

$$\mathcal{L} = i\bar{\psi}\gamma^\mu\partial_\mu\psi - m\bar{\psi}\psi - e\mu^{2-\frac{n}{2}}A_\mu\bar{\psi}\gamma^\mu\psi$$
$$- \frac{1}{4}(\partial_\mu A_\nu - \partial_\nu A_\mu)^2 - \frac{1}{2\xi}(\partial^\mu A_\mu)^2. \tag{5.477}$$

We need also generalize the Dirac's matrices γ^μ to n-dimension. The algebra of the Dirac's matrices in Eq.(2.234) is now defined in an n-dimensional spacetime. We have

$$\delta^\mu_\mu = n, \tag{5.478}$$

which leads to

$$\gamma^\mu\gamma_\mu = n, \tag{5.479a}$$
$$\gamma_\mu\gamma_\nu\gamma^\mu = (2 - n)\gamma_\nu. \tag{5.479b}$$

We have also the following trace relations of the gamma matrices γ_μ

$$\mathrm{Tr}(\gamma^\mu) = 0, \tag{5.480a}$$
$$\mathrm{Tr}(\gamma^{\mu_1}\gamma^{\mu_2}\cdots\gamma^{\mu_n}) = 0 \quad \textbf{(n is an odd number)}, \tag{5.480b}$$
$$\mathrm{Tr}(\mathbb{I}) = f(n), \tag{5.480c}$$
$$\mathrm{Tr}(\gamma^\mu\gamma^\nu) = f(n)\eta^{\mu\nu}, \tag{5.480d}$$
$$\mathrm{Tr}(\gamma^\mu\gamma^\kappa\gamma^\nu\gamma^\lambda) = f(n)(\eta^{\mu\kappa}\eta^{\nu\lambda} - \eta^{\mu\nu}\eta^{\kappa\lambda} + \eta^{\mu\lambda}\eta^{\kappa\nu}), \tag{5.480e}$$

where $f(n)$ is a function of dimension n with $f(4) = 4$. The explicit form of $f(n)$ is irrelevant to the final results after the limit $\varepsilon = 4 - n \to 0$ is taken.

Now we calculate the fermion self-energy $\Sigma(p)$. After generalizing Eq.(5.469) to n-dimension, we obtain

$$\Sigma(p) = -ie^2\mu^{4-n}\int\frac{d^nk}{(2\pi)^n}\gamma^\mu\frac{i}{\not{p} - \not{k} - m}\frac{-i\eta_{\mu\nu}}{k^2}\gamma^\nu$$
$$= -ie^2\mu^{4-n}\int\frac{d^nk}{(2\pi)^n}\frac{\gamma^\mu(\not{p} - \not{k} + m)\gamma_\mu}{[(p-k)^2 - m^2]k^2}. \tag{5.481}$$

Using the integration identity
$$\frac{1}{ab} = \int_0^1 \frac{dz}{[az + b(1-z)]^2},$$
(5.482)
we have
$$\Sigma(p) = -ie^2\mu^{4-n}\int_0^1 dz \int \frac{d^n k}{(2\pi)^n} \frac{\gamma^\mu(\not{p} - \not{k} + m)\gamma_\mu}{[(p-k)^2 z - m^2 z + k^2(1-z)]^2}.$$
(5.483)
Changing the integral variable k to $k' = k - pz$, we obtain
$$\Sigma(p) = -\int_0^1 dz \int \frac{d^n k'}{(2\pi)^n} \frac{i\mu^{4-n}e^2\gamma^\mu(\not{p} - \not{p}z - \not{k}' + m)\gamma_\mu}{[(p(1-z) - k')^2 z - m^2 z + (k' + pz)^2(1-z)]^2}$$
$$= -i\mu^{4-n}e^2\int_0^1 dz \int \frac{d^n k'}{(2\pi)^n} \frac{\gamma^\mu(\not{p} - \not{p}z - \not{k}' + m)\gamma_\mu}{[k'^2 - m^2 z + p^2 z(1-z)]^2}.$$
(5.484)
The term linear in k' contributes zero after integration. Thus Eq.(5.484) becomes
$$\Sigma(p) = -i\mu^{4-n}e^2\int_0^1 dz\gamma^\mu(\not{p} - \not{p}z + m)\gamma_\mu$$
$$\times \int \frac{d^n k'}{(2\pi)^n} \frac{1}{[k'^2 - m^2 z + pz(1-z)]^2}.$$
(5.485)
Using Eq.(F.35), we have
$$\Sigma(p) = (-1)^{\frac{n}{2}}\mu^{4-n}e^2\frac{\Gamma\left(2 - \frac{n}{2}\right)}{(4\pi)^{\frac{n}{2}}}\int_0^1 dz\gamma^\mu[\not{p}(1-z) + m]\gamma_\mu$$
$$\times [-m^2 z + p^2 z(1-z)]^{\frac{n}{2}-2}.$$
(5.486)
Using the mathematical identity
$$\Gamma\left(\frac{\varepsilon}{2}\right) = \frac{2}{\varepsilon} - \gamma + \mathcal{O}(\varepsilon)$$
(5.487)
and Eq.(5.479), we obtain
$$\Sigma(p) = -\frac{e^2}{16\pi^2}\Gamma\left(\frac{\varepsilon}{2}\right)\int_0^1 dz\{2\not{p}(1-z) - 4m + \varepsilon[\not{p}(1-z) + m]\}$$
$$\times \left[\frac{m^2 z - p^2 z(1-z)}{4\pi\mu^2}\right]^{-\frac{\varepsilon}{2}}$$
$$= -\frac{e^2}{16\pi^2}\left(\frac{2}{\varepsilon} - \gamma\right)\int_0^1 dz\{2\not{p}(1-z) - 4m + \varepsilon[\not{p}(1-z) + m]\}$$
$$\times \exp\left[-\frac{\varepsilon}{2}\ln\frac{m^2 z - p^2 z(1-z)}{4\pi\mu^2}\right]$$
(5.488)
$$= \frac{e^2}{8\pi^2\varepsilon}(-\not{p} + 4m) + \frac{e^2}{16\pi^2}\left\{\not{p}(-1+\gamma) - 2m(1+2\gamma)\right.$$
$$\left. + 2\int_0^1 dz[\not{p}(1-z) - 2m]\ln\frac{m^2 z - p^2 z(1-z)}{4\pi\mu^2}\right\}$$
$$= \frac{e^2}{8\pi^2\varepsilon}(-\not{p} + 4m) + \text{finite}.$$

To calculate the vacuum polarization $\Pi_{\mu\nu}(k)$, we generalize Eq.(5.473) to n-dimension, which gives

$$
\begin{aligned}
\Pi_{\mu\nu} &= i\mu^{4-n}e^2 \int \frac{d^n p}{(2\pi)^n} \mathrm{Tr}\left(\gamma_\mu \frac{1}{\not{p}-m}\gamma_\nu \frac{1}{\not{p}-\not{k}-m}\right) \\
&= i\mu^{4-n}e^2 \int \frac{d^n p}{(2\pi)^n} \frac{\mathrm{Tr}[\gamma_\mu(\not{p}+m)\gamma_\nu(\not{p}-\not{k}+m)]}{(p^2-m^2)[(p-k)^2-m^2]}.
\end{aligned}
\tag{5.489}
$$

Using the integration identity Eq.(5.482), we have

$$
\begin{aligned}
\Pi_{\mu\nu} &= i\mu^{4-n}e^2 \int_0^1 dz \int \frac{d^n p}{(2\pi)^n} \\
&\quad \frac{\mathrm{Tr}[\gamma_\mu(\not{p}+m)\gamma_\nu(\not{p}-\not{k}+m)]}{\{(p^2-m^2)(1-z)+[(p-k)^2-m^2]z\}^2}.
\end{aligned}
\tag{5.490}
$$

Changing the integral variable from p to $p' = p - kz$ yields

$$
\begin{aligned}
\Pi_{\mu\nu} &= i\mu^{4-n}e^2 \int_0^1 dz \int \frac{d^n p'}{(2\pi)^n} \\
&\quad \frac{\mathrm{Tr}[\gamma_\mu(\not{p}'+\not{k}z+m)\gamma_\nu(\not{p}'-\not{k}(1-z)+m)]}{[p'^2-m^2+k^2z(1-z)]^2}.
\end{aligned}
\tag{5.491}
$$

Using Eq.(5.480) and discarding the terms odd in p' which are integrated to zero, Eq.(5.491) becomes

$$
\begin{aligned}
\Pi_{\mu\nu} &= i\mu^{4-n}e^2 \int_0^1 dz \int \frac{d^n p'}{(2\pi)^n} \\
&\quad \frac{[p'^\kappa p'^\lambda - k^\kappa k^\lambda z(1-z)]\mathrm{Tr}(\gamma_\mu\gamma_\kappa\gamma_\nu\gamma_\lambda)+m^2\mathrm{Tr}(\gamma_\mu\gamma_\nu)}{[p'^2-m^2+k^2z(1-z)]^2}.
\end{aligned}
\tag{5.492}
$$

Using Eq.(5.480), we have

$$
\begin{aligned}
\Pi_{\mu\nu} &= i\mu^{4-n}e^2 f(n) \int_0^1 dz \int \frac{d^n p}{(2\pi)^n} \\
&\quad \frac{[2p_\mu p_\nu - 2z(1-z)(k_\mu k_\nu - k^2\eta_{\mu\nu})-\eta_{\mu\nu}[p^2-m^2+k^2z(1-z)]}{[p^2-m^2+k^2z(1-z)]^2} \\
&= ie^2\mu^{4-n}f(n) \int_0^1 dz \int \frac{d^n p}{(2\pi)^n}\left\{ \frac{2p_\mu p_\nu}{[p^2-m^2+k^2z(1-z)]^2} \right. \\
&\quad \left. - \frac{2z(1-z)(k_\mu k_\nu - k^2\eta_{\mu\nu})}{[p^2-m^2+k^2z(1-z)]^2} - \frac{\eta_{\mu\nu}}{p^2-m^2+k^2z(1-z)} \right\}.
\end{aligned}
\tag{5.493}
$$

Using Eqs.(F.35) and (F.47), we can see that the contributions of the first and third terms in the integrand cancel. Then Eq.(5.493) becomes

$$
\Pi_{\mu\nu} = -ie^2\mu^{4-n}f(n) \int_0^1 dz \int \frac{d^n p}{(2\pi)^n} \frac{2z(1-z)(k_\mu k_\nu - k^2\eta_{\mu\nu})}{[p^2-m^2+k^2z(1-z)]^2}.
\tag{5.494}
$$

Using Eq.(F.35), we have

$$\Pi_{\mu\nu} = (-1)^{\frac{n}{2}} e^2 \mu^{4-n} \frac{\Gamma\left(2 - \frac{n}{2}\right)}{(4\pi)^{\frac{n}{2}}} f(n) \int_0^1 dz \frac{2z(1-z)(k_\mu k_\nu - k^2 \eta_{\mu\nu})}{[-m^2 z + k^2 z(1-z)]^{2-\frac{n}{2}}}.$$

(5.495)

Using Eq.(5.487), we obtain

$$\begin{aligned}
\Pi_{\mu\nu}(k) &= \frac{e^2}{16\pi^2} \Gamma\left(\frac{\varepsilon}{2}\right) f(n) \int_0^1 dz\, 2z(1-z)(k_\mu k_\nu - k^2 \eta_{\mu\nu}) \\
&\quad \times \left[\frac{m^2 z - k^2 z(1-z)}{4\pi\mu^2}\right]^{-\frac{\varepsilon}{2}} \\
&= \frac{e^2 f(n)}{16\pi^2} \left(\frac{2}{\varepsilon} - \gamma\right) (k_\mu k_\nu - k^2 \eta_{\mu\nu}) \\
&\quad \times \int_0^1 dz\, 2z(1-z) \exp\left[-\frac{\varepsilon}{2} \ln \frac{m^2 z - k^2 z(1-z)}{4\pi\mu^2}\right] \\
&= \frac{e^2}{2\pi^2} (k_\mu k_\nu - k^2 \eta_{\mu\nu}) \\
&\quad \times \left\{\frac{1}{3\varepsilon} - \frac{\gamma}{6} - \int_0^1 dz\, z(1-z) \ln\left[\frac{m^2 z - k^2 z(1-z)}{4\pi\mu^2}\right]\right\} \\
&= (k_\mu k_\nu - k^2 \eta_{\mu\nu}) \left(\frac{e^2}{6\pi^2 \varepsilon} + \text{finite}\right).
\end{aligned}$$

(5.496)

The finite part is a function of k^2. When k^2 is small, we have

$$\begin{aligned}
\Pi_{\mu\nu}(k) &= \frac{e^2}{2\pi^2} (k_\mu k_\nu - k^2 \eta_{\mu\nu}) \\
&\quad \times \left\{\frac{1}{3\varepsilon} - \frac{\gamma}{6} - \int_0^1 dz\, z(1-z) \left[1 - \frac{k^2}{m^2} z(1-z)\right] + \cdots\right\} \\
&= \frac{e^2}{2\pi^2} (k_\mu k_\nu - k^2 \eta_{\mu\nu}) \\
&\quad \times \left[\frac{1}{3\varepsilon} + c' + \frac{k^2}{m^2} \int_0^1 dz\, z^2(1-z)^2 + \mathcal{O}(k^4)\right] \\
&= \frac{e^2}{2\pi^2} (k_\mu k_\nu - k^2 \eta_{\mu\nu}) \left[\frac{1}{3\varepsilon} + c' + \frac{k^2}{30m^2} + \mathcal{O}(k^4)\right],
\end{aligned}$$

(5.497)

where c' is a finite constant which can be merged with $\frac{1}{3\varepsilon}$ and thus be neglected. Therefore, we obtain

$$\Pi_{\mu\nu}(k) = \frac{e^2}{6\pi^2} (k_\mu k_\nu - k^2 \eta_{\mu\nu}) \left[\frac{1}{\varepsilon} + \frac{k^2}{10m^2} + \mathcal{O}(k^4)\right],$$

(5.498)

Now we evaluate the vertex function Λ_μ. we generalize Eq.(5.476) to n-dimension, which gives

$$
\begin{aligned}
-&ie\mu^{2-\frac{n}{2}}\Lambda_\mu(p,q,p') \\
&= (-ie\mu^{2-\frac{n}{2}})^3 \int \frac{d^n k}{(2\pi)^n} \gamma_\nu \frac{i}{\not{p}' - \not{k} - m} \gamma_\mu \frac{i}{\not{p} - \not{k} - m} \gamma_\kappa \frac{-i\eta^{\nu\kappa}}{k^2} \\
&= -(e\mu^{2-\frac{n}{2}})^3 \int \frac{d^n k}{(2\pi)^n} \frac{\gamma_\nu(\not{p}' - \not{k} + m)\gamma_\mu(\not{p} - \not{k} + m)\gamma^\nu}{k^2[(p-k)^2 - m^2][(p'-k)^2 - m^2]}.
\end{aligned}
\tag{5.499}
$$

To evaluate the integral in Eq.(5.499), we use the following mathematical formula

$$
\frac{1}{abc} = 2 \int_0^1 dx \int_0^{1-x} dy \frac{1}{[a(1-x-y) + bx + cy]^3},
\tag{5.500}
$$

which is called the 2-parameter Feynman formula. Then Eq.(5.499) becomes

$$
\Lambda_\mu(p,q,p') = -i\frac{2e^2\mu^{4-n}}{(2\pi)^n} \int_0^1 dx \int_0^{1-x} dy \int d^n k
$$
$$
\frac{\gamma_\nu(\not{p}' - \not{k} + m)\gamma_\mu(\not{p} - \not{k} + m)\gamma^\nu}{[k^2 - m^2(x+y) - 2k(px + p'y) + p^2 x + p'^2 y]^3}.
\tag{5.501}
$$

Changing the integral variable from k to $k' = k - px - p'y$, we obtain

$$
\Lambda_\mu(p,q,p') = -i\frac{2e^2\mu^{4-n}}{(2\pi)^n} \int_0^1 dx \int_0^{1-x} dy \int d^n k'
$$
$$
\frac{\gamma_\nu[\not{p}'(1-y) - \not{p}x - \not{k}' + m]\gamma_\mu[\not{p}(1-x) - \not{p}'y - \not{k}' + m]\gamma^\nu}{[k'^2 - m^2(x+y) + p^2 x(1-x) + p'^2 y(1-y) - 2p\cdot p'xy]^3}.
\tag{5.502}
$$

The divergent part comes from the terms quadratic in k in the numerator. The terms linear in k contribute to zero and the other terms give the convergent integrands. We use $\Lambda_\mu^{(1)}$ to denote the divergent part and $\Lambda_\mu^{(2)}$ to denote the convergent part. Then we have

$$
\Lambda_\mu^{(1)}(p,q,p') = -i\frac{2e^2\mu^{4-n}}{(2\pi)^n} \int_0^1 dx \int_0^{1-x} dy \int d^n k
$$
$$
\frac{k^\kappa k^\lambda \gamma_\nu \gamma_\kappa \gamma_\mu \gamma_\lambda \gamma^\nu}{[k^2 - m^2(x+y) + p^2 x(1-x) + p'^2 y(1-y) - 2p\cdot p'xy]^3}.
\tag{5.503}
$$

Using Eq.(F.47), we obtain

$$
\Lambda_\mu^{(1)}(p,q,p') = \frac{e^2}{2}\mu^{4-n} \left(\frac{1}{4\pi}\right)^n \Gamma\left(2 - \frac{n}{2}\right) \int_0^1 dx \int_0^{1-x} dy
$$
$$
\frac{\gamma_\nu \gamma_\kappa \gamma_\mu \gamma^\kappa \gamma^\nu}{[m^2(x+y) - p^2 x(1-x) - p'^2 y(1-y) + 2p\cdot p'xy]^{2-\frac{n}{2}}}.
\tag{5.504}
$$

Using Eq.(2.234), one can easily prove

$$\gamma_\nu \gamma_\kappa \gamma_\mu \gamma_\lambda \gamma^\nu = (2-n)\gamma_\kappa \gamma_\mu \gamma_\lambda + 2(\gamma_\mu \gamma_\lambda \gamma_\kappa - \gamma_\kappa \gamma_\lambda \gamma_\mu), \qquad (5.505)$$

which gives

$$\gamma_\nu \gamma_\kappa \gamma_\mu \gamma^\kappa \gamma^\nu = (2-n)^2 \gamma_\mu. \qquad (5.506)$$

Using Eqs.(5.506) and (5.487), we have

$$\Lambda_\mu^{(1)}(p,q,p') = \frac{e^2}{2}\frac{1}{16\pi^2}\Gamma\left(\frac{\varepsilon}{2}\right)\int_0^1 dx \int_0^{1-x} dy (2-n)^2 \gamma_\mu$$
$$\times \left[\frac{m^2(x+y) - p^2 x(1-x) - p'^2 y(1-y) + 2p\cdot p' xy}{4\pi\mu^2}\right]^{-\frac{\varepsilon}{2}}$$
$$= \frac{e^2}{2}\frac{1}{4\pi^2}\left(\frac{2}{\varepsilon} - \gamma\right)\gamma_\mu \int_0^1 dx \int_0^{1-x} dy$$
$$\exp\left[-\frac{\varepsilon}{2}\ln\frac{m^2(x+y) - p^2 x(1-x) - p'^2 y(1-y) + 2p\cdot p' xy}{4\pi\mu^2}\right]$$
$$= \frac{e^2}{8\pi^2}\gamma_\mu\left[\frac{1}{\varepsilon} - \frac{\gamma}{2} - \int_0^1 dx \int_0^{1-x} dy\right.$$
$$\left.\ln\frac{m^2(x+y) - p^2 x(1-x) - p'^2 y(1-y) + 2p\cdot p' xy}{4\pi\mu^2}\right]$$
$$= \frac{e^2}{8\pi^2\varepsilon}\gamma_\mu + \text{finite}. \qquad (5.507)$$

The convergent part $\Lambda_\mu^{(2)}$ comes from the other terms in Eq.(5.502)

$$\Lambda_\mu^{(2)}(p,q,p') = -\frac{e^2}{16\pi^2}\int_0^1 dx \int_0^{1-x} dy$$
$$\frac{\gamma_\nu[\not{p}'(1-y) - \not{p}x + m]\gamma_\mu[\not{p}(1-x) - \not{p}'y + m]\gamma^\nu}{m^2(x+y) - p^2 x(1-x) - p'^2 y(1-y) + 2p\cdot p' xy}. \qquad (5.508)$$

Summing up the contributing from Eqs.(5.507) and (5.508), we have

$$\Lambda_\mu(p,q,p') = \Lambda_\mu^{(1)}(p,q,p') + \Lambda_\mu^{(2)}(p,q,p'). \qquad (5.509)$$

5.16 Renormalization of QED

We have separated the divergent parts from the finite ones in the self-energy, vacuum polarization and vertex function in the previous section. Similar to Section 5.6 for the scalar field, we can use the renormalization procedure to eliminate the divergent parts.

First we add a mass counter term $\mathcal{L}_{\delta m}$ to the Lagrangian density of spinor fermions

$$\mathcal{L}_F = i\bar{\psi}\slashed{\partial}\psi - m\bar{\psi}\psi. \tag{5.510}$$

The mass counter term has the form

$$\mathcal{L}_{\delta m} = A\bar{\psi}\psi \tag{5.511}$$

with

$$A = -\frac{3me^2}{8\pi^2\varepsilon}. \tag{5.512}$$

Then we make a transformation

$$\psi \to \sqrt{Z_2}\psi \tag{5.513}$$

with

$$Z_2 = 1 - \frac{e^2}{8\pi^2\varepsilon}. \tag{5.514}$$

The transformation given by Eq.(5.513) is called the wave function renormalization. Eq.(5.513) is equivalent to adding a counter term $i(Z_2-1)\bar{\psi}\slashed{\partial}\psi$ to the kinetic term (the first term in Eq.(5.510)). Then we have the bare Lagrangian density for spinor fermions

$$\begin{aligned}
\mathcal{L}_{FB} &= iZ_2\bar{\psi}\slashed{\partial}\psi - Z_2(m+A)\bar{\psi}\psi \\
&= i\bar{\psi}_B\slashed{\partial}\psi_B - m_B\bar{\psi}_B\psi_B
\end{aligned} \tag{5.515}$$

with

$$\psi_B = \sqrt{Z_2}\psi \tag{5.516}$$

and

$$m_B = m + A = m\left(1 - \frac{3e^2}{8\pi^2\varepsilon}\right). \tag{5.517}$$

ψ_B is called the bare field. Thus the effective counter Lagrangian density has the form

$$\mathcal{L}_{FC} = iB\bar{\psi}\slashed{\partial}\psi - \delta m\bar{\psi}\psi \tag{5.518}$$

with

$$B = Z_2 - 1 = -\frac{e^2}{8\pi^2\varepsilon}. \tag{5.519}$$

and

$$\delta m = Z_2(m + A) - m$$

$$= m \left(1 - \frac{3e^2}{8\pi^2\varepsilon}\right) \left(1 - \frac{e^2}{8\pi^2\varepsilon}\right) - m$$

$$= m \left(1 - \frac{e^2}{2\pi^2\varepsilon}\right) - m \qquad (5.520)$$

$$= -\frac{e^2 m}{2\pi^2\varepsilon}.$$

Using Eq.(5.488)), the new self-energy term of fermions becomes

$$\Sigma^r(p) = \Sigma(p) + \delta m - B\slashed{p}$$

$$= \frac{e^2}{8\pi^2\varepsilon}(-\slashed{p} + 4m) + \text{finite terms} + \delta m - B\slashed{p} \qquad (5.521)$$

$$= \text{finite terms}.$$

Thus the renormalized self-energy $\Sigma^r(p)$ contains only the convergent parts. The renormalized vertex function $\Gamma^{(2)}(p)$ then becomes

$$\Gamma^{(2)}(p) = (S_F'(p))^{-1}$$

$$= S_F^{-1}(p) - \Sigma(p) \qquad (5.522)$$

$$= \slashed{p} - m + \text{finite terms}.$$

For the massless vector bosons, we make a transformation

$$A^\mu \rightarrow \sqrt{Z_3} A^\mu \qquad (5.523)$$

with

$$Z_3 = 1 - \frac{e^2}{6\pi^2\varepsilon}. \qquad (5.524)$$

Eq.(5.524) is a wave function renormalization. Then the Lagrangian density for gauge bosons becomes

$$\mathcal{L}_{\text{BB}} = -\frac{1}{4} Z_3 F_{\mu\nu} F^{\mu\nu} + \frac{E}{2}(\partial_\mu A^\mu)^2$$

$$= -\frac{1}{4} Z_3 (\partial_\mu A_\nu - \partial_\nu A_\mu)^2 + \text{gauge terms} \qquad (5.525)$$

$$= -\frac{1}{4}(\partial_\mu A_{\text{B}\nu} - \partial_\nu A_{\text{B}\mu})^2 + \text{gauge terms}$$

with

$$A_{\text{B}}^\mu = \sqrt{Z_3} A^\mu. \qquad (5.526)$$

$\frac{E}{2}(\partial_\mu A^\mu)^2$ in Eq.(5.525) is the gauge term where E can be chosen to be any value. The transformation Eq.(5.523) is equivalent to adding a counter term

$$\mathcal{L}_{\text{BC}} = -\frac{C}{4}F_{\mu\nu}F^{\mu\nu} \tag{5.527}$$

with

$$C = Z_3 - 1 = -\frac{e^2}{6\pi^2\varepsilon}. \tag{5.528}$$

According to Eq.(5.498), the vacuum polarization $\Pi_{\mu\nu}(k)$ has the form

$$\Pi_{\mu\nu}(k) = (k_\mu k_\nu - k^2\eta_{\mu\nu})\Pi(k^2) \tag{5.529}$$

with

$$\begin{aligned}
\Pi(k^2) &= \frac{e^2}{6\pi^2}\left(\frac{1}{\varepsilon} + \frac{k^2}{10m^2}\right) + \mathcal{O}(k^4) \\
&= \frac{e^2}{6\pi^2\varepsilon} + \Pi_f(k^2),
\end{aligned} \tag{5.530}$$

where $\Pi_f(k^2)$ is finite and approaches to zero as $k^2 \to 0$. Eq.(5.529) satisfies the gauge invariance condition

$$k_\mu\Pi^{\mu\nu}(k) = 0. \tag{5.531}$$

Using Eq.(5.474), we have

$$\begin{aligned}
D'_{\text{F}\mu\nu}(k) = D_{\text{F}\mu\nu}(k) \\
- D_{\text{F}\mu\alpha}(k)(k^\alpha k^\beta - k^2\eta^{\alpha\beta})\Pi(k^2)D_{\text{F}\beta\nu}(k) + \cdots ,
\end{aligned} \tag{5.532}$$

which gives

$$D'_{\text{F}\mu\nu}(k) = \frac{1}{k^2[1 + \Pi(k^2)]}\left[-\eta_{\mu\nu} - \frac{k_\mu k_\nu}{k^2}\Pi(k^2)\right]. \tag{5.533}$$

The term containing $\frac{k_\mu k_\nu}{k^2}$ is the gauge term. Thus the propagator can be written as

$$\begin{aligned}
D'_{\text{F}\mu\nu}(k) &= \frac{-\eta_{\mu\nu}}{k^2[1 + \Pi(k^2)]} + \text{gauge terms} \\
&= \frac{-\eta_{\mu\nu}}{k^2\left[1 + \dfrac{e^2}{6\pi^2\varepsilon} + \Pi_f(k^2)\right]} + \text{gauge terms} \\
&= \frac{-Z_3\eta_{\mu\nu}}{k^2[1 + \Pi_f(k^2)]} + \text{gauge terms}.
\end{aligned} \tag{5.534}$$

Using Eq.(5.526), we obtain the renormalized propagator

$$
\begin{aligned}
D^r_{\text{F}\mu\nu}(k) &= \langle 0|T(A_\mu A_\nu)|0\rangle \\
&= Z_3^{-1}\langle 0|T(A_{\text{B}\mu} A_{\text{B}\nu})|0\rangle \\
&= Z_3^{-1}\frac{-Z_3\eta_{\mu\nu}}{k^2[1+\Pi_f(k^2)]} + \text{gauge terms} \\
&= \frac{-\eta_{\mu\nu}}{k^2[1+\Pi_f(k^2)]} + \text{gauge terms} \\
&= \frac{-\eta_{\mu\nu}}{k^2}\left(1 - \frac{e^2}{60\pi^2}\frac{k^2}{m^2} + \mathcal{O}(k^4)\right) + \text{gauge terms},
\end{aligned}
\tag{5.535}
$$

which contains only finite terms. The correction term in Eq.(5.535)

$$
D^U_{\text{F}\mu\nu}(k) = \frac{e^2}{60\pi^2}\frac{\eta_{\mu\nu}}{m^2}
\tag{5.536}
$$

is called the Uehling term, which contributes a correction to the Coulomb interaction. Using Eq.(5.536), the modified Coulomb interaction takes the form

$$
U^r_{\text{coul}} = \frac{1}{2}\int d^3x\frac{\rho(\mathbf{x}',t)\rho(\mathbf{x},t)}{4\pi|\mathbf{x}-\mathbf{x}'|} + \frac{\rho(\mathbf{x}',t)\rho(\mathbf{x},t)}{60\pi^2 m^2}\delta^3(\mathbf{x}-\mathbf{x}').
\tag{5.537}
$$

This correction to the Coulomb interaction gives a contribution to the Lamb shift.

In order to eliminate the divergent term of the vertex function $\Lambda^{(1)}_\mu$, we introduce an interaction counter term

$$
\mathcal{L}_{\text{IC}} = -De\mu^{2-\frac{n}{2}}\bar{\psi}A\!\!\!/\psi
\tag{5.538}
$$

with

$$
D = -\frac{e^2}{8\pi^2\varepsilon}.
\tag{5.539}
$$

The corresponding interaction part of the Lagrangian density reads

$$
\begin{aligned}
\mathcal{L}_{\text{IB}} &= -(1+D)e\mu^{\frac{\varepsilon}{2}}A^\mu\bar{\psi}\gamma_\mu\psi \\
&= -Z_1 e\mu^{\frac{\varepsilon}{2}}A^\mu\bar{\psi}\gamma_\mu\psi
\end{aligned}
\tag{5.540}
$$

with

$$
Z_1 = 1 - \frac{e^2}{8\pi^2\varepsilon}.
\tag{5.541}
$$

Using Eqs.(5.516) and (5.526), Eq.(5.540) becomes

$$
\mathcal{L}_{\text{IB}} = -\frac{Z_1}{Z_3^{\frac{1}{2}}Z_2}e\mu^{\frac{\varepsilon}{2}}A^\mu_{\text{B}}\bar{\psi}_{\text{B}}\gamma_\mu\psi_{\text{B}}.
\tag{5.542}
$$

We introduce the bare charge e_{B} defined by

$$e_{\mathrm{B}} \equiv \frac{Z_1}{Z_3^{\frac{1}{2}} Z_2} e\mu^{\frac{\varepsilon}{2}} = e\mu^{\frac{\varepsilon}{2}} Z_3^{-\frac{1}{2}}. \tag{5.543}$$

In the derivation of Eq.(5.543), we have used

$$Z_1 = Z_2. \tag{5.544}$$

It should be noted that $Z_1 = Z_2$ given by Eq.(5.544) is an exact formula due to the Ward identity Eq.(5.461) and

$$\langle 0|T(\bar{\psi}_{\mathrm{B}}\psi_{\mathrm{B}})|0\rangle = Z_2\langle 0|T(\bar{\psi}\psi)|0\rangle. \tag{5.545}$$

In terms of bare charge e_{B}, $\mathcal{L}_{\mathrm{IB}}$ can be expressed as

$$\mathcal{L}_{\mathrm{IB}} = -e_{\mathrm{B}} A_{\mathrm{B}}^\mu \bar{\psi}_{\mathrm{B}} \gamma_\mu \psi_{\mathrm{B}}. \tag{5.546}$$

Using Eqs.(5.509) and (5.463), we have

$$\begin{aligned}
\Lambda_\mu(p, q, p') &= (1 + D)\gamma_\mu + \frac{e^2}{8\pi^2\varepsilon}\gamma_\mu + \text{finite terms} \\
&= \gamma_\mu + \text{finite terms},
\end{aligned} \tag{5.547}$$

which contains only the convergent part.

Adding Eq.(5.515), Eq.(5.525) and Eq.(5.546), we obtain the total bare Lagrangian density

$$\begin{aligned}
\mathcal{L}_{\mathrm{B}} &= i\bar{\psi}_{\mathrm{B}}\partial\!\!\!/\psi_{\mathrm{B}} - m_{\mathrm{B}}\bar{\psi}_{\mathrm{B}}\psi_{\mathrm{B}} - e_{\mathrm{B}}A_{\mathrm{B}}^\mu\bar{\psi}_{\mathrm{B}}\gamma_\mu\psi_{\mathrm{B}} \\
&\quad - \frac{1}{4}(\partial_\mu A_{\mathrm{B}\nu} - \partial_\nu A_{\mathrm{B}\mu})^2 + \text{gauge terms} \\
&= iZ_2\bar{\psi}\partial\!\!\!/\psi - Z_2(m + A)\bar{\psi}\psi - Z_1 e\mu^{\frac{\varepsilon}{2}}A^\mu\bar{\psi}\gamma_\mu\psi \\
&\quad - \frac{Z_3}{4}(\partial_\mu A_\nu - \partial_\nu A_\mu)^2 + \text{gauge terms}
\end{aligned} \tag{5.548}$$

Using the bare Lagrangian density $\mathcal{L}_{\mathrm{B}}$, we can eliminate the divergent parts and obtain the finite renormalized quantities for the quasi-particles.

5.17   Renormalization group

In Section 5.6, we have shown how to use the renormalization method to eliminate the infinite terms in the integrations of scalar boson field with the ϕ^4 interaction term. Due to the wave function renormalization Eq.(5.161), the renormalized two-point function $\Gamma_r^{(2)}$ can be written as

$$\Gamma_r^{(2)}(p, g, m, \mu) = Z(g, m, \mu)\Gamma^{(2)}(p, g_0, m_0), \tag{5.549}$$

where g and m are the renormalized coupling constant and mass which are the functions of bare coupling constant g_0, bare mass m_0 and μ. Expressed in terms of g_0 and m_0, $\Gamma^{(2)}(p, g_0, m_0)$ is independent of μ. Eq.(5.549) can also be written as

$$\Gamma^{(2)}(p, g_0, m_0) = Z^{-1}(g, m, \mu)\Gamma_r^{(2)}(p, g, m, \mu). \tag{5.550}$$

Similarly, for the renormalized n-particles vertex functions, we have

$$\Gamma_r^{(n)}(p, g, m, \mu) = Z^{\frac{n}{2}}(g, m, \mu)\Gamma^{(n)}(p, g_0, m_0), \tag{5.551}$$

or

$$\Gamma^{(n)}(p, g_0, m_0) = Z^{-\frac{n}{2}}(g, m, \mu)\Gamma_r^{(n)}(p, g, m, \mu). \tag{5.552}$$

Since the unrenormalized vertex function $\Gamma^{(n)}$ is independent of μ, it is invariant under the transformation

$$\mu \to e^{\alpha}\mu. \tag{5.553}$$

The transformation forms a group, which is called the *renormalization group*. Thus we have

$$\frac{\partial \Gamma^{(n)}}{\partial \mu} = 0 \tag{5.554}$$

or

$$\mu\frac{\partial}{\partial \mu}\Gamma^{(n)} = 0. \tag{5.555}$$

Inserting Eq.(5.552) into Eq.(5.555), we have

$$\mu\frac{\partial}{\partial \mu}[Z^{-\frac{n}{2}}(\mu)\Gamma_r^{(n)}(p, g, m, \mu)] = 0. \tag{5.556}$$

In Eq.(5.556), g and m depend on μ. Performing the differentiation gives

$$-\frac{n}{2}\left(Z^{-\frac{n}{2}-1}\mu\frac{\partial}{\partial \mu}Z\right)\Gamma_r^{(n)}$$
$$+ Z^{-\frac{n}{2}}\left(\mu\frac{\partial}{\partial \mu} + \mu\frac{\partial g}{\partial \mu}\frac{\partial}{\partial g} + \mu\frac{\partial m}{\partial \mu}\frac{\partial}{\partial m}\right)\Gamma_r^{(n)} = 0. \tag{5.557}$$

Multiplying Eq.(5.557) by $Z^{-\frac{n}{2}}$, we have

$$\left(-n\mu\frac{\partial}{\partial \mu}\ln\sqrt{Z} + \mu\frac{\partial}{\partial \mu} + \mu\frac{\partial g}{\partial \mu}\frac{\partial}{\partial g} + \mu\frac{\partial m}{\partial \mu}\frac{\partial}{\partial m}\right)\Gamma_r^{(n)} = 0. \tag{5.558}$$

We introduce the following quantities:

$$\gamma \equiv \frac{\partial}{\partial \mu} \ln \sqrt{Z}, \tag{5.559a}$$

$$\beta \equiv \mu \frac{\partial g}{\partial \mu}, \tag{5.559b}$$

$$\gamma_m \equiv \frac{\mu}{m} \frac{\partial m}{\partial \mu}. \tag{5.559c}$$

In terms of the parameters γ, β and γ_m in Eq.(5.559), Eq.(5.558) is expressed as

$$\left(\mu \frac{\partial}{\partial \mu} + \beta \frac{\partial}{\partial g} - n\gamma + m\gamma_m \frac{\partial}{\partial m} \right) \Gamma_r^{(n)} = 0. \tag{5.560}$$

Eq.(5.560) is called the renormalization group equation (RG equation), which determines the variation of the renormalized $\Gamma_r^{(n)}$ with the change of the regularization parameter μ. μ scales as a mass. The renormalization procedure can be considered as a scale transformation. Thus we can express RG equation Eq.(5.560) as a transformation under a change of scale. We introduce a scale transformation

$$p \to tp. \tag{5.561}$$

For a $n_\varepsilon = 4 - \varepsilon$ dimension spacetime, the Lagrangian density scales as

$$[\mathcal{L}] = l^{-n_\varepsilon}, \tag{5.562}$$

where l has a length dimension because $S = \int d^{n_\varepsilon} x \mathcal{L}$ is dimensionless. Momentum dimension Λ scales inversely with l. We have

$$[\mathcal{L}] = \Lambda^{n_\varepsilon}. \tag{5.563}$$

The kinetic energy term is $\partial_\mu \phi \partial^\mu \phi$. Thus $[\partial_\mu \phi \partial^\mu \phi] = \Lambda^{n_\varepsilon}$. Since $[\partial_\mu] = \Lambda$, we have

$$[\phi] = \Lambda^{\frac{n_\varepsilon}{2} - 1}. \tag{5.564}$$

Thus we have

$$[G^{(n)}(x_1, \cdots, x_n)] = \Lambda^{n\left(\frac{n_\varepsilon}{2} - 1\right)} \tag{5.565}$$

and

$$\begin{aligned} [G^{(n)}(p_1, \cdots, p_n)] &= \Lambda^{-(n-1)n_\varepsilon + n\left(\frac{n_\varepsilon}{2} - 1\right)} \\ &= \Lambda^{-n\left(\frac{n_\varepsilon}{2} + 1\right) + n_\varepsilon}. \end{aligned} \tag{5.566}$$

Using Eq.(5.116), we have

$$[\Gamma^{(n)}(x_1, \cdots, x_n)] = \Lambda^{n\left(\frac{n_\varepsilon}{2} + 1\right)} \tag{5.567}$$

and

$$[\Gamma^{(n)}(p_1, \cdots, p_n)] = \Lambda^{-(n-1)n_\varepsilon + n\left(\frac{n_\varepsilon}{2}+1\right)}. \tag{5.568}$$

Thus the vertex function $\Gamma^{(n)}$ has a mass dimension

$$D = n_\varepsilon + n\left(1 - \frac{n_\varepsilon}{2}\right) = n_\varepsilon + n\left(\frac{\varepsilon}{2} - 1\right). \tag{5.569}$$

Under the scale transformation $p \to tp$, $\Gamma^{(n)}$ transforms as

$$\Gamma_r^{(n)}(tp, g, m, \mu) = t^D \Gamma_r^{(n)}(p, g, t^{-1}m, t^{-1}\mu). \tag{5.570}$$

Performing the differentiation with respect to t gives

$$\left(t\frac{\partial}{\partial t} + m\frac{\partial}{\partial m} + \mu\frac{\partial}{\partial \mu} - D\right)\Gamma_r^{(n)} = 0. \tag{5.571}$$

Using Eq.(5.560) to eliminate $\mu\frac{\partial \Gamma_r^{(n)}}{\partial \mu}$, we have

$$\left(-t\frac{\partial}{\partial t} + \beta\frac{\partial}{\partial g} - n\gamma + m(\gamma_m - 1)\frac{\partial}{\partial m} + D\right)\Gamma_r^{(n)}(tp, g, m, \mu) = 0. \tag{5.572}$$

Eq.(5.572) determines the variation of $\Gamma_r^{(n)}$ when the momentum p is scaled up by a factor t.

If we fix μ as a constant, Eq.(5.572) can be considered as a differential equation with t as the variable. The coupling constant g and mass m should be functions of t.

We expect that Γ satisfying the differential equation Eq.(5.572) should fulfill the following relation

$$\Gamma_r^{(n)}(tp, g, m, \mu) = f(t)\Gamma_r^{(n)}(p, g(t), m(t), \mu), \tag{5.573}$$

which states that a change in t can be offset by the changes in m and g, and an overall factor $f(t)$. Differentiating Eq.(5.573), we have

$$t\frac{\partial}{\partial t}\Gamma_r^{(n)}(tp, g, m, \mu)$$

$$= t\frac{df}{dt}\Gamma_r^{(n)}(p, g(t), m(t), \mu) + tf(t)\left(\frac{\partial m}{\partial t}\frac{\partial \Gamma_r^{(n)}}{\partial m} + \frac{\partial g}{\partial t}\frac{\partial \Gamma_r^{(n)}}{\partial g}\right)$$

$$= \left(t\frac{df}{dt} + tf(t)\frac{\partial m}{\partial t}\frac{\partial}{\partial m} + tf(t)\frac{\partial g}{\partial t}\frac{\partial}{\partial g}\right)\Gamma_r^{(n)}(p, g(t), m(t), \mu)$$

$$= \left(t\frac{df}{dt} + tf(t)\frac{\partial m}{\partial t}\frac{\partial}{\partial m} + tf(t)\frac{\partial g}{\partial t}\frac{\partial}{\partial g}\right)\frac{1}{f(t)}\Gamma_r^{(n)}(tp, g, m, \mu), \tag{5.574}$$

which leads to

$$\left(-t\frac{\partial f}{\partial t} + \frac{t}{f}\frac{df}{dt} + t\frac{\partial m}{\partial t}\frac{\partial}{\partial m} + t\frac{\partial g}{\partial t}\frac{\partial}{\partial g}\right)\frac{1}{f(t)}\Gamma_r^{(n)}(tp, g, m, \mu) = 0. \tag{5.575}$$

Comparing Eq.(5.575) with Eq.(5.572), we can see that the two equations have the same form. In order to make the two equations equivalent, we should have the equal coefficients of $\frac{\partial}{\partial g}$, which gives

$$t\frac{\partial g(t)}{\partial t} = \beta(g(t)). \tag{5.576}$$

The coefficients of $\frac{\partial}{\partial m}$ should also be equal, which leads to

$$t\frac{\partial m}{\partial t} = m(\gamma_m - 1). \tag{5.577}$$

Comparing the remaining terms, we have

$$\frac{t}{f}\frac{df}{dt} = D - n\gamma. \tag{5.578}$$

Eq.(5.576) dictates the change of the effective coupling constant. Its solution $g(t)$ is called the running coupling constant. The asymptotic limit of $g(t)$ as $t \to \infty$ can be determined if we have solved the above differential equations. $m(t)$ is the solution of Eq.(5.577). $f(t)$ can be obtained by solving Eq.(5.578). Integrating Eq.(5.578), we obtain

$$f(t) = t^D \exp\left[-\int_1^t \frac{n\gamma(g(t))}{t}dt\right]. \tag{5.579}$$

Inserting Eq.(5.579) into Eq.(5.573), we have

$$\Gamma_r^{(n)}(tp, g, m, \mu) = t^D \exp\left[-n\int_1^t \frac{\gamma(g(t))}{t}dt\right]$$
$$\Gamma_r^{(n)}(p, g(t), m(t), \mu), \tag{5.580}$$

where $D = 4 - n$ in the limit $\varepsilon \to 0$. The exponential factor is called the anomalous dimension. Eq.(5.576), Eq.(5.577), and Eq.(5.579) determine how the vertex function $\Gamma_r^{(n)}$ varies with the change of the scaling factor t. Of particular interest is the large and small momentum behaviors of the quantum fields.

5.18   Asymptotic behavior of coupling constant

5.18.1   *Stable fixed point*

Now we analyse some possible asymptotic behavior of the running coupling constant $g(t)$. $\beta(g)$ determines the behavior of the running coupling constant $g(t)$. Two typical function forms of $\beta(g)$ are shown in Fig.5.27. The zeros of β in Fig.5.27 are called the fixed points. Usually we have a zero of β

at $g = 0$. $\beta(g)$ could also have a zero at $g = g_0$. In Fig.5.27(a), $\beta > 0$ when $g < g_0$ and $\beta < 0$ when $g > g_0$. When $g < g_0$, according to Eq.(5.576), g increases with the increase of t. g is driven towards g_0 if g_0 is smaller than g_0. On the other hand, g decreases with the increase of t when $g > g_0$. g is driven back towards g_0 when g is larger than g_0. Thus $g(t \to \infty) \to g_0$. g_0 is called an ultra-violet stable fixed point. Similar argument leads to $g(t \to 0) \to g = 0$. Thus $g(0) = 0$ is called an infra-red stable fixed point. For the function form $\beta(g)$ shown in Fig.5.27(b), $\beta < 0$ when $g < g_0$ and $\beta > 0$ when $g > g_0$. The sign of β is reversed as compared with $\beta(g)$ shown in Fig.5.27(a). In this case, $g = g_0$ becomes an infra-red stable fixed point and $g = 0$ is an ultra-violet stable fixed point. Thus, when we increase the momentum (t increases), the running coupling constant $g(t)$ decreases. In the infinite momentum (or energy) limit, the effective interaction vanishes, which is called the asymptotic freedom.

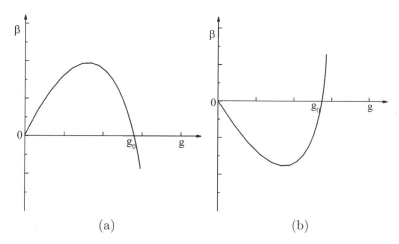

Fig. 5.27 Two possible function forms of $\beta(g)$. (a) $g = 0$ is an infra-red stable fixed point and $g = g_0$ is an ultra-violet stable fixed point. (b) $g = 0$ is an ultra-violet stable fixed point and $g = g_0$ is an infra-red stable fixed point.

5.18.2   *Asymptotic behavior of scalar field*

Now we consider the asymptotic behavior of the scalar boson field with the ϕ^4 interaction. Using Eq.(5.178c), we have

$$
\begin{aligned}
g &= g_0 \mu^{-\varepsilon} \frac{Z^2}{Z_g} \\
&= g_0 \mu^{-\varepsilon} \left(1 - \frac{3g}{16\pi^2 \varepsilon} \right).
\end{aligned}
\tag{5.581}
$$

Differentiating Eq.(5.581) with respect to μ, we obtain

$$
\begin{aligned}
\mu \frac{\partial g}{\partial \mu} &= g_0 \mu^{-\varepsilon} (-\varepsilon) \left(1 - \frac{3g}{16\pi^2 \varepsilon} \right) \\
&\quad + g_0 \mu^{-\varepsilon} \left(- \frac{3\mu \frac{\partial g}{\partial \mu}}{16\pi^2 \varepsilon} \right).
\end{aligned}
\tag{5.582}
$$

Solving Eq.(5.582) for $\mu \frac{\partial g}{\partial \mu}$, we have

$$
\mu \frac{\partial g}{\partial \mu} = -\varepsilon g + \frac{3g^2}{16\pi^2} + \mathcal{O}(g^3),
\tag{5.583}
$$

where we have used $g = g_0 \mu^{-\varepsilon} + \mathcal{O}(g^2)$. We will neglect the $\mathcal{O}(g^3)$ terms. Using Eq.(5.559) and taking the limit $\varepsilon \to 0$, we have

$$
\beta(g) = \lim_{\varepsilon \to 0} \mu \frac{\partial g}{\partial \mu} = \frac{3g^2}{16\pi^2} > 0.
\tag{5.584}
$$

Thus the running coupling constant increases with increasing t, which means that the higher momentum, the stronger interaction. The solution of Eq.(5.576) is given by

$$
g = \frac{g_0}{1 - \frac{3}{16\pi^2} g_0 \ln \left(\frac{t}{t_0} \right)},
\tag{5.585}
$$

which shows that g increases with increasing t.

5.18.3   *Asymptotic behavior of QED*

We can estimate the asymptotic behavior of QED using Eq.(5.543). Inserting Eq.(5.524) into Eq.(5.543), we have

$$
e_{\mathrm{B}} = e\mu^{\frac{\varepsilon}{2}} \left(1 + \frac{e^2}{12\pi^2 \varepsilon} \right).
\tag{5.586}
$$

The bare charge e_B is independent of μ. Differentiating Eq.(5.586) with respect to μ, we obtain

$$\mu \frac{\partial e}{\partial \mu} \mu^{\frac{\varepsilon}{2}} \left(1 + \frac{e^2}{12\pi^2\varepsilon} \right) + e\mu^{\frac{\varepsilon}{2}} \frac{\varepsilon}{2} \left(1 + \frac{e^2}{12\pi^2\varepsilon} \right) + e\mu^{\frac{\varepsilon}{2}} \frac{2e\mu \frac{\partial e}{\partial \mu}}{12\pi^2\varepsilon} = 0, \qquad (5.587)$$

which can be rewritten as

$$\mu \frac{\partial e}{\partial \mu} \left(1 + \frac{3e^2}{12\pi^2\varepsilon} \right) = -e \frac{\varepsilon}{2} \left(1 + \frac{e^2}{12\pi^2\varepsilon} \right) \qquad (5.588)$$

or

$$\begin{aligned}
\mu \frac{\partial e}{\partial \mu} &= -e \frac{\varepsilon}{2} \left(1 + \frac{e^2}{12\pi^2\varepsilon} \right) \left(1 + \frac{3e^2}{12\pi^2\varepsilon} \right)^{-1} \\
&= -e \frac{\varepsilon}{2} \left(1 - \frac{e^2}{6\pi^2\varepsilon} \right) + \mathcal{O}(e^5) \qquad (5.589) \\
&= -\frac{\varepsilon}{2} e + \frac{e^2}{12\pi^2} + \mathcal{O}(e^5).
\end{aligned}$$

Taking limit $\varepsilon \to 0$, Eq.(5.589) becomes

$$\mu \frac{\partial e}{\partial \mu} = \frac{e^3}{12\pi^2}. \qquad (5.590)$$

According to Eq.(5.559), we have

$$\beta = \frac{e^3}{12\pi^2}. \qquad (5.591)$$

Since $\beta > 0$, the running coupling constant increases with the increase of t. The solution of Eq.(5.576) for QED is

$$e^2(t) = \frac{e^2(t_0)}{1 - \frac{e^2(t_0)}{6\pi^2} \ln \left(\frac{t}{t_0} \right)}, \qquad (5.592)$$

which shows that e increases with increasing t. When

$$t = t_0 \exp \left(\frac{6\pi^2}{e^2} \right), \qquad (5.593)$$

the charge e approaches to ∞, which is called the Landau singularity. The increased e with increasing t can be understood as follows: A charge is screened by the opposite charge of the antiparticles. As t increases, the momentum increases, which means that the charge is measured more locally. Then the nonlocal screening effect is weaker, which leads to larger charge value.

Chapter 6

From quantum field theory to quantum mechanics

We are now ready to deduce some approximate formalisms of physics which are important for the applications. One is the formalism of quantum mechanics which is applicable in microscopic scale and low energy. The other is the formalism of classical fields, which is applicable in macroscopic scale where the fluctuation and correlation are small. First we consider the systems with low energy where the quantum mechanics is used. When we say something is small or low, we should have a reference point. Here the reference energy is the mass m of quasi-particles. When the energy variation is much smaller than the mass of quasi-particles, the energy is said to be small. This is also called the non-relativistic limit. In this case, the number of quasi-particles is conserved because the creation of one quasi-particle costs an energy of m, which is much larger than the available energy. We only use quantum mechanics to describe the massive particles. For massless field, we do not have the mass of particle as a gauge energy and the conservation of particle number. Massless field is approximated directly as the classics massless field, such as electromagnetic field in the case of photons. In the following, we will not use the natural units so that we write out the Planck constant $\hbar$ and the speed of light c explicitly in the discussions of the non-relativistic and classical limits. We will also call quasi-particle simply particle.

6.1   Non-relativistic limit of Klein-Gordon equation

First we consider the scalar boson field described by the Klein-Gordon equation.

$$\frac{1}{c^2}\ddot{\hat{\phi}}(\mathbf{x},t) = \left(\nabla^2 - \frac{m^2c^2}{\hbar^2}\right)\hat{\phi}(\mathbf{x},t). \tag{6.1}$$

In order to derive the non-relativistic limit of the free Klein-Gordon equation, we make an ansatz

$$\hat{\phi}(\mathbf{x},t) = \hat{\varphi}(\mathbf{x},t)\exp\left(-\frac{i}{\hbar}mc^2t\right). \tag{6.2}$$

We have split the time dependence of $\hat{\phi}$ into two terms: the fast oscillating term $\exp\left(-\frac{i}{\hbar}mc^2t\right)$ and the slow changing term $\hat{\varphi}(\mathbf{x},t)$.

In the non-relativistic limit, the difference of the energy E of the particle and the mass m is small. We define

$$E' \equiv E - mc^2. \tag{6.3}$$

In the non-relativistic limit, $E' \ll E \approx mc^2$. Thus

$$\left\langle i\hbar\frac{\partial\hat{\varphi}}{\partial t}\right\rangle \approx \langle E'\hat{\varphi}\rangle \ll \langle mc^2\hat{\varphi}\rangle. \tag{6.4}$$

Using the ansatz Eq.(6.2), we have

$$\begin{aligned}
\frac{\partial\hat{\phi}}{\partial t} &= \left(\frac{\partial\hat{\varphi}}{\partial t} - i\frac{mc^2}{\hbar}\hat{\varphi}\right)\exp\left(-\frac{i}{\hbar}mc^2t\right) \\
&\approx -i\frac{mc^2}{\hbar}\hat{\varphi}\exp\left(-\frac{i}{\hbar}mc^2t\right)
\end{aligned} \tag{6.5}$$

and

$$\begin{aligned}
\frac{\partial^2\hat{\phi}}{\partial t^2} &= \frac{\partial}{\partial t}\left[\left(\frac{\partial\hat{\varphi}}{\partial t} - i\frac{mc^2}{\hbar}\hat{\varphi}\right)\exp\left(-\frac{i}{\hbar}mc^2t\right)\right] \\
&\approx \left(-i\frac{mc^2}{\hbar}\frac{\partial\hat{\varphi}}{\partial t} - i\frac{mc^2}{\hbar}\frac{\partial\hat{\varphi}}{\partial t} - \frac{m^2c^4}{\hbar^2}\hat{\varphi}\right)\exp\left(-\frac{i}{\hbar}mc^2t\right) \\
&= -\left(i\frac{2mc^2}{\hbar}\frac{\partial\hat{\varphi}}{\partial t} + \frac{m^2c^4}{\hbar^2}\hat{\varphi}\right)\exp\left(-\frac{i}{\hbar}mc^2t\right).
\end{aligned} \tag{6.6}$$

Inserting Eq.(6.6) into Eq.(6.1), we have

$$\begin{aligned}
&-\frac{1}{c^2}\left(i\frac{2mc^2}{\hbar}\frac{\partial\hat{\varphi}}{\partial t} + \frac{m^2c^4}{\hbar^2}\hat{\varphi}\right)\exp\left(-\frac{i}{\hbar}mc^2t\right) \\
&= \left(\nabla^2 - \frac{m^2c^2}{\hbar^2}\right)\hat{\varphi}\exp\left(-\frac{i}{\hbar}mc^2t\right).
\end{aligned} \tag{6.7}$$

Eliminating the fast oscillating term $\exp\left(-\frac{i}{\hbar}mc^2t\right)$, Eq.(6.7) becomes

$$i\hbar\frac{\partial\hat{\varphi}}{\partial t} = -\frac{\hbar^2}{2m}\nabla^2\hat{\varphi}. \tag{6.8}$$

Eq.(6.8) is the Schrödinger equation in operator form for scalar bosons.

6.2 Non-relativistic limit of the Dirac equation

Now we consider the non-relativistic limit of the Dirac equation Eq.(2.247)

$$(i\hbar\gamma^\mu\partial_\mu - mc)\hat{\psi} = 0. \tag{6.9}$$

or

$$i\hbar\frac{\partial\hat{\psi}}{\partial t} = -i\hbar c\boldsymbol{\alpha} \cdot \boldsymbol{\nabla}\hat{\psi} + \beta mc^2\hat{\psi}. \tag{6.10}$$

The coupling of the Dirac fermion field with the photon field should maintain the gauge invariance. We introduce the covariant derivative $D_\mu = \partial_\mu - i\frac{e}{\hbar c}\hat{A}_\mu$ to replace the ordinary derivative ∂_μ to include this interaction term with the photon field. In the classical limit, $\hat{A}_\mu$ is replaced by its classical value and becomes the electromagnetic four-potential

$$A^\mu = \{A_0(x), \mathbf{A}(x)\}. \tag{6.11}$$

Then the Dirac equation in the electromagnetic potentials is given by

$$i\hbar\frac{\partial\hat{\psi}}{\partial t} = [c\boldsymbol{\alpha} \cdot (-i\hbar\mathbf{D}) + eA_0 + \beta mc^2]\hat{\psi}. \tag{6.12}$$

Since particle-antiparticle creation and annihilation are negligible in low energy, we can consider particles and antiparticles separately. Thus the four-component spinor $\hat{\psi}$ is decomposed into two-component spinors

$$\hat{\psi} = \begin{pmatrix} \hat{\hat{\varphi}} \\ \hat{\hat{\chi}} \end{pmatrix}. \tag{6.13}$$

Then the Dirac equation Eq.(6.12) reads

$$i\hbar\frac{\partial}{\partial t}\begin{pmatrix} \hat{\hat{\varphi}} \\ \hat{\hat{\chi}} \end{pmatrix} = \begin{pmatrix} c\boldsymbol{\sigma} \cdot (-i\hbar\mathbf{D})\hat{\hat{\chi}} \\ c\boldsymbol{\sigma} \cdot (-i\hbar\mathbf{D})\hat{\hat{\varphi}} \end{pmatrix} + eA_0\begin{pmatrix} \hat{\hat{\varphi}} \\ \hat{\hat{\chi}} \end{pmatrix} + mc^2\begin{pmatrix} \hat{\hat{\varphi}} \\ -\hat{\hat{\chi}} \end{pmatrix}. \tag{6.14}$$

In the derivation of Eq.(6.14), we have used the explicit representation of Dirac's matrices

$$\alpha_i = \begin{pmatrix} 0 & \sigma_i \\ \sigma_i & 0 \end{pmatrix}, \quad \beta = \begin{pmatrix} I & 0 \\ 0 & -I \end{pmatrix}, \tag{6.15}$$

where σ_i are Pauli's 2×2 matrices given by Eq.(2.260) and I is the 2×2 unit matrix.

Similar to the case of the non-relativistic limit of the Klein-Gordon equation, we use the ansatz

$$\begin{pmatrix} \hat{\hat{\varphi}} \\ \hat{\hat{\chi}} \end{pmatrix} = \begin{pmatrix} \hat{\varphi} \\ \hat{\chi} \end{pmatrix}\exp\left(-\frac{imc^2}{\hbar}t\right). \tag{6.16}$$

Inserting Eq.(6.16) into Eq.(6.14), we have

$$i\hbar\frac{\partial}{\partial t}\begin{pmatrix}\hat{\varphi}\\\hat{\chi}\end{pmatrix} = \begin{pmatrix}c\boldsymbol{\sigma}\cdot(-i\hbar\mathbf{D})\hat{\chi}\\c\boldsymbol{\sigma}\cdot(-i\hbar\mathbf{D})\hat{\varphi}\end{pmatrix} + eA_0\begin{pmatrix}\hat{\varphi}\\\hat{\chi}\end{pmatrix} - 2mc^2\begin{pmatrix}0\\-\hat{\chi}\end{pmatrix}. \quad (6.17)$$

When the kinetic energy and potential energy are much smaller than the rest energy mc^2, i.e.

$$\left\langle i\hbar\frac{\partial\hat{\chi}}{\partial t}\right\rangle \ll \langle mc^2\hat{\chi}\rangle \quad (6.18)$$

and

$$\langle eA_0\hat{\chi}\rangle \ll \langle mc^2\hat{\chi}\rangle, \quad (6.19)$$

we have

$$\hat{\chi} = \frac{\boldsymbol{\sigma}\cdot(-i\hbar\mathbf{D})}{2mc}\hat{\varphi} \quad (6.20)$$

for the lower component of Eq.(6.17). This means that $\hat{\chi}$ is the small component of the field operator $\hat{\psi}$ and $\hat{\varphi}$ is the large component of the field operator $\hat{\psi}$. Inserting Eq.(6.20) into the upper equation of Eq.(6.17), we have

$$i\hbar\frac{\partial\hat{\varphi}}{\partial t} = \frac{[\boldsymbol{\sigma}\cdot(-i\hbar\mathbf{D})][\boldsymbol{\sigma}\cdot(-i\hbar\mathbf{D})]}{2m}\hat{\varphi} + eA_0\hat{\varphi}. \quad (6.21)$$

Using the relation

$$(\boldsymbol{\sigma}\cdot\mathbf{A})(\boldsymbol{\sigma}\cdot\mathbf{B}) = \mathbf{A}\cdot\mathbf{B} + i\boldsymbol{\sigma}\cdot(\mathbf{A}\times\mathbf{B}), \quad (6.22)$$

we have

$$\begin{aligned}&[\boldsymbol{\sigma}\cdot(-i\hbar\mathbf{D})][\boldsymbol{\sigma}\cdot(-i\hbar\mathbf{D})]\\&= \left(-i\hbar\boldsymbol{\nabla}-\frac{e}{c}\mathbf{A}\right)^2 + i\boldsymbol{\sigma}\cdot\left[\left(-i\hbar\boldsymbol{\nabla}-\frac{e}{c}\mathbf{A}\right)\times\left(-i\hbar\boldsymbol{\nabla}-\frac{e}{c}\mathbf{A}\right)\right]\\&= \left(-i\hbar\boldsymbol{\nabla}-\frac{e}{c}\mathbf{A}\right)^2 - \frac{e}{c}\hbar\boldsymbol{\sigma}\cdot(\boldsymbol{\nabla}\times\mathbf{A})\\&= \left(-i\hbar\boldsymbol{\nabla}-\frac{e}{c}\mathbf{A}\right)^2 - \frac{e\hbar}{c}\boldsymbol{\sigma}\cdot\mathbf{B}.\end{aligned} \quad (6.23)$$

Thus, Eq.(6.21) becomes

$$i\hbar\frac{\partial\hat{\varphi}}{\partial t} = \left[\frac{1}{2m}\left(-i\hbar\boldsymbol{\nabla}-\frac{e}{c}\mathbf{A}\right)^2 - \frac{e\hbar}{2mc}\boldsymbol{\sigma}\cdot\mathbf{B} + eA_0\right]\hat{\varphi}. \quad (6.24)$$

This is the *Pauli equation* in operator version. The two components of $\hat{\varphi}$ describe the spin degrees of freedom.

In the case of a homogeneous magnetic field $\mathbf{B}_0$,

$$\mathbf{A} = \frac{1}{2}\mathbf{B}_0\times\mathbf{r}. \quad (6.25)$$

When the magnetic field is weak, we have

$$
\left(-i\hbar\boldsymbol{\nabla} - \frac{e}{c}\mathbf{A}\right)^2 = \left(-i\hbar\boldsymbol{\nabla} - \frac{e}{2c}\mathbf{B}_0 \times \mathbf{r}\right)^2
$$

$$
\approx (-i\hbar\boldsymbol{\nabla})^2 - \frac{e}{c}(\mathbf{B}_0 \times \mathbf{r}) \cdot (-i\hbar\boldsymbol{\nabla}) \qquad (6.26)
$$

$$
= (-i\hbar\boldsymbol{\nabla})^2 - \frac{e}{c}(\mathbf{B}_0 \cdot \mathbf{L}),
$$

where

$$
\mathbf{L} \equiv \mathbf{r} \times (-i\hbar\boldsymbol{\nabla}) \qquad (6.27)
$$

is defined as the *operator of orbital angular momentum*. In the derivation of Eq.(6.26), we have neglected the quadratic terms of $\mathbf{A}$. We define

$$
\mathbf{S} \equiv \frac{1}{2}\hbar\boldsymbol{\sigma} \qquad (6.28)
$$

as *spin operator*. Then we obtain the *Schrödinger equation* in the operator form

$$
i\hbar\frac{\partial\hat{\varphi}}{\partial t} = \left[\frac{1}{2m}(-i\hbar\boldsymbol{\nabla})^2 - \frac{e}{2mc}(\mathbf{L} + 2\mathbf{S}) \cdot \mathbf{B}_0 + eA_0\right]\hat{\varphi}. \qquad (6.29)
$$

The factor 2 before $\mathbf{S}$ is the g factor for spin. When the relativistic effect is considered, the g factor for spin has a little deviation from 2. Since the spin degeneracy g_s for spin-$\frac{1}{2}$ fermions is 2, we often use the same notation g_s for them.

6.3 Generator of space rotation transformation

$\mathbf{L} = \frac{1}{i}\mathbf{r} \times \hbar\boldsymbol{\nabla}$ can be shown to be the generator of the space rotation transformation. Thus the conservation of angular momentum $\mathbf{L}$ is related to the spatial isotropy. Now we consider a rotation around the z-axis by a small angle $\delta\theta$ as shown in Fig.6.1. Under such a rotation, a point P at x and y is at $x' = x + \delta\theta \cdot y$ and $y' = y - \delta\theta \cdot x$ in the new coordinate reference. A function ψ in the new coordinate reference can be written as

$$
\begin{aligned}
\psi(x', y', z') &= \psi(x + \delta\theta \cdot y, y - \delta\theta \cdot x, z) \\
&= \psi(x, y, z) + \delta\theta \cdot y\frac{\partial\psi}{\partial x} - \delta\theta \cdot x\frac{\partial\psi}{\partial y} \\
&= (1 + \frac{i}{\hbar}\delta\theta L_z)\psi(x, y, z) \\
&= R_z(\delta\theta)\psi(x, y, z)
\end{aligned} \qquad (6.30)
$$

with

$$R_z(\delta\theta) = e^{\frac{i}{\hbar}\delta\theta L_z}. \tag{6.31}$$

Thus $L_z = \frac{1}{i}(\mathbf{r} \times \hbar\boldsymbol{\nabla})_z$ is the generator of the space rotation transformation around the z-axis. Similarly, L_z and L_y are the generators of the space rotation transformations around the x-axis and y-axis, respectively. The infinitesimal rotation transformation around the i-axis with the rotational angle $\delta\theta_i$ is given by

$$R_i(\delta\theta_i) = 1 + \frac{i}{\hbar}L_i\delta\theta_i. \tag{6.32}$$

The matrix of a finite rotation around an axis $\mathbf{n}$ with an angle θ is given by

$$R_{\mathbf{n}}(\theta) = e^{\frac{i}{\hbar}\mathbf{L}\cdot\mathbf{n}\theta} = e^{\frac{i}{\hbar}\mathbf{L}\cdot\boldsymbol{\theta}}. \tag{6.33}$$

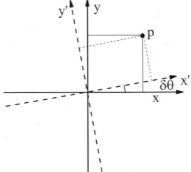

Fig. 6.1   Rotation around the z-axis by a small angle $\delta\theta$.

Using the matrix R for rotation, we write the spatial rotation in the form of

$$\begin{pmatrix} x' \\ y' \\ z' \end{pmatrix} = (R) \begin{pmatrix} x \\ y \\ z \end{pmatrix}. \tag{6.34}$$

The rotation matrix around the z-axis is given by (see Eq.(A.61))

$$R_z(\theta) = \begin{pmatrix} \cos(\theta) & \sin(\theta) & 0 \\ -\sin(\theta) & \cos(\theta) & 0 \\ 0 & 0 & 1 \end{pmatrix}. \tag{6.35}$$

Similarly the matrices for the rotations around the x- and y-axes are given by

$$R_x(\phi) = \begin{pmatrix} 1 & 0 & 0 \\ 0 & \cos(\phi) & \sin(\phi) \\ 0 & -\sin(\phi) & \cos(\phi) \end{pmatrix} \tag{6.36}$$

and

$$R_y(\varphi) = \begin{pmatrix} \cos(\varphi) & 0 & -\sin(\varphi) \\ 0 & 1 & 0 \\ \sin(\varphi) & 0 & \cos(\varphi) \end{pmatrix}, \tag{6.37}$$

respectively.

For an active rotation around the z-axis, which rotates a vector $\mathbf{A}$ around the z-axis, we have

$$\begin{pmatrix} A'_x \\ A'_y \\ A'_z \end{pmatrix} = \begin{pmatrix} \cos(\theta) & \sin(\theta) & 0 \\ -\sin(\theta) & \cos(\theta) & 0 \\ 0 & 0 & 1 \end{pmatrix} \begin{pmatrix} A_x \\ A_y \\ A_z \end{pmatrix}. \tag{6.38}$$

When we rotate the axes, leaving the vector fixed, the rotation is called the passive rotation, which is described by the matrix with minus rotational angle.

The operator of orbital angular momentum $\mathbf{L}$, as the generator for the matrix of rotation R, is given by

$$L_x = \frac{\hbar}{i} \frac{dR_x(\phi)}{d\phi}\Big|_{\phi=0} = \begin{pmatrix} 0 & 0 & 0 \\ 0 & 0 & -i \\ 0 & i & 0 \end{pmatrix} \hbar, \tag{6.39}$$

$$L_y = \frac{\hbar}{i} \frac{dR_y(\varphi)}{d\varphi}\Big|_{\varphi=0} = \begin{pmatrix} 0 & 0 & i \\ 0 & 0 & 0 \\ -i & 0 & 0 \end{pmatrix} \hbar, \tag{6.40}$$

$$L_z = \frac{\hbar}{i} \frac{dR_z(\theta)}{d\theta}\Big|_{\theta=0} = \begin{pmatrix} 0 & -i & 0 \\ i & 0 & 0 \\ 0 & 0 & 0 \end{pmatrix} \hbar. \tag{6.41}$$

Eqs.(6.39)-(6.41) are the matrix form of $\mathbf{L}$, which relates $\mathbf{L}$ with the rotation transformation R. It is easy to verify the following commutation relation for $\mathbf{L}$.

$$[L_i, L_j] = i\hbar\epsilon^{ijk}L_k. \tag{6.42}$$

6.4 Spin-orbital coupling

In the derivation of Eq.(6.20), we have neglected the term $i\hbar\frac{\partial\hat{\chi}}{\partial t}$ and $eA_0\hat{\chi}$ in the lower equation of Eq.(6.17). We can maintain the first order terms of

$i\hbar\frac{\partial\hat\chi}{\partial t}$ and $eA_0\hat\chi$ and obtain a more accurate equation. The lower equation of Eq.(6.17) has the form

$$i\hbar\frac{\partial}{\partial t}\hat\chi - eA_0\hat\chi + 2mc^2\hat\chi = c\boldsymbol{\sigma}\cdot(-i\hbar\mathbf{D})\hat\varphi. \qquad (6.43)$$

We consider the field A_0 as time independent. We can write the solution of Eq.(6.43) in the following form

$$\hat\chi = [i\hbar\partial_t - eA_0 + 2mc^2]^{-1}c\boldsymbol{\sigma}\cdot(-i\hbar\mathbf{D})\hat\varphi. \qquad (6.44)$$

Inserting Eq.(6.44) into the upper equation of Eq.(6.17), we have

$$(i\hbar\partial_t - eA_0)\hat\varphi - c\boldsymbol{\sigma}\cdot(-i\hbar\mathbf{D})\frac{1}{i\hbar\partial_t - eA_0 + 2mc^2}c\boldsymbol{\sigma}\cdot(-i\hbar\mathbf{D})\hat\varphi = 0. \quad (6.45)$$

We expand the operator $[i\hbar\partial_t - eA_0 + 2mc^2]^{-1}$ and keep the first order terms of $i\hbar\frac{\partial}{\partial t}$ and eA_0.

$$\begin{aligned}
\frac{1}{i\hbar\partial_t - eA_0 + 2mc^2} &= \frac{1}{2mc^2}\left(1 + \frac{i\hbar\partial_t - eA_0}{2mc^2}\right)^{-1} \\
&= \frac{1}{2mc^2}\left(1 - \frac{i\hbar\partial_t - eA_0}{2mc^2}\right) \\
&= \frac{1}{2mc^2} - \frac{i\hbar\partial_t - eA_0}{4m^2c^4}.
\end{aligned} \qquad (6.46)$$

Then Eq.(6.45) becomes

$$\begin{aligned}
(i\hbar\partial_t - eA_0)\hat\varphi = &\frac{[\boldsymbol{\sigma}\cdot(-i\hbar\mathbf{D})]^2}{2m}\hat\varphi \\
&- \frac{\boldsymbol{\sigma}\cdot(-i\hbar\mathbf{D})(i\hbar\partial_t - eA_0)\boldsymbol{\sigma}\cdot(-i\hbar\mathbf{D})}{4m^2c^2}\hat\varphi.
\end{aligned} \qquad (6.47)$$

In the following, we will neglect the $\mathbf{A}$ term for the first order correction. Keeping the lowest order terms, the second term on the right hand side can be rewritten as

$$\begin{aligned}
&\boldsymbol{\sigma}\cdot(-i\hbar\mathbf{D})(i\hbar\partial_t - eA_0)\boldsymbol{\sigma}\cdot(-i\hbar\mathbf{D})\hat\varphi \\
=&\boldsymbol{\sigma}\cdot(-i\hbar\mathbf{D})\boldsymbol{\sigma}\cdot(-i\hbar\mathbf{D})(i\hbar\partial_t - eA_0)\hat\varphi \\
&+ \boldsymbol{\sigma}\cdot(-i\hbar\mathbf{D})[i\hbar\partial_t - eA_0, \boldsymbol{\sigma}\cdot(-i\hbar\mathbf{D})]\hat\varphi \\
=&\frac{[\boldsymbol{\sigma}\cdot(-i\hbar\mathbf{D})]^4}{2m}\hat\varphi + \boldsymbol{\sigma}\cdot(-i\hbar\mathbf{D})[\boldsymbol{\sigma}\cdot(-i\hbar\mathbf{D}), eA_0]\hat\varphi.
\end{aligned} \qquad (6.48)$$

Now we evaluate the commutator in Eq.(6.48),

$$\begin{aligned}
[\boldsymbol{\sigma}\cdot(-i\hbar\mathbf{D}), eA_0] &= \boldsymbol{\sigma}\cdot(-i\hbar\boldsymbol{\nabla})eA_0 - eA_0\boldsymbol{\sigma}\cdot(-i\hbar\boldsymbol{\nabla}) \\
&= -ie\hbar\boldsymbol{\sigma}\cdot\boldsymbol{\nabla}A_0 \\
&= ie\hbar\boldsymbol{\sigma}\cdot\mathbf{E}.
\end{aligned} \qquad (6.49)$$

Then

$$\boldsymbol{\sigma} \cdot (-i\hbar\boldsymbol{\nabla})[\boldsymbol{\sigma} \cdot (-i\hbar\mathbf{D}), eA_0]$$
$$= \boldsymbol{\sigma} \cdot (-i\hbar\boldsymbol{\nabla})(ie\hbar\boldsymbol{\sigma} \cdot \mathbf{E}) \tag{6.50}$$
$$= e\hbar^2(\boldsymbol{\nabla} \cdot \mathbf{E} + \mathbf{E} \cdot \boldsymbol{\nabla}) + e\hbar\boldsymbol{\sigma} \cdot (i\hbar\boldsymbol{\nabla} \times \mathbf{E} - i\hbar\mathbf{E} \times \boldsymbol{\nabla}).$$

In the case of the static electromagnetic field, $\boldsymbol{\nabla} \times \mathbf{E} = 0$. Thus Eq.(6.45) becomes

$$\begin{aligned}
i\hbar\frac{\partial}{\partial_t}\hat{\varphi} = \Big[& \frac{1}{2m}(-i\hbar\boldsymbol{\nabla})^2 - \frac{e}{2mc}(\mathbf{L} + 2\mathbf{S}) \cdot \mathbf{B}_0 + eA_0 \\
& - \frac{(-i\hbar\boldsymbol{\nabla})^4}{8m^3c^2} - \frac{e\hbar^2}{4m^2c^2}(\boldsymbol{\nabla} \cdot \mathbf{E} + \mathbf{E} \cdot \boldsymbol{\nabla}) \\
& + \frac{ie\hbar\mathbf{S} \cdot (\mathbf{E} \times \boldsymbol{\nabla})}{2m^2c^2} \Big]\hat{\varphi}.
\end{aligned} \tag{6.51}$$

In the above equation, the $\frac{(-i\hbar\boldsymbol{\nabla})^4}{8m^3c^2}$ term is the relativistic kinetic correction. $\frac{e\hbar^2}{4m^2c^2}(\boldsymbol{\nabla} \cdot \mathbf{E} + \mathbf{E} \cdot \boldsymbol{\nabla})$ is the Darwin term. Since it contains a non-hermitian term $\mathbf{E} \cdot \boldsymbol{\nabla}$, a further transformation is usually performed to make the Darwin term hermitian when the Darwin term is used in a Hamiltonian. The last term is the spin-orbital coupling term which is denoted as H_{so}. For a spherical potential A_0, the last term in Eq.(6.51) becomes

$$\begin{aligned}
H_{so} &= \frac{-e}{2m^2c^2}\mathbf{S} \cdot [\mathbf{E} \times (-i\hbar\boldsymbol{\nabla})] \\
&= \frac{e}{2m^2c^2}\frac{1}{r}\frac{\partial A_0}{\partial r}\mathbf{S} \cdot [\mathbf{r} \times (-i\hbar\boldsymbol{\nabla})] \\
&= \frac{e}{2m^2c^2}\frac{1}{r}\frac{\partial A_0}{\partial r}(\mathbf{S} \cdot \mathbf{L}),
\end{aligned} \tag{6.52}$$

which shows clearly that it describes the spin-orbital interaction.

6.5 Operator of time translation transformation in quantum mechanics

When we inspect the Schrödinger equation, we can see that the left hand side is $i\hbar\partial_t\hat{\varphi}$ and the right hand side contains no time derivative.

Since $\hat{\pi} = i\hat{\psi}^\dagger$, according to Eqs.(2.65) and (2.66), we have

$$\{\hat{\varphi}_\alpha(\mathbf{x}, t), \hat{\varphi}_\beta^\dagger(\mathbf{x}', t)\} = \delta_{\alpha\beta}\delta(\mathbf{x} - \mathbf{x}'), \tag{6.53a}$$

$$\{\hat{\varphi}_\alpha(\mathbf{x}, t), \hat{\varphi}_\beta(\mathbf{x}', t)\} = \{\hat{\varphi}_\alpha^\dagger(\mathbf{x}, t), \hat{\varphi}_\beta^\dagger(\mathbf{x}', t)\} = 0. \tag{6.53b}$$

Thus $\hat{\varphi}(\mathbf{x}, t)$ and $\hat{\varphi}^\dagger(\mathbf{x}, t)$ behave similarly as annihilation and creation operators. It should be noted that we have used the complex field operator for the Dirac fermions and thus we have the complex $\hat{\varphi}$ and $\hat{\varphi}^\dagger$. The operators $\hat{\varphi}(\mathbf{x}, t)$ and $\hat{\varphi}^\dagger(\mathbf{x}, t)$ can be considered as complex annihilation and creation operators. We often use $\hat{a}$ and $\hat{a}^\dagger$ to denote $\hat{\varphi}$ and $\hat{\varphi}^\dagger$, respectively, as an indication of their properties.

The Schrödinger equation Eq.(6.29) in operator form can be written as

$$i\hbar \frac{\partial \hat{\varphi}}{\partial t} = [\hat{\varphi}, \hat{H}] \tag{6.54}$$

with

$$\hat{H} \equiv \int d^3x \left\{ \frac{1}{2m} (i\hbar \boldsymbol{\nabla} \hat{\varphi}^\dagger) \cdot (-i\hbar \boldsymbol{\nabla} \hat{\varphi}) \right. \\ \left. + \left[-\frac{e}{2mc} (\mathbf{L} + 2\mathbf{S}) \cdot \mathbf{B}_0 + eA_0 \right] \hat{\varphi}^\dagger \hat{\varphi} \right\}, \tag{6.55}$$

When we use the notation of annihilation and creation operators, Eq.(6.55) becomes

$$\hat{H} = \int d^3x \left\{ \frac{1}{2m} (i\hbar \boldsymbol{\nabla} \hat{a}^\dagger(\mathbf{x}, t)) \cdot (-i\hbar \boldsymbol{\nabla} \hat{a}(\mathbf{x}, t)) \right. \\ \left. + \left[-\frac{e}{2mc} (\mathbf{L} + 2\mathbf{S}) \cdot \mathbf{B}_0 + eA_0 \right] \hat{a}^\dagger(\mathbf{x}, t) \hat{a}(\mathbf{x}, t) \right\}, \tag{6.56}$$

Since $\hat{\varphi}$ and $\hat{\varphi}^\dagger$ obey the same equation of motion, we have also

$$i\hbar \frac{\partial \hat{\varphi}^\dagger}{\partial t} = [\hat{\varphi}^\dagger, \hat{H}]. \tag{6.57}$$

Using the notation of annihilation and creation operators, we have

$$i\hbar \frac{\partial \hat{a}^\dagger}{\partial t} = [\hat{a}^\dagger, \hat{H}], \tag{6.58a}$$

$$i\hbar \frac{\partial \hat{a}}{\partial t} = [\hat{a}, \hat{H}]. \tag{6.58b}$$

$\hat{H}$ is then the generator of time translation transformation. Thus, according to Eq.(2.322), $\hat{H}$ is the Hamiltonian of the system.

The Hamiltonian can be written as

$$\hat{H} = \hat{T} + \hat{U} \tag{6.59}$$

with

$$\hat{T} = -\frac{\hbar^2}{2m} \int d^3x \hat{a}^\dagger(\mathbf{x}, t) \boldsymbol{\nabla}^2 \hat{a}(\mathbf{x}, t) \tag{6.60}$$

and

$$\hat{U} = \int d^3x U(\mathbf{x}) \hat{a}^\dagger(\mathbf{x}, t) \hat{a}(\mathbf{x}, t). \tag{6.61}$$

$\hat{T}$ is the kinetic energy operator and $\hat{U}$ is the potential operator. $\hat{U}$ in Eq.(6.61) is the local one-body potential operator. In the later sections, we will show that $\hat{U}$ can be many-body potential operator.

6.6   Transformation of basis

We consider a system of N-particles. We denote $\mathcal{H}_N$ the Hilbert space of states for a system of N identical particles.

The creation and annihilation operators can operate in different bases. Of particular important are the state vectors $|x\rangle = |\mathbf{x}, t\rangle$. The meaning of $|x\rangle = \hat{a}^\dagger(\mathbf{x}, t)|0\rangle$ is that there is a particle at position $\mathbf{x}$.

Creation and annihilation operators in another basis can be derived as follows: Inserting the completeness relation, we obtain a transformation which transforms the orthonormal basis $\{|\alpha\rangle\}$ into another basis $\{|\tilde{\alpha}\rangle\}$

$$|\tilde{\alpha}\rangle = \sum_\alpha |\alpha\rangle\langle\alpha|\tilde{\alpha}\rangle. \tag{6.62}$$

By the definition of the creation operators $\hat{a}^\dagger_{\tilde{\alpha}}$ and $\hat{a}^\dagger_\alpha$, we have

$$
\begin{aligned}
\hat{a}^\dagger_{\tilde{\alpha}}|\tilde{\alpha}_1, \tilde{\alpha}_2, \cdots \tilde{\alpha}_n\rangle &= |\tilde{\alpha}, \tilde{\alpha}_1, \tilde{\alpha}_2, \cdots \tilde{\alpha}_n\rangle \\
&= \sum_\alpha \langle\alpha|\tilde{\alpha}\rangle|\alpha, \tilde{\alpha}_1, \tilde{\alpha}_2, \cdots \tilde{\alpha}_n\rangle \\
&= \sum_\alpha \langle\alpha|\tilde{\alpha}\rangle\hat{a}^\dagger_\alpha|\tilde{\alpha}_1, \tilde{\alpha}_2, \cdots \tilde{\alpha}_n\rangle.
\end{aligned}
\tag{6.63}
$$

Since Eq.(6.63) is valid for any state $|\tilde{\alpha}_1, \tilde{\alpha}_2, \cdots \tilde{\alpha}_n\rangle$, we obtain the operator relation

$$\hat{a}^\dagger_{\tilde{\alpha}} = \sum_\alpha \langle\alpha|\tilde{\alpha}\rangle\hat{a}^\dagger_\alpha. \tag{6.64}$$

The annihilation operators satisfy the adjoint equation

$$\hat{a}_{\tilde{\alpha}} = \sum_\alpha \langle\tilde{\alpha}|\alpha\rangle\hat{a}_\alpha. \tag{6.65}$$

The commutation and anticommutation for $\hat{a}^\dagger_{\tilde{\alpha}}$ and $\hat{a}_{\tilde{\alpha}'}$ can be obtained straightforwardly from Eqs.(6.64) and (6.65),

$$
\begin{aligned}
[\hat{a}_{\tilde{\alpha}'}, \hat{a}^\dagger_{\tilde{\alpha}}]_\pm &= \sum_{\alpha\alpha'} \langle\tilde{\alpha}'|\alpha\rangle\langle\alpha|\tilde{\alpha}\rangle[\hat{a}_{\alpha'}, \hat{a}^\dagger_\alpha]_\pm \\
&= \sum_\alpha \langle\tilde{\alpha}'|\alpha\rangle\langle\alpha|\tilde{\alpha}\rangle \\
&= \langle\tilde{\alpha}'|\tilde{\alpha}\rangle \\
&= \delta_{\tilde{\alpha}'\tilde{\alpha}}.
\end{aligned}
\tag{6.66}
$$

Thus we can get the expansion of the operators $\hat{a}^{\dagger}(\mathbf{x}, t)$ and $\hat{a}(\mathbf{x}, t)$ on other basis $\{|\alpha\rangle\}$

$$\hat{a}^{\dagger}(\mathbf{x}) = \sum_{\alpha} \langle \alpha | \mathbf{x} \rangle \hat{a}_{\alpha}^{\dagger} = \sum_{\alpha} \varphi_{\alpha}^{*}(\mathbf{x}) \hat{a}_{\alpha}^{\dagger}, \tag{6.67a}$$

$$\hat{a}(\mathbf{x}) = \sum_{\alpha} \langle \mathbf{x} | \alpha \rangle \hat{a}_{\alpha} = \sum_{\alpha} \varphi_{\alpha}(\mathbf{x}) \hat{a}_{\alpha}, \tag{6.67b}$$

where $\varphi_{\alpha}(\mathbf{x}) \equiv \langle \mathbf{x}, t | \alpha, t \rangle = \langle \mathbf{x} | \alpha \rangle$ is called the *wave function* in the coordinate representation of the state $|\alpha\rangle$. Wave function was named to reflect that a state satisfies the wave equation. In quantum mechanics, we usually use another definition for wave functions

$$\varphi_{\alpha}(\mathbf{x}, t) \equiv \langle \mathbf{x}, 0 | \alpha, t \rangle. \tag{6.68}$$

which contains the time evolution information of the state.

Since any operator can be expressed as a linear combination of the set of all product of the operators $(\hat{a}_{\alpha}^{\dagger}, \hat{a}_{\alpha})$, we can discuss the properties of any operator in terms of creation and annihilation operators.

If $\{|\alpha\rangle\}$ is an orthonormal basis of $\mathcal{H}$ describing single-particle states, the canonical orthonormal basis of $\mathcal{H}_N$ is the tensor products

$$|\alpha_1 \cdots \alpha_N\rangle \equiv |\alpha_1\rangle \otimes |\alpha_2\rangle \otimes \cdots \otimes |\alpha_N\rangle. \tag{6.69}$$

These basis states have the wave functions:

$$\begin{aligned}
\varphi_{\alpha_1 \alpha_2 \cdots \alpha_N}&(\mathbf{x}_1, \cdots \mathbf{x}_N) \\
&= \langle \mathbf{x}_1, \cdots \mathbf{x}_N | \alpha_1, \cdots \alpha_N \rangle \\
&= ((\langle \mathbf{x}_1 | \otimes \langle \mathbf{x}_2 | \otimes \cdots \otimes \langle \mathbf{x}_N |)(|\alpha_1\rangle \otimes |\alpha_2\rangle \otimes \cdots \otimes |\alpha_N\rangle)) \\
&= \varphi_{\alpha_1}(\mathbf{x}_1) \varphi_{\alpha_2}(\mathbf{x}_2) \cdots \varphi_{\alpha_N}(\mathbf{x}_N).
\end{aligned} \tag{6.70}$$

The overlap of two basis states is given by

$$\begin{aligned}
\langle \alpha_1, \alpha_2, &\cdots \alpha_N | \alpha_1', \alpha_2', \cdots \alpha_N' \rangle \\
&= ((\langle \alpha_1 | \otimes \langle \alpha_2 | \otimes \cdots \otimes \langle \alpha_N |)(|\alpha_1'\rangle \otimes |\alpha_2'\rangle \otimes \cdots \otimes |\alpha_N'\rangle)) \\
&= \langle \alpha_1 | \alpha_1' \rangle \langle \alpha_2 | \alpha_2' \rangle \cdots \langle \alpha_N | \alpha_N' \rangle.
\end{aligned} \tag{6.71}$$

The completeness of the basis is given by

$$\sum_{\alpha_1, \alpha_2, \cdots \alpha_N} |\alpha_1, \alpha_2, \cdots \alpha_N\rangle \langle \alpha_1, \alpha_2, \cdots \alpha_N| = 1. \tag{6.72}$$

The wave function of N bosons is symmetric and satisfies

$$\varphi(\mathbf{x}_{P_1}, \mathbf{x}_{P_2}, \cdots \mathbf{x}_{P_N}) = \varphi(\mathbf{x}_1, \mathbf{x}_2, \cdots \mathbf{x}_N), \tag{6.73}$$

where $(P_1, P_2, \cdots, P_N)$ is a permutation of the set $(1, 2, \cdots, N)$. The wave function of N fermions is antisymmetric under the exchange of any pair of particles and satisfies

$$\varphi(\mathbf{x}_{P_1}, \mathbf{x}_{P_2}, \cdots \mathbf{x}_{P_N}) = (-1)^{S_P} \varphi(\mathbf{x}_1, \mathbf{x}_2, \cdots \mathbf{x}_N), \tag{6.74}$$

where $(-1)^{S_P}$ denotes the sign or parity of the permutation P. S_P is the number of exchanges of two numbers that bring the permutation $(P_1, P_2, \cdots, P_N)$ to the original form $(1, 2, \cdots, N)$. For convenience, we adopt a unified notation for both bosons and fermions

$$\varphi(\mathbf{x}_{P_1}, \mathbf{x}_{P_2}, \cdots \mathbf{x}_{P_N}) = \xi^{S_P} \varphi(\mathbf{x}_1, \mathbf{x}_2, \cdots \mathbf{x}_N), \tag{6.75}$$

where $\xi = 1$ or -1 for bosons or fermions respectively. These symmetries pose the restrictions on the Hilbert space of identical particle systems. When a wave function $\varphi(\mathbf{x}_1, \mathbf{x}_2, \cdots \mathbf{x}_N)$ is symmetric under a permutation of particles, it belongs to the Hilbert space of N bosons B_N. When a wave function $\varphi(\mathbf{x}_1, \mathbf{x}_2, \cdots \mathbf{x}_N)$ is antisymmetric under the permutations, it belongs to the Hilbert space of N fermions F_N.

We use the symmetrization operator P_B and the anti-symmetrization operator P_F in $\mathcal{H}_N$ to obtain the symmetrized wave functions

$$P_\alpha \varphi(\mathbf{x}_1, \mathbf{x}_2, \cdots \mathbf{x}_N) = \frac{1}{N!} \sum_P \xi^{S_P} \varphi(\mathbf{x}_{P_1}, \mathbf{x}_{P_2}, \cdots \mathbf{x}_{P_N}), \tag{6.76}$$

where $\alpha = B, F$. For any wave function φ, we have

$$\begin{aligned}
P_\alpha^2 \varphi(\mathbf{x}_1, \mathbf{x}_2, \cdots \mathbf{x}_N) \\
= \frac{1}{N!} \frac{1}{N!} \sum_P \sum_{P'} \xi^{S_{P'}} \xi^{S_P} \varphi(\mathbf{x}_{P'P_1}, \mathbf{x}_{P'P_2}, \cdots \mathbf{x}_{P'P_N}) \\
= \frac{1}{N!} \sum_P \left(\frac{1}{N!} \sum_{Q=P'P} \xi^{S_Q} \varphi(\mathbf{x}_{Q_1}, \mathbf{x}_{Q_2}, \cdots \mathbf{x}_{Q_N}) \right) \\
= \frac{1}{N!} \sum_P P_\alpha \varphi(\mathbf{x}_1, \mathbf{x}_2, \cdots \mathbf{x}_N)) \\
= P_\alpha \varphi(\mathbf{x}_1, \mathbf{x}_2, \cdots \mathbf{x}_N)).
\end{aligned} \tag{6.77}$$

The symmetrized wave function corresponds to the symmetrized state $|\alpha_1, \alpha_2, \cdots \alpha_N\rangle_S$ with one particle in state α_1, one particle in state α_2, $\cdots$, and one particle in state α_N defined by

$$\begin{aligned}
P_\alpha |\alpha_1, \alpha_2, \cdots \alpha_N\rangle = \frac{1}{N!} \sum_P \xi^{S_{P'}} |\alpha_{P_1}\rangle \otimes |\alpha_{P_2}\rangle \otimes \cdots \otimes |\alpha_{P_N}\rangle \\
= \frac{1}{\sqrt{N!}} |\alpha_1, \alpha_2, \cdots \alpha_N\rangle_S.
\end{aligned} \tag{6.78}$$

$P_\alpha|\alpha_1, \alpha_2, \cdots \alpha_N\rangle$ is the basis of B_N or F_N with the completeness relation in B_N or F_N given by

$$
\sum_{\alpha_1, \alpha_2, \cdots \alpha_N} P_\alpha|\alpha_1, \alpha_2, \cdots \alpha_N\rangle\langle\alpha_1, \alpha_2, \cdots \alpha_N|P_\alpha
$$
$$
= \frac{1}{N!} \sum_{\alpha_1, \alpha_2, \cdots \alpha_N} |\alpha_1, \alpha_2, \cdots \alpha_N\rangle_{SS}\langle\alpha_1, \alpha_2, \cdots \alpha_N| = 1.
$$

(6.79)

Similar to Eq.(2.32), we have

$$
{}_S\langle\alpha_1', \alpha_2', \cdots \alpha_N'|\alpha_1, \alpha_2, \cdots \alpha_N\rangle_S = \prod_\alpha n_\alpha!.
$$

(6.80)

The orthonormal basis for the Hilbert space B_N or F_N has the form

$$
|\alpha_1, \alpha_2, \cdots \alpha_N\rangle_{SN}
$$
$$
= \frac{1}{\sqrt{\prod_\alpha n_\alpha!}}|\alpha_1, \alpha_2, \cdots \alpha_N\rangle_S
$$
$$
= \frac{1}{\sqrt{N! \prod_\alpha n_\alpha!}} \sum_P \xi^{S_P}|\alpha_{P_1}\rangle \otimes |\alpha_{P_2}\rangle \otimes \cdots \otimes |\alpha_{P_N}\rangle.
$$

(6.81)

The overlap of a state $|\beta_1, \beta_2, \cdots \beta_N\rangle$ constructed from an orthonormal basis $|\beta\rangle$ and the state $|\alpha_1, \alpha_2, \cdots \alpha_N\rangle_{SN}$ reads

$$
\langle\beta_1, \beta_2, \cdots \beta_N|\alpha_1, \alpha_2, \cdots \alpha_N\rangle_{SN}
$$
$$
= \frac{1}{\sqrt{N! \prod_\alpha n_\alpha!}} \sum_P \xi^{S_P}\langle\beta_1|\alpha_{P_1}\rangle\langle\beta_2|\alpha_{P_2}\rangle \cdots \langle\beta_N|\alpha_{P_N}\rangle
$$
$$
\equiv \frac{1}{\sqrt{N! \prod_\alpha n_\alpha!}} S(\langle\beta_i|\alpha_i\rangle),
$$

(6.82)

where $S(M_{ij})$ is the permanent for bosons

$$
\text{Per}\{M_{ij}\} \equiv \sum_P M_{1P_1} M_{2P_2} \cdots M_{NP_N}
$$

(6.83)

and the determinant for fermions

$$
\det(M_{ij}) \equiv \sum_P (-1)^{S_P} M_{1P_1} M_{2P_2} \cdots M_{NP_N}.
$$

(6.84)

In the coordinate representation, we have a basis consisting of the permanents of wave functions for bosons

$$
\varphi_{\alpha_1\alpha_2\cdots\alpha_N}(\mathbf{x}_1, \cdots \mathbf{x}_N) = \langle\mathbf{x}_1, \cdots \mathbf{x}_N|\alpha_1, \cdots \alpha_N\rangle_{SN}
$$
$$
= \frac{1}{\sqrt{N! \prod_\alpha n_\alpha!}}\text{Per}(\varphi_{\alpha_i}(\mathbf{x}_i))
$$

(6.85)

and a basis consisting of the Slater determinants for fermions

$$\varphi_{\alpha_1\alpha_2\cdots\alpha_N}(\mathbf{x}_1,\cdots\mathbf{x}_N) = \langle\mathbf{x}_1,\cdots\mathbf{x}_N|\alpha_1,\cdots\alpha_N\rangle_{SN}$$
$$= \frac{1}{\sqrt{N!}}\det(\varphi_{\alpha_i}(\mathbf{x}_i)). \tag{6.86}$$

The overlap of two normalized states reads

$$_{SN}\langle\beta_1,\beta_2,\cdots\beta_N|\alpha_1,\alpha_2,\cdots\alpha_N\rangle_{SN} = \frac{1}{\sqrt{\prod_\beta n_\beta!\prod_\alpha n_\alpha!}}S(\langle\beta_i|\alpha_i\rangle). \tag{6.87}$$

Inserting Eq.(6.81) into Eq.(6.79), we have the completeness relation for the states $|\alpha_1,\alpha_2,\cdots\alpha_N\rangle_{SN}$

$$\sum_{\alpha_1,\alpha_2,\cdots\alpha_N}\frac{\prod_\alpha n_\alpha!}{N!}|\alpha_1,\alpha_2,\cdots\alpha_N\rangle_{SN}\ _{SN}\langle\alpha_1,\alpha_2,\cdots\alpha_N\rangle| = 1. \tag{6.88}$$

By the definition of creation operator, we have

$$\hat{a}_\alpha^\dagger|\alpha_1,\cdots\alpha_N\rangle = |\alpha,\alpha_1,\cdots\alpha_N\rangle$$
$$= |\alpha\rangle \otimes |\alpha_1,\cdots\alpha_N\rangle. \tag{6.89}$$

Thus we can also write

$$\hat{a}_\alpha^\dagger = |\alpha\rangle \otimes. \tag{6.90}$$

Since $\hat{a}_\alpha$ is the adjoint of the creation operator $\hat{a}_\alpha^\dagger$, we can write

$$\hat{a}_\alpha = \otimes\langle\alpha|. \tag{6.91}$$

The creation operator $\hat{a}_\alpha^\dagger$ does not operate within one space B_N or F_N. They transform states in the space B_N or F_N to those in B_{N+1} or F_{N+1} and thus operate within the Fock space B or F, which is defined as the direct sum of the boson or fermion spaces.

$$B = B_0 \otimes B_1 \otimes B_2 \otimes \cdots = \otimes_{n=0}^\infty B_n, \tag{6.92a}$$
$$F = F_0 \otimes F_1 \otimes F_2 \otimes \cdots = \otimes_{n=0}^\infty F_n \tag{6.92b}$$

with $B_0 = F_0 = |0\rangle$ and $B_1 = F_1 = \mathcal{H}_1$. $|0\rangle, |\alpha\rangle, |\alpha_1,\alpha_2\rangle,\cdots$ form the basis for the Fock space. The completeness relation in the Fock space is

$$|0\rangle\langle0| + \sum_{N=1}^\infty\frac{1}{N!}\sum_{\alpha_1,\alpha_2,\cdots\alpha_N}|\alpha_1,\alpha_2,\cdots\alpha_N\rangle_{SS}\langle\alpha_1,\alpha_2,\cdots\alpha_N|$$
$$= |0\rangle\langle0| + \sum_{N=1}^\infty\frac{1}{N!}\sum_{\alpha_1,\alpha_2,\cdots\alpha_N}(\prod_\alpha n_\alpha!) \tag{6.93}$$
$$\times |\alpha_1,\alpha_2,\cdots\alpha_N\rangle_{SN}\ _{SN}\langle\alpha_1,\alpha_2,\cdots\alpha_N|$$
$$= 1.$$

6.7   One-body operators

A convenient technique is to use the basis in which an operator is diagonal. An operator $\hat{U}$ is diagonal when the operator $\hat{U}$ is expressed as

$$\hat{U} = \sum_{\alpha} \hat{a}_{\alpha}^{\dagger} U_{\alpha} \hat{a}_{\alpha}. \tag{6.94}$$

Eq.(6.94) can also be expressed as

$$\hat{U} = \sum_{\alpha} |\alpha\rangle U_{\alpha} \langle\alpha|. \tag{6.95}$$

When we calculate $\langle\alpha|\hat{U}|\alpha\rangle$, we have

$$\langle\alpha|\hat{U}|\alpha\rangle = \sum_{\alpha'} \langle\alpha|\alpha'\rangle U_{\alpha'} \langle\alpha'|\alpha\rangle = U_{\alpha}. \tag{6.96}$$

Using Eq.(6.78), we have

$$\begin{aligned}
&_S\langle\alpha_1', \alpha_2', \cdots \alpha_N'|\hat{U}|\alpha_1, \alpha_2, \cdots \alpha_N\rangle_S \\
&= \sum_{P} \xi^{S_P} \sum_{i=1}^{N} \prod_{k\neq i} \langle\alpha_{P_k}'|\alpha_k\rangle\langle\alpha_{P_i}'|\hat{U}|\alpha_i\rangle \\
&= \left(\sum_{i=1}^{N} U_{\alpha_i}\right) {}_S\langle\alpha_1', \alpha_2', \cdots \alpha_N'|\alpha_1, \alpha_2, \cdots \alpha_N\rangle_S.
\end{aligned} \tag{6.97}$$

Using Eqs.(6.64) and (6.65), we may transform the diagonal representation of $\hat{U}$ to a representation with a general basis

$$\begin{aligned}
\hat{U} &= \sum_{\alpha\lambda\mu} U_{\alpha}\langle\lambda|\alpha\rangle\langle\alpha|\mu\rangle\hat{a}_{\lambda}^{\dagger}\hat{a}_{\mu} \\
&= \sum_{\lambda\mu} \langle\lambda|\hat{U}|\mu\rangle\hat{a}_{\lambda}^{\dagger}\hat{a}_{\mu}.
\end{aligned} \tag{6.98}$$

where

$$\begin{aligned}
\langle\lambda|\hat{U}|\mu\rangle &= \sum_{\alpha} \langle\lambda|\alpha\rangle U_{\alpha}\langle\alpha|\mu\rangle \\
&= \int d^3x d^3y \sum_{\alpha} \langle\lambda|\mathbf{x}\rangle\langle\mathbf{x}|\alpha\rangle U_{\alpha}\langle\alpha|\mathbf{y}\rangle\langle\mathbf{y}|\mu\rangle \\
&= \int d^3x d^3y \varphi_{\lambda}^*(x)\langle x|\hat{U}|y\rangle\varphi_{\mu}(y)
\end{aligned} \tag{6.99}$$

with $\varphi_{\lambda}^*(x) = \langle\lambda|x\rangle$ and $\varphi_{\mu}(y) = \langle y|\mu\rangle$.

In the $\{\mathbf{x}\}$ representation, the kinetic energy operator $\hat{T}$ reads

$$\begin{aligned}
\hat{T} &= -\frac{\hbar^2}{2m} \int d^3x \hat{a}^\dagger(\mathbf{x}) \nabla^2 \hat{a}(\mathbf{x}) \\
&= -\frac{\hbar^2}{2m} \int d^3x |\mathbf{x}\rangle \nabla^2 \langle\mathbf{x}|.
\end{aligned} \tag{6.100}$$

A local one-body operator $\hat{U}$ can be written as

$$\begin{aligned}
\hat{U} &= \int d^3x U(\mathbf{x}) \hat{a}^\dagger(\mathbf{x}) \hat{a}(\mathbf{x}) \\
&= \int d^3x U(\mathbf{x}) \hat{n}(\mathbf{x}) \\
&= \int d^3x U(\mathbf{x}) |\mathbf{x}\rangle\langle\mathbf{x}|.
\end{aligned} \tag{6.101}$$

6.8 Schrödinger equation

Since $\hat{H}$ is the generator of time translation transformation, for a state $|\mathbf{x}_1, \mathbf{x}_2, \cdots \mathbf{x}_N\rangle$, the time evolution is given by

$$|\mathbf{x}_1, \mathbf{x}_2, \cdots \mathbf{x}_N, t\rangle = e^{\frac{i}{\hbar}\hat{H}t}|\mathbf{x}_1, \mathbf{x}_2, \cdots \mathbf{x}_N, 0\rangle \tag{6.102}$$

or

$$-i\hbar \frac{\partial}{\partial t}|\mathbf{x}_1, \mathbf{x}_2, \cdots \mathbf{x}_N, t\rangle = \hat{H}|\mathbf{x}_1, \mathbf{x}_2, \cdots \mathbf{x}_N, 0\rangle. \tag{6.103}$$

Now we consider the wave functions in the $\mathbf{x}$ representation. We use the definition Eq.(6.68) of the wave functions for quantum mechanics.

$$\begin{aligned}
\varphi_{\alpha_1, \alpha_2, \cdots \alpha_N}(\mathbf{x}_1, \mathbf{x}_2, \cdots \mathbf{x}_N, t) &= \langle\mathbf{x}_1, \mathbf{x}_2, \cdots \mathbf{x}_N | \alpha_1, \alpha_2, \cdots \alpha_N, t\rangle_{SN} \\
&= \langle\mathbf{x}_1, \mathbf{x}_2, \cdots \mathbf{x}_N, t | \alpha_1, \alpha_2, \cdots \alpha_N\rangle_{SN} \quad (6.104) \\
&= \langle\mathbf{x}_1, \mathbf{x}_2, \cdots \mathbf{x}_N | e^{-\frac{i}{\hbar}\hat{H}t} | \alpha_1, \alpha_2, \cdots \alpha_N\rangle_{SN}.
\end{aligned}$$

When we express $\hat{H}$ as diagonal in the bases $|\mathbf{x}\rangle$, we have

$$\begin{aligned}
&\varphi_{\alpha_1, \alpha_2, \cdots \alpha_N}(\mathbf{x}_1, \mathbf{x}_2, \cdots \mathbf{x}_N, t) \\
&= \langle\mathbf{x}_1, \mathbf{x}_2, \cdots \mathbf{x}_N | e^{-\frac{i}{\hbar}\hat{H}t} \int d^3x_1' d^3x_2' \cdots d^3x_N' \\
&\quad \frac{1}{N!} |\mathbf{x}_1', \mathbf{x}_2', \cdots \mathbf{x}_N'\rangle_{SS}\langle\mathbf{x}_1', \mathbf{x}_2', \cdots \mathbf{x}_N' | \alpha_1, \alpha_2, \cdots \alpha_N\rangle_{SN} \quad (6.105) \\
&= \int d^3x_1' d^3x_2' \cdots d^3x_N' \frac{1}{N!} {}_S\langle\mathbf{x}_1, \mathbf{x}_2, \cdots \mathbf{x}_N | e^{-\frac{i}{\hbar}\hat{H}t} | \mathbf{x}_1', \mathbf{x}_2', \cdots \mathbf{x}_N'\rangle_S \\
&\quad \langle\mathbf{x}_1', \mathbf{x}_2', \cdots \mathbf{x}_N' | \alpha_1, \alpha_2, \cdots \alpha_N\rangle_{SN}.
\end{aligned}$$

In the derivation of Eq.(6.105), we have used $P_\alpha^2 = P_\alpha$. Thus

$$i\hbar\frac{\partial}{\partial t}\varphi_{\alpha_1,\alpha_2,\cdots\alpha_N}(\mathbf{x}_1,\mathbf{x}_2,\cdots\mathbf{x}_N,t)$$

$$= \int d^3x_1' d^3x_2'\cdots d^3x_N' \frac{1}{N!}{}_S\langle\mathbf{x}_1,\mathbf{x}_2,\cdots\mathbf{x}_N|\hat{H}|\mathbf{x}_1',\mathbf{x}_2',\cdots\mathbf{x}_N'\rangle_S \qquad (6.106)$$

$$\varphi_{\alpha_1,\alpha_2,\cdots\alpha_N}(\mathbf{x}_1',\mathbf{x}_2',\cdots\mathbf{x}_N',t).$$

Using Eq.(6.97), we have

$$i\hbar\frac{\partial}{\partial t}\varphi_{\alpha_1,\alpha_2,\cdots\alpha_N}(\mathbf{x}_1,\mathbf{x}_2,\cdots\mathbf{x}_N,t)$$

$$= \int d^3x_1' d^3x_2'\cdots d^3x_N' \frac{1}{N!}\sum_{i=1}^{N} H_{\mathbf{x}_i'}{}_S\langle\mathbf{x}_1,\mathbf{x}_2,\cdots\mathbf{x}_N|\mathbf{x}_1',\mathbf{x}_2',\cdots\mathbf{x}_N'\rangle_S$$

$$\varphi_{\alpha_1,\alpha_2,\cdots\alpha_N}(\mathbf{x}_1',\mathbf{x}_2',\cdots\mathbf{x}_N',t)$$

$$= \int d^3x_1' d^3x_2'\cdots d^3x_N' \sum_{i=1}^{N} H_{\mathbf{x}_i'}\delta(\mathbf{x}_1-\mathbf{x}_1')\delta(\mathbf{x}_2-\mathbf{x}_2')\cdots\delta(\mathbf{x}_N-\mathbf{x}_N') \qquad (6.107)$$

$$\varphi_{\alpha_1,\alpha_2,\cdots\alpha_N}(\mathbf{x}_1',\mathbf{x}_2',\cdots\mathbf{x}_N',t)$$

$$= \sum_{i=1}^{N} H_{\mathbf{x}_i}\varphi_{\alpha_1,\alpha_2,\cdots\alpha_N}(\mathbf{x}_1,\mathbf{x}_2,\cdots\mathbf{x}_N,t).$$

where

$$H_{\mathbf{x}_i} = \frac{-\hbar^2}{2m}\nabla_{\mathbf{x}_i}^2 + U(\mathbf{x}_i). \qquad (6.108)$$

Eq.(6.107) is called the *Schrödinger equation*. We introduce the total Hamiltonian

$$H = \sum_{i=1}^{N}\left[\frac{-\hbar^2}{2m}\nabla_{\mathbf{x}_i}^2 + U(\mathbf{x}_i)\right]. \qquad (6.109)$$

Then Eq.(6.107) becomes

$$i\hbar\frac{\partial}{\partial t}\varphi(\mathbf{x}_1,\mathbf{x}_2,\cdots\mathbf{x}_N,t) = H\left(\left\{\frac{\hbar}{i}\nabla_{\mathbf{x}_i}\right\},\{\mathbf{x}_i\}\right)\varphi(\mathbf{x}_1,\mathbf{x}_2,\cdots\mathbf{x}_N,t),$$

$$(6.110)$$

which is the Schrödinger equation for an N-particle system. In Eq.(6.110), for simplicity, we have omitted the subscript $\alpha_1,\alpha_2,\cdots\alpha_N$, which is important only when the initial configuration matters. $\varphi(\mathbf{x}_1,\mathbf{x}_2,\cdots\mathbf{x}_N,t)$ is called the wave function for N-particle non-relativistic quantum system. It should be noted that in a system of quantum mechanics, the particle number N is conserved.

We can introduce the operators for the physical observables of particles in quantum mechanics. $|\mathbf{x}\rangle = \hat{a}^\dagger(\mathbf{x})|0\rangle$ has the meaning that there is a particle at position $\mathbf{x}$. When an operator $\hat{A}$ for the physical observable A of particles acts on the state $|\mathbf{x}\rangle$, it should give the value of the physical observable A of the particle at position $\mathbf{x}$.

$$\hat{A}|\mathbf{x}\rangle = A(\mathbf{x})|\mathbf{x}\rangle. \tag{6.111}$$

In particular, the position operator $\hat{\mathbf{x}}$ acts on $|\mathbf{x}\rangle$ to give the value of the position $\mathbf{x}$ of the particle.

$$\hat{\mathbf{x}}|\mathbf{x}\rangle = \mathbf{x}|\mathbf{x}\rangle. \tag{6.112}$$

Thus $|\mathbf{x}\rangle$ is also the eigenstate of the position operator. For any function of position operator $f(\hat{\mathbf{x}})$, we have $f(\hat{\mathbf{x}})|\mathbf{x}\rangle = f(\mathbf{x})|\mathbf{x}\rangle$. $f(\mathbf{x})$ is the value of $f(\hat{\mathbf{x}})$ at $\mathbf{x}$.

We define the Hamiltonian operator $\hat{H}(\hat{p}, \hat{q})$ in quantum mechanics as

$$\hat{H}(\hat{p}, \hat{q}) \equiv \sum_{i=1}^{N} \left[\frac{\hat{p}_i^2}{2m} + \hat{U}(\hat{q}_i) \right]. \tag{6.113}$$

with

$$\hat{p}_i = -i\hbar \nabla_{\hat{\mathbf{x}}_i}. \tag{6.114}$$

and

$$\hat{q}_i = \hat{\mathbf{x}}_i. \tag{6.115}$$

$\hat{p}_i \equiv -i\hbar \nabla_{\hat{\mathbf{x}}_i}$ is called the momentum operator in quantum mechanics. The momentum operator $\hat{p}$ and position operator $\hat{q}$ obey the following commutation relation

$$[\hat{q}, \hat{p}] = [\hat{q}, -i\hbar \nabla_{\hat{q}}] = i\hbar. \tag{6.116}$$

Similar to $|\mathbf{x}\rangle$, $|\mathbf{p}\rangle = \hat{a}_{\mathbf{p}}^\dagger|0\rangle$ is a state that there is a quanta with momentum $\mathbf{p}$. When the momentum operator $\hat{p}$ of a quanta acts on $|\mathbf{p}\rangle$, it gives the value of the momentum $\mathbf{p}$ of the quanta.

$$\hat{\mathbf{p}}|\mathbf{p}\rangle = \mathbf{p}|\mathbf{p}\rangle. \tag{6.117}$$

$|\mathbf{p}\rangle$ is thus also the eigenstate of the momentum operator.

The Schrödinger equation Eq.(6.110) for an N-particle system can be expressed as the operator form of quantum mechanics

$$i\hbar \frac{\partial}{\partial t}|\alpha_1, \alpha_2, \cdots \alpha_N\rangle_{SN} = \hat{H}(\hat{p}, \hat{q})|\alpha_1, \alpha_2, \cdots \alpha_N\rangle_{SN}. \tag{6.118}$$

Composite fermions can behave as bosons. When two fermions are strongly bound, they can be considered as one identity and a pair of fermion field operators are used as one operator. The operators composed of a pair of anti-commuted operators obeys the commutation relations of bosons. Therefore, a pair of bound fermions can be considered as a boson. In this case, the Schrödinger equation Eq.(6.110) can also be used for the composite Dirac fermions that behave as bosons.

Chapter 7

Electromagnetic field

7.1 Current density

We consider the photon field in the classical limit (also called the electromagnetic field) coupled to a spinor fermion field (also called Dirac fermion field). The Lagrangian density of the photon field coupled with the spinor fermion field is given by Eq.(2.579) with the form

$$
\begin{aligned}
\mathcal{L} &= \mathcal{L}_{Dirac} + \mathcal{L}_{photon} + \mathcal{L}_{int} \\
&= \bar{\psi}(i\gamma^\mu \partial_\mu - m)\psi - \frac{1}{4}F_{\mu\nu}F^{\mu\nu} - e\bar{\psi}\gamma^\mu\psi A_\mu.
\end{aligned}
\tag{7.1}
$$

The coupling term in Eq.(7.1) is $j_e^\mu A_\mu$ with

$$
j_e^\mu \equiv e\bar{\psi}\gamma^\mu\psi.
\tag{7.2}
$$

j_e^μ is called the *Dirac current*. In terms of j_e^μ, the Lagrangian density of the massless vector boson field with the coupling term can be expressed as

$$
\mathcal{L} = -\frac{1}{4}F_{\mu\nu}F^{\mu\nu} - j_e^\mu A_\mu = \mathcal{L}_0 + \mathcal{L}_{int}.
\tag{7.3}
$$

As the coupled field, the field operator $\hat{\psi}$ of the spinor fermion field satisfies the Dirac equation Eq.(2.247),

$$
i\hbar\frac{\partial\hat{\psi}}{\partial t} = \left(\frac{\hbar c}{i}\boldsymbol{\alpha}\cdot\boldsymbol{\nabla} + \beta mc^2\right)\hat{\psi}
\tag{7.4}
$$

with

$$
\hat{\psi} = \begin{pmatrix} \hat{\psi}_1(\mathbf{x},t) \\ \hat{\psi}_2(\mathbf{x},t) \\ \hat{\psi}_3(\mathbf{x},t) \\ \hat{\psi}_4(\mathbf{x},t) \end{pmatrix}.
\tag{7.5}
$$

Since we deal with the classical limit, we have used the units with the Planck constant $\hbar$ and the speed of light c written out explicitly.

Now we construct the four-current and the equation of continuity. Multiplying Eq.(7.4) from the left by $\hat{\psi}^\dagger = (\hat{\psi}_1^\dagger(\mathbf{x}, t), \hat{\psi}_2^\dagger(\mathbf{x}, t), \hat{\psi}_3^\dagger(\mathbf{x}, t), \hat{\psi}_4^\dagger(\mathbf{x}, t))$, we obtain

$$i\hbar\hat{\psi}^\dagger \frac{\partial}{\partial t}\hat{\psi} = \frac{\hbar c}{i} \sum_{k=1}^{3} \hat{\psi}^\dagger \alpha_k \frac{\partial}{\partial x^k}\hat{\psi} + mc^2\hat{\psi}^\dagger\beta\hat{\psi}. \tag{7.6}$$

We further use the hermitian conjugate of Eq.(7.4)

$$-i\hbar\frac{\partial\hat{\psi}^\dagger}{\partial t} = -\frac{\hbar c}{i} \sum_{k=1}^{3} \frac{\partial\hat{\psi}^\dagger}{\partial x^k}\alpha_k^\dagger + mc^2\hat{\psi}^\dagger\beta^\dagger. \tag{7.7}$$

Multiplying the equation from the right by $\hat{\psi}$ and taking into consideration the hermiticity of Dirac's matrices $(\alpha_i^\dagger = \alpha_i, \beta_i^\dagger = \beta_i)$, we have

$$-i\hbar\frac{\partial\hat{\psi}^\dagger}{\partial t}\hat{\psi} = -\frac{\hbar c}{i} \sum_{k=1}^{3} \frac{\partial\hat{\psi}^\dagger}{\partial x^k}\alpha_k\hat{\psi} + m_0c^2\hat{\psi}^\dagger\beta\hat{\psi}. \tag{7.8}$$

Subtracting Eq.(7.8) from Eq.(7.6), we obtain

$$i\hbar\frac{\partial}{\partial t}(\hat{\psi}^\dagger\hat{\psi}) = \frac{\hbar c}{i} \sum_{k=1}^{3} \frac{\partial}{\partial x^k}(\hat{\psi}^\dagger\alpha_k\hat{\psi}). \tag{7.9}$$

We define a positive definite density operator of the form

$$\hat{\rho}(\mathbf{x}) \equiv \hat{\psi}^\dagger(\mathbf{x})\hat{\psi}(\mathbf{x}) = (\hat{\psi}_1^\dagger, \hat{\psi}_2^\dagger, \hat{\psi}_3^\dagger, \hat{\psi}_4^\dagger) \begin{pmatrix} \hat{\psi}_1 \\ \hat{\psi}_2 \\ \hat{\psi}_3 \\ \hat{\psi}_4 \end{pmatrix} = \sum_{i=1}^{4} \hat{\psi}_i^\dagger(\mathbf{x})\hat{\psi}_i(\mathbf{x}). \tag{7.10}$$

and the current density operator $\hat{\mathbf{j}}$

$$\hat{\mathbf{j}} \equiv c\hat{\psi}^\dagger\boldsymbol{\alpha}\hat{\psi}. \tag{7.11}$$

Then Eq.(7.9) becomes the equation of continuity

$$\frac{\partial\hat{\rho}}{\partial t} + \boldsymbol{\nabla}\cdot\hat{\mathbf{j}} = 0. \tag{7.12}$$

We obtain the conservation law directly from Eq.(7.12)

$$\frac{\partial}{\partial t}\int_V d^3x\hat{\psi}^\dagger(\mathbf{x})\hat{\psi}(\mathbf{x}) = -\int_V \boldsymbol{\nabla}\cdot\hat{\mathbf{j}}d^3x = -\int_s \hat{\mathbf{j}}\cdot d\mathbf{s} = 0, \tag{7.13}$$

where V denotes the volume of the system and s is the surface of the volume V. $c\hat{\rho}$ and $\hat{\mathbf{j}}$ form a four-vector, which reads

$$\{\hat{j}^\mu\} \equiv \{c\hat{\rho}, \hat{\mathbf{j}}\} \equiv \{\hat{j}^0, \hat{\mathbf{j}}\} = \{c\hat{\psi}^\dagger\hat{\psi}, c\hat{\psi}^\dagger\boldsymbol{\alpha}\hat{\psi}\} = \{c\hat{\psi}^\dagger\gamma^0\gamma^\mu\hat{\psi}\} \tag{7.14}$$

or

$$\hat{j}^\mu(x) = c\hat{\psi}^\dagger(x)\gamma^0\gamma^\mu\hat{\psi}(x). \tag{7.15}$$

According to Eq.(2.300), $\hat{j}^\mu(x)$ transforms under the Lorentz transformation as a four-vector.

Since $\hat{j}^\mu(x)$ is a four-vector, we can write the equation of continuity in the Lorentz covariant form

$$\frac{\partial \hat{j}^\mu}{\partial x^\mu} = 0. \tag{7.16}$$

Comparing with Eq.(7.2), we have

$$j_e^\mu = \frac{e}{c}j^\mu. \tag{7.17}$$

7.2 Classical limit

Now we consider the photon field. It is easy to deduce the classical limit using the path integral formalism

$$Z = \int DA \exp\left[\frac{i}{\hbar} \int d^4x \mathcal{L}(A)\right], \tag{7.18}$$

where the Lagrangian density of photon field $\mathcal{L}$ is given by Eq.(7.3). In the classical limit, the action is much larger than $\hbar$, we can calculate the path integral using the stationary phase approximation. In the limit $\hbar \to 0$, the path integral is given by the value of the integrand at the extremum of $S = \int d^4x \mathcal{L}(A_c)$, where A_c is determined by the Euler-Lagrange equation. The Euler-Lagrange variational procedure is often called the principle of least action.

In the following, we will use the Gaussian units for the electromagnetic field. In the Gaussian units, the action has the form

$$S = -\frac{1}{16\pi c} \int F_{\mu\nu}F^{\mu\nu}d^4x - \frac{1}{c^2} \int j_e^\mu A_\mu d^4x. \tag{7.19}$$

The variation of the action gives

$$\delta S = -\int \frac{1}{c}\left(\frac{1}{c}j_e^\mu \delta A_\mu + \frac{1}{8\pi}F^{\mu\nu}\delta F_{\mu\nu}\right)d^4x = 0. \tag{7.20}$$

Inserting $F_{\mu\nu} = \frac{\partial A_\nu}{\partial x^\mu} - \frac{\partial A_\mu}{\partial x^\nu}$, we have

$$\begin{aligned}
\delta S &= -\int \frac{1}{c}\left(\frac{1}{c}j_e^\mu \delta A_\mu + \frac{1}{8\pi}F^{\mu\nu}\frac{\delta\partial A_\nu}{\partial x^\mu} - \frac{1}{8\pi}F^{\mu\nu}\frac{\delta\partial A_\mu}{\partial x^\nu}\right)d^4x \\
&= -\int \frac{1}{c}\left(\frac{1}{c}j_e^\mu \delta A_\mu - \frac{1}{4\pi}F^{\mu\nu}\frac{\delta\partial A_\mu}{\partial x^\nu}\right)d^4x.
\end{aligned} \tag{7.21}$$

Integrating by parts and using Gauss's theorem, we have

$$\delta S = - \int \frac{1}{c} \left(\frac{1}{c} j_e^\mu + \frac{1}{4\pi} \frac{\partial F^{\mu\nu}}{\partial x^\nu} \right) \delta A_\mu d^4 x$$
$$- \frac{1}{4\pi c} \int F^{\mu\nu} \delta \partial A_\mu dS_\nu. \tag{7.22}$$

Neglecting the surface integration, we obtain

$$- \int \left(\frac{1}{c} j_e^\mu + \frac{1}{4\pi} \frac{\partial F^{\mu\nu}}{\partial x^\nu} \right) \delta A_\mu d^4 x = 0. \tag{7.23}$$

Since the variations δA_μ are arbitrary, the coefficients of δA_μ should be zero, which gives

$$\frac{\partial F^{\mu\nu}}{\partial x^\nu} = - \frac{4\pi}{c} j_e^\mu. \tag{7.24}$$

We can display $F^{\mu\nu}$ in matrix form in terms of $\mathbf{E}$ and $\mathbf{B}$.

$$(F^{\mu\nu}) = \begin{pmatrix} 0 & -E^1 & -E^2 & -E^3 \\ E^1 & 0 & -B^3 & B^2 \\ E^2 & B^3 & 0 & -B^1 \\ E^3 & -B^2 & B^1 & 0 \end{pmatrix}. \tag{7.25}$$

$F^{\mu\nu}$ is thus also called the electromagnetic field tensor or field strength tensor.

7.3   Maxwell equations

Expressing Eq.(7.24) in terms of $\mathbf{E}$ and $\mathbf{B}$, and also using $j_e^\mu = \{c\rho_e, \mathbf{j}_e\}$, we have

$$\mathbf{\nabla} \times \mathbf{B} = \frac{1}{c} \frac{\partial \mathbf{E}}{\partial t} + \frac{4\pi}{c} \mathbf{j}_e, \tag{7.26a}$$

$$\mathbf{\nabla} \cdot \mathbf{E} = 4\pi \rho_e. \tag{7.26b}$$

The equation of $\nu = 0$ for Eq.(7.24) is

$$\partial_1 F^{10} + \partial_2 F^{20} + \partial_3 F^{30} = 4\pi \rho_e, \tag{7.27}$$

which gives Eq.(7.26b). The equation of $\nu = 1$ for Eq.(7.24) is

$$\frac{4\pi}{c} j_e^1 = \partial_0 F^{01} + \partial_2 F^{21} + \partial_3 F^{31}$$
$$= -\frac{\partial E^1}{\partial ct} + \frac{\partial B^3}{\partial x_2} - \frac{\partial B^2}{\partial x_3} \tag{7.28}$$
$$= -\frac{1}{c} \frac{\partial E^1}{\partial t} + (\mathbf{\nabla} \times \mathbf{B})^1.$$

We can obtain the similar equations for other components. Thus we have Eq.(7.26a)

According to the definition of $\mathbf{E}$ and $\mathbf{B}$ given by Eqs.(2.559) and (2.560), we have $\mathbf{B} = \boldsymbol{\nabla} \times \mathbf{A}$ and $\mathbf{E} = -\frac{1}{c}\frac{\partial \mathbf{A}}{\partial t} - \boldsymbol{\nabla} A_0$. Taking the divergence of both sides of the equation $\mathbf{B} = \boldsymbol{\nabla} \times \mathbf{A}$, we have

$$\boldsymbol{\nabla} \cdot \mathbf{B} = 0. \tag{7.29}$$

Evaluating $\boldsymbol{\nabla} \times \mathbf{E}$ gives

$$\begin{aligned}
\boldsymbol{\nabla} \times \mathbf{E} &= -\frac{1}{c}\frac{\partial}{\partial t}\boldsymbol{\nabla} \times \mathbf{A} - \boldsymbol{\nabla} \times \boldsymbol{\nabla} A_0 \\
&= -\frac{1}{c}\frac{\partial \mathbf{B}}{\partial t}.
\end{aligned} \tag{7.30}$$

Altogether we have the following four equations

$$\boldsymbol{\nabla} \cdot \mathbf{E} = 4\pi \rho_e, \tag{7.31a}$$

$$\boldsymbol{\nabla} \cdot \mathbf{B} = 0, \tag{7.31b}$$

$$\boldsymbol{\nabla} \times \mathbf{E} = -\frac{1}{c}\frac{\partial \mathbf{B}}{\partial t}, \tag{7.31c}$$

$$\boldsymbol{\nabla} \times \mathbf{B} = \frac{1}{c}\frac{\partial \mathbf{E}}{\partial t} + \frac{4\pi}{c}\mathbf{j}_e, \tag{7.31d}$$

which are called the *Maxwell equations*. When j_e^μ is replaced by its classical values, Eq.(7.31) is the classical Maxwell equations.

Historically, Eq.(7.31a) is called Gauss's law. Eq.(7.31c) is known as Faraday's law. Eq.(7.31d) is Ampere's law when $\frac{\partial E}{\partial t}$ is neglected. $\frac{\partial E}{\partial t}$ was added by Maxwell.

Eq.(7.31a) and Eq.(7.31d) are the inhomogeneous equation with the source terms, which are equivalent to Eq.(7.24). Eq.(7.31b) and Eq.(7.31c) are the homogeneous equations without source terms. We can also express Eq.(7.31b) and Eq.(7.31c) in terms of $F^{\mu\nu}$. We introduce an antisymmetric tensor $\tilde{F}^{\mu\nu}$ dual to $F^{\mu\nu}$ by

$$\tilde{F}^{\mu\nu} = \frac{1}{2}\epsilon^{\mu\nu\rho\sigma}F_{\rho\sigma}, \tag{7.32}$$

where $\epsilon^{\mu\nu\rho\sigma}$ is the antisymmetric Levi-Civita symbol in four dimension. $\tilde{F}^{\mu\nu}$ is called the dual field strength tensor. We can easily check that the matrix form of $\tilde{F}^{\mu\nu}$ in terms of $\mathbf{E}$ and $\mathbf{B}$ is

$$(\tilde{F}^{\mu\nu}) = \begin{pmatrix} 0 & -B^1 & -B^2 & -B^3 \\ B^1 & 0 & E^3 & -E^2 \\ B^2 & -E^3 & 0 & E^1 \\ B^3 & E^2 & -E^1 & 0 \end{pmatrix}. \tag{7.33}$$

One can easily verify that the homogenous equations Eq.(7.31b) and Eq.(7.31c) are equivalent to the equation

$$\partial_\mu \tilde{F}^{\mu\nu} = 0. \tag{7.34}$$

Thus the Maxwell equations can be rewritten in a compact form as

$$\partial_\mu F^{\mu\nu} = \frac{4\pi}{c} j_e^\nu, \tag{7.35a}$$

$$\partial_\mu \tilde{F}^{\mu\nu} = 0. \tag{7.35b}$$

Due to the antisymmetry of $\epsilon^{\mu\nu\rho\sigma}$, Eq.(7.34) also leads to

$$\partial^\lambda F^{\mu\nu} + \partial^\mu F^{\nu\lambda} + \partial^\nu F^{\lambda\mu} = 0, \tag{7.36}$$

which reflects the antisymmetric property of $F^{\mu\nu}$.

7.4   Gauge invariance

Usually we denote A_0 as φ. $\mathbf{A}$ and φ are also called the *vector potential* and *scalar potential*, respectively. The Lagrangian density for the photon field is invariant under the gauge transformation

$$\mathbf{A} \to \mathbf{A}' = \mathbf{A} + \nabla\Lambda, \tag{7.37a}$$

$$\varphi \to \varphi' = \varphi - \frac{1}{c}\frac{\partial\Lambda}{\partial t}. \tag{7.37b}$$

The electric field $\mathbf{E}$ and magnetic field $\mathbf{B}$ are also invariant under the gauge transformation Eq.(7.37).

$$\nabla \times \mathbf{A}' = \nabla \times \mathbf{A} = \mathbf{B}, \tag{7.38a}$$

$$-\frac{1}{c}\frac{\partial\mathbf{A}'}{\partial t} - \nabla\varphi' = -\frac{1}{c}\frac{\partial\mathbf{A}}{\partial t} - \nabla\varphi = \mathbf{E}. \tag{7.38b}$$

Thus $\mathbf{E}$ and $\mathbf{B}$ are independent of the gauge type.

7.5   Radiation of electromagnetic waves

Inserting $\mathbf{B} = \nabla \times \mathbf{A}$ and $\mathbf{E} = -\frac{1}{c}\frac{\partial\mathbf{A}}{\partial t} - \nabla\varphi$ into Eq.(7.26), we have

$$\nabla \times (\nabla \times \mathbf{A}) = -\frac{1}{c^2}\frac{\partial^2\mathbf{A}}{\partial t^2} - \frac{1}{c}\frac{\partial}{\partial t}\nabla\varphi + \frac{4\pi}{c}\mathbf{j}_e, \tag{7.39a}$$

$$-\frac{1}{c}\frac{\partial}{\partial t}\nabla \cdot \mathbf{A} - \nabla^2\varphi = 4\pi\rho_e. \tag{7.39b}$$

Eq.(7.39) can be reformulated into the form

$$\nabla^2 \mathbf{A} - \frac{1}{c^2}\frac{\partial^2 \mathbf{A}}{\partial t^2} - \nabla\left(\nabla \cdot \mathbf{A} + \frac{1}{c}\frac{\partial \varphi}{\partial t}\right) = -\frac{4\pi}{c}\mathbf{j}_e, \tag{7.40a}$$

$$\nabla^2 \varphi + \frac{1}{c}\frac{\partial}{\partial t}\nabla \cdot \mathbf{A} = -4\pi\rho_e. \tag{7.40b}$$

We introduce the *Lorentz gauge*

$$\nabla \cdot \mathbf{A} + \frac{1}{c}\frac{\partial \varphi}{\partial t} = 0, \tag{7.41}$$

which can be satisfied by appropriate selection of the gauge transformation Eq.(7.37). Using the Lorentz gauge Eq.(7.41), Eq.(7.40) becomes

$$\nabla^2 \mathbf{A} - \frac{1}{c^2}\frac{\partial^2 \mathbf{A}}{\partial t^2} = -\frac{4\pi}{c}\mathbf{j}_e, \tag{7.42a}$$

$$\nabla^2 \varphi - \frac{1}{c^2}\frac{\partial^2 \varphi}{\partial t^2} = -4\pi\rho_e. \tag{7.42b}$$

Eqs.(7.42a) and (7.42b) are the d'Alembert equations for $\mathbf{A}$ and φ, respectively. They are wave equations. The solutions of the inhomogeneous wave equations Eq.(7.42) can be obtained in the following way.

First we consider the solution of the equation

$$\nabla^2 \varphi - \frac{1}{c^2}\frac{\partial^2 \varphi}{\partial t^2} = Q(t)\delta^3(\mathbf{r}), \tag{7.43}$$

which is the wave equation Eq.(7.42b) for a source of a point-like charge at the origin of coordinates.

Outside the origin $r = 0$, we have

$$\nabla^2 \varphi - \frac{1}{c^2}\frac{\partial^2 \varphi}{\partial t^2} = 0. \tag{7.44}$$

The source $Q(t)\delta(r)$ is spherically symmetric. We formulate the Laplacian operator in the spherical coordinates. Eq.(7.44) becomes

$$\frac{1}{r^2}\frac{\partial}{\partial r}\left(r^2\frac{\partial \varphi}{\partial r}\right) - \frac{1}{c^2}\frac{\partial^2 \varphi}{\partial t^2} = 0. \tag{7.45}$$

We introduce

$$u(r,t) = \varphi(r,t)r. \tag{7.46}$$

Inserting Eq.(7.46) into Eq.(7.45), we have

$$\frac{\partial^2 u}{\partial r^2} - \frac{1}{c^2}\frac{\partial^2 u}{\partial t^2} = 0. \tag{7.47}$$

Eq.(7.47) is a one-dimensional wave equation, which has the solution of the form

$$u(r,t) = f_1\left(t - \frac{r}{c}\right) + f_2\left(t + \frac{r}{c}\right). \tag{7.48}$$

We choose only $f_1\left(t - \frac{r}{c}\right)$ because the solution $f_2\left(t + \frac{r}{c}\right)$ does not satisfy the causality principle. Thus the solution of Eq.(7.44) has the form

$$\varphi(r,t) = \frac{f_1\left(t - \frac{r}{c}\right)}{r}. \tag{7.49}$$

When $r \to 0$, the potential $Q(t)\delta^3(\mathbf{r})$ approaches to infinity and thus the spatial derivatives become much larger than the time derivative. The second term in Eq.(7.43) can be neglected when $r \to 0$. Using the formula

$$\triangle\left(\frac{1}{r}\right) = -4\pi\delta^3(\mathbf{r}), \tag{7.50}$$

we obtain the solution of Eq.(7.43)

$$\varphi(r,t) = \frac{Q\left(t - \frac{r}{c}\right)}{r}. \tag{7.51}$$

For an arbitrary distribution $\rho_e(\mathbf{x}',t)$, we replace $Q(t)$ by $\rho_e d^3x'$ and integrate over the whole space, which gives the solution of Eq.(7.42b)

$$\varphi(\mathbf{x},t) = \int \frac{\rho_e\left(\mathbf{x}',t - \frac{r}{c}\right)}{r} d^3x' \tag{7.52}$$

with $r = |\mathbf{x}' - \mathbf{x}|$. Eq.(7.52) shows that the scalar potential $\varphi(\mathbf{x},t)$ at $\mathbf{x}$ and at the time t is the sum of the spherical waves originated from the source elements at $\mathbf{x}'$ and at times $t - \frac{r}{c}$. Similarly we have the solution of Eq.(7.42a)

$$A(\mathbf{x},t) = \frac{1}{c}\int \frac{\mathbf{j}_e\left(\mathbf{x}',t - \frac{r}{c}\right)}{r} d^3x'. \tag{7.53}$$

In the region outside the source, we have

$$\nabla^2\mathbf{A} - \frac{1}{c^2}\frac{\partial^2\mathbf{A}}{\partial t^2} = 0, \tag{7.54a}$$

$$\nabla^2\varphi - \frac{1}{c^2}\frac{\partial^2\varphi}{\partial t^2} = 0. \tag{7.54b}$$

The solutions of Eq.(7.54) are the superpositions of plane waves

$$\mathbf{A} = \mathbf{A}_0 e^{i(\mathbf{k}\cdot\mathbf{x}-\omega t)}, \tag{7.55a}$$

$$\varphi = \varphi_0 e^{i(\mathbf{k}\cdot\mathbf{x}-\omega t)} \tag{7.55b}$$

with

$$k = \frac{\omega}{c}. \tag{7.56}$$

ω and $\mathbf{k}$ are called the frequency and wave vector of the plane wave, respectively. Using the Lorentz gauge, we have

$$\varphi_0 = \frac{c}{\omega} \mathbf{k} \cdot \mathbf{A}_0. \tag{7.57}$$

Eq.(7.55) shows that the propagation speed of the wave is c.

The differential equations for the electromagnetic field $\mathbf{E}$ and $\mathbf{B}$ outside the source have the similar form as Eq.(7.54). Since $\mathbf{B} = \boldsymbol{\nabla} \times \mathbf{A}$, when we take the curl of Eq.(7.54b), we have

$$\boldsymbol{\nabla} \times (\nabla^2 \mathbf{A}) - \frac{1}{c^2} \frac{\partial^2 \boldsymbol{\nabla} \times \mathbf{A}}{\partial t^2} = 0. \tag{7.58}$$

Interchanging the order of the Laplacian and curl operators gives

$$\boldsymbol{\nabla} \times (\nabla^2 \mathbf{A}) = \nabla^2 (\boldsymbol{\nabla} \times \mathbf{A}) = \nabla^2 \mathbf{B}. \tag{7.59}$$

Thus we get the differential equation for $\mathbf{B}$

$$\nabla^2 \mathbf{B} - \frac{1}{c^2} \frac{\partial^2 \mathbf{B}}{\partial t^2} = 0. \tag{7.60}$$

Similarly, using $\mathbf{E} = -\frac{1}{c} \frac{\partial \mathbf{A}}{\partial t} - \boldsymbol{\nabla}\varphi$, we obtain the wave equation for the electric field $\mathbf{E}$

$$\nabla^2 \mathbf{E} - \frac{1}{c^2} \frac{\partial^2 \mathbf{E}}{\partial t^2} = 0. \tag{7.61}$$

Eq.(7.60) and Eq.(7.61) are the wave equations in free space for the electromagnetic field $\mathbf{E}$ and $\mathbf{B}$. Their solutions can be build by the superposition of the plane waves.

In free space, according to Eq.(7.31), we have

$$\boldsymbol{\nabla} \times \mathbf{E} = -\frac{1}{c} \frac{\partial \mathbf{B}}{\partial t}, \tag{7.62a}$$

$$\boldsymbol{\nabla} \times \mathbf{B} = -\frac{1}{c} \frac{\partial \mathbf{E}}{\partial t}. \tag{7.62b}$$

From Eq.(7.62), we can see that $\mathbf{E}$ and $\mathbf{B}$ described by a plane wave should be in a direction perpendicular to each other. Both of them are also perpendicular to the direction of the wave vector $\mathbf{k}$.

7.6   Proca equation

For a massive vector boson field with the Lagrangian density Eq.(2.439) coupled to a spinor fermion field, we have

$$\mathcal{L} = -\frac{1}{4}F_{\mu\nu}F^{\mu\nu} + \frac{1}{2}m^2 A_\mu A^\mu - j_\mu A^\mu. \tag{7.63}$$

The Euler-Lagrange equation for it reads

$$\partial_\mu F^{\mu\nu} + m^2 A^\nu = j^\nu, \tag{7.64}$$

which is called the Proca equations. Taking the divergence of Eq.(7.64) gives

$$\partial_\nu A^\nu = 0. \tag{7.65}$$

The Lorentz gauge condition holds always and we do not have the freedom of the gauge transformations as we have for the Maxwell equations. When Eq.(7.65) is used, Eq.(7.64) becomes

$$(\Box + m^2)A^\mu = j^\mu. \tag{7.66}$$

In Section 4.4 on the Higgs mechanism, we have shown that massless vector bosons acquire mass at the expense of the Goldstone bosons. A massless vector boson with two components becomes a massive vector boson with three components while the Goldstone boson disappears. The Higgs mechanism transforms the massless vector bosons into the massive vector bosons and breaks the gauge symmetry which is no more needed for the massive vector bosons.

From Eqs.(2.273), (7.42) and (7.66), when the source terms are not included, we can see that each component of the spinor and vector fields satisfies a Klein-Gordon equation ($m = 0$ for photons). This is not strange because the Klein-Gordon equation is an equation for particles with one component. The Dirac, Maxwell and Proca equations describe the relations between the spin components. When we do not consider the relations between the spin components, the equations reduce to the simple Klein-Gordon equation for particles with one component.

7.7   Poisson equation

For a static electric field, the Maxwell equations have the form

$$\nabla \cdot \mathbf{E} = 4\pi\rho_e, \tag{7.67a}$$

$$\nabla \times \mathbf{E} = 0. \tag{7.67b}$$

The electric field $\mathbf{E}$ is expressed by the relation

$$\mathbf{E} = -\boldsymbol{\nabla}\varphi. \tag{7.68}$$

Substituting Eq.(7.68) into Eq.(7.67a), we have

$$\Delta\varphi = -4\pi\rho_e. \tag{7.69}$$

Eq.(7.69) is called the *Poisson equation*.

In vacuum, $\rho_e = 0$, the scalar potential φ satisfies the Laplace equation.

$$\Delta\varphi = 0. \tag{7.70}$$

We define the Green's function $G(\mathbf{x} - \mathbf{x}')$ of the Laplace equation by

$$\Delta G(\mathbf{x} - \mathbf{x}') = \delta^3(\mathbf{x} - \mathbf{x}'). \tag{7.71}$$

$G(\mathbf{x} - \mathbf{x}')$ has the form

$$G(\mathbf{x} - \mathbf{x}') = -\frac{1}{4\pi}\frac{1}{|\mathbf{x} - \mathbf{x}'|}. \tag{7.72}$$

Using Eq.(7.50), one can easily check that the function $G(\mathbf{x} - \mathbf{x}')$ given by Eq.(7.72) is the solution of Eq.(7.71). Thus the scalar potential φ determined by Eq.(7.69) takes the form

$$\varphi = \int \frac{\rho_e}{r} dV. \tag{7.73}$$

In Section 2.6.2.3, we have set $\varphi = 0$ for the massless vector boson field. Now we have $\varphi \neq 0$ for $\mathbf{E}$. $\varphi \neq 0$ is contributed by the charge source. The electric field comes from the two sources: One $(-\frac{1}{c}\frac{\partial \mathbf{A}}{\partial t})$ from $\mathbf{A}$ contributed by the photon field and the other $(-\nabla\varphi)$ from the interaction of the photon field with the spinor field. Using the gauge transformations, one can realize the division of $\mathbf{E}$ into the contribution from the interaction and the one from photons while the Lagrangian is kept unchanged.

7.8 Electrostatic energy of charges

Now we calculate the energy of the electromagnetic field coupled with a source $j_e^\mu(x)$. The canonical energy-momentum tensor $\Theta^{\mu\nu}$ reads

$$\Theta^{\mu\nu} = \frac{\partial \mathcal{L}}{\partial(\partial_\mu A_\sigma)}\partial^\nu A_\sigma - \eta^{\mu\nu}\mathcal{L}, \tag{7.74}$$

where $\mathcal{L}$ is given by Eq.(7.3). Using the relation

$$\frac{\partial F_{\alpha\beta}F^{\alpha\beta}}{\partial(\partial_\mu A_\sigma)} = 4F^{\mu\sigma}, \tag{7.75}$$

we have

$$\Theta^{\mu\nu} = \frac{1}{16\pi}\eta^{\mu\nu}F_{\alpha\beta}F^{\alpha\beta} - \frac{1}{4\pi}F^{\mu\sigma}\partial^{\nu}A_{\sigma} + \frac{1}{c}\eta^{\mu\nu}j_e^{\sigma}A_{\sigma}. \qquad (7.76)$$

We introduce the symmetric energy-momentum tensor

$$T^{\mu\nu} = \Theta^{\mu\nu} + \partial_{\kappa}\chi^{\kappa\mu\nu} \qquad (7.77)$$

with

$$\chi^{\kappa\mu\nu} \equiv -\frac{1}{4\pi}F^{\kappa\mu}A^{\nu} = \frac{1}{4\pi}F^{\mu\kappa}A^{\nu}. \qquad (7.78)$$

Using Eq.(7.24), we find

$$\begin{aligned}
T^{\mu\nu} = &\frac{1}{16\pi}\eta^{\mu\nu}F_{\alpha\beta}F^{\alpha\beta} + \frac{1}{4\pi}F^{\mu\sigma}F_{\sigma}{}^{\nu} \\
&+ \frac{1}{c}\eta^{\mu\nu}j_e^{\sigma}A_{\sigma} - \frac{1}{c}j_e^{\mu}A^{\nu}.
\end{aligned} \qquad (7.79)$$

Then the energy is given by

$$\begin{aligned}
H &= \int d^3x T_0^0 \\
&= \int d^3x \left(\frac{1}{16\pi}F_{\alpha\beta}F^{\alpha\beta} + \frac{1}{4\pi}F^{0\sigma}F_{\sigma 0} + \frac{1}{c}j_e^{\sigma}A_{\sigma} - \frac{1}{c}j_e^0 A_0\right) \qquad (7.80) \\
&= \int d^3x \left[\frac{1}{8\pi}(\mathbf{E}^2 + \mathbf{B}^2) - \frac{1}{c}\mathbf{j}_e \cdot \mathbf{A}\right].
\end{aligned}$$

Now we determine the electrostatic energy of a system with charges. In this case, $\mathbf{j}_e = 0$ and $\mathbf{B} = 0$. The electrostatic energy of charges is given by

$$U = \frac{1}{8\pi}\int E^2 dV. \qquad (7.81)$$

Using $\mathbf{E} = -\boldsymbol{\nabla}\varphi$, we obtain

$$\begin{aligned}
U &= -\frac{1}{8\pi}\int \mathbf{E} \cdot \boldsymbol{\nabla}\varphi dV \\
&= -\frac{1}{8\pi}\int \boldsymbol{\nabla} \cdot (\varphi\mathbf{E})dV + \frac{1}{8\pi}\int \varphi\boldsymbol{\nabla} \cdot \mathbf{E}dV.
\end{aligned} \qquad (7.82)$$

Using Gauss's theorem, the first term in Eq.(7.82) can be changed into a surface integration. Neglecting the surface integration, we have

$$\begin{aligned}
U &= \frac{1}{2}\int \rho_e\varphi dV \\
&= \frac{1}{2}\int \frac{\rho_e(\mathbf{x})\rho_e(\mathbf{x}')}{r}d^3x d^3x',
\end{aligned} \qquad (7.83)$$

which is called the Coulomb energy. U can be considered as an effective interaction term for Dirac fermions and added to the Lagrangian for Dirac fermions when we study the Dirac fermion field.

7.9 Many-body operators

When we use the operator form $\hat{\rho}_e = e\hat{\psi}^\dagger\hat{\psi}$ for ρ_e, we obtain the interaction term of the Coulomb type in the Hamiltonian operator

$$\hat{U} = \frac{1}{2}\int d^3x d^3x' \frac{e^2}{|\mathbf{x} - \mathbf{x}'|}\hat{\psi}^\dagger(\mathbf{x})\hat{\psi}^\dagger(\mathbf{x}')\hat{\psi}(\mathbf{x}')\hat{\psi}(\mathbf{x}), \qquad (7.84)$$

$\hat{U}$ is a two-body operator. A two-body operator $\hat{U}$ can be expressed in the following form using the basis in which $\hat{U}$ is diagonal

$$\hat{U} = \frac{1}{2}\sum_{\alpha\beta}U_{\alpha\beta}|\alpha\beta\rangle\langle\alpha\beta| = \frac{1}{2}\sum_{\alpha\beta}U_{\alpha\beta}\hat{a}_\alpha^\dagger\hat{a}_\beta^\dagger\hat{a}_\beta\hat{a}_\alpha \qquad (7.85)$$

with

$$U_{\alpha\beta} = \langle\alpha\beta|\hat{U}|\alpha\beta\rangle. \qquad (7.86)$$

We can evaluate a general matrix element similar to the case for a one-body operator (see Eq.(6.97))

$$_S\langle\alpha_1', \alpha_2', \cdots \alpha_N'|\hat{U}|\alpha_1, \alpha_2, \cdots \alpha_N\rangle_S$$

$$= \sum_P \xi^{S_P} \sum_{i\neq j}^N \prod_{k\neq i,j} \langle\alpha_{P_k}'|\alpha_k\rangle\langle\alpha_{P_i}'\alpha_{P_j}'|\hat{U}|\alpha_i\alpha_j\rangle \qquad (7.87)$$

$$= (\frac{1}{2}\sum_{i\neq j}^N U_{\alpha_i\alpha_j})_S\langle\alpha_1', \alpha_2', \cdots \alpha_N'|\alpha_1, \alpha_2, \cdots \alpha_N\rangle_S.$$

The factor $\frac{1}{2}\sum_{i\neq j}^N U_{\alpha_i\alpha_j}$ is the sum over all distinct pairs of particles in the state $|\alpha_1, \alpha_2, \cdots \alpha_N\rangle$. If $|\alpha\rangle$ and $|\beta\rangle$ are different, the number of pairs is $n_\alpha n_\beta$. If $|\alpha\rangle = |\beta\rangle$, the number of pairs is $n_\alpha(n_\alpha - 1)$. To help counting, we define an operator $\hat{P}_{\alpha\beta}$ which counts the number of the particle pairs in the states $|\alpha\rangle$ and $|\beta\rangle$.

$$\hat{P}_{\alpha\beta} \equiv \hat{n}_\alpha\hat{n}_\beta - \delta_{\alpha\beta}\hat{n}_\alpha, \qquad (7.88)$$

$\hat{P}_{\alpha\beta}$ can be expressed in terms of the creation and annihilation operators

$$\hat{P}_{\alpha\beta} = \hat{a}_\alpha^\dagger\hat{a}_\alpha\hat{a}_\beta^\dagger\hat{a}_\beta - \delta_{\alpha\beta}\hat{a}_\alpha^\dagger\hat{a}_\alpha$$

$$= \hat{a}_\alpha^\dagger\xi\hat{a}_\beta^\dagger\hat{a}_\alpha\hat{a}_\beta \qquad (7.89)$$

$$= \hat{a}_\alpha^\dagger\hat{a}_\beta^\dagger\hat{a}_\beta\hat{a}_\alpha.$$

Using the operator $\hat{P}_{\alpha\beta}$, Eq.(7.87) becomes

$$_S\langle\alpha_1', \alpha_2', \cdots \alpha_N'|\hat{U}|\alpha_1, \alpha_2, \cdots \alpha_N\rangle_S$$

$$= {}_S\langle\alpha_1', \alpha_2', \cdots \alpha_N'|\frac{1}{2}\sum_{\alpha\beta}U_{\alpha\beta}\hat{P}_{\alpha\beta}|\alpha_1, \alpha_2, \cdots \alpha_N\rangle_S. \qquad (7.90)$$

$\hat{U}$ can also be expressed in terms of the operator $\hat{P}_{\alpha\beta}$ by

$$
\begin{aligned}
\hat{U} &= \frac{1}{2} \sum_{\alpha\beta} U_{\alpha\beta} \hat{P}_{\alpha\beta} \\
&= \frac{1}{2} \sum_{\alpha\beta} \langle \alpha\beta | \hat{U} | \alpha\beta \rangle \hat{a}_\alpha^\dagger \hat{a}_\beta^\dagger \hat{a}_\beta \hat{a}_\alpha .
\end{aligned}
\tag{7.91}
$$

We can transform the diagonal representation to that of an arbitrary basis, which gives the general expression for a two-body potential

$$
\hat{U} = \frac{1}{2} \sum_{\lambda\mu\nu\rho} \langle \lambda\mu | \hat{U} | \nu\rho \rangle \hat{a}_\lambda^\dagger \hat{a}_\mu^\dagger \hat{a}_\rho \hat{a}_\nu .
\tag{7.92}
$$

Using symmetrized states, Eq.(7.92) becomes

$$
\begin{aligned}
\hat{U} &= \frac{1}{4} \sum_{\lambda\mu\nu\rho} \langle \lambda\mu | \hat{U} (|\nu\rho\rangle + \xi|\rho\nu\rangle) \hat{a}_\lambda^\dagger \hat{a}_\mu^\dagger \hat{a}_\rho \hat{a}_\nu \\
&= \frac{1}{4} \sum_{\lambda\mu\nu\rho} {}_S\langle \lambda\mu | \hat{U} | \nu\rho \rangle_S \hat{a}_\lambda^\dagger \hat{a}_\mu^\dagger \hat{a}_\rho \hat{a}_\nu .
\end{aligned}
\tag{7.93}
$$

The Coulomb interaction $\hat{U}$ in Eq.(7.84) is an interaction that is diagonal in the $\{\mathbf{x}\}$ representation.

Generally, for a two-body interaction $\hat{U}$ diagonal in the $\{\mathbf{x}\}$ representation, we can express $\hat{U}$ in the following form

$$
\hat{U} = \frac{1}{2} \int d^3x d^3y \upsilon(\mathbf{x} - \mathbf{y}) \hat{\psi}^\dagger(\mathbf{x}) \hat{\psi}^\dagger(\mathbf{y}) \hat{\psi}(\mathbf{y}) \hat{\psi}(\mathbf{x}) .
\tag{7.94}
$$

We can generalize the two-body interaction to the n-body interaction described by an n-body operator

$$
\begin{aligned}
\hat{U}_n =& \frac{1}{n!} \sum_{\lambda_1\cdots\lambda_n} \sum_{\mu_1\cdots\mu_n} \langle \lambda_1\cdots\lambda_n | \hat{U}_n | \mu_1\cdots\mu_n \rangle \\
& \times \hat{a}_{\lambda_1}^\dagger \cdots \hat{a}_{\lambda_n}^\dagger \hat{a}_{\mu_1} \cdots \hat{a}_{\mu_n} .
\end{aligned}
\tag{7.95}
$$

In the expression of Eq.(7.95), the normal ordered form is used, in which all the creation operators are on the left of all the annihilation operators.

7.10   Potentials of charge particles in classical limit

In a classical system, when the distances between the particles are large and thus the particles can be considered as point-like particles, we have

$$
\rho = \Sigma_i e_i \delta(\mathbf{x} - \mathbf{x}_i),
\tag{7.96}
$$

where the sum is over all the charges. $\mathbf{x}_i$ is the position of the particles with charge e_i. Then we have

$$A_0(\mathbf{x}) = \int \frac{\rho}{r} dV = \sum_i \frac{e_i}{|\mathbf{x} - \mathbf{x}_i|} \tag{7.97}$$

and

$$U = \frac{1}{2} \sum_{i \neq j} \frac{e_i e_j}{|\mathbf{x}_i - \mathbf{x}_j|}. \tag{7.98}$$

In particular, the interaction potential of two charges is

$$U = \frac{e_1 e_2}{|\mathbf{x}_1 - \mathbf{x}_2|}. \tag{7.99}$$

Eq.(7.99) is called the *Coulomb interaction* for the point-like charged particles.

7.11 Multipole expansion

According to Eq.(7.73), the scalar potential has the form

$$\varphi = \int \frac{\rho(\mathbf{x}')}{|\mathbf{x} - \mathbf{x}'|} d^3 x'. \tag{7.100}$$

Eq.(7.100) can be expressed as an expansion in spherical harmonics using the mathematical formula

$$\frac{1}{|\mathbf{x} - \mathbf{x}'|} = 4\pi \sum_{l=0}^{\infty} \sum_{m=-l}^{l} \frac{1}{2l+1} \frac{r_<^l}{r_>^{2l+1}} Y_{lm}^*(\theta', \phi') Y_{lm}(\theta, \phi), \tag{7.101}$$

where $r_<$ is the smaller of $|\mathbf{x}|$ and $|\mathbf{x}'|$, and $r_>$ is the larger of $|\mathbf{x}|$ and $|\mathbf{x}'|$. Inserting Eq.(7.101) into Eq.(7.100) we have

$$\varphi = 4\pi \sum_{l=0}^{\infty} \sum_{m=-l}^{l} \frac{1}{2l+1} \left[\int Y_{lm}^*(\theta', \phi') r'^l \rho(\mathbf{x}') d^3 x' \right] \frac{Y_{lm}(\theta, \phi)}{r^{2l+1}}. \tag{7.102}$$

In the derivation of Eq.(7.102), we have set $r_< = r'$ and $r_> = r$ because we consider the potential outside the charge distribution.

We introduce the multipole moments q_{lm} defined by

$$q_{lm} \equiv \int Y_{lm}^*(\theta', \phi') r'^l \rho(\mathbf{x}') d^3 x'. \tag{7.103}$$

In terms of q_{lm}, the scalar potential is expressed by an expansion formula

$$\varphi = 4\pi \sum_{l=0}^{\infty} \sum_{m=-l}^{l} \frac{1}{2l+1} q_{lm} Y_{lm}(\theta, \phi) \frac{1}{r^{2l+1}}. \tag{7.104}$$

The expansion can also be expressed in terms of Cartesian coordinates as

$$\varphi(\mathbf{x}) = \frac{q}{r} + \frac{\mathbf{p} \cdot \mathbf{x}}{r^3} + \frac{1}{2} \sum_{ij} Q_{ij} \frac{x_i x_j}{r^5} + \cdots, \tag{7.105}$$

where

$$q \equiv \int \rho(\mathbf{x}')d^3x', \tag{7.106a}$$

$$\mathbf{p} \equiv \int \mathbf{x}'\rho(\mathbf{x}')d^3x', \tag{7.106b}$$

$$Q_{ij} \equiv \int (3x_i'x_j' - r'^2\delta_{ij})\rho(\mathbf{x}')d^3x'. \tag{7.106c}$$

q is the total charge, which is also called the monopole moment. $\mathbf{p}$ is called the electric dipole moment. q_{ij} is called the quadrupole moment tensor.

Using Eq.(7.68), the electric field $\mathbf{E}$ generated by the total charge q is given by

$$\mathbf{E}_q = \frac{q}{r^2}\mathbf{n}. \tag{7.107}$$

and the electric field $\mathbf{E}$ due to the dipole $\mathbf{p}$ has the form

$$\mathbf{E_p} = \frac{3\mathbf{n}(\mathbf{p} \cdot \mathbf{n}) - \mathbf{p}}{r^3}, \tag{7.108}$$

where $\mathbf{n}$ is the unit vector in the direction of $\mathbf{r}$.

7.11.1   *Interaction energy of dipole moments*

Now we consider the interaction energy U of a localized charge distribution $\rho(\mathbf{x})$ placed in an external electric field $\mathbf{E}$ generated by the scalar potential $\varphi(\mathbf{x})$. From Eq.(7.83), neglecting the self-interaction terms, the interaction energy is given by

$$U = \int \rho(x)\varphi(\mathbf{x})d^3x. \tag{7.109}$$

We expand the potential in a Taylor series.

$$\varphi(\mathbf{x}) = \varphi(0) + \mathbf{x} \cdot \nabla\varphi(\mathbf{x})\Big|_{\mathbf{x}=0} + \frac{1}{2}\sum_i \sum_j x_i x_j \frac{\partial^2\varphi(\mathbf{x})}{\partial x_i \partial x_j}\Big|_{\mathbf{x}=0} + \cdots$$

$$= \varphi(0) - \mathbf{x} \cdot \mathbf{E} - \frac{1}{2}\sum_i \sum_j x_i x_j \frac{\partial E_j}{\partial x_i}\Big|_{\mathbf{x}=0} + \cdots. \tag{7.110}$$

We can subtract $\frac{1}{6}r^2\nabla\cdot\mathbf{E}$ from the last term of Eq.(7.110) because $\nabla\cdot\mathbf{E} = 0$. Thus we find

$$\varphi(\mathbf{x}) = \varphi(0) - \mathbf{x}\cdot\mathbf{E}(0)$$

$$- \frac{1}{6}\sum_i\sum_j(3x_ix_j - r^2\delta_{ij})\frac{\partial E_j(\mathbf{x})}{\partial x_i}\bigg|_{\mathbf{x}=0} + \cdots . \qquad (7.111)$$

Inserting Eq.(7.111) into Eq.(7.109), we have

$$U = q\varphi(0) - \mathbf{p}\cdot\mathbf{E}(0) - \frac{1}{6}\sum_i\sum_j Q_{ij}\frac{\partial E_j(\mathbf{x})}{\partial x_i}\bigg|_{\mathbf{x}=0} + \cdots . \qquad (7.112)$$

The electric field generated by a dipole is given by Eq.(7.108). When we insert the electric field given by Eq.(7.108) into the second term of the above equation, we obtain the interaction energy U_{12} of two dipoles $\mathbf{p}_1$ and $\mathbf{p}_2$.

$$U_{12} = \frac{\mathbf{p}_1\cdot\mathbf{p}_2 - 3(\mathbf{n}\cdot\mathbf{p}_1)(\mathbf{n}\cdot\mathbf{p}_2)}{|\mathbf{x}_1 - \mathbf{x}_2|^3}, \qquad (7.113)$$

where $\mathbf{x}_1$ and $\mathbf{x}_2$ are the position vectors of $\mathbf{p}_1$ and $\mathbf{p}_2$, respectively. $\mathbf{n}$ is the unit vector in the direction of $\mathbf{x}_1 - \mathbf{x}_2$.

7.11.2 *Magnetic moment*

We consider the magnetic effect of a stationary current. According to Eq.(7.53), the vector potential for a stationary current is given by

$$A(\mathbf{x}, t) = \frac{1}{c}\int\frac{\mathbf{j}(\mathbf{x}')}{|\mathbf{x} - \mathbf{x}'|}d^3x'. \qquad (7.114)$$

Now we consider the case that $\mathbf{j}$ is a localized current distribution. Using the expansion formula

$$\frac{1}{|\mathbf{x} - \mathbf{x}'|} = \frac{1}{|\mathbf{x}|} + \frac{\mathbf{x}\cdot\mathbf{x}'}{|\mathbf{x}|^3} + \cdots , \qquad (7.115)$$

we obtain

$$A_i(\mathbf{x}, t) = \frac{1}{c}\frac{1}{|\mathbf{x}|}\int j_i(\mathbf{x}')d^3x' + \frac{1}{c}\frac{\mathbf{x}}{|\mathbf{x}|^3}\cdot\int j_i(\mathbf{x}')\mathbf{x}'d^3x' + \cdots . \qquad (7.116)$$

Since $\mathbf{j}$ is a stationary current, we have

$$\nabla'\cdot\mathbf{j} = 0. \qquad (7.117)$$

Then

$$\int j_id^3x' = \int\mathbf{j}\cdot\nabla'x_i'd^3x' = -\int x_i'\nabla'\cdot\mathbf{j}d^3x' = 0. \qquad (7.118)$$

Thus the first term in Eq.(7.116) vanishes. Integrating by parts, we obtain

$$\int x'_i \mathbf{j} \cdot \nabla' x'_j d^3 x' = -\int x'_j \nabla' \cdot (x'_i \mathbf{j}) d^3 x'$$
$$= -\int x'_j \mathbf{j} \cdot \nabla' x'_i d^3 x'. \tag{7.119}$$

Eq.(7.119) can be rewritten as

$$\int (x'_i j_j + x'_j j_i) d^3 x' = 0. \tag{7.120}$$

Using Eq.(7.120), we can rewrite Eq.(7.116) as

$$
\begin{aligned}
A_i(\mathbf{x}, t) &= \frac{1}{c} \frac{\mathbf{x}}{|\mathbf{x}|^3} \cdot \int \mathbf{x}' j_i(\mathbf{x}') d^3 x' \\
&= \frac{1}{c|\mathbf{x}|^3} \sum_j x_j \int x'_j j_i(\mathbf{x}') d^3 x' \\
&= -\frac{1}{c|\mathbf{x}|^3} \frac{1}{2} \sum_j x_j \int [x'_i j_j(\mathbf{x}') - x'_j j_i(\mathbf{x}')] d^3 x' \\
&= -\frac{1}{c|\mathbf{x}|^3} \frac{1}{2} \sum_{jk} \epsilon^{ijk} x_j \int (\mathbf{x}' \times \mathbf{j})_k d^3 x' \\
&= -\frac{1}{c|\mathbf{x}|^3} \frac{1}{2} \left[\mathbf{x} \times \int (\mathbf{x}' \times \mathbf{j}) \right]_i d^3 x'.
\end{aligned}
\tag{7.121}
$$

We introduce the magnetic moment density defined by

$$\mathbf{\mathcal{M}}(\mathbf{x}) \equiv \frac{1}{2c} \mathbf{x} \times \mathbf{j}(\mathbf{x}) \tag{7.122}$$

and magnetic moment defined by

$$\mathbf{m} \equiv \frac{1}{2c} \int \mathbf{x}' \times \mathbf{j}(\mathbf{x}') d^3 x'. \tag{7.123}$$

Eq.(7.121) becomes

$$\mathbf{A}(\mathbf{x}) = \frac{\mathbf{m} \times \mathbf{x}}{|\mathbf{x}|^3}. \tag{7.124}$$

Using $\mathbf{B} = \nabla \times \mathbf{A}$, we obtain the magnetic field generated by a magnetic moment.

$$\mathbf{B}(\mathbf{x}) = \frac{3\mathbf{n}(\mathbf{n} \cdot \mathbf{m}) - \mathbf{m}}{|\mathbf{x}|^3}, \tag{7.125}$$

where $\mathbf{n}$ is the unit vector in the direction of $\mathbf{x}$. When the current is generated by the point-like charges q_i with the masses m_i and velocities $\mathbf{v}_i$, the current density $\mathbf{j}$ reads

$$\mathbf{j} = \sum_i q_i \mathbf{v}_i \delta^3(\mathbf{x} - \mathbf{x}_i). \tag{7.126}$$

Then the magnetic moment has the form

$$\mathbf{m} = \frac{1}{2c} \sum_i q_i (\mathbf{x}_i \times \mathbf{v}_i). \tag{7.127}$$

Expressing $\mathbf{m}$ in terms of the orbital angular momentum

$$\mathbf{L}_i = m_i (\mathbf{x}_i \times \mathbf{v}_i) \tag{7.128}$$

of the ith particle, we have

$$\mathbf{m} = \sum_i \frac{q_i}{2m_i c} \mathbf{L}_i. \tag{7.129}$$

We can see that the magnetic moment is proportional to the orbital angular momentum of a charge.

7.12 Maxwell equations in a media

Now we consider the electromagnetic fields in a media. The media can be a gas, a liquid or a solid. The matter of the media consists of atoms or molecules. When an electromagnetic field is applied, local electric dipoles and magnetic moments can be induced. According to Eq.(7.105), the scalar potential has the form

$$\varphi(\mathbf{x}) = \int d^3 x' \left[\frac{\rho(\mathbf{x}')}{|\mathbf{x} - \mathbf{x}'|} + \frac{\mathbf{P}(\mathbf{x}') \cdot (\mathbf{x} - \mathbf{x}')}{|\mathbf{x} - \mathbf{x}'|^3} \right], \tag{7.130}$$

where $\mathbf{P}(\mathbf{x}')$ is the induced electric dipole density, which is also called the polarization vector. Eq.(7.130) can be rewritten as

$$\begin{aligned}
\varphi(\mathbf{x}) &= \int d^3 x' \left[\frac{\rho(\mathbf{x}')}{|\mathbf{x} - \mathbf{x}'|} + \mathbf{P}(\mathbf{x}') \cdot \nabla' \left(\frac{1}{|\mathbf{x} - \mathbf{x}'|} \right) \right] \\
&= \int d^3 x' \frac{1}{|\mathbf{x} - \mathbf{x}'|} \left[\rho(\mathbf{x}') - \nabla' \cdot \mathbf{P}(\mathbf{x}') \right].
\end{aligned} \tag{7.131}$$

In the derivation of Eq.(7.131), we have performed integration by parts for the second term.

Using $\mathbf{E} = -\nabla \varphi$, we have

$$\nabla \cdot \mathbf{E} = 4\pi (\rho - \nabla \cdot \mathbf{P}). \tag{7.132}$$

The dipole moment can be considered to contribute an effective charge density

$$\rho_{\mathrm{P}} = -\nabla \cdot \mathbf{P}. \tag{7.133}$$

To simplify Eq.(7.132), we introduce the electric displacement $\mathbf{D}$ defined by

$$\mathbf{D} \equiv \mathbf{E} + 4\pi\mathbf{P}. \tag{7.134}$$

Then Eq.(7.132) becomes

$$\nabla \cdot \mathbf{D} = 4\pi\rho. \tag{7.135}$$

When there are induced magnetic moments, according to Eq.(7.121), the vector potential $\mathbf{A}$ has the form

$$\mathbf{A}(\mathbf{x}) = \int d^3x' \left[\frac{\mathbf{j}(\mathbf{x}')}{c|\mathbf{x} - \mathbf{x}'|} + \frac{\mathbf{M}(\mathbf{x}') \times (\mathbf{x} - \mathbf{x}')}{|\mathbf{x} - \mathbf{x}'|^3} \right], \tag{7.136}$$

where $\mathbf{M}(\mathbf{x}')$ is the induced magnetic moment, which is also called the macroscopic magnetization. Integrating by parts for the second term, we can rewrite Eq.(7.136) as

$$\begin{aligned} \mathbf{A}(\mathbf{x}) &= \int d^3x' \left[\frac{\mathbf{j}(\mathbf{x}')}{c|\mathbf{x} - \mathbf{x}'|} + \mathbf{M}(\mathbf{x}') \times \nabla' \left(\frac{1}{|\mathbf{x} - \mathbf{x}'|} \right) \right] \\ &= \int d^3x' \frac{1}{c|\mathbf{x} - \mathbf{x}'|} \left[\mathbf{j}(\mathbf{x}') + c\nabla' \times \mathbf{M}(\mathbf{x}') \right]. \end{aligned} \tag{7.137}$$

Using $\mathbf{B} = \nabla \times \mathbf{A}$, we have

$$\nabla \times \mathbf{B} = \frac{4\pi}{c}(\mathbf{j} + c\nabla \times \mathbf{M}). \tag{7.138}$$

The magnetization $\mathbf{M}$ can be considered to contribute an effective current density

$$\mathbf{j}_\mathrm{M} = c\nabla \times \mathbf{M}. \tag{7.139}$$

We introduce a macroscopic quantity defined by

$$\mathbf{H} \equiv \mathbf{B} - 4\pi\mathbf{M}. \tag{7.140}$$

Conventionally, $\mathbf{H}$ is called the magnetic field and $\mathbf{B}$ is called magnetic induction. Physically, it is more natural to call $\mathbf{B}$ as magnetic field. However, we need to respect the history convention. Since $\mathbf{B} = \mathbf{H}$ in the vacuum, we do not need to distinguish $\mathbf{B}$ with $\mathbf{H}$ for the vacuum,, and we have also called $\mathbf{B}$ magnetic field in the case of vacuum. In terms of $\mathbf{H}$, Eq.(7.138) reads

$$\nabla \times \mathbf{H} = \frac{4\pi}{c}\mathbf{j}. \tag{7.141}$$

When the electromagnetic field is time-dependent, we need use Eq.(7.31d)

$$\nabla \times \mathbf{B} = \frac{1}{c}\frac{\partial \mathbf{E}}{\partial t} + \frac{4\pi}{c}\mathbf{j}_\mathrm{t}. \tag{7.142}$$

The total current $\mathbf{j}_t$ consists of three parts.

$$\mathbf{j}_t = \mathbf{j} + \mathbf{j}_P + \mathbf{j}_M, \tag{7.143}$$

where $\mathbf{j}_P$ is the induced current due to the variation of polarization $\mathbf{P}$ and $\mathbf{j}_M$ is the induced current due to the induced magnetization $\mathbf{M}$. From equation

$$\rho_P = -\nabla \cdot \mathbf{P}, \tag{7.144}$$

we can see that polarization $\mathbf{P}$ comes from the polarization charges. When $\mathbf{P}$ changes with time, there is motion of the polarization charge ρ_P, which leads to the current $\mathbf{j}_P$. The current $\mathbf{j}_P$ reads

$$\mathbf{j}_P = nq_P \mathbf{v} = nq_P \frac{d\mathbf{x}}{dt}, \tag{7.145}$$

where n is the charge number per unit volume, q_P is the charge of particles and $\mathbf{v}$ is the velocity of the oscillating charges. In terms of $\mathbf{P}$, Eq.(7.145) can be expressed as

$$\mathbf{j}_P = nq_P \frac{d\mathbf{x}}{dt} = \frac{d(nq_P\mathbf{x})}{dt} = \frac{d\mathbf{P}}{dt}. \tag{7.146}$$

Inserting Eq.(7.146) and Eq.(7.139) into Eq.(7.143), we have

$$\mathbf{j}_t = \mathbf{j} + \frac{d\mathbf{P}}{dt} + c\nabla \times \mathbf{M}, \tag{7.147}$$

Using Eq.(7.147), Eq.(7.142) becomes

$$\nabla \times \mathbf{B} = \frac{1}{c}\frac{\partial \mathbf{E}}{\partial t} + \frac{4\pi}{c}\frac{\partial \mathbf{P}}{\partial t} + 4\pi\nabla \times \mathbf{M} + \frac{4\pi}{c}\mathbf{j}. \tag{7.148}$$

or

$$\nabla \times (\mathbf{B} - 4\pi\mathbf{M}) = \frac{1}{c}\frac{\partial}{\partial t}(\mathbf{E} + 4\pi\mathbf{P}) + \frac{4\pi}{c}\mathbf{j}. \tag{7.149}$$

Using $\mathbf{H} = \mathbf{B} - 4\pi\mathbf{M}$ and $\mathbf{D} = \mathbf{E} + 4\pi\mathbf{P}$, Eq.(7.149) becomes

$$\nabla \times \mathbf{H} = \frac{1}{c}\frac{\partial \mathbf{D}}{\partial t} + \frac{4\pi}{c}\mathbf{j}. \tag{7.150}$$

In summary, the Maxwell equations in a media takes the form

$$\nabla \cdot \mathbf{D} = 4\pi\rho, \tag{7.151a}$$

$$\nabla \cdot \mathbf{B} = 0, \tag{7.151b}$$

$$\nabla \times \mathbf{E} = -\frac{1}{c}\frac{\partial \mathbf{B}}{\partial t}, \tag{7.151c}$$

$$\nabla \times \mathbf{H} = \frac{1}{c}\frac{\partial \mathbf{D}}{\partial t} + \frac{4\pi}{c}\mathbf{j}, \tag{7.151d}$$

7.13   Magnetic moment due to spin

In addition to the contribution of the local current to the magnetic moment, the spin of a particle also contributes to a local magnetic moment. According Eq.(7.2), the current density for spinor fermions is given by

$$j_e^\mu = e\bar{\psi}\gamma^\mu\psi. \tag{7.152}$$

Using the Dirac equation in an electromagnetic field Eq.(6.12), Eq.(7.152) can be rewritten as

$$
\begin{aligned}
j_e^\mu &= \frac{e}{2}(\bar{\psi}\gamma^\mu\psi + \bar{\psi}\gamma^\mu\psi) \\
&= -\frac{ie\hbar}{2mc}\Big\{ -\bar{\psi}\gamma^\mu\gamma^\nu\left(\frac{\partial}{\partial x^\nu} - \frac{ie}{\hbar c}A_\nu\right)\psi \\
&\qquad + \left[\left(\frac{\partial}{\partial x^\nu} + \frac{ie}{\hbar c}A_\nu\right)\bar{\psi}\right]\gamma^\nu\gamma^\mu\psi \Big\} \\
&= j_1^\mu + j_2^\mu
\end{aligned}
\tag{7.153}
$$

with

$$j_1^\mu \equiv -\frac{ie\hbar}{2mc}\left(\frac{\partial\bar{\psi}}{\partial x_\mu}\psi - \bar{\psi}\frac{\partial\psi}{\partial x_\mu}\right) + \frac{e^2}{mc^2}A^\mu\bar{\psi}\psi \tag{7.154}$$

and

$$
\begin{aligned}
j_2^\mu &\equiv -\frac{ie\hbar}{2mc}\Big(-\bar{\psi}\gamma^\mu\gamma^\nu\frac{\partial\psi}{\partial x^\nu} + \frac{\partial\bar{\psi}}{\partial x^\nu}\gamma^\nu\gamma^\mu\psi \\
&\qquad + \frac{ie}{\hbar c}A_\nu\bar{\psi}\gamma^\mu\gamma^\nu\psi + \frac{ie}{\hbar c}A_\nu\bar{\psi}\gamma^\nu\gamma^\mu\psi\Big)_{\mu\neq\nu} \\
&= -\frac{e\hbar}{2mc}\frac{\partial}{\partial x^\nu}(\bar{\psi}\sigma^{\nu\mu}\psi).
\end{aligned}
\tag{7.155}
$$

j_1^μ corresponds to the terms of $\mu = \nu$ and j_2^μ corresponds to the terms of $\mu \neq \nu$. The decomposition of j_e^μ into j_1^μ and j_2^μ is the Gordon decomposition which we will also discuss in Section 7.15. j_1^μ does not contain gamma matrix and corresponds to the current density in Eq.(8.54) in the nonrelativistic case. When we use time-dependence factor $\exp\left(-\frac{i}{\hbar}mc^2 t\right)$ given in Eq.(6.16) and $|eA_\mu| \ll mc^2$, we have

$$j_1^0 = e\varphi^\dagger\varphi, \tag{7.156}$$

which is the charge density. For j_2^μ in the non-relativistic limit, $\bar{\psi}\sigma^{\mu 0}\psi = -\bar{\psi}\sigma^{0\mu}\psi$ can be ignored. From Eqs.(2.333) and (2.268b), we can see that

$$\frac{1}{2}\sigma_{ij} = \epsilon^{ijk}\mathcal{S}_k = \epsilon^{ijk}\Sigma_k \tag{7.157}$$

with

$$\mathcal{S}^k = \frac{1}{4}\epsilon^{ijk}\sigma_{ij} \tag{7.158}$$

corresponds to the spin of the particle. Then we have

$$\begin{aligned}
(\mathbf{j}_2)_i &= -\frac{e\hbar}{2m}\frac{\partial}{\partial x_j}(\bar{\psi}\sigma_{ij}\psi) \\
&= -\frac{e\hbar}{m}\left(\boldsymbol{\nabla}\times\bar{\psi}\boldsymbol{S}\psi\right)_i.
\end{aligned} \tag{7.159}$$

Thus spin contributes an intrinsic magnetic moment density

$$\boldsymbol{M}_s = -\frac{e\hbar}{mc}\bar{\psi}\boldsymbol{S}\psi \tag{7.160}$$

in view of Eq.(7.139). Thus we should also include the magnetization due to the spin of particles in Eq.(7.138). Comparing Eq.(7.160) with Eq.(7.129), we can see that there is a factor difference of 2. This factor is defined as the gyromagnetic ratio (g factor) and we have $g = 2$, which is consistent with Eq.(6.29).

7.14 Interaction energy of magnetic moment with magnetic field

According to Eq.(7.3), the interaction energy for the charged spinor fermions in an electromagnetic field is given by

$$H_{\text{int}} = \int j_e^\mu A_\mu d^3x. \tag{7.161}$$

The contribution from j_2^μ reads

$$\begin{aligned}
H_s &= \int j_2^\mu A_\mu d^3x \\
&= -\int \frac{e\hbar}{2mc}\frac{\partial}{\partial x^\nu}\left(\bar{\psi}\sigma^{\nu\mu}\psi\right)A_\mu d^3x \\
&= \int \frac{e\hbar}{2mc}\frac{\partial A_\mu}{\partial x^\nu}(\bar{\psi}\sigma^{\nu\mu}\psi)d^3x \\
&= \int \frac{e\hbar}{2mc}\left[\frac{1}{2}\frac{\partial A_\mu}{\partial x^\nu}(\bar{\psi}\sigma^{\nu\mu}\psi) + \frac{1}{2}\frac{\partial A_\nu}{\partial x^\mu}(\bar{\psi}\sigma^{\mu\nu}\psi)\right]d^3x \\
&= -\int \frac{e\hbar}{2mc}\left(\frac{1}{2}F^{\nu\mu}\bar{\psi}\sigma^{\nu\mu}\psi\right)d^3x.
\end{aligned} \tag{7.162}$$

In the derivation of the third line, we have integrated by parts and dropped the divergence term because the integration for the divergence terms can be transformed into surface integration and vanishes.

In the non-relativistic limit, we can neglect the smaller component of $\psi = \left(\begin{smallmatrix}\varphi\\\chi\end{smallmatrix}\right)$. Using $B_k = (\boldsymbol{\nabla} \times \mathbf{A})_k = \varepsilon^{ijk}\nabla_i A_j$ and $\sigma_{ij} = \varepsilon^{ijk}\Sigma_k$, Eq.(7.162) becomes

$$H_{\mathrm{s}} = -\frac{e\hbar}{2mc}\mathbf{B} \cdot \int \varphi^\dagger \boldsymbol{\sigma}\varphi d^3x = -\frac{e\hbar}{2mc}\mathbf{B} \cdot \mathbf{m}_s \tag{7.163}$$

with

$$\mathbf{m}_s = \int \varphi^\dagger \boldsymbol{\sigma}\varphi d^3x. \tag{7.164}$$

$\mathbf{m}_s$ is the magnetic moment of spin. Eq.(7.163) describes the interaction of the magnetic moment of spin with a magnetic field.

7.15   The anomalous magnetic moment

In the Lagrangian density of QED, we have an interaction term

$$\mathcal{L}_{\mathrm{I}} = -eA^\mu\bar\psi\gamma_\mu\psi. \tag{7.165}$$

When we consider interaction between quasi-particles, the interaction should be a renormalized one.

$$\mathcal{L}_{\mathrm{IR}} = -e_p A^\mu \bar\psi_{p'}\Gamma_\mu(p',p)\psi_p, \tag{7.166}$$

which corresponds to an interaction energy

$$\begin{aligned}
H_{\mathrm{I}} &= e_p \int d^3x A^\mu(x)\bar\psi_{p'}(x)\Gamma_\mu(p',p)\psi_p(x)\\
&= \frac{e_p}{V}\sqrt{\frac{m^2}{\omega_{\mathbf{p}'}\omega_{\mathbf{p}}}} \int d^3x e^{-i(p'-p)\cdot x}A^\mu(x)\bar u(p')\Gamma_\mu(p',p)u(p),
\end{aligned} \tag{7.167}$$

where V is the normalization volume. In Eq.(7.167), we have made the normalization for the states $|\psi_p\rangle$ in such a way that a box with volume V contains only one particle. Using Eq.(5.475), we have

$$\bar u(p')\Gamma_\mu u(p) = \bar u(p')\gamma_\mu u(p) + \bar u(p')\Lambda_\mu u(p). \tag{7.168}$$

Now we show that $\bar u(p')\Lambda_\mu u(p)$ has the form

$$\bar u(p')\Lambda_\mu u(p) = \bar u(p')\left[D\gamma_\mu - \frac{e^2}{16\pi^2 m}(p_\mu + p'_\mu)\right]u(p), \tag{7.169}$$

where D is a function of q^2 whose detailed form we do not necessarily have to know for the calculations of the magnetic interaction. According to Eq.(5.509), Λ_μ is the addition of $\Lambda_\mu^{(1)}$ and $\Lambda_\mu^{(2)}$. From Eq.(5.507), we can see that $\Lambda_\mu^{(1)}$ makes a contribution only to $D\gamma_\mu$ in Eq.(7.169). Thus the second term in Eq.(7.169) comes from $\Lambda_\mu^{(2)}$. $u(p)$ satisfies the Dirac equation Eq.(2.429). We have

$$\gamma_\mu p^\mu u(p) = m u(p), \qquad (7.170\text{a})$$

$$\bar{u}(p')\gamma_\mu p'^\mu = m\bar{u}(p'). \qquad (7.170\text{b})$$

Using

$$\gamma^\mu \gamma^\nu = \frac{1}{2}\{\gamma^\mu, \gamma^\nu\} + \frac{1}{2}[\gamma^\mu, \gamma^\nu] = \eta^{\mu\nu} - i\sigma^{\mu\nu}, \qquad (7.171)$$

Eq.(7.170) can be written as

$$\gamma_\mu u(p) = \frac{1}{m}(p_\mu - i\sigma_{\mu\nu}p^\nu)u(p), \qquad (7.172\text{a})$$

$$\bar{u}(p')\gamma_\mu = \frac{1}{m}\bar{u}(p')(p'_\mu + i\sigma_{\mu\nu}p'^\nu). \qquad (7.172\text{b})$$

Combining above two equations, we have

$$\begin{aligned}
\bar{u}(p')\gamma_\mu u(p) &= \frac{1}{2}\bar{u}(p')[\gamma_\mu u(p)] + \frac{1}{2}[\bar{u}(p')\gamma_\mu]u(p) \\
&= \frac{1}{2m}\bar{u}(p')[(p_\mu + p'_\mu) + i\sigma_{\mu\nu}q^\nu]u(p),
\end{aligned} \qquad (7.173)$$

where $q = p' - p$. Eq.(7.173) is called the Gordon decomposition.

Now we consider the integration in Eq.(5.508). Since the initial and final states are free particle states, we have $p^2 = p'^2 = m^2$. In the static limit, $q^2 = (p' - p)^2 \to 0$. Then the denominator of the integrand in Eq.(5.508) takes the value

$$m^2(x+y) - p^2 x(1-x) - p'^2 y(1-y) + 2p \cdot p' xy = m^2(x+y)^2. \qquad (7.174)$$

To evaluate the numerator, we note that $p\!\!\!/'$ on the left and $p\!\!\!/$ on the right can be replaced by m according to the Dirac equation Eq.(7.170). Using the following gamma matrix identities,

$$\gamma_\mu \gamma_\nu + \gamma_\nu \gamma_\mu = 2\eta_{\mu\nu}, \qquad (7.175\text{a})$$

$$\gamma_\nu \gamma^\alpha \gamma^\beta \gamma^\nu = 4\eta^{\alpha\beta}, \qquad (7.175\text{b})$$

$$\gamma_\nu \gamma_\alpha \gamma_\mu \gamma^\beta \gamma^\nu = -2\gamma^\beta \gamma_\mu \gamma_\alpha, \qquad (7.175\text{c})$$

the numerator of the integrand in Eq.(5.508) can be rewritten as

$$4m(y - xy - x^2)p_\mu + 4m(x - xy - y^2)p'_\mu + A\gamma_\mu. \qquad (7.176)$$

The term $A\gamma_\mu$ makes a contribution to $D\gamma_\mu$ in Eq.(7.169). The integration of the remaining terms gives

$$
\begin{aligned}
\Lambda'^{(2)}_\mu &= -\frac{e_p^2}{4\pi^2 m}\int_0^1 dx \int_0^{1-x} dy \frac{(y-xy-x^2)p_\mu + (x-xy-y^2)p'_\mu}{(x+y)^2}\\
&= -\frac{e_p^2}{4\pi^2 m}\int_0^1 dx \int_0^{1-x} dy \frac{[y-x(x+y)]p_\mu + [x-y(x+y)]p'_\mu}{(x+y)^2}\\
&= -\frac{e_p^2}{4\pi^2 m}\left(\frac{1}{4}p_\mu + \frac{1}{4}p'_\mu\right)\\
&= -\frac{e_p^2}{16\pi^2 m}\left(p_\mu + p'_\mu\right).
\end{aligned}
\tag{7.177}
$$

Thus we obtain Eq.(7.169).

Inserting Eq.(7.169) into Eq.(7.168), we obtain

$$
\begin{aligned}
\bar{u}(p')\Gamma_\mu(p',p)u(p) = \bar{u}(p')\Big\{ &\gamma_\mu[F_1(q^2) + F_2(q^2)]\\
&- \frac{1}{2m}(p'+p)_\mu F_2(q^2)\Big\}u(p)
\end{aligned}
\tag{7.178}
$$

with

$$
F_2(q^2) = \frac{e_p^2}{8\pi^2}
\tag{7.179}
$$

and

$$
F_1(q^2) = 1 + D - \frac{e_p^2}{8\pi^2}
\tag{7.180}
$$

in the static limit $q^2 \to 0$. Using the Gordon decomposition, Eq.(7.178) becomes

$$
\bar{u}(p')\Gamma_\mu(p',p)u(p) = \bar{u}(p')\left[\gamma_\mu F_1(q^2) + i\frac{1}{2m}F_2(q^2)q^\nu \sigma_{\mu\nu}\right]u(p).
\tag{7.181}
$$

Let us analyse the meaning of $F_1(q^2)$ and $F_2(q^2)$ in Eq.(7.181). First, we consider the effect of pure static electric field. In this case, $\mathbf{A} = 0$. The interaction energy given by Eq.(7.167) becomes

$$
\begin{aligned}
H_I &= \frac{e_p}{V}\sqrt{\frac{m^2}{\omega_{\mathbf{p}'}\omega_{\mathbf{p}}}}\int d^3 x e^{-i(p'-p)\cdot x}A^0(x)\bar{u}(p')\\
&\quad \times \left\{\gamma_0[F_1(q^2) + F_2(q^2)] - \frac{1}{2m}(p'+p)_0 F_2(q^2)\right\}u(p)\\
&= \frac{e_p}{V}\sqrt{\frac{m^2}{\omega_{\mathbf{p}'}\omega_{\mathbf{p}}}}\int d^3 x e^{-i(p'-p)\cdot x}A^0(x)\bar{u}(p')\\
&\quad \times \left[\gamma_0 F_1(q^2) + \frac{2m\gamma_0 - \omega_{\mathbf{p}'} - \omega_{\mathbf{p}}}{2m}F_2(q^2)\right]u(p).
\end{aligned}
\tag{7.182}
$$

In the non-relativistic limit, we can neglect the lower components of the Dirac spinors. We have

$$\bar{u}u = 1 \qquad (7.183)$$

and

$$\bar{u}\gamma_0 u = 1. \qquad (7.184)$$

Thus, in the static limit $q^2 \to 0$, Eq.(7.182) becomes

$$H_{\mathrm{I}} = e_p F_1(0) \frac{1}{V} \int d^3x e^{-i\mathbf{q}\cdot\mathbf{x}} A^0(x)$$
$$= e_p F_1(0) A^0. \qquad (7.185)$$

This is the interaction energy which should be $e_p A^0$ as given by Eq.(6.29) when e_p is the renormalized electric charge. Thus we have

$$F_1(0) = 1. \qquad (7.186)$$

Then we consider the effect of magnetic field. Using Eq.(7.181) and Gordon decomposition, we find

$$A^k(x)\bar{u}(p')\Gamma_k(p',p)u(p) = A^k(x)\bar{u}(p')\left[\gamma_k F_1(q^2) + i\frac{1}{2m}F_2(q^2)q^\nu \sigma_{k\nu}\right]u(p)$$
$$= A^k(x)\bar{u}(p')\left\{\frac{1}{2m}(p'_k + p_k)F_1(q^2)\right. \qquad (7.187)$$
$$\left. + i\frac{1}{2m}[F_1(q^2) + F_2(q^2)]q^\nu \sigma_{k\nu}\right\}u(p).$$

Using Eq.(2.268b), which gives

$$\sigma^{ij} = \epsilon^{ijk}\Sigma_k \qquad (7.188)$$

with

$$\Sigma = \begin{pmatrix} \boldsymbol{\sigma} & 0 \\ 0 & \boldsymbol{\sigma} \end{pmatrix}, \qquad (7.189)$$

we have

$$A^k(x)\bar{u}(p')\Gamma_k(p',p)u(p)$$
$$= -\bar{u}(p')\left[\mathbf{A}\cdot\frac{\mathbf{P}}{m}F_1(q^2) + i\frac{1}{2m}(F_1(q^2) + F_2(q^2))\mathbf{q}\times\mathbf{A}\cdot\Sigma\right]u(p). \qquad (7.190)$$

Thus, in the static limit $q^2 \to 0$, we obtain

$$H_{\mathrm{I}} = \frac{e_p}{V}\int d^3x e^{-i\mathbf{q}\cdot\mathbf{x}}\left\{-F_1(q^2)\mathbf{A}\cdot\frac{\mathbf{P}}{m}\bar{u}(p')u(p)\right.$$
$$\left. + \frac{1}{2m}[F_1(q^2) + F_2(q^2)](-i\mathbf{q}\times\mathbf{A})\cdot\bar{u}(p')\Sigma u(p)\right\}$$
$$= -e_p\bar{u}(p')u(p)\frac{1}{V}\int d^3x e^{-i\mathbf{q}\cdot\mathbf{x}}\frac{\mathbf{P}}{m}\cdot\mathbf{A}(x) \qquad (7.191)$$
$$- \frac{e_p}{2m}(1 + F_2(0))\bar{u}(p')\Sigma u(p)\cdot\frac{1}{V}\int d^3x e^{-i\mathbf{q}\cdot\mathbf{x}}\boldsymbol{\nabla}\times\mathbf{A}(x)$$
$$= -e_p\frac{\mathbf{P}}{m}\cdot\mathbf{A}(x) - \frac{e_p}{2m}\left(1 + \frac{e_p^2}{8\pi^2}\right)2\mathbf{S}\cdot\mathbf{B}.$$

The first term in Eq.(7.191) is the interaction energy of a moving charge in a magnetic field. For a constant magnetic field,

$$\mathbf{A} = \frac{1}{2}\mathbf{B} \times \mathbf{x}. \tag{7.192}$$

Thus

$$-\frac{e_p}{m}\mathbf{p} \cdot \mathbf{A} = -\frac{e_p}{2m}\mathbf{p} \cdot (\mathbf{B} \times \mathbf{x})$$
$$= -\frac{e_p}{2m}\mathbf{L} \cdot \mathbf{B}, \tag{7.193}$$

which is the orbital term in Eq.(6.29). The second term in Eq.(7.191) is the interaction energy of a spinor fermion with spin-$\frac{1}{2}$ in a magnetic field, which can be expressed as $-\frac{e_p g}{2m}\mathbf{S} \cdot \mathbf{B}$. The gyromagnetic ratio g is given by

$$g = 2\left(1 + \frac{e_p^2}{8\pi^2}\right)$$
$$= 2\left(1 + \frac{\alpha}{2\pi}\right) \tag{7.194}$$

with

$$\alpha = \frac{e_p^2}{4\pi}. \tag{7.195}$$

α is called the *fine structure constant*. For electron, $\alpha = \frac{1}{137}$ is small and we often neglect the contribution of $F_2(0)$ and take $g = 2$. Thus the magnetic moment $\boldsymbol{m_s}$ of a spinor fermion with spin-$\frac{1}{2}$ is given by

$$\boldsymbol{m_s} = \frac{e_p g}{2m}\mathbf{S}. \tag{7.196}$$

Historically, theoretical value of g was first found to be 2, which was not consistent with the precise experimental value. This inconsistency is considered as 'anomalous' and thus $\boldsymbol{m_s}$ of electron is called anomalous magnetic moment. The accurate theoretical value of g was calculated by Schwinger in 1949, which was one of the most remarkable masterpiece in the theoretical calculations.

Chapter 8

Quantum mechanics

8.1 Equations of motion for operators in quantum mechanics

We consider an operator $\hat{A}$ diagonal in the $\{\mathbf{x}\}$ representation. We define the mean value of the operator as

$$\overline{\hat{A}} \equiv \langle \alpha | \hat{A} | \alpha \rangle = \int d^3x \int d^3x' \langle \alpha | \mathbf{x} \rangle \langle \mathbf{x} | \hat{A} | \mathbf{x}' \rangle \langle \mathbf{x}' | \alpha \rangle = \int \varphi^* A \varphi d^3x \equiv \bar{A}. \quad (8.1)$$

Let us calculate the temporal variation of $\hat{A}$.

$$\frac{d}{dt} \overline{\hat{A}} = \int \varphi^* \frac{dA}{dt} \varphi d^3x + \int \left(\frac{d\varphi^*}{dt} A\varphi + \varphi^* A \frac{d\varphi}{dt} \right) d^3x. \quad (8.2)$$

The second integral can be simplified with the aid of the Schrödinger equation

$$\frac{\partial \varphi}{\partial t} = -\frac{i}{\hbar} H\varphi \quad (8.3)$$

and

$$\frac{\partial \varphi^*}{\partial t} = \frac{i}{\hbar} H^* \varphi^* = \frac{i}{\hbar} H\varphi^*. \quad (8.4)$$

We have used the hermiticity of H in the derivation of Eq.(8.4). Then we have

$$\begin{aligned}
\frac{d}{dt} \overline{\hat{A}} &= \int \varphi^* \frac{dA}{dt} \varphi d^3x + \frac{i}{\hbar} \int \varphi^* [H, A] \varphi d^3x \\
&= \frac{\overline{\partial \hat{A}}}{\partial t} + \frac{i}{\hbar} \overline{[\hat{H}, \hat{A}]}.
\end{aligned} \quad (8.5)$$

If we define the mean value of $\frac{d\hat{A}}{dt}$ as the temporal derivative of the mean value $\bar{\hat{A}}$

$$\overline{\frac{d\hat{A}}{dt}} \equiv \frac{d\overline{\hat{A}}}{dt}, \quad (8.6)$$

we have

$$\frac{d\hat{A}}{dt} = \frac{\partial \hat{A}}{\partial t} + \frac{i}{\hbar}[\hat{H}, \hat{A}], \tag{8.7}$$

which is called the equation of motion for operators in quantum mechanics.

8.1.1   *Ehrenfest's theorem*

We use a Hamiltonian of a particle in a potential $U(\mathbf{x})$. According to Eq.(8.7), the time derivatives of the position and momentum operators are given by

$$\frac{d\hat{\mathbf{x}}}{dt} = \frac{i}{\hbar}[\hat{H}, \hat{\mathbf{x}}], \tag{8.8a}$$

$$\frac{d\hat{\mathbf{p}}}{dt} = \frac{i}{\hbar}[\hat{H}, \hat{\mathbf{p}}]. \tag{8.8b}$$

We evaluate the commutators.

$$[\hat{H}, \hat{\mathbf{x}}] = \left[\frac{1}{2m}\hat{\mathbf{p}}^2, \hat{\mathbf{x}}\right] = \frac{\hbar}{i}\frac{\hat{\mathbf{p}}}{m}. \tag{8.9}$$

and

$$[\hat{H}, \hat{\mathbf{p}}] = [\hat{U}(\hat{\mathbf{x}}), \hat{\mathbf{p}}] = -\frac{\hbar}{i}\frac{\partial \hat{U}}{\partial \hat{\mathbf{x}}}. \tag{8.10}$$

Thus Eqs.(8.8) becomes

$$\frac{d\hat{\mathbf{x}}}{dt} = \frac{\hat{\mathbf{p}}}{m}, \tag{8.11a}$$

$$\frac{d\hat{\mathbf{p}}}{dt} = -\frac{\partial \hat{U}}{\partial \hat{\mathbf{x}}}. \tag{8.11b}$$

Taking the mean values of Eqs.(8.11), we have

$$\bar{\hat{\mathbf{p}}} = m\overline{\frac{d\hat{\mathbf{x}}}{dt}} \equiv m\bar{\mathbf{v}}, \tag{8.12a}$$

$$\frac{d\bar{\hat{\mathbf{p}}}}{dt} = -\overline{\frac{\partial \hat{U}}{\partial \hat{\mathbf{x}}}} \equiv \mathbf{F}. \tag{8.12b}$$

where $\mathbf{v} \equiv \bar{\mathbf{v}} = \overline{\frac{d\hat{\mathbf{x}}}{dt}}$ is called the velocity and $\mathbf{F} = -\overline{\frac{\partial \hat{U}}{\partial \hat{\mathbf{x}}}}$ is called the force. This is *Ehrenfest's theorem*. Since the mean values are equal to the most probable values in the classical limit, we could use both the mean values and the most probable values to describe the observables of the classical particles. Eq.(8.12) are the quantum version of the *Newton's equations* (*Newton's second law*) written as

$$\mathbf{p} = m\mathbf{v} = m\frac{d\mathbf{x}}{dt}, \tag{8.13a}$$

$$\frac{d\mathbf{p}}{dt} = \mathbf{F} = -\frac{\partial U(\mathbf{x})}{\partial \mathbf{x}}, \tag{8.13b}$$

where $\mathbf{x}$, $\mathbf{p}$ and $\mathbf{v}$ are the position, momentum and velocity vectors of a classical particle, respectively.

8.1.2 *Euler-Lagrange equation of quantum mechanics*

The Hamiltonian operator for one-particle systems is given by

$$\hat{H} = \frac{\hat{\mathbf{p}}^2}{2m} + V(\hat{\mathbf{x}}). \tag{8.14}$$

The momentum and position operators $\hat{\mathbf{p}}$ and $\hat{\mathbf{x}}$ obey the commutation relation Eq.(6.116)

$$[\hat{x}_i, \hat{p}_i] = i\hbar. \tag{8.15}$$

From commutation relation, we have

$$[\hat{x}_i, \hat{p}_i{}^n] = i\hbar n \hat{p}_i{}^{n-1}, \tag{8.16a}$$

$$[\hat{p}_i, \hat{x}_i{}^n] = -i\hbar n \hat{x}_i{}^{n-1}. \tag{8.16b}$$

Since $\hat{H}(\hat{\mathbf{p}}, \hat{\mathbf{x}})$ does not contain the time derivatives of $\hat{\mathbf{p}}$ and $\hat{\mathbf{x}}$, using Eq.(8.16), we have

$$[\hat{x}_i, \hat{H}(\hat{\mathbf{p}}, \hat{\mathbf{x}})] = i\hbar\frac{\partial \hat{H}}{\partial \hat{p}_i}, \tag{8.17a}$$

$$[\hat{p}_i, \hat{H}(\hat{\mathbf{p}}, \hat{\mathbf{x}})] = -i\hbar\frac{\partial \hat{H}}{\partial \hat{x}_i}. \tag{8.17b}$$

We introduce the Lagrangian function operator $\mathcal{L}$ defined as

$$\mathcal{L} \equiv \hat{\mathbf{p}} \cdot \frac{d\hat{\mathbf{x}}}{dt} - \hat{H}(\hat{\mathbf{p}}, \hat{\mathbf{x}}). \tag{8.18}$$

Comparing Eq.(8.17) with Eq.(8.8), we have

$$\frac{d\hat{\mathbf{x}}}{dt} = \frac{\partial \hat{H}}{\partial \hat{\mathbf{p}}} \tag{8.19a}$$

$$\frac{d\hat{\mathbf{p}}}{dt} = -\frac{\partial \hat{H}}{\partial \hat{\mathbf{x}}}. \tag{8.19b}$$

The total differential of the Lagrangian function operator as an operator function of $\hat{\mathbf{x}}$ and $\dot{\hat{\mathbf{x}}}$ is

$$d\mathcal{L} = \frac{\partial \mathcal{L}}{\partial \hat{\mathbf{x}}} \cdot d\hat{\mathbf{x}} + \frac{\partial \mathcal{L}}{\partial \dot{\hat{\mathbf{x}}}} \cdot d\dot{\hat{\mathbf{x}}}. \tag{8.20}$$

Using Eq.(8.18), we have

$$d\mathcal{L} = \hat{\mathbf{p}} \cdot d\dot{\hat{\mathbf{x}}} + \dot{\hat{\mathbf{x}}} \cdot d\hat{\mathbf{p}} - d\hat{H}$$
$$= \hat{\mathbf{p}} \cdot d\dot{\hat{\mathbf{x}}} + \dot{\hat{\mathbf{x}}} \cdot d\hat{\mathbf{p}} - \frac{\partial \hat{H}}{\partial \hat{\mathbf{x}}} \cdot d\hat{\mathbf{x}} - \frac{\partial \hat{H}}{\partial \hat{\mathbf{p}}} \cdot d\hat{\mathbf{p}}. \tag{8.21}$$

Inserting Eq.(8.19), we obtain

$$d\mathcal{L} = \hat{\mathbf{p}} \cdot d\dot{\hat{\mathbf{x}}} + \dot{\hat{\mathbf{x}}} \cdot d\hat{\mathbf{p}} + \dot{\hat{\mathbf{p}}} \cdot d\hat{\mathbf{x}} - \dot{\hat{\mathbf{x}}} \cdot d\hat{\mathbf{p}}$$
$$= \hat{\mathbf{p}} \cdot d\dot{\hat{\mathbf{x}}} + \dot{\hat{\mathbf{p}}} \cdot d\hat{\mathbf{x}}. \tag{8.22}$$

Comparing Eq.(8.20) with Eq.(8.22), we find

$$\dot{\hat{\mathbf{p}}} = \frac{\partial \mathcal{L}}{\partial \hat{\mathbf{x}}}, \quad \hat{\mathbf{p}} = \frac{\partial \mathcal{L}}{\partial \dot{\hat{\mathbf{x}}}}, \tag{8.23}$$

which leads to the Euler-Lagrange equation in operator form for quantum mechanics

$$\frac{d}{dt} \left(\frac{\partial \mathcal{L}}{\partial \dot{\hat{\mathbf{x}}}} \right) - \frac{\partial \mathcal{L}}{\partial \hat{\mathbf{x}}} = 0. \tag{8.24}$$

Therefore, similar to quantum field theory, quantum mechanics can also be expressed in the frame of the Euler-Lagrange equation. Both quantum mechanics and quantum field theory have the similar mathematical structures, which we call the canonical formalism.

8.1.3   *Constants of motion*

When an operator $\hat{A}$ commutes with the Hamiltonian operator and is not time dependent explicitly, we have

$$\frac{d\hat{A}}{dt} = \frac{\partial \hat{A}}{\partial t} + \frac{i}{\hbar}[\hat{H}, \hat{A}] = 0. \tag{8.25}$$

Thus $\hat{A}$ is a constant of motion. The Hamiltonian operator $\hat{H}$ apparently commutes with itself. It is a constant of motion, which is the law of the conservation of energy. When $\frac{\partial U}{\partial x} = 0$, we have $\hat{\mathbf{p}} =$const.. In the classical limit, it is *Newton's first law*.

8.1.4 *Conservation of angular momentum*

For a central potential, the potential is only a function of the radius r. There is a constant of motion related with the *angular momentum operator* defined as

$$\hat{\mathbf{L}} \equiv \hat{\mathbf{x}} \times \hat{\mathbf{p}} = -i\hbar \hat{\mathbf{x}} \times \boldsymbol{\nabla}. \tag{8.26}$$

In the Cartesian coordinates, the components of $\hat{\mathbf{L}}$ read

$$\hat{L}_x = \hat{y}\hat{p}_z - \hat{z}\hat{p}_y = -i\hbar \left(\hat{y}\frac{\partial}{\partial \hat{z}} - \hat{z}\frac{\partial}{\partial \hat{y}} \right), \tag{8.27a}$$

$$\hat{L}_y = \hat{z}\hat{p}_x - \hat{x}\hat{p}_z = -i\hbar \left(\hat{z}\frac{\partial}{\partial \hat{x}} - \hat{x}\frac{\partial}{\partial \hat{z}} \right), \tag{8.27b}$$

$$\hat{L}_z = \hat{x}\hat{p}_y - \hat{y}\hat{p}_x = -i\hbar \left(\hat{x}\frac{\partial}{\partial \hat{y}} - \hat{y}\frac{\partial}{\partial \hat{x}} \right). \tag{8.27c}$$

Using Eq.(8.27), we obtain the following commutation relations of the angular momentum components by straightforward calculations

$$[\hat{L}_y, \hat{L}_z] = \hat{L}_y\hat{L}_z - \hat{L}_z\hat{L}_y = i\hbar\hat{L}_x, \tag{8.28a}$$

$$[\hat{L}_z, \hat{L}_x] = \hat{L}_z\hat{L}_x - \hat{L}_x\hat{L}_z = i\hbar\hat{L}_y, \tag{8.28b}$$

$$[\hat{L}_x, \hat{L}_y] = \hat{L}_x\hat{L}_y - \hat{L}_y\hat{L}_x = i\hbar\hat{L}_z. \tag{8.28c}$$

For example,

$$\begin{aligned}
[\hat{L}_x, \hat{L}_y] &= [\hat{y}\hat{p}_z - \hat{z}\hat{p}_y, \hat{z}\hat{p}_x - \hat{x}\hat{p}_z] \\
&= [\hat{y}\hat{p}_z, \hat{z}\hat{p}_x] + [\hat{z}\hat{p}_y, \hat{x}\hat{p}_z] \\
&= \hat{y}\hat{p}_x[\hat{p}_z, \hat{z}] + \hat{p}_y\hat{x}[\hat{z}, \hat{p}_z] \\
&= i\hbar(\hat{x}\hat{p}_y - \hat{y}\hat{p}_x) \\
&= i\hbar\hat{L}_z.
\end{aligned} \tag{8.29}$$

Eq.(8.28) can be written in a compact form

$$[\hat{L}_i, \hat{L}_j] = i\hbar\varepsilon^{ijk}\hat{L}_k, \tag{8.30}$$

where ε^{ijk} is the antisymmetric Levi-Civita symbol in three dimensions defined by Eq.(F.3) in Appendix F. The commutation relation Eq.(8.30) is equivalent with the operator relation

$$\hat{\mathbf{L}} \times \hat{\mathbf{L}} = i\hbar\hat{\mathbf{L}}. \tag{8.31}$$

For the spin operator, we have the similar commutation relations. Using the relation for Pauli's matrices

$$\sigma_i\sigma_j = \delta_{ij} + i\varepsilon^{ijk}\sigma_k, \tag{8.32}$$

we have

$$[\frac{\sigma_i}{2}, \frac{\sigma_j}{2}] = i\varepsilon^{ijk}\frac{\sigma_k}{2}. \tag{8.33}$$

Using Eq.(6.28), we find

$$[\hat{S}_i, \hat{S}_j] = i\hbar\varepsilon^{ijk}\hat{S}_k, \tag{8.34}$$

or

$$\hat{\mathbf{S}} \times \hat{\mathbf{S}} = i\hbar\hat{\mathbf{S}}. \tag{8.35}$$

The spin operator obeys the same commutation relations as the orbital angular momentum operator.

The square of angular momentum operator is given by

$$\hat{\mathbf{L}}^2 = \hat{L}_x^2 + \hat{L}_y^2 + \hat{L}_z^2. \tag{8.36}$$

$\hat{\mathbf{L}}^2$ commutes with all components of the angular momentum operator

$$[\hat{\mathbf{L}}^2, \hat{L}_x] = [\hat{\mathbf{L}}^2, \hat{L}_y] = [\hat{\mathbf{L}}^2, \hat{L}_z] = 0. \tag{8.37}$$

Eq.(8.37) can be verified by straightforward calculations. For example,

$$\begin{aligned}
[\hat{\mathbf{L}}^2, \hat{L}_x] &= [\hat{L}_x\hat{L}_x + \hat{L}_y\hat{L}_y + \hat{L}_z\hat{L}_z, \hat{L}_x] \\
&= \hat{L}_y[\hat{L}_y, \hat{L}_x] + [\hat{L}_y, \hat{L}_x]\hat{L}_y \\
&\quad + \hat{L}_z[\hat{L}_z, \hat{L}_x] + [\hat{L}_z, \hat{L}_x]\hat{L}_z \\
&= \hat{L}_y(-i\hbar\hat{L}_z) + (-i\hbar\hat{L}_z)\hat{L}_y \\
&\quad + \hat{L}_z(i\hbar\hat{L}_y) + (i\hbar\hat{L}_y)\hat{L}_z \\
&= 0.
\end{aligned} \tag{8.38}$$

In a system with spherical symmetry, it is convenient to write the angular momentum in the spherical coordinates

$$r^2 = x^2 + y^2 + z^2, \quad \cos\theta = \frac{z}{r}, \quad \tan\varphi = \frac{y}{x}. \tag{8.39}$$

or

$$x = r\sin\theta\cos\varphi, \quad y = r\sin\theta\sin\varphi, \quad z = r\cos\theta. \tag{8.40}$$

In the spherical coordinates, Eq.(8.27) becomes

$$L_x = i\hbar\left(\sin\varphi\frac{\partial}{\partial\theta} + \cot\theta\cos\varphi\frac{\partial}{\partial\varphi}\right), \tag{8.41a}$$

$$L_y = i\hbar\left(-\cos\varphi\frac{\partial}{\partial\theta} + \cot\theta\sin\varphi\frac{\partial}{\partial\varphi}\right), \tag{8.41b}$$

$$L_z = -i\hbar\frac{\partial}{\partial\varphi}. \tag{8.41c}$$

Inserting Eq.(8.41) into Eq.(8.36), we have

$$\mathbf{L}^2 = -\hbar^2 \left\{ \frac{1}{\sin\theta} \frac{\partial}{\partial\theta} \left(\sin\theta \frac{\partial}{\partial\theta} \right) + \frac{1}{\sin^2\theta} \frac{\partial^2}{\partial\varphi^2} \right\} \equiv -\hbar^2 \triangle_{\theta,\varphi}. \quad (8.42)$$

The operator $\hat{\mathbf{L}}^2$ commutes with $\hat{U}(\hat{r})$.

Since Hamiltonian operator with a central potential can be written as

$$\hat{H} = \hat{T}_r + \frac{\hat{L}^2}{2m\hat{r}^2} + \hat{U}(\hat{r}) \quad (8.43)$$

with

$$T_r = -\frac{\hbar^2}{2m} \frac{1}{r^2} \frac{\partial}{\partial r} \left(r^2 \frac{\partial}{\partial r} \right), \quad (8.44)$$

we have

$$[\hat{H}, \hat{L}^2] = 0. \quad (8.45)$$

This is the *law of conservation of angular momentum* (Kepler's second law). Because $[L^2, \hat{L}_z] = 0$ and thus $[\hat{H}, \hat{L}_z] = 0$, the z component of angular momentum is also conserved.

8.2 Elementary aspects of Schrödinger equation

We consider a system of N-particles with Hamiltonian operator given by

$$\hat{H} = \sum_{i=1}^{N} \left(\frac{\hat{\mathbf{p}}_i}{2m_i} + \hat{V}_i(\hat{\mathbf{x}}_i, t) \right) + \sum_{i \neq j} \hat{V}_{ij}(\hat{\mathbf{x}}_i, \hat{\mathbf{x}}_j), \quad (8.46)$$

where $\hat{V}_i(\hat{\mathbf{x}}_i, t)$ is the external potential, which is the one particle potential. $\hat{V}_{ij}(\hat{\mathbf{x}}_i, \hat{\mathbf{x}}_j)$ is the interaction potential between two particles i and j, which is the two particle potential, such as Coulomb potential.

The Schrödinger equation reads

$$i\hbar \frac{\partial\varphi}{\partial t} = H\varphi, \quad (8.47)$$

where φ is the wave function.

Let us derive the equation of continuity for the wave function. First we consider one particle case. The wave function is a function in three-dimensional space. We define $W \equiv \varphi\varphi^* = |\varphi|^2$. Since

$$\varphi_\alpha \varphi_\alpha^* = \langle \mathbf{x}|\alpha, t\rangle \langle \alpha, t|\mathbf{x}\rangle = \langle \mathbf{x}|\hat{a}_\alpha^\dagger(t)\hat{a}_\alpha(t)|\mathbf{x}\rangle = \langle \mathbf{x}|\hat{n}_\alpha(t)|\mathbf{x}\rangle, \quad (8.48)$$

the meaning of

$$W(\mathbf{x}; t)dV = \varphi\varphi^* dV \quad (8.49)$$

can be interpreted as the probability of the particle occurring in the volume element dV at the position $\mathbf{x}$ and time t for the state $|\alpha, t\rangle$.

For an N-particle system, the wave function is a function in $3N$ dimensional space, which is called the configuration space of the system. We denote an infinitesimal small volume element in the configuration space as dV

$$dV = dV_1 dV_2 \cdots dV_N = d^3 x_1 d^3 x_2 \cdots d^3 x_N. \qquad (8.50)$$

Then

$$W(\mathbf{x}_1, \mathbf{x}_2, \cdots, \mathbf{x}_N; t) dV = \varphi^* \varphi dV \qquad (8.51)$$

is the probability of the particle 1 occurring in the volume element dV_1 at $\mathbf{x}_1, \cdots$ and Nth particle occurring in dV_N at $\mathbf{x}_N$ at time t.

Integrating Eq.(8.51) with respect to the coordinates of all particles, excluding the particle k, we obtain

$$W(\mathbf{x}_k) dV_k = dV_k \int \varphi^* \varphi d\Omega_k, \qquad (8.52)$$

where $d\Omega_k$ is defined by $dV = dV_k d\Omega_k$. $W(\mathbf{x}_k) dV_k$ is the probability of a particle occurring in dV_k at $\mathbf{x}_k$.

Using the Schrödinger equation, we can obtain the equation of continuity for the probability W in the configuration space. Multiplying Eq.(8.47) by φ^* and then subtracting the corresponding complex-conjugated equation, we have

$$i\hbar \frac{\partial}{\partial t}(\varphi^* \varphi) = -\frac{\hbar^2}{2} \sum_{i=1}^{N} \frac{1}{m_i}(\varphi^* \nabla_i^2 \varphi - \varphi \nabla_i^2 \varphi^*). \qquad (8.53)$$

We define

$$\mathbf{j}_i \equiv \frac{i\hbar}{2m_i}(\varphi \nabla_i \varphi^* - \varphi^* \nabla_i^2 \varphi), \qquad (8.54)$$

which is called the current density of the particle i. Eq.(8.53) becomes

$$\frac{\partial W}{\partial t} + \sum_{i=1}^{N} \boldsymbol{\nabla} \cdot \mathbf{j}_i(\mathbf{x}_1, \mathbf{x}_2, \cdots, \mathbf{x}_N; t) = 0. \qquad (8.55)$$

Eq.(8.55) is the equation of continuity for the probability W. When we integrate Eq.(8.55), we have

$$\int \frac{\partial}{\partial t} W(\mathbf{x}_1, \mathbf{x}_2, \cdots, \mathbf{x}_N; t) d\Omega_i = \frac{\partial}{\partial t} W(\mathbf{x}_i, t). \qquad (8.56)$$

The second term in Eq.(8.55) becomes

$$\sum_{i'=1}^{N} \int \boldsymbol{\nabla}_{i'} \cdot \mathbf{j}_{i'} d\Omega_i = \int \boldsymbol{\nabla}_i \cdot \mathbf{j}_i d\Omega_i + \sum_{i' \neq i}^{N} \int \boldsymbol{\nabla}_{i'} \cdot \mathbf{j}_{i'} d\Omega_i. \qquad (8.57)$$

The integral $\int \boldsymbol{\nabla}_{i'} \cdot \mathbf{j}_{i'} d\Omega_i$ for $i' \neq i$ is zero because it can be transformed into surface integrals. Thus we obtain the equation of continuity for each particle

$$\frac{\partial W(\mathbf{x}_i, t)}{\partial t} + \boldsymbol{\nabla} \cdot \mathbf{j}_i(\mathbf{x}_i, t) = 0. \qquad (8.58)$$

8.3 Newton's law

The total momentum operator $\hat{\mathbf{p}}$ of the N-particle system is given by

$$\hat{\mathbf{p}} = \sum_{i=1}^{N} \hat{\mathbf{p}}_i = -i\hbar \sum_{i=1}^{N} \hat{\boldsymbol{\nabla}}_i. \qquad (8.59)$$

Let us consider the time derivative of the momentum operator $\hat{\mathbf{p}}$.

$$\frac{d\hat{\mathbf{p}}}{dt} = \frac{i}{\hbar}(\hat{H}\hat{\mathbf{p}} - \hat{\mathbf{p}}\hat{H}). \qquad (8.60)$$

Inserting $\hat{H}$ in Eq.(8.46) into Eq.(8.60), we have

$$\begin{aligned}
\frac{d\hat{\mathbf{p}}}{dt} = \Big[\Big(\sum_{k=1}^{N} \hat{V}_k + \sum_{k \neq j}^{N} \hat{V}_{kj} \Big) \Big(\sum_{i=1}^{N} \hat{\boldsymbol{\nabla}}_i \Big) \\
- \Big(\sum_{i=1}^{N} \hat{\boldsymbol{\nabla}}_i \Big) \Big(\sum_{k=1}^{N} \hat{V}_k + \sum_{k \neq j}^{N} \hat{V}_{kj} \Big) \Big].
\end{aligned} \qquad (8.61)$$

We use the following formula to simplify Eq.(8.61),

$$\hat{V}_k \Big(\sum_{i=1}^{N} \hat{\boldsymbol{\nabla}}_i \Big) - \Big(\sum_{i=1}^{N} \hat{\boldsymbol{\nabla}}_i \Big) \hat{V}_k = -\hat{\boldsymbol{\nabla}}_k \hat{V}_k(\hat{\mathbf{x}}). \qquad (8.62)$$

When V_{kj} is only a function of the distance between particles, we have

$$\hat{\boldsymbol{\nabla}}_k \hat{V}_{kj} = \frac{d\hat{V}_{kj}}{d\hat{r}_{kj}} \hat{\boldsymbol{\nabla}}_k \hat{r}_{kj} = \frac{d\hat{V}_{kj}}{d\hat{r}_{kj}} \frac{\hat{\mathbf{r}}_{kj}}{\hat{r}_{kj}}, \qquad (8.63a)$$

$$\hat{\boldsymbol{\nabla}}_j \hat{V}_{kj} = \frac{d\hat{V}_{kj}}{d\hat{r}_{kj}} \hat{\boldsymbol{\nabla}}_j \hat{r}_{kj} = -\frac{d\hat{V}_{kj}}{d\hat{r}_{kj}} \frac{\hat{\mathbf{r}}_{kj}}{\hat{r}_{kj}}, \qquad (8.63b)$$

which gives

$$\hat{\boldsymbol{\nabla}}_k \hat{V}_{kj} = -\hat{\boldsymbol{\nabla}}_j \hat{V}_{kj}. \qquad (8.64)$$

We define

$$\mathbf{F}_{kj} \equiv \overline{\hat{\boldsymbol{\nabla}}_k \hat{V}_{kj}}. \tag{8.65}$$

$\mathbf{F}_{kj}$ is called the *force* exerted by the particle j at $\mathbf{x}_j$ on the particle k at $\mathbf{x}_k$. Then Eq.(8.64) becomes

$$\mathbf{F}_{kj} = -\mathbf{F}_{jk}. \tag{8.66}$$

Eq.(8.66) is the so-called *Newton's third law* of classical mechanics, which states that the action is equal to the minus reaction.

Using Eqs.(8.62) and (8.64), we find

$$\frac{d\hat{\mathbf{p}}}{dt} = -\sum_{i=1}^{N} \hat{\boldsymbol{\nabla}}_i \hat{V}_i(\hat{\mathbf{x}}_i, t) \equiv \sum_{i=1}^{N} \hat{F}_i \tag{8.67}$$

with

$$\hat{F}_i = -\hat{\boldsymbol{\nabla}}_i \hat{V}_i(\hat{\mathbf{x}}_i, t). \tag{8.68}$$

Using Eq.(8.12), we have in the classical limit

$$\frac{d\mathbf{p}}{dt} = \sum_i m_i \frac{d\mathbf{v}_i}{dt} = \sum_i \mathbf{F}_i, \tag{8.69}$$

which is Newton's second law for the whole system. When $V_i = 0$, we have

$$\frac{d\hat{\mathbf{p}}}{dt} = 0, \tag{8.70}$$

which is the *law of momentum conservation*. If we consider $\hat{\mathbf{p}}_i$, we have

$$\frac{d\hat{\mathbf{p}}_i}{dt} = -\sum_{k=1}^{N} \hat{\boldsymbol{\nabla}}_i \hat{V}_i(\hat{\mathbf{x}}_i, t) - \sum_{j \neq i}^{N} \hat{\boldsymbol{\nabla}}_i \hat{V}_{ij}(\hat{\mathbf{r}}_{ij}). \tag{8.71}$$

Using Eq.(8.12), we have in the classical limit

$$m_i \frac{d\mathbf{v}_i}{dt} = -\sum_{k=1}^{N} \boldsymbol{\nabla}_i V_i(\mathbf{x}_i, t) - \sum_{j \neq i}^{N} \boldsymbol{\nabla}_i V_{ij}(\mathbf{r}_{ij}). \tag{8.72}$$

which is *Newton's second law* for the particle i. When there are no forces ($V_i = 0, V_{ij} = 0$), we get $\frac{d\mathbf{v}_i}{dt} = 0$. This is *Newton's first law*, which states that the velocity of particle remains constant if there is no force acting on it.

8.4 The theorem of mass center

For an object consisting of several particles, we have a simple theorem to describe the motion of the object as a whole disregarding the details of the interactions between the constituent particles. Eq.(8.69) can be rewritten as

$$\mathbf{F} \equiv \sum_i \mathbf{F}_i = \sum_i m_i \frac{d\mathbf{v}_i}{dt} = \frac{d^2(\sum_i m_i \mathbf{r}_i)}{dt^2}, \tag{8.73}$$

where $\mathbf{F}$ is the total force acting on the object. We introduce a vector $\mathbf{R}$ defined by

$$\mathbf{R} = \frac{1}{M} \sum_i m_i \mathbf{r}_i, \tag{8.74}$$

where

$$M = \sum_i m_i \tag{8.75}$$

is the total mass of the object. Then Eq.(8.73) takes the simple form.

$$\mathbf{F} = \frac{d^2 M \mathbf{R}}{dt^2} = M \frac{d^2 \mathbf{R}}{dt^2}. \tag{8.76}$$

Thus the external force is the total mass times the acceleration of a mean point located at $\mathbf{R}$. $\mathbf{R}$ is called the mass center of the body, which is a kind of average of $\mathbf{r}_i$ in which different $\mathbf{r}_i$ have weights proportional to the mass m_i of the particle at $\mathbf{r}_i$. Eq.(8.76) is the theorem of mass center. The theorem of mass center has an important implication. It shows that if Newton's law is applicable to the constituent particles, it would also applicable to the larger objects consisting of these particles. Thus Newton's law $\mathbf{F} = m\mathbf{a}$ is a law reproducing itself on a larger scale. When the classical Newton's law is applicable in a certain scale, it can be applied in a larger and larger scale, and becomes more and more accurate. In this way, we can divide the scale into macroscopic scale in which the classical mechanics is a good approximation and the microscopic scale in which quantum field theory and quantum mechanics should be used.

8.5 Lorentz force

We consider the case of a particle in the electromagnetic field. The corresponding Hamiltonian operator of a particle in quantum mechanics for

Eq.(6.56) is given by

$$\hat{H} = \frac{\hat{\mathbf{p}}^2}{2m} - \frac{e}{2mc}(\hat{\mathbf{L}} + 2\hat{\mathbf{S}}) \cdot \hat{\mathbf{B}}_0 + e\hat{A}_0 \tag{8.77}$$

with $\hat{\mathbf{L}} = \hat{\mathbf{x}} \times \hat{\mathbf{p}}$. Using Eq.(8.8) and mathematical relation $\mathbf{a} \cdot (\mathbf{b} \times \mathbf{c}) = \mathbf{b} \cdot (\mathbf{c} \times \mathbf{a}) = \mathbf{c} \cdot (\mathbf{a} \times \mathbf{b})$, we have

$$\begin{aligned}
\frac{d\hat{\mathbf{x}}}{dt} &= \frac{i}{\hbar}[\hat{H}, \hat{\mathbf{x}}] \\
&= \frac{\hat{\mathbf{p}}}{m} + \frac{i}{\hbar}(-i\hbar)\frac{e}{2mc}(\hat{\mathbf{x}} \times \hat{\mathbf{B}}_0) \\
&= \frac{\hat{\mathbf{p}}}{m} + \frac{e}{2mc}(\hat{\mathbf{x}} \times \hat{\mathbf{B}}_0)
\end{aligned} \tag{8.78}$$

and

$$\begin{aligned}
\frac{d\hat{\mathbf{p}}}{dt} &= \frac{i}{\hbar}[\hat{H}, \hat{\mathbf{p}}] \\
&= \frac{e}{2mc}(\hat{\mathbf{p}} \times \hat{\mathbf{B}}_0) + e\hat{\mathbf{E}}.
\end{aligned} \tag{8.79}$$

Eq.(8.78) can be rewritten as

$$\hat{\mathbf{p}} = m\frac{d\hat{\mathbf{x}}}{dt} - \frac{e}{2c}(\hat{\mathbf{x}} \times \hat{\mathbf{B}}_0). \tag{8.80}$$

Inserting Eq.(8.80) into Eq.(8.79), we have

$$m\frac{d^2\hat{\mathbf{x}}}{dt^2} - \frac{e}{2c}\left(\frac{d\hat{\mathbf{x}}}{dt} \times \hat{\mathbf{B}}_0\right) = \frac{e}{2c}\left(\frac{d\hat{\mathbf{x}}}{dt} \times \hat{\mathbf{B}}_0\right) + e\hat{\mathbf{E}}. \tag{8.81}$$

Expressing in terms of $\hat{\mathbf{v}} = \frac{d\hat{\mathbf{x}}}{dt}$, Eq.(8.81) becomes

$$\begin{aligned}
m\frac{d\hat{\mathbf{v}}}{dt} &= \frac{e}{c}(\hat{\mathbf{v}} \times \hat{\mathbf{B}}_0) + e\hat{\mathbf{E}} \\
&= \hat{\mathbf{f}}_L
\end{aligned} \tag{8.82}$$

with

$$\mathbf{f}_L \equiv e\mathbf{E} + \frac{e}{c}(\mathbf{v} \times \mathbf{B}_0). \tag{8.83}$$

$\mathbf{f}_L$ is called the *Lorentz force*, which is the force that an electromagnetic field exerts on a particle with a charge e.

8.6   Path integral formalism for quantum mechanics

8.6.1   *Feymann's path integral for one-particle systems*

The observable operator $\hat{A}$ is a function of momentum operator $\hat{p}$ and position operator $\hat{q}$. Let us describe the dynamics of a non-relativistic system in path integral formalism, as we did in the quantum field theory. First we consider the simplest case of a particle moving in a potential in one-dimensional space. The commutation relation of the momentum operator $\hat{p}$ and position operator $\hat{q}$ is given by

$$[\hat{q}, \hat{p}] = i\hbar. \tag{8.84}$$

The eigenstates of these operators span the Hilbert space. Their eigen equations are

$$\hat{q}|q\rangle = q|q\rangle, \tag{8.85a}$$
$$\hat{p}|p\rangle = p|p\rangle. \tag{8.85b}$$

The state vectors are normalized by

$$\langle q'|q\rangle = \delta(q' - q), \tag{8.86a}$$
$$\langle p'|p\rangle = \delta(p' - p) \tag{8.86b}$$

and obey the completeness relations

$$\int dq|q\rangle\langle q| = 1, \tag{8.87a}$$

$$\int dp|p\rangle\langle p| = 1. \tag{8.87b}$$

According to Eq.(6.114), $\hat{p} = -i\hbar \frac{d}{d\hat{q}}$. We apply $\hat{p}$ to the eigenstate $|p\rangle$ and then project it onto $\langle q|$. We obtain

$$\langle q|\hat{p}|p\rangle = p\langle q|p\rangle = -i\hbar\frac{d}{dq}\langle q|p\rangle. \tag{8.88}$$

Solving the differential equation, we have

$$\langle q|p\rangle = \frac{1}{(2\pi\hbar)^{\frac{1}{2}}}e^{\frac{i}{\hbar}pq}. \tag{8.89}$$

Thus the coordinate representation of the momentum eigenstate is a plane wave.

Using the relation of the wave function of particle $\varphi_\alpha(q,t)$ with the quantum state $|\alpha, t\rangle$

$$\varphi_\alpha(q,t) = \langle q|\alpha, t\rangle, \tag{8.90}$$

we can expand the quantum state $|\alpha, t\rangle$ of the system by

$$|\alpha, t\rangle = \int dq' \varphi_\alpha(q', t)|q'\rangle. \tag{8.91}$$

The wave function of one particle satisfies the Schrödinger equation in the non-relativistic limit.

$$i\hbar \frac{\partial}{\partial t} \varphi_\alpha(q, t) = H(p, q)\varphi_\alpha(q, t). \tag{8.92}$$

The formal solution of this equation is

$$\varphi_\alpha(q, t) = e^{-\frac{i}{\hbar}Ht}\varphi_\alpha(q, 0). \tag{8.93}$$

In Eq.(8.90), $|q\rangle$ forms a basis in the Hilbert space. Since $|q\rangle$ does not change with the time, $|q\rangle$ is considered as a rest basis in the Hilbert space. We can also define a time dependent basis $|q, t\rangle_b$ by

$$|q, t\rangle_b \equiv e^{\frac{i}{\hbar}\hat{H}t}|q\rangle. \tag{8.94}$$

$|q, t\rangle_b$ plays the role of a moving basis in the Hilbert space. For simplicity of notation, we usually omit the subscript 'b'. Since

$$\begin{aligned} |\alpha, t\rangle &= \int dq \varphi_\alpha(q, t)|q\rangle \\ &= \int dq e^{-\frac{i}{\hbar}Ht}\varphi_\alpha(q, 0)|q\rangle, \end{aligned} \tag{8.95}$$

we have

$$\begin{aligned} \varphi_\alpha(q, t) &= \langle q|\alpha, t\rangle \\ &= \int dq' \langle q|e^{-\frac{i}{\hbar}Ht}\varphi_\alpha(q', 0)|q'\rangle \\ &= \int dq' \langle q, t|\varphi_\alpha(q', 0)|q'\rangle \\ &= \langle q, t|\alpha, 0\rangle \\ &= \langle q, t|\alpha\rangle_H, \end{aligned} \tag{8.96}$$

where $|\alpha\rangle_H \equiv |\alpha, 0\rangle$ is called the Heisenberg state vector. Meanwhile $|\alpha\rangle_S \equiv |\alpha, t\rangle$ is called the Schrödinger state vector.

Since Eq.(8.94) is a unitary transformation, the orthonormality and completeness relations remain valid for the time-dependent states. We have $|q, t\rangle$

$$\begin{aligned} \langle q', t|q, t\rangle &= \langle q'|e^{-\frac{i}{\hbar}\hat{H}t}e^{\frac{i}{\hbar}\hat{H}t}|q\rangle \\ &= \langle q'|q\rangle \\ &= \delta(q' - q) \end{aligned} \tag{8.97}$$

and

$$\int dq|q,t\rangle\langle q,t| = \int dq e^{\frac{i}{\hbar}\hat{H}t}|q\rangle\langle q|e^{-\frac{i}{\hbar}\hat{H}t}$$

$$= e^{\frac{i}{\hbar}\hat{H}t}e^{-\frac{i}{\hbar}\hat{H}t} \tag{8.98}$$

$$= 1.$$

According to Eq.(8.8a), the time evolution of the coordinate operator is given by

$$\hat{q}(t) = e^{\frac{i}{\hbar}\hat{H}t}\hat{q}e^{-\frac{i}{\hbar}\hat{H}t}. \tag{8.99}$$

$|q,t\rangle = e^{\frac{i}{\hbar}\hat{H}t}|q\rangle$ is the eigenstates of $\hat{q}(t)$ because

$$\hat{q}(t)|q,t\rangle = e^{\frac{i}{\hbar}\hat{H}t}\hat{q}e^{-\frac{i}{\hbar}\hat{H}t}e^{\frac{i}{\hbar}\hat{H}t}|q\rangle$$

$$= e^{\frac{i}{\hbar}\hat{H}t}q|q\rangle \tag{8.100}$$

$$= q|q,t\rangle.$$

Now we consider the transition amplitude

$$\langle q',t'|q,t\rangle = \langle q'|e^{-\frac{i}{\hbar}\hat{H}(t'-t)}|q\rangle. \tag{8.101}$$

$\langle q',t'|q,t\rangle$ is also called the Feynman kernel in quantum mechanics, which is similar to the Feynman kernel in quantum field theory. The Feynman kernel contains all the information one can get by solving the Schrödinger equation Eq.(8.92). We can obtain the time development of the wave function at arbitrary t' by the integration

$$\varphi_\alpha(q',t') = \langle q',t'|\alpha\rangle_H$$

$$= \int dq\langle q',t'|q,t\rangle\langle q,t|\alpha\rangle_H \tag{8.102}$$

$$= \int dq\langle q',t'|q,t\rangle\varphi_\alpha(q,t).$$

In order to construct the path integral formalism, we divide the time interval (t,t') into many small slices with equal length.

$$t_n = t + n\epsilon \tag{8.103}$$

with

$$\epsilon = \frac{t'-t}{N}. \tag{8.104}$$

We insert a complete set of basis states $|q_n,t_n\rangle$ at each of the grid points t_n $(n=1,\cdots,N-1)$ in the Feynman kernel

$$\langle q',t'|q,t\rangle = \int dq_{N-1}\cdots\int dq_2\int dq_1$$

$$\langle q',t'|q_{N-1},t_{N-1}\rangle\cdots\langle q_2,t_2|q_1,t_1\rangle\langle q_1,t_1|q,t\rangle. \tag{8.105}$$

Using Eq.(8.94), each of the kernel elements under the integral can be written as

$$\langle q_{n+1}, t_{n+1} | q_n, t_n \rangle = \langle q_{n+1} | e^{-\frac{i}{\hbar} \hat{H}(\hat{p}, \hat{q}) \epsilon} | q_n \rangle. \tag{8.106}$$

This kernel element is also called the transfer matrix, which is denoted as $T(q_{n+1}, q_n)$. When ϵ is small, the time-evolution operator can be approximated by a Taylor expansion

$$\langle q_{n+1}, t_{n+1} | q_n, t_n \rangle = \langle q_{n+1} | \left[1 - \frac{i\epsilon}{\hbar} \hat{H}(\hat{p}, \hat{q}) \right] | q_n \rangle + \mathcal{O}(\epsilon^2). \tag{8.107}$$

Since the Hamiltonian depends on $\hat{p}$ and $\hat{q}$, we also insert a complete set of the momentum eigenstates

$$\langle q_{n+1} | \hat{H}(\hat{p}, \hat{q}) | q_n \rangle = \int dp_n \langle q_{n+1} | p_n \rangle \langle p_n | \hat{H}(\hat{p}, \hat{q}) | q_n \rangle. \tag{8.108}$$

The operators $\hat{p}$ and $\hat{q}$ can act to the left or to the right on their eigenstates, we have

$$\langle p_n | \hat{H}(\hat{p}, \hat{q}) | q_n \rangle = \langle p_n | q_n \rangle H(p_n, q_n). \tag{8.109}$$

One can also use a more symmetric prescription, the so-called Weyl's operator ordering. $\langle p_n | q_n \rangle H(p_n, q_n)$ in Eq.(8.109) can be replaced by $\langle p_n | q_n \rangle H(p_n, \frac{1}{2}(q_{n+1} + q_n))$. We will use the notation $H(p_n, \bar{q}_n)$ in the following so that we can choose $\bar{q}_n = q_n$ or $\bar{q}_n = \frac{1}{2}(q_{n+1} + q_n)$ in the derivations. Using Eq.(8.89), we have

$$\langle q_{n+1}, t_{n+1} | q_n, t_n \rangle = \int \frac{dp_n}{2\pi\hbar} \exp\left(\frac{i}{\hbar} p_n(q_{n+1} - q_n) \right)$$
$$\times \left[1 - \frac{i\epsilon}{\hbar} H(p_n, \bar{q}_n) \right] + \mathcal{O}(\epsilon^2). \tag{8.110}$$

Taking the limit $\epsilon \to 0$ or $N \to \infty$, we have

$$\langle q', t' | q, t \rangle = \lim_{N \to \infty} \int \prod_{n=1}^{N-1} dq_n \prod_{n=0}^{N-1} \frac{dp_n}{2\pi\hbar} \exp\left(\frac{i\epsilon}{\hbar} p_n \frac{q_{n+1} - q_n}{\epsilon} \right)$$
$$\prod_{n=0}^{N-1} \left[1 - \frac{i\epsilon}{\hbar} H(p_n, \bar{q}_n) \right]. \tag{8.111}$$

We can rewrite Eq.(8.111) using the representation of the exponential function

$$\prod_{n=0}^{N-1} \left(1 + \frac{x_n}{N} \right) = \exp\left(\lim_{N \to \infty} \frac{1}{N} \sum_{n=0}^{N-1} x_n \right). \tag{8.112}$$

Then Eq.(8.111) becomes

$$\langle q', t'|q, t\rangle = \lim_{N \to \infty} \int \prod_{n=1}^{N-1} dq_n \prod_{n=0}^{N-1} \frac{dp_n}{2\pi\hbar}$$
$$\exp\left(\frac{i\epsilon}{\hbar} \sum_{n=0}^{N-1} \left[p_n \frac{q_{n+1} - q_n}{\epsilon} - H(p_n, \bar{q}_n)\right]\right). \quad (8.113)$$

In the limit $N \to \infty$, the sample values become continuous. The integral becomes the functional integral, which is also called the path integral physically. We introduce the notation of path integral.

$$\int \prod_{n=1}^{N-1} dq_n \to \int Dq \quad \text{and} \quad \int \prod_{n=0}^{N-1} dp_n \to \int Dp. \quad (8.114)$$

In the limit $\epsilon \to 0$,

$$\frac{q_{n+1} - q_n}{\epsilon} \to \dot{q}(t_n), \quad \epsilon \sum_{n=0}^{N-1} f(t_n) \to \int_t^{t'} d\tau f(\tau). \quad (8.115)$$

Then we obtain the path integral expression for the Feynman kernel

$$\langle q', t'|q, t\rangle = \int Dq Dp \frac{1}{2\pi\hbar} \exp\left(\frac{i}{\hbar} \int_t^{t'} d\tau [p\dot{q} - H(p, q)]\right). \quad (8.116)$$

The path integral is over all function $p(t)$ in the momentum space and $q(t)$ in the position space with the boundary conditions

$$q(t) = q \quad \text{and} \quad q(t') = q'. \quad (8.117)$$

8.6.2 *Lagrangian function in quantum mechanics*

The Hamiltonian of an one-dimensional non-relativistic system with one particle has the following standard form

$$H(p, q) = \frac{1}{2m}p^2 + V(q). \quad (8.118)$$

The first term is called the kinetic energy term and the second term is called the potential term. Since the dependence on p is quadratic form, the integration over p can be carried out explicitly. We use the Gaussian integral formula

$$\int_{-\infty}^{\infty} dx e^{-\frac{1}{2}ax^2 + Jx} = \left(\frac{2\pi}{a}\right)^{\frac{1}{2}} e^{\frac{J^2}{2a}}. \quad (8.119)$$

The path integral expression for the Feynman kernel Eq.(8.116) becomes

$$
\begin{aligned}
\langle q', t' | q, t \rangle &= \int DqDp \frac{1}{2\pi\hbar} \exp\left(\frac{i}{\hbar} \int_t^{t'} d\tau \left[p\dot{q} - \frac{1}{2m}p^2 - V(q) \right]\right) \\
&= \mathcal{N} \int Dq \exp\left(\frac{i}{\hbar} \int_t^{t'} d\tau \left[\frac{m}{2}\dot{q}^2 - V(q) \right]\right),
\end{aligned}
\tag{8.120}
$$

where $\mathcal{N}$ is the normalization constant. We introduce

$$
S \equiv \int_t^{t'} d\tau L(q, \dot{q})
\tag{8.121}
$$

with

$$
L(q, \dot{q}) = \frac{m}{2}\dot{q}^2 - V(q) = p\dot{q} - H(p, q).
\tag{8.122}
$$

S is called the *action functional* and L is the *Lagrangian function* in quantum mechanics. Eq.(8.120) becomes

$$
\langle q', t' | q, t \rangle = \mathcal{N} \int Dq \exp\left(\frac{i}{\hbar} S[q, \dot{q}]\right).
\tag{8.123}
$$

8.6.3   Hamilton's equations

For a classical system, the action is much larger than the Planck constant $\hbar$. We can use stationary phase approximation. The path integral is approximated with the extreme value of the integrand. The extreme condition of the action is given by

$$
\delta S = 0,
\tag{8.124}
$$

which leads to the Euler-Lagrange equation

$$
\frac{d}{dt}\frac{\partial L}{\partial \dot{q}} - \frac{\partial L}{\partial q} = 0.
\tag{8.125}
$$

Inserting the Lagrangian function in Eq.(8.122) into the Euler-Lagrange equation Eq.(8.125), we have

$$
m\ddot{q} = -\frac{\partial V}{\partial q},
\tag{8.126}
$$

which is the Newton's equation of motion. $\ddot{q}$ is called the acceleration of particle.

The momentum can be obtained by

$$
p = \frac{\partial L}{\partial \dot{q}}.
\tag{8.127}
$$

We use p as an independent variable to replace $\dot{q}$ in the Lagrangian and call it the canonically conjugate momentum here. Instead of the variables q and $\dot{q}$, we express the Lagrangian in terms of q and p. The Hamiltonian is then obtained from the Lagrangian as a Legendre's transformation

$$H(q, p) = p\dot{q}(p) - L(q, \dot{q}(p)). \tag{8.128}$$

The equation of motion (Euler-Lagrange equation) is equivalent to the following equations

$$\dot{p} = -\frac{\partial H}{\partial q}, \tag{8.129a}$$

$$\dot{q} = \frac{\partial H}{\partial p}. \tag{8.129b}$$

Eq.(8.129) is called Hamilton's equations and is equivalent to Newton's equation.

We introduce a notation for the following derivatives of two functions A and B.

$$\{A, B\}_{PB} = \frac{\partial A}{\partial q}\frac{\partial B}{\partial p} - \frac{\partial A}{\partial p}\frac{\partial B}{\partial q}, \tag{8.130}$$

which is called the Poisson bracket. With the help of the Poisson bracket, the time evolution of a physical quantity A can be evaluated with the following formula

$$\begin{aligned}
\frac{dA}{dt} &= \frac{\partial A}{\partial t} + \frac{\partial A}{\partial p}\dot{p} + \frac{\partial A}{\partial q}\dot{q} \\
&= \frac{\partial A}{\partial t} - \frac{\partial A}{\partial p}\frac{\partial H}{\partial q} + \frac{\partial A}{\partial q}\frac{\partial H}{\partial p} \\
&= \frac{\partial A}{\partial t} + \{A, H\}_{PB}.
\end{aligned} \tag{8.131}$$

In particular, when H does not depend on time explicitly, we have

$$\frac{dH}{dt} = \frac{\partial H}{\partial t} + \{H, H\}_{PB} = 0. \tag{8.132}$$

Thus H is a constant of motion for a classical system, which means that the energy is conserved.

8.6.4 *Path integral formalism for multi-particle systems*

We can easily extend the path integral formalism of one particle in one-dimensional space to general systems of multi-particles in three dimensional space. We use the simplified notations. $(\mathbf{q}_1, \mathbf{q}_2, \cdots, \mathbf{q}_N)$ is denoted as (q)

and $(\mathbf{p}_1, \mathbf{p}_2, \cdots, \mathbf{p}_N)$ as (p). The path integral for the transition amplitude (Feynman kernel) is given by

$$
\langle \mathbf{q}'_1, \mathbf{q}'_2, \cdots, \mathbf{q}'_N, t' | \mathbf{q}_1, \mathbf{q}_2, \cdots, \mathbf{q}_N, t \rangle
$$

$$
= \int \prod_{\alpha=1}^{N} D\mathbf{q}_\alpha \prod_{\beta=1}^{N} D\mathbf{p}_\beta \frac{1}{(2\pi\hbar)^3} \tag{8.133}
$$

$$
\times \exp\left\{ \frac{i}{\hbar} \int_t^{t'} d\tau \left[\sum_{\alpha=1}^{N} \mathbf{p}_\alpha \cdot \dot{\mathbf{q}}_\alpha - H(p, q) \right] \right\}.
$$

The integration is over all functions $p(t)$ in the momentum space and over $q(t)$ in the position space with the boundary conditions $q(t) = q$ and $q'(t) = q'$.

We use the general Hamiltonian form

$$
H(q, p) = \frac{1}{2} p^T M^{-1} p + V(q). \tag{8.134}
$$

The path integral Eq.(8.133) can be derived from the amplitude $\langle q_{l+1}, t_{l+1} | q_l, t_l \rangle$ in a similar way as for the one particle system.

$$
\langle q', t' | q, t \rangle
$$

$$
= \lim_{L \to \infty} \int \prod_{\alpha=1}^{N} \prod_{l=1}^{L-1} d^3 q_{\alpha l} \langle q', t' | q_{L-1}, t_{L-1} \rangle \cdots \langle q_2, t_2 | q_1, t_1 \rangle \langle q_1, t_1 | q, t \rangle. \tag{8.135}
$$

Inserting the completeness relation of the momentum eigenstates $|p_l\rangle$, similar to Eq.(8.110), we have

$$
\langle q'_{l+1}, t'_{l+1} | q_l, t_l \rangle
$$

$$
= \int \prod_{\alpha=1}^{N} \frac{dp_{\alpha l}}{(2\pi\hbar)^3} \exp\left(\frac{i\epsilon}{\hbar} [p_l^T \dot{q}_l - H(p_l, q_l)] \right) \tag{8.136}
$$

$$
= \int \prod_{\alpha=1}^{N} \frac{dp_{\alpha l}}{(2\pi\hbar)^3} \exp\left(\frac{i\epsilon}{\hbar} \left[-\frac{1}{2} p_l^T M^{-1} p_l + p_l^T \dot{q}_l - V(q_l) \right] \right),
$$

which leads to Eq.(8.133).

Using the Gaussian integration formula

$$
\int d^d x \exp\left(-\frac{1}{2} x^T A x + x^T y \right)
$$

$$
= (2\pi)^{\frac{d}{2}} \exp\left(-\frac{1}{2} \mathrm{Tr} \ln A \right) \exp\left(\frac{1}{2} y^T A^{-1} y \right), \tag{8.137}
$$

We obtain

$$\langle q'_{l+1}, t'_{l+1} | q_l, t_l \rangle$$

$$= (2\pi\hbar)^{-3}(2\pi)^{\frac{3}{2}} \exp\left[-\frac{1}{2}\text{Tr}\ln\left(\frac{i\epsilon}{\hbar}M^{-1}\right)\right]$$

$$\times \exp\left[\frac{1}{2}\left(\frac{i\epsilon}{\hbar}\dot{q}_l^T\right)\left(\frac{i\epsilon}{\hbar}M^{-1}\right)^{-1}\left(\frac{i\epsilon}{\hbar}\dot{q}_l\right)\right]\exp\left[\frac{i\epsilon}{\hbar}(-V(q_l))\right]$$

$$= (2\pi\hbar)^{-3}(2\pi)^{\frac{3}{2}} \exp\left[-\frac{1}{2}\text{Tr}\ln\left(\frac{i\epsilon}{\hbar}I\right)\right]\exp\left(-\frac{1}{2}\text{Tr}\ln M^{-1}\right) \tag{8.138}$$

$$\times \exp\left(\frac{1}{2}\frac{i\epsilon}{\hbar}\dot{q}_l^T M\dot{q}_l\right)\exp\left[\frac{i\epsilon}{\hbar}(-V(q_l))\right]$$

$$= (2\pi i\hbar\epsilon)^{-\frac{3}{2}} \exp\left(\frac{1}{2}\text{Tr}\ln M\right)\exp\left[\frac{i\epsilon}{\hbar}\left(\frac{1}{2}\dot{q}_l^T M\dot{q}_l - V(q_l)\right)\right].$$

We define the Lagrangian function by

$$L(q_l, \dot{q}_l) \equiv \frac{1}{2}\dot{q}_l^T M\dot{q}_l - V(q_l). \tag{8.139}$$

Then the Feynman kernel can be written as

$$\langle q', t' | q, t \rangle = \lim_{L\to\infty}(2\pi i\hbar\epsilon)^{-\frac{3LN}{2}} \int \prod_{\alpha=1}^{N}\prod_{l=1}^{L-1} d^3 q_{\alpha l} \exp\left(\frac{1}{2}\text{Tr}\ln M\right)$$

$$\times \exp\left[\frac{i\epsilon}{\hbar}\sum_{l=1}^{L-1}L(q_l, \dot{q}_l)\right]. \tag{8.140}$$

We absorb the extra factor $\exp\left(\frac{1}{2}\text{Tr}\ln M\right)$ into the normalization constant $\mathcal{N}$. The Feynman kernel becomes

$$\langle q', t' | q, t \rangle = \mathcal{N}\int Dq \exp\left[\frac{i}{\hbar}\int d\tau L(q, \dot{q})\right]. \tag{8.141}$$

8.7 Three representations

In the calculations of quantum mechanics, there are three types of formalisms. We call them Schrödinger, Heisenberg and interaction representations, which will be discussed in the following.

8.7.1 *Schrödinger representation*

The elementary equation of quantum mechanics is the Schrödinger equation

$$i\hbar\frac{\partial}{\partial t}\varphi_\alpha(t) = H\varphi_\alpha(t), \tag{8.142}$$

which has the formal solution

$$\varphi_\alpha(t) = e^{-\frac{i}{\hbar}Ht}\varphi_\alpha(0). \tag{8.143}$$

The operators $\hat{\mathbf{p}} = \frac{\hbar}{i}\nabla_{\hat{\mathbf{x}}}$ and $\hat{\mathbf{x}}$ do not contain the time t as an explicit variable. The average value of an operator $\hat{O}(\hat{\mathbf{p}}, \hat{\mathbf{x}})$ is given by

$$\begin{aligned}
\bar{\hat{O}} &= \langle \alpha | \hat{O}(\hat{\mathbf{p}}, \hat{\mathbf{x}}) | \alpha \rangle \\
&= \int \varphi_\alpha^* O\left(\frac{\hbar}{i}\nabla, \mathbf{x}\right) \varphi_\alpha d^3 x.
\end{aligned} \tag{8.144}$$

This is the formalism of the so-called *Schrödinger representation*. In the Schrödinger representation, we have time dependent wave function $\varphi_\alpha(t)$ and time independent operators.

8.7.2 Heisenberg representation

Since the experimental measured properties of quantum system are manifested in the average values of operators given by Eq.(8.144), we can transform the average value formula Eq.(8.144) into the following form

$$\begin{aligned}
\bar{\hat{O}} &= {}_S\langle \alpha, t | \hat{O}_S | \alpha, t \rangle_S \\
&= {}_S\langle \alpha, 0 | e^{\frac{i}{\hbar}\hat{H}t} \hat{O}_S(0) e^{-\frac{i}{\hbar}\hat{H}t} | \alpha, 0 \rangle_S.
\end{aligned} \tag{8.145}$$

We have used the subscript S to denote the Schrödinger representation. We now define a new representation of state vectors and operators

$$|\alpha\rangle_H \equiv |\alpha, 0\rangle_S \tag{8.146}$$

and

$$\hat{O}_H \equiv e^{\frac{i}{\hbar}\hat{H}t} \hat{O}_S(0) e^{-\frac{i}{\hbar}\hat{H}t}. \tag{8.147}$$

This representation denoted by the subscript H is called the *Heisenberg representation*. In the Heisenberg representation, we have the same formula to calculate the average value of operators

$$\begin{aligned}
\bar{\hat{O}} &= {}_S\langle \alpha | \hat{O}_S | \alpha \rangle_S \\
&= {}_S\langle \alpha, 0 | e^{\frac{i}{\hbar}\hat{H}t} \hat{O}_S(0) e^{-\frac{i}{\hbar}\hat{H}t} | \alpha, 0 \rangle_S \\
&= {}_H\langle \alpha | \hat{O}_H | \alpha \rangle_H.
\end{aligned} \tag{8.148}$$

Therefore, the Schrödinger representation formalism, in which the wave function $\varphi(t)$ is time dependent and operators are time independent, can be transformed into an equivalent formalism, in which the wave function is time independent and operators are time dependent.

In the Heisenberg representation, operators are time dependent. The equation of motion which has the solution of Eq.(8.147) can be easily found to have the form

$$i\hbar\frac{\partial}{\partial t}\hat{O}(t) = [\hat{O}(t), \hat{H}].\tag{8.149}$$

This equation has the same form as the Heisenberg's equations of motion in quantum field theory (for example Eq.(2.199)). In quantum field theory, the operators are the field operators. The Heisenberg's equations of motion are the physical time evolution equation for quantum fields. Here Eq.(8.149) is an equation defined to facilitate the calculations.

8.7.3 *Interaction representation*

Many-body Schrödinger equations are often difficult to be solved. The usual approach is the perturbation expansion, which can be dealt more easily in the interaction representation. We divide the Hamiltonian operator into two parts

$$\hat{H} = \hat{H}_0 + \hat{V},\tag{8.150}$$

where $\hat{H}_0$ is a Hamiltonian that its Schrödinger equations can be solved exactly. Generally $\hat{H}_0$ is taken to be the Hamiltonian without interactions. In some cases, $\hat{H}_0$ is chosen to be a solvable Hamiltonian including some specific interactions. The term $\hat{V}$ is the remaining parts of $\hat{H}$. The principle of choosing $\hat{H}_0$ and $\hat{V}$ is to make the effect of $\hat{V}$ small while maintaining $\hat{H}_0$ solvable.

In the interaction representation, both operators and state vectors are time dependent. They are defined by the following formulas:

$$\hat{O}_I(t) \equiv e^{\frac{i}{\hbar}\hat{H}_0 t}\hat{O}_S e^{-\frac{i}{\hbar}\hat{H}_0 t},\tag{8.151}$$

and

$$|\alpha, t\rangle_I \equiv e^{\frac{i}{\hbar}\hat{H}_0 t}e^{-\frac{i}{\hbar}\hat{H}t}|\alpha, 0\rangle_I$$

$$= e^{\frac{i}{\hbar}\hat{H}_0 t}e^{-\frac{i}{\hbar}\hat{H}t}|\alpha\rangle_H\tag{8.152}$$

$$= e^{\frac{i}{\hbar}\hat{H}_0 t}|\alpha, t\rangle_S,$$

where we have used subscript I to denote the interaction representation. In the interaction representation, the average value of the operator $\hat{O}$ is given by

$$\overline{\hat{O}} = {}_S\langle\alpha|\hat{O}_S|\alpha\rangle_S$$

$$= {}_S\langle\alpha|e^{-\frac{i}{\hbar}\hat{H}_0 t}\hat{O}_I e^{\frac{i}{\hbar}\hat{H}_0 t}|\alpha\rangle_S\tag{8.153}$$

$$= {}_I\langle\alpha|\hat{O}_I|\alpha\rangle_I.$$

We have the same formula in the interaction representation to calculate the average values of operators as in the other representations. The time dependence of the operators is governed by the unperturbed Hamiltonian $\hat{H}_0$

$$i\hbar\frac{\partial}{\partial t}\hat{O}_I(t) = [\hat{O}_I(t), \hat{H}_0]. \tag{8.154}$$

Differentiating Eq.(8.152), we have

$$
\begin{aligned}
\frac{\partial}{\partial t}|\alpha, t\rangle_I &= \frac{i}{\hbar}e^{\frac{i}{\hbar}\hat{H}_0 t}(\hat{H}_0 - \hat{H})e^{-\frac{i}{\hbar}\hat{H}t}|\alpha, 0\rangle_I \\
&= -\frac{i}{\hbar}e^{\frac{i}{\hbar}\hat{H}_0 t}\hat{V}e^{-\frac{i}{\hbar}\hat{H}t}|\alpha, 0\rangle_I \\
&= -\frac{i}{\hbar}e^{\frac{i}{\hbar}\hat{H}_0 t}\hat{V}e^{-\frac{i}{\hbar}\hat{H}_0 t}\left[e^{\frac{i}{\hbar}\hat{H}_0 t}e^{-\frac{i}{\hbar}\hat{H}t}|\alpha, 0\rangle_I\right] \\
&= -\frac{i}{\hbar}\hat{V}_I(t)|\alpha, t\rangle_I.
\end{aligned}
\tag{8.155}
$$

The time dependence of the state vectors is governed by the interaction $\hat{V}$.

We introduce an operator $\hat{U}_I(t)$ defined by

$$\hat{U}_I(t) \equiv e^{\frac{i}{\hbar}\hat{H}_0 t}e^{-\frac{i}{\hbar}\hat{H}t}, \tag{8.156}$$

which has the meaning of the operator of time translation transformation for states in the interaction representation. From Eq.(8.152), we have

$$|\alpha, t\rangle_I = \hat{U}_I(t)|\alpha, 0\rangle_I \tag{8.157}$$

with $\hat{U}_I(0) = 1$. The time derivative of $\hat{U}_I(t)$ reads

$$
\begin{aligned}
\frac{\partial}{\partial t}\hat{U}_I(t) &= \frac{i}{\hbar}e^{\frac{i}{\hbar}\hat{H}_0 t}(\hat{H}_0 - \hat{H})e^{-\frac{i}{\hbar}\hat{H}t} \\
&= -\frac{i}{\hbar}e^{\frac{i}{\hbar}\hat{H}_0 t}\hat{V}e^{-\frac{i}{\hbar}\hat{H}t} \\
&= -\frac{i}{\hbar}e^{\frac{i}{\hbar}\hat{H}_0 t}\hat{V}e^{-\frac{i}{\hbar}\hat{H}_0 t}\left[e^{\frac{i}{\hbar}\hat{H}_0 t}e^{-\frac{i}{\hbar}\hat{H}t}\right] \\
&= -\frac{i}{\hbar}\hat{V}_I(t)\hat{U}_I(t).
\end{aligned}
\tag{8.158}
$$

We can solve this equation in the following way. Integrating both sides of the equation, we obtain

$$\hat{U}_I(t) - \hat{U}_I(0) = -\frac{i}{\hbar}\int_0^t dt_1 \hat{V}_I(t)\hat{U}_I(t). \tag{8.159}$$

Since $\hat{U}_I(0) = 1$, we have

$$\hat{U}_I(t) = 1 - \frac{i}{\hbar}\int_0^t dt_1 \hat{V}_I(t)\hat{U}_I(t). \tag{8.160}$$

Repeating the iteration procedures, we obtain

$$
\begin{aligned}
\hat{U}_I(t) =& 1 - \frac{i}{\hbar} \int_0^t dt_1 \hat{V}_I(t_1) \\
& + \left(\frac{-i}{\hbar}\right)^2 \int_0^t dt_1 \int_0^{t_1} dt_2 \hat{V}_I(t_1)\hat{V}_I(t_2) + \cdots \\
=& \sum_{n=0}^{\infty} \left(\frac{-i}{\hbar}\right)^n \int_0^t dt_1 \int_0^{t_1} dt_2 \cdots \int_0^{t_{n-1}} dt_n \\
& \times \hat{V}_I(t_1)\hat{V}_I(t_2) \cdots \hat{V}_I(t_n).
\end{aligned}
\tag{8.161}
$$

Using the time-ordering operator T, we can change Eq.(8.161) into the following form

$$
\begin{aligned}
\hat{U}_I(t) =& \sum_{n=0}^{\infty} \frac{1}{n!} \left(\frac{-i}{\hbar}\right)^n \int_0^t dt_1 \int_0^t dt_2 \cdots \int_0^t dt_n \\
& T[\hat{V}_I(t_1)\hat{V}_I(t_2) \cdots \hat{V}_I(t_n)] \\
=& T \exp\left[-i \int_0^t dt_1 \hat{V}_I(t_1)\right].
\end{aligned}
\tag{8.162}
$$

Thus, the interaction representation formalism can be used to do perturbation calculations.

In terms of the operator $\hat{U}_I(t)$, we introduce another important operator $\hat{S}_I$ (often called S matrix)

$$
\hat{S}_I(t,t') \equiv \hat{U}_I(t)\hat{U}_I^\dagger(t').
\tag{8.163}
$$

We can also express the time evolution of the states in the interaction representation in terms of the S matrix.

$$
\begin{aligned}
|\alpha, t\rangle_I &= \hat{U}_I(t)|\alpha, 0\rangle_I \\
&= \hat{S}_I(t,t')\hat{U}_I(t')|\alpha, 0\rangle_I \\
&= \hat{S}_I(t,t')|\alpha, t'\rangle_I.
\end{aligned}
\tag{8.164}
$$

Thus the S matrix changes the wave function $\varphi_I(t)$ at time t into the wave function $\varphi_I(t')$ at time t'. One can easily check that $\hat{S}_I$ operator has the following properties:

$$
\hat{S}_I(t,t) = 1,
\tag{8.165a}
$$

$$
\hat{S}_I^\dagger(t,t') = \hat{U}_I(t')\hat{U}_I^\dagger(t) = \hat{S}_I(t',t),
\tag{8.165b}
$$

$$
\hat{S}_I(t,t')\hat{S}_I(t',t'') = \hat{S}_I(t,t'').
\tag{8.165c}
$$

8.8   De Broglie waves

Although we start from the constituent principle of identical particles, all the equations of motion are wave equations, we have the particle-wave duality naturally. When there is no interaction between particles, a particle state with momentum $\mathbf{p}$ is a plane wave state with the wave function given by

$$\varphi(\mathbf{x}, t) = A \exp\left[i(\mathbf{k} \cdot \mathbf{x} - \omega_{\mathbf{k}} t)\right]. \tag{8.166}$$

where A is an amplitude factor. $\mathbf{k}$ and $\omega_{\mathbf{k}}$ are the wave number and frequency respectively (as an example, see Eq.(2.400) for Dirac fermions). The momentum $\mathbf{p}$ is given by

$$\mathbf{p} = \hbar \mathbf{k} = \frac{h}{\lambda} \frac{\mathbf{k}}{|\mathbf{k}|}, \tag{8.167}$$

where $h = 2\pi\hbar$. Both h and $\hbar$ are called the Planck constant. $\hbar$ is also called the reduced Planck constant or Dirac constant. But we will not make a difference of their names and call both the Planck constant in this book. The energy of the state is given by

$$E = \hbar \omega_{\mathbf{k}}. \tag{8.168}$$

Eqs.(8.167) and (8.168) are called the *De Broglie relations*.

For massless photons,

$$E = \hbar \omega_{\mathbf{k}} = \hbar k c. \tag{8.169}$$

For massive particles,

$$E = \hbar \omega_{\mathbf{k}} = \sqrt{m^2 c^4 + \hbar^2 k^2 c^2} = \sqrt{m^2 c^4 + p^2 c^2}. \tag{8.170}$$

In the nonrelativistic limit, expanding Eq.(8.170) gives

$$E = mc^2 + \frac{p^2}{2m} + \cdots \doteq \frac{p^2}{2m} + mc^2. \tag{8.171}$$

When the constant factor mc^2 is subtracted, the energy becomes

$$E' = E - mc^2 = \frac{p^2}{2m}. \tag{8.172}$$

For a plane wave, we have two types of velocities. One is the phase velocity v_p defined by

$$v_p \equiv \frac{\omega_{\mathbf{k}}}{k} = \frac{E}{p} = \frac{\sqrt{m^2 c^4 + p^2 c^2}}{p}. \tag{8.173}$$

Another is the group velocity v_g defined by

$$v_g \equiv \frac{d\omega_{\mathbf{k}}}{dk} = \frac{dE}{dp} = \frac{pc^2}{\sqrt{m^2c^4 + p^2c^2}}. \tag{8.174}$$

The phase velocity is larger than the speed of light in vacuum, while the group velocity is smaller than the speed of light. A plane wave is not a local state. The plane wave state extends the whole space. When we talk about the motion of a particle, we actually talk about the energy transport and thus the particle should be in a localized state. Therefore, the phase velocity can not be assigned as a velocity of the particle. We can show that the group velocity corresponds to the velocity of the energy transport. A free particle in a localized state is described by a finite wave packet. For simplicity, we assign the direction of the wave vector as x direction. A group of waves propagating in the x direction can be described by the following localized wave packet state.

$$\varphi(x,t) = \int_{k_0 - \triangle k}^{k_0 + \triangle k} c(k) \exp\left[i(kx - \omega(k)t)\right]dk, \tag{8.175}$$

where k_0 is the mean wave vector of the group and $\triangle k$ measures the extension of the wave packet. We consider a localized wave packet with $\triangle k \ll k_0$. We can expand the frequency ω in a Taylor series around k_0.

$$\omega(k) = \omega(k_0) + \left(\frac{d\omega}{dk}\right)_{k=k_0} (k - k_0)$$
$$+ \frac{1}{2}\left(\frac{d^2\omega}{dk^2}\right)_{k=k_0} (k - k_0)^2 + \cdots. \tag{8.176}$$

Inserting the expansion Eq.(8.176) into Eq.(8.175) and taking $k' = k - k_0$ as the new integration variable, we obtain

$$\varphi(x,t) = \exp\left[i(k_0x - \omega(k_0)t)\right]$$
$$\times \int_{-\triangle k}^{\triangle k} \exp\left[i(x - v_g t)k'\right]C(k_0 + k')dk'. \tag{8.177}$$

Since $k' \ll k_0$, we have $C(k_0 + k') \approx C(k_0)$. Then Eq.(8.177) becomes

$$\varphi(x,t) = \exp\left[i(k_0x - \omega(k_0)t)\right]C(k_0) \int_{-\triangle k}^{\triangle k} \exp\left[i(x - v_g t)k'\right]dk'$$
$$= 2C(k_0)\frac{\sin[\triangle k(x - v_g t)]}{x - v_g t} \exp\left[i(k_0x - \omega(k_0)t)\right] \tag{8.178}$$
$$= \mathcal{C}(x,t) \exp\left[i(k_0x - \omega(k_0)t)\right]$$

with

$$C(x,t) = 2C(k)\frac{\sin[\triangle k(x - v_g t)]}{x - v_g t}.$$
(8.179)

$C(x,t)$ is the amplitude of the wave packet. $C(x,t)$ has its maximum at

$$v_g t - x = 0,$$
(8.180)

which means that the maximum of the amplitude moves with a velocity

$$v = \frac{dx}{dt} = v_g.$$
(8.181)

The velocity of the wave packet is the one that the amplitude maximum of the wave packet propagates with. Thus the velocity of a particle state is the group velocity, which can only be smaller than the speed of light in vacuum. For massless photons, its group velocity is given by

$$v_c = \frac{d\omega}{dk} = \frac{d(ck)}{dk} = c,$$
(8.182)

which is the reason we call c as the speed of light in vacuum. Any massless particles have the velocity of c. Since all massive particles have smaller group velocities than c, photons in vacuum have the maximum velocity in nature.

Since we can add any constant E_0 to $\omega(k)$ in Eq.(8.175) without changing the configuration of the wave packet state, we have a freedom to choose the energy zero in our calculations.

Eq.(8.174) can be rewritten as

$$p^2 c^4 = v_g^2 (m^2 c^4 + p^2 c^2)$$
(8.183)

or

$$p^2 c^2 (c^2 - v_g^2) = v_g^2 m^2 c^4.$$
(8.184)

Solving for $\mathbf{p}$ from Eq.(8.184), we find

$$\mathbf{p} = \frac{mc\mathbf{v}_g}{\sqrt{c^2 - v_g^2}} = \frac{m\mathbf{v}_g}{\sqrt{1 - \frac{v_g^2}{c^2}}},$$
(8.185)

which shows that

$$m' = \frac{m}{\sqrt{1 - \frac{v_g^2}{c^2}}}$$
(8.186)

can be considered as a mass for a moving particle with the velocity of $\mathbf{v}_g$. The mass is thus increased by a factor of $\frac{1}{\sqrt{1 - \frac{v_g^2}{c^2}}}$ for a moving particle.

8.9 Regions of past, present and future in spacetime

Since the massless photons have the largest velocity c, we can see that photons travel along the light cone as shown in Fig.8.1. Other particles have smaller velocity than the speed of light. They can only travel along a line inside the time-like region. Therefore, we can separate the spacetime into time-like regions and space-like regions as shown in Fig.8.1. A signal or event in the time-like region 1 would eventually reaches the point O at a speed smaller than the speed of light. Thus we call the time-like region 1 the region of past. A signal or event happened at the point O will reach the points in the time-like region 2. Thus we call the time-like region 2 the region of future. The event at the point O are affected by the events in the region of past and the events in the region of future are affected by the events at the point O. However, the events in the space-like region will not affect the events in at the point O because the interval between the point O and a point in the space-like region is space-like. The quantum version of this statement is proved in Section 5.1.2. The causality principle prohibits the simultaneous nonlocal interaction. Here we can see clearly the locality principle and the causality principle are correlated with each other. Suppose there is a nonlocal simultaneous interaction. Then an event at a point O with the position $\mathbf{x}_0$ and time t_0 would affect the events in the space-like region, which eventually affects the future of the event happened at the position $\mathbf{x}_0$ and time t_0. Then a person at O would somehow can affect the future and manipulate the event in the future and cause the paradox. Therefore, the causality principle needs the locality principle for escorting.

8.10 Statistical interpretation of wave functions

As for the meaning of the wave function $\varphi(\mathbf{x}, t)$, according to Eq.(8.48), $\varphi(\mathbf{x}, t)\varphi^*(\mathbf{x}, t) = \langle x|\hat{n}_\alpha|x \rangle$ is the probability that there is the particle at $\mathbf{x}$. This is the statistical interpretation given by Max Born. However, one may ask why there is only a probability of finding a particle at a position $\mathbf{x}$ instead of finding it definitely. A particle should be in one position at a time. If a particle is in one place, how does it manage to move to another place in distance simultaneously. This will give a nonlocal existence for a particle. The interpretation is that a particle can be in a state which is nonlocal. This state could be characterized by some guiding fields as

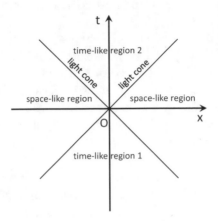

Fig. 8.1   The time-like and space-like regions in spacetime.

termed by Born. The guiding quantities could be energy, momentum, or spin. These quantities are generally conserved due to the symmetry, which makes the corresponding states stable. However, the particles themselves are local. We can only find a particle when the particle is at **x**. The particles are always created and annihilated everywhere with a probability. Due to the conservation of energy, the number of particles is conserved. For a single-particle state, although particles create and annihilate everywhere for all time. there is only one particle in total at any moment, which occurs in someplace. If we make a measurement, we could pick up the particle only when it is created. The probability of creating a particle at **x** is $|\varphi|^2$. Thus we have only a probability of $|\varphi|^2$ to pick up the particle at **x**. After we pick up the particle, the particle can not be annihilated back as usual. This is the irreversibility of the measurement, which leads to the quantum collapse to the state that the measurement selects. A path of a particle in a classical limit is only the path of the localized state guided by the equations of motion and conservation laws.

8.11   Heisenberg uncertainty principle

In the microscopic scale, we are not able to measure the exact position and momentum of a particle simultaneously, which is called the Heisenberg uncertainty principle. We will show in the following that the Heisenberg uncertainty principle is just a property of the position and momentum operators due to the particle-wave duality.

First let us make a qualitative analysis, we consider a one-dimensional wave packet. From Eq.(8.179), we can see that the distance $\triangle x$ of the first minimum of the wave packet amplitude at $x = x_m$ from the maximum at $x = 0$ is determined by the factor $\sin(\triangle k \cdot x_m) = \sin(\triangle k \cdot \triangle x)$. This distance can be characterized as the extension of the wave packet. We obtain

$$\triangle k \triangle x = \pi. \tag{8.187}$$

Using de Broglie relations, the momentum extension is determined by $\hbar \triangle k$. We have

$$\triangle p \triangle x \cong \pi \hbar. \tag{8.188}$$

This relation shows that the position and momentum of a particle state can not be determined exactly at the same time.

Now we derive the uncertainty principle. The average values of the momentum and position operators are given by

$$\bar{p}_x = \int \varphi^*(x) \left(-i\hbar \frac{\partial}{\partial x} \right) \varphi(x) dx, \tag{8.189a}$$

$$\bar{x} = \int \varphi^*(x) x \varphi(x) dx. \tag{8.189b}$$

The deviation from the average value is characterized by the mean-square deviations $\overline{(\triangle p_x)^2}$ and $\overline{(\triangle x)^2}$ defined by

$$\overline{(\triangle p_x)^2} \equiv \overline{(p_x - \bar{p}_x)^2} = \overline{p_x^2} - \bar{p}_x^2, \tag{8.190a}$$

$$\overline{(\triangle x)^2} \equiv \overline{(x - \bar{x})^2} = \overline{x^2} - \bar{x}^2. \tag{8.190b}$$

$\overline{p_x^2}$ and $\overline{x^2}$ can be evaluated by

$$\overline{p_x^2} = \int \varphi^*(x) \left(-\hbar^2 \frac{\partial^2}{\partial x^2} \right) \varphi(x) dx, \tag{8.191a}$$

$$\overline{x^2} = \int \varphi^*(x) x^2 \varphi(x) dx. \tag{8.191b}$$

To establish the connection between $\overline{\triangle p_x^2}$ and $\overline{\triangle x^2}$, we consider the integral

$$I(\alpha) = \int_{-\infty}^{\infty} \left| \alpha \triangle x \varphi(x) + \frac{i}{\hbar} \triangle p_x \varphi(x) \right|^2 dx \quad \alpha \in \mathbb{R}. \tag{8.192}$$

Since the integrand is positive, we have

$$I(\alpha) \geq 0. \tag{8.193}$$

Expanding the integrand, we have

$$I(\alpha) = \alpha^2 \int_{-\infty}^{\infty} \Delta x^2 |\varphi(x)|^2 dx$$
$$+ \alpha \int_{-\infty}^{\infty} \Delta x \left[\left(\frac{i}{\hbar} \Delta p_x \varphi^*(x) \right) \varphi(x) + \varphi^*(x) \left(\frac{i}{\hbar} \Delta p_x \varphi(x) \right) \right] dx$$
$$+ \int_{-\infty}^{\infty} \left(\frac{i}{\hbar} \Delta p_x \varphi^*(x) \right) \left(\frac{i}{\hbar} \Delta p_x \varphi(x) \right) dx.$$

$$(8.194)$$

The second term on the right hand side of Eq.(8.194) can be simplified using integration by parts.

$$\int_{-\infty}^{\infty} \Delta x \left[\left(\frac{i}{\hbar} \Delta p_x \varphi^*(x) \right) \varphi(x) + \varphi^*(x) \left(\frac{i}{\hbar} \Delta p_x \varphi(x) \right) \right] dx$$
$$= \int_{-\infty}^{\infty} \Delta x \left[\frac{d\varphi^*(x)}{dx} \varphi(x) + \varphi^*(x) \frac{d\varphi(x)}{dx} \right]$$
$$- 2 \int_{-\infty}^{\infty} \Delta x \left[\frac{i}{\hbar} \bar{p}_x \varphi^*(x) \varphi(x) \right] dx$$
$$= \int_{-\infty}^{\infty} \Delta x \frac{d}{dx} [\varphi^*(x) \varphi(x)] dx$$
$$= \Delta x \varphi^*(x) \varphi(x) \Big|_{-\infty}^{\infty} - \int_{-\infty}^{\infty} \varphi^*(x) \varphi(x) dx$$
$$= -1.$$

$$(8.195)$$

The third term can be evaluated as follows:

$$\int_{-\infty}^{\infty} \left(\frac{i}{\hbar} \Delta p_x \varphi^*(x) \right) \left(\frac{i}{\hbar} \Delta p_x \varphi(x) \right) dx$$
$$= \int_{-\infty}^{\infty} \frac{d\varphi^*(x)}{dx} \frac{d\varphi(x)}{dx} dx - \frac{1}{\hbar^2} \bar{p}_x^2$$
$$= \varphi^*(x) \frac{d\varphi(x)}{dx} \Big|_{-\infty}^{\infty} - \int_{-\infty}^{\infty} \varphi^*(x) \frac{d^2\varphi(x)}{dx^2} dx - \frac{1}{\hbar^2} \bar{p}_x^2 \qquad (8.196)$$
$$= \frac{1}{\hbar^2} \int_{-\infty}^{\infty} \varphi^*(x) \left(-\hbar^2 \frac{d^2}{dx^2} \right) \varphi(x) dx - \frac{1}{\hbar^2} \bar{p}_x^2$$
$$= \frac{1}{\hbar^2} \overline{(\Delta p_x)^2}.$$

Then we have

$$I(\alpha) = \overline{(\Delta x)^2} \alpha^2 - \alpha + \frac{1}{\hbar^2} \overline{(\Delta p_x)^2} \geq 0. \qquad (8.197)$$

α can be any real number. Eq.(8.197) shows that the minimum of $I(\alpha)$ should be larger than or equal to zero. The minimum $I(\alpha_m)$ of $I(\alpha)$ is given by

$$I(\alpha_m) = \frac{1}{\hbar^2}\overline{(\triangle p_x)^2} - \frac{1}{4\overline{(\triangle x)^2}} \geq 0, \tag{8.198}$$

which leads to

$$\overline{\triangle p_x^2 \triangle x^2} \geq \frac{\hbar^2}{4}. \tag{8.199}$$

Eq.(8.199) is called the *Heisenberg uncertainty relation* for momentum and position. The Heisenberg uncertainty principle is not limited to the position and momentum operators. We can derive the Heisenberg uncertainty relations for arbitrary observables.

Let us consider two hermitian operators $\hat{A}$ and $\hat{B}$. The commutator of the two operators has the form

$$[\hat{A}, \hat{B}] = i\hat{C}. \tag{8.200}$$

$\hat{C}$ is called the remainder of commutation (or commutation rest). When $\hat{A}$ and $\hat{B}$ commute, $\hat{C}$ is zero. It is easy to see that $\hat{C}$ is also a hermitian operator. The deviation of the operators from the mean values is defined by

$$\triangle\hat{A} \equiv \hat{A} - \overline{\hat{A}}, \tag{8.201a}$$

$$\triangle\hat{B} \equiv \hat{B} - \overline{\hat{B}}. \tag{8.201b}$$

It is easy to prove that $\triangle\hat{A}$ and $\triangle\hat{B}$ obey the same commutation relation as that for $\hat{A}$ and $\hat{B}$

$$[\triangle\hat{A}, \triangle\hat{B}] = i\hat{C}. \tag{8.202}$$

Similar to the discussions for $\hat{p}_x$ and $\hat{x}$, we define an integral

$$I(\alpha) \equiv \int |(\alpha\triangle A - i\triangle B)\varphi|^2 \, dx \geq 0 \quad \alpha \in \mathbb{R}. \tag{8.203}$$

We can evaluate $I(\alpha)$ in the following way

$$I(\alpha) = \int (\alpha\triangle A - i\triangle B)^*\varphi^*(\alpha\triangle A - i\triangle B)\varphi dx$$

$$= \int \varphi^*(\alpha\triangle A + i\triangle B)(\alpha\triangle A - i\triangle B)\varphi dx$$

$$= \int \varphi^* \left[\alpha^2(\triangle A)^2 + i\alpha(\triangle B\triangle A - \triangle A\triangle B) + (\triangle B)^2\right]\varphi dx \quad (8.204)$$

$$= \int \varphi^* \left[\alpha^2(\triangle A)^2 + \alpha C + (\triangle B)^2\right]\varphi dx$$

$$= \alpha^2\overline{(\triangle\hat{A})^2} + \alpha\overline{\hat{C}} + \overline{(\triangle\hat{B})^2}.$$

Since $I(\alpha) \geq 0$ for any real number, the minimum $I(\alpha_m)$ of $I(\alpha)$ should also be larger than or equal to zero.

$$I(\alpha_m) = \overline{(\Delta\hat{B})^2} - \frac{\overline{\hat{C}}^2}{4\overline{(\Delta\hat{A})^2}} \geq 0, \qquad (8.205)$$

which leads to

$$\overline{(\Delta\hat{A})^2}\,\overline{(\Delta\hat{B})^2} \geq \frac{\overline{\hat{C}}^2}{4}. \qquad (8.206)$$

Eq.(8.206) is the Heisenberg uncertainty principle in its most general form. In particular, according to the Schrödinger equation, $\hat{E} = i\hbar\frac{\partial}{\partial t}$. The energy operator $\hat{E}$ in quantum mechanics is equal to $i\hbar\frac{\partial}{\partial t}$. The commutation relation between energy and time operators is given by

$$[\hat{E}, \hat{t}] = i\hbar. \qquad (8.207)$$

Thus the uncertainty relation for energy and time reads

$$\overline{(\Delta\hat{E})^2}\,\overline{(\Delta\hat{t})^2} \geq \frac{\hbar^2}{4}. \qquad (8.208)$$

This relation means that a particle state with short life time will experience large energy change. This phenomenon is related to the conservation of energy. If one wants to transform a state with a definite energy (an eigenstate of Hamiltonian) to a state with a different energy, one needs apply external disturbance. Otherwise, it would not change. The change can be achieved either in a short time by applying a large external disturbance or in a long time by a small disturbance.

8.12   Stationary states

A state which does not change with the time in average is called the stationary state. We will show that it is the eigenstate of Hamiltonian.

When the Hamiltonian operator $\hat{H}$ is not time-dependent explicitly, we have

$$\frac{d\hat{H}}{dt} = [\hat{H}, \hat{H}] = 0. \qquad (8.209)$$

The energy is a constant of motion. In this case, we can separate the variables x and t of the time-dependent Schrödinger equation

$$i\hbar\frac{\partial}{\partial t}\varphi(\mathbf{x}, t) = H\varphi(\mathbf{x}, t) \qquad (8.210)$$

with the separated form of solutions

$$\varphi(\mathbf{x}, t) = \varphi(\mathbf{x}) f(t). \qquad (8.211)$$

We have

$$i\hbar \varphi(\mathbf{x}) \frac{\partial}{\partial t} f(t) = H\varphi(\mathbf{x}) f(t). \qquad (8.212)$$

After separating the variables, we have

$$i\hbar \frac{\dot{f}(t)}{f(t)} = \frac{H\varphi(\mathbf{x})}{\varphi(\mathbf{x})} = \text{const} = E, \qquad (8.213)$$

which gives the time dependent part as

$$f(t) = f_0 \exp\left(-i\frac{Et}{\hbar}\right). \qquad (8.214)$$

The function with the spatial argument obeys the stationary Schrödinger equation

$$H\varphi(\mathbf{x}) = E\varphi(\mathbf{x}). \qquad (8.215)$$

This equation is also called the Schrödinger equation and is used mostly because H does not depends on time explicitly in most applications. Mathematically, Eq.(8.215) is an eigenvalue equation of Hamiltonian. E is the energy eigenvalue, which is real because the Hamiltonian is a hermitian operator. Generally the eigenvalue equation Eq.(8.215) has a set of solutions $\varphi_n(\mathbf{x})$ characterized by n. n is called the *quantum number*. The energy eigenvalue E_n is also numbered using n. The solution $\varphi_n(\mathbf{x}, t)$ then has the form

$$\varphi_n(\mathbf{x}, t) = \varphi_n(\mathbf{x}) \exp\left(-i\frac{E_n t}{\hbar}\right), \qquad (8.216)$$

which is an oscillatory function in time, with the phase factor $\exp\left(-i\frac{E_n t}{\hbar}\right)$. We generally normalize the solutions with

$$\int \varphi_n(\mathbf{x}, t)^* \varphi_n(\mathbf{x}, t) dV = \int \varphi_n(\mathbf{x})^* \varphi_n(\mathbf{x}) dV = 1. \qquad (8.217)$$

This normalization condition means that a state contains one particle. It can be shown that $\varphi_n(\mathbf{x})$ are orthonormal, i.e. $\langle \varphi_n | \varphi_m \rangle = \delta_{nm}$. The general solution of the time-dependent Schrödinger equation is a superposition of all $\varphi_n(\mathbf{x}, t)$.

$$\varphi(\mathbf{x}, t) = \sum_n C_n(0) \varphi_n(\mathbf{x}) e^{-i\frac{E_n}{\hbar} t}$$

$$= \sum_n \left[\int \varphi(\mathbf{x}', 0) \varphi_n^*(\mathbf{x}') d^3 x' \right] \varphi_n(\mathbf{x}) e^{-i\frac{E_n}{\hbar} t}. \qquad (8.218)$$

Chapter 9

Applications of quantum mechanics

In this chapter, we will apply the methods of quantum mechanics to several simple physical systems. These systems are important for both understanding the quantum phenomena and applications of quantum mechanics.

9.1 Harmonic oscillator

9.1.1 *Classical solution*

We consider a one-dimensional system. When the interaction potential $V(x)$ has a local minimum at x_0, we can expand the potential $V(x)$ in a Taylor series about the minimum

$$
\begin{aligned}
V(x) &= V(x_0) + V'(x_0)(x - x_0) + \frac{1}{2}V''(x_0)(x - x_0)^2 + \cdots \\
&= V(x_0) + \frac{1}{2}k(x - x_0)^2 + \cdots
\end{aligned}
\tag{9.1}
$$

with

$$
k = V''(x_0).
\tag{9.2}
$$

When energy is small, we can neglected the higher order term. $V(x_0)$ is a constant term and can be dropped since it only affects the reference energy. Thus the potential in Eq.(9.1) can be written as

$$
V(x) = \frac{1}{2}m\omega^2 x^2.
\tag{9.3}
$$

with

$$
\omega = \sqrt{\frac{k}{m}},
\tag{9.4}
$$

where m is the mass of particle.

For a potential given by Eq.(9.3), Newton's equation is

$$m\frac{d^2x}{dt^2} = F = -\frac{\partial V(x)}{\partial x} = -kx. \tag{9.5}$$

Eq.(9.5) is also called *Hook's law*. The solution of Eq.(9.5) is

$$x(t) = A\sin(\omega t) + B\cos(\omega t), \tag{9.6}$$

which consists of the harmonic functions. This is the reason we call this system the *harmonic oscillator*.

9.1.2   *Hamiltonian operator in terms of $\hat{a}^\dagger$ and $\hat{a}$*

The Hamiltonian of a harmonic oscillator is given by

$$H = -\frac{\hbar^2}{2m}\nabla^2 + \frac{1}{2}m\omega^2 x^2. \tag{9.7}$$

The Schrödinger equation for the harmonic oscillator has the form

$$-\frac{\hbar^2}{2m}\frac{d^2\varphi}{dx^2} + \frac{1}{2}m\omega^2 x^2\varphi = E\varphi. \tag{9.8}$$

Instead of expressing the Hamiltonian operator in terms of the momentum operator $\hat{p}$ and position operator $\hat{x}$, we define two non-hermitian operators

$$\hat{a} \equiv \sqrt{\frac{m\omega}{2\hbar}}\left(\hat{x} + \frac{i\hat{p}}{m\omega}\right), \tag{9.9a}$$

$$\hat{a}^\dagger \equiv \sqrt{\frac{m\omega}{2\hbar}}\left(\hat{x} - \frac{i\hat{p}}{m\omega}\right). \tag{9.9b}$$

We will show that they have the same properties as the annihilation and creation operators we introduced previously in the quantum field theory. Historically, physicists first introduced the hermitian operators $\hat{p}$ and $\hat{x}$, and found that $\hat{p}$ and $\hat{x}$ satisfied the canonical commutation relation $[\hat{x}, \hat{p}] = i\hbar$. Then using Eq.(9.9) to introduce the annihilation and creation operators $\hat{a}$ and $\hat{a}^\dagger$. The second procedure is thus called the second quantization. In this book, we use the annihilation and creation operators to derive the canonical commutation relations in the quantum field theory. The procedure is reversed in the quantum field theory.

Using the commutation relation of $\hat{p}$ and $\hat{x}$, we have

$$[\hat{a}, \hat{a}^\dagger] = \frac{1}{2\hbar}(-i[\hat{x}, \hat{p}] + i[\hat{p}, \hat{x}]) = 1. \tag{9.10}$$

Eq.(9.10) shows that $\hat{a}$ and $\hat{a}^\dagger$ have the same commutation relation as the annihilation and creation operators. Similarly, we can also define the number operator

$$\hat{N} \equiv \hat{a}^\dagger \hat{a}. \tag{9.11}$$

Using the definition of $\hat{a}$ and $\hat{a}^\dagger$ in Eq.(9.9), we have

$$\begin{aligned}
\hat{a}^\dagger \hat{a} &= \frac{m\omega}{2\hbar}\left(\hat{x}^2 + \frac{\hat{p}^2}{m^2\omega^2}\right) + \frac{i}{2\hbar}[\hat{x},\hat{p}] \\
&= \frac{\hat{H}}{\hbar\omega} - \frac{1}{2}.
\end{aligned} \tag{9.12}$$

Thus we can express the Hamiltonian operator in terms of $\hat{a}$ and $\hat{a}^\dagger$ as

$$\hat{H} = \hbar\omega\left(\hat{a}^\dagger \hat{a} + \frac{1}{2}\right) = \hbar\omega\left(\hat{N} + \frac{1}{2}\right). \tag{9.13}$$

9.1.3 *Eigenvalues and eigenstates*

Since the Hamiltonian operator $\hat{H}$ commutes with $\hat{N}$, $\hat{H}$ and $\hat{N}$ can be diagonalized simultaneously. We introduce $|n\rangle$ to denote the eigenstate of $\hat{N}$ with the eigenvalue n.

$$\hat{N}|n\rangle = n|n\rangle. \tag{9.14}$$

Using Eq.(9.13), we have

$$\hat{H}|n\rangle = \hbar\omega\left(n + \frac{1}{2}\right)|n\rangle. \tag{9.15}$$

Thus the energy eigenvalues are given by

$$E_n = \hbar\omega\left(n + \frac{1}{2}\right). \tag{9.16}$$

Now we show that n is a nonnegative integer. First we note that

$$\begin{aligned}
\hat{N}\hat{a}^\dagger|n\rangle &= ([\hat{N},\hat{a}^\dagger] + \hat{a}^\dagger\hat{N})|n\rangle \\
&= (\hat{a}^\dagger[\hat{a},\hat{a}^\dagger] + \hat{a}^\dagger\hat{N})|n\rangle \\
&= (n+1)\hat{a}^\dagger|n\rangle
\end{aligned} \tag{9.17}$$

and

$$\begin{aligned}
\hat{N}\hat{a}|n\rangle &= ([\hat{N},\hat{a}] + \hat{a}\hat{N})|n\rangle \\
&= (\hat{a}^\dagger[\hat{a},\hat{a}] + [\hat{a}^\dagger,\hat{a}]\hat{a} + \hat{a}\hat{N})|n\rangle \\
&= (n-1)\hat{a}|n\rangle.
\end{aligned} \tag{9.18}$$

Thus $\hat{a}^\dagger|n\rangle$ and $\hat{a}|n\rangle$ are also the eigenstates of $\hat{N}$ with eigenvalue $n + 1$ (increased by one) and $n - 1$ (decreased by one), respectively. This is the reason we call $\hat{a}^\dagger$ and $\hat{a}$ the creation (or raising) operator and annihilation (or lowering) operator.

Since $\hat{a}|n\rangle$ is the eigenstate of $\hat{N}$ with eigenvalue $n - 1$, we have

$$\hat{a}|n\rangle = c|n - 1\rangle, \tag{9.19}$$

where c is a normalized constant, which can be determined by

$$\langle n|\hat{a}^\dagger\hat{a}|n\rangle = |c|^2\langle n - 1|n - 1\rangle = |c|^2. \tag{9.20}$$

Thus

$$n = |c|^2 \geq 0. \tag{9.21}$$

Eq.(9.21) shows that n is a nonnegative number. Therefore, we can write Eq.(9.19) in the following form

$$\hat{a}|n\rangle = e^{i\delta}\sqrt{n}|n - 1\rangle, \tag{9.22}$$

where δ is a phase parameter. Similarly, we have

$$\hat{a}^\dagger|n\rangle = e^{-i\delta}\sqrt{n + 1}|n + 1\rangle. \tag{9.23}$$

Applying the annihilation operator $\hat{a}$ to Eq.(9.22) consecutively, we have

$$\hat{a}^k|n\rangle = e^{ik\delta}\sqrt{n(n - 1)\cdots(n - k + 1)}|n - k\rangle. \tag{9.24}$$

If n is not an integer, we will have an eigenstate $|n'\rangle = |n - k\rangle$ with n' a negative number for $k > n$, which contradicts with Eq.(9.21) demanding that n' has to be a nonnegative number. Thus n can only be a nonnegative integer. When n is a nonnegative integer, the sequence $|n\rangle$ terminates at $n = 0$.

Since the smallest value of n is zero, the ground state $|0\rangle$ of the harmonic oscillator has the energy

$$E_0 = \frac{1}{2}\hbar\omega, \tag{9.25}$$

which is called the zero-point energy. Other eigenstates can be obtained by applying the creation operator $\hat{a}^\dagger$ successively

$$|n\rangle = \frac{(\hat{a}^\dagger)^n}{\sqrt{n!}}|0\rangle. \tag{9.26}$$

Thus $\hat{H}$ has the eigenstate $|n\rangle$ $(n = 0, 1, 2, \cdots)$ with the eigenvalue

$$E_n = \hbar\omega\left(n + \frac{1}{2}\right). \tag{9.27}$$

The orthonormality requires

$$\langle n'|\hat{a}|n\rangle = e^{i\delta}\sqrt{n}\langle n'|n-1\rangle$$
$$= e^{i\delta}\sqrt{n}\delta_{n',n-1} \tag{9.28}$$

and

$$\langle n'|\hat{a}^{\dagger}|n\rangle = e^{-i\delta}\sqrt{n+1}\langle n'|n+1\rangle$$
$$= e^{-i\delta}\sqrt{n+1}\delta_{n',n+1}. \tag{9.29}$$

Expressing $\hat{x}$ and $\hat{p}$ in terms of $\hat{a}$ and $\hat{a}^{\dagger}$, we obtain from Eq.(9.9)

$$\hat{x} = \sqrt{\frac{\hbar}{2m\omega}}\left(\hat{a}+\hat{a}^{\dagger}\right), \tag{9.30a}$$

$$\hat{p} = i\sqrt{\frac{\hbar m\omega}{2}}\left(-\hat{a}+\hat{a}^{\dagger}\right). \tag{9.30b}$$

Thus we have the matrix elements of the operators $\hat{x}$ and $\hat{p}$

$$\langle n'|\hat{x}|n\rangle = \sqrt{\frac{\hbar}{2m\omega}}\left(e^{i\delta}\sqrt{n}\delta_{n',n-1}+e^{-i\delta}\sqrt{n+1}\delta_{n',n+1}\right), \tag{9.31a}$$

$$\langle n'|\hat{p}|n\rangle = i\sqrt{\frac{\hbar m\omega}{2}}\left(-e^{i\delta}\sqrt{n}\delta_{n',n-1}+e^{-i\delta}\sqrt{n+1}\delta_{n',n+1}\right). \tag{9.31b}$$

9.1.4   *Wave functions*

Now let us derive the energy eigenstate in the x representation. For the ground state, we have

$$\hat{a}|0\rangle = 0. \tag{9.32}$$

Multiplying Eq.(9.32) with $\langle x|$, we have

$$\langle x|\hat{a}|0\rangle = \sqrt{\frac{m\omega}{2\hbar}}\langle x|\left(\hat{x}+\frac{i\hat{p}}{m\omega}\right)|0\rangle$$
$$= \sqrt{\frac{m\omega}{2\hbar}}\left(x+\frac{\hbar}{m\omega}\frac{d}{dx}\right)\langle x|0\rangle \tag{9.33}$$
$$= 0.$$

Eq.(9.33) is the differential equation for the wave function $\varphi_0(x) \equiv \langle x|0\rangle$ of the ground state, which has the following solution

$$\varphi_0(x) = \langle x|0\rangle = \frac{1}{\pi^{\frac{1}{4}}\sqrt{x_0}}\exp\left[-\frac{1}{2}\left(\frac{x}{x_0}\right)^2\right], \tag{9.34}$$

where

$$x_0 \equiv \sqrt{\frac{\hbar}{m\omega}}. \tag{9.35}$$

In the normalization of the wave function, we have taken the irrelevant phase parameter δ to be zero.

The wave functions of other states read

$$\begin{aligned}
\varphi_n(x) &= \langle x|n\rangle \\
&= \langle x| \left[\frac{(\hat{a}^\dagger)^n}{\sqrt{n!}} \right] |0\rangle \\
&= \frac{1}{\pi^{\frac{1}{4}}\sqrt{2^n n!}} \frac{1}{x_0^{n+\frac{1}{2}}} \left(x - x_0^2 \frac{d}{dx} \right)^n \exp\left[-\frac{1}{2}\left(\frac{x}{x_0} \right)^2 \right].
\end{aligned} \tag{9.36}$$

Using the Hermite polynomials defined by the Rodriguez formula

$$H_n(\xi) = (-1)^n e^{\xi^2} \left(\frac{d}{d\xi} \right)^n e^{-\xi^2}, \tag{9.37}$$

we can rewrite Eq.(9.36) as

$$\varphi_n(x) = \left(\frac{m\omega}{\pi\hbar} \right)^{\frac{1}{4}} \frac{1}{\sqrt{2^n n!}} H_n\left(\frac{x}{x_0} \right) e^{-\frac{1}{2}\left(\frac{x}{x_0} \right)^2}. \tag{9.38}$$

Eq.(9.38) can be derived from Eq.(9.36) using the following recursion relations for the Hermite polynomials

$$H_{n+1}(\xi) = 2\xi H_n(\xi) - 2n H_{n-1}(\xi) \tag{9.39}$$

and

$$\frac{dH_n}{d\xi} = 2n H_{n-1}(\xi). \tag{9.40}$$

9.2 Schrödinger equation for a central potential

9.2.1 *Schrödinger equation in spherical coordinates*

In order to treat the problem of central potential, we express the Schrödinger equation in the spherical coordinates. The Schrödinger equation for a particle in a central potential $V(r)$ reads

$$-\frac{\hbar^2}{2m}\nabla^2\psi + V(r)\psi = E\psi. \tag{9.41}$$

In Eq.(9.41), we have used ψ, instead of φ, to represent the wave function in order to avoid the confusion with the angular coordinate φ used in the spherical coordinates.

In the spherical coordinates, the Laplacian operator ∇^2 has the form

$$\nabla^2 = \frac{1}{r^2}\frac{\partial}{\partial r}\left(r^2\frac{\partial}{\partial r}\right) + \frac{1}{r^2\sin\theta}\frac{\partial}{\partial \theta}\left(\sin\theta\frac{\partial}{\partial \theta}\right) + \frac{1}{r^2\sin^2\theta}\frac{\partial^2}{\partial \varphi^2}. \quad (9.42)$$

The Schrödinger equation Eq.(9.41) becomes

$$-\frac{\hbar^2}{2m}\left[\frac{1}{r^2}\frac{\partial}{\partial r}\left(r^2\frac{\partial \psi}{\partial r}\right) + \frac{1}{r^2\sin\theta}\frac{\partial}{\partial \theta}\left(\sin\theta\frac{\partial \psi}{\partial \theta}\right) + \frac{1}{r^2\sin^2\theta}\frac{\partial^2 \psi}{\partial \varphi^2}\right]$$
$$+ V(r)\psi = E\psi. \quad (9.43)$$

9.2.2 *Separation of variables*

Using the following separation of variables

$$\psi(r,\theta,\varphi) = R(r)Y(\theta,\varphi), \quad (9.44)$$

we can separate Eq.(9.43) into a radial and an angular part. Inserting Eq.(9.44) into Eq.(9.43), we have

$$-\frac{\hbar^2}{2m}\left[\frac{Y}{r^2}\frac{d}{dr}\left(r^2\frac{dR}{dr}\right) + \frac{R}{r^2\sin\theta}\frac{\partial}{\partial \theta}\left(\sin\theta\frac{\partial Y}{\partial \theta}\right) + \frac{R}{r^2\sin^2\theta}\frac{\partial^2 Y}{\partial \varphi^2}\right]$$
$$+ V(r)RY = ERY. \quad (9.45)$$

Eq.(9.45) can be rewritten as

$$\left\{\frac{1}{R}\frac{d}{dr}\left(r^2\frac{dR}{dr}\right) - \frac{2mr^2}{\hbar^2}[V(r) - E]\right\}$$
$$+ \frac{1}{Y}\left\{\frac{1}{\sin\theta}\frac{\partial}{\partial \theta}\left(\sin\theta\frac{\partial Y}{\partial \theta}\right) + \frac{1}{\sin^2\theta}\frac{\partial^2 Y}{\partial \varphi^2}\right\} = 0. \quad (9.46)$$

The terms in the first curly bracket depend only on r and the remaining terms depend only on θ and φ. Thus, each must be a constant. We write the separation constant in the form of $l(l+1)$. Then Eq.(9.46) becomes two equations

$$\frac{1}{R}\frac{d}{dr}\left(r^2\frac{dR}{dr}\right) - \frac{2mr^2}{\hbar^2}[V(r) - E] = l(l+1), \quad (9.47a)$$

$$\frac{1}{Y}\left\{\frac{1}{\sin\theta}\frac{\partial}{\partial \theta}\left(\sin\theta\frac{\partial Y}{\partial \theta}\right) + \frac{1}{\sin^2\theta}\frac{\partial^2 Y}{\partial \varphi^2}\right\} = -l(l+1). \quad (9.47b)$$

Eq.(9.47b) for the eigenvalues and eigenstates of the angular part can be written as

$$-\hbar^2\left\{\frac{1}{\sin\theta}\frac{\partial}{\partial \theta}\left(\sin\theta\frac{\partial Y}{\partial \theta}\right) + \frac{1}{\sin^2\theta}\frac{\partial^2 Y}{\partial \varphi^2}\right\} = l(l+1)\hbar^2 Y \quad (9.48)$$

or

$$\mathbf{L}^2 Y = l(l+1)\hbar^2 Y. \quad (9.49)$$

Thus Y is the eigenstate of $\mathbf{L}^2$ and $l(l+1)\hbar^2$ is the eigenvalue of $\mathbf{L}^2$.

9.2.3   *Angular momentum operators*

Now we derive the eigenstates and eigenvalues of the angular momentum operator. For the applications to more generalized cases, we consider the total angular momentum operator

$$\hat{\mathbf{J}} = \hat{\mathbf{L}} + \hat{\mathbf{S}}. \tag{9.50}$$

$\hat{\mathbf{J}}^2$ is given by

$$\hat{\mathbf{J}}^2 = \hat{J}_x^2 + \hat{J}_y^2 + \hat{J}_z^2. \tag{9.51}$$

It commutes with each component $\hat{J}_i$ of $\hat{\mathbf{J}}$

$$[\hat{\mathbf{J}}^2, \hat{J}_i] = 0. \tag{9.52}$$

Thus $\hat{\mathbf{J}}^2$ and $\hat{J}_z$ can be diagonalized simultaneously. We denote the eigenvalues of $\hat{\mathbf{J}}^2$ and $\hat{J}_z$ by a and b, respectively. We have

$$\hat{\mathbf{J}}^2 |a, b\rangle = a|a, b\rangle, \tag{9.53a}$$

$$\hat{J}_z |a, b\rangle = b|a, b\rangle. \tag{9.53b}$$

We introduce two non-hermitian operators

$$\hat{J}_\pm = \hat{J}_x \pm i\hat{J}_y. \tag{9.54}$$

$\hat{J}_\pm$ are called the *ladder operators*. They satisfy the following commutation relations

$$[\hat{J}_+, \hat{J}_-] = 2\hbar \hat{J}_z, \tag{9.55a}$$

$$[\hat{J}_z, \hat{J}_\pm] = \pm\hbar \hat{J}_\pm, \tag{9.55b}$$

$$[\hat{\mathbf{J}}^2, \hat{J}_\pm] = 0. \tag{9.55c}$$

Using above commutation relations, we have

$$\begin{aligned}
\hat{J}_z(\hat{J}_\pm|a, b\rangle) &= ([\hat{J}_z, \hat{J}_\pm] + \hat{J}_\pm \hat{J}_z)|a, b\rangle \\
&= (b \pm \hbar)(\hat{J}_\pm|a, b\rangle).
\end{aligned} \tag{9.56}$$

Eq.(9.56) shows that $\hat{J}_\pm|a, b\rangle$ are the eigenstates of $\hat{J}_z$ with the eigenvalues $b \pm \hbar$. When we apply $\hat{J}_+(\hat{J}_-)$ to an eigenstate $|a, b\rangle$ of $\hat{J}_z$, we obtain an eigenstate of $\hat{J}_z$ with its eigenvalue increased (decreased) by one unit of $\hbar$. This is the reason why $\hat{J}_\pm$ are called the ladder operators.

Applying $\hat{\mathbf{J}}^2$ to $\hat{J}_\pm|a, b\rangle$, we have

$$\hat{\mathbf{J}}^2(\hat{J}_\pm|a, b\rangle) = \hat{J}_\pm \hat{\mathbf{J}}^2 |a, b\rangle = a\hat{J}_\pm|a, b\rangle. \tag{9.57}$$

Eq.(9.57) shows that $\hat{J}_\pm|a, b\rangle$ are also the eigenstates of $\hat{\mathbf{J}}^2$ with the eigenvalue a. Thus

$$\hat{J}_\pm|a, b\rangle = C_\pm|a, b \pm \hbar\rangle, \tag{9.58}$$

where $C_\pm$ are the normalization constant.

9.2.4   *Eigenvalues of* $\hat{\mathbf{J}}^2$ *and* $\hat{J}_z$

First we prove that the eigenvalue b of $\hat{J}_z$ has an upper limit for a given eigenvalue a of $\hat{\mathbf{J}}^2$. We use the following formula

$$\hat{\mathbf{J}}^2 - \hat{J}_z^2 = \frac{1}{2}(\hat{J}_+\hat{J}_- + \hat{J}_-\hat{J}_+)$$
$$= \frac{1}{2}(\hat{J}_+\hat{J}_+^\dagger + \hat{J}_+^\dagger\hat{J}_+). \tag{9.59}$$

Thus

$$\langle a,b|(\hat{\mathbf{J}}^2 - \hat{J}_z^2)|a,b\rangle = \frac{1}{2}\langle a,b|(\hat{J}_+\hat{J}_+^\dagger + \hat{J}_+^\dagger\hat{J}_+)|a,b\rangle$$
$$= \frac{1}{2}(|C^+|^2 + |C^-|^2) \geq 0, \tag{9.60}$$

which means

$$\langle a,b|(\hat{\mathbf{J}}^2 - \hat{J}_z^2)|a,b\rangle = a - b^2 \geq 0. \tag{9.61}$$

Eq.(9.61) is equivalent to

$$-a^{\frac{1}{2}} \leq b \leq a^{\frac{1}{2}}. \tag{9.62}$$

Therefore, b has an upper limit for a given a. When we apply $\hat{J}_+$ successively to $|a,b\rangle$, we could obtain the eigenstates of $\hat{\mathbf{J}}^2$ and $\hat{J}_z$ with increased eigenvalues of $\hat{J}_z$ until the upper limit of the eigenvalues of $\hat{J}_z$ is reached. We denote the maximum eigenvalue of $\hat{J}_z$ as b_{max}. Then

$$\hat{J}_+|a,b_{max}\rangle = 0. \tag{9.63}$$

Otherwise $\hat{J}_+|a,b_{max}\rangle$ would be the eigenstate of $\hat{J}_z$ with the eigenvalue $b_{max} + \hbar$, which contradicts with the statement that b_{max} is the maximum eigenvalue. Applying Eq.(9.63) with $\hat{J}_-$ gives

$$\hat{J}_-\hat{J}_+|a,b_{max}\rangle = 0. \tag{9.64}$$

$\hat{J}_-\hat{J}_+$ can be rewritten as

$$\hat{J}_-\hat{J}_+ = \hat{J}_x^2 + \hat{J}_y^2 - i(\hat{J}_y\hat{J}_x - \hat{J}_x\hat{J}_y)$$
$$= \hat{\mathbf{J}}^2 - \hat{J}_z^2 - \hbar\hat{J}_z. \tag{9.65}$$

Inserting Eq.(9.65) into Eq.(9.64), we have

$$(\hat{\mathbf{J}}^2 - \hat{J}_z^2 - \hbar\hat{J}_z)|a,b_{max}\rangle = (a - b_{max}^2 - \hbar b_{max})|a,b_{max}\rangle$$
$$= 0. \tag{9.66}$$

Since $|a,b_{max}\rangle$ is not a null state, we have

$$a - b_{max}^2 - \hbar b_{max} = 0. \tag{9.67}$$

Solving a gives

$$a = b_{max}(b_{max} + \hbar). \tag{9.68}$$

Eq.(9.62) also shows that there is a lower limit of the eigenvalues b. We denote the minimum value of b as b_{min}. Then

$$\hat{J}_-|a, b_{min}\rangle = 0. \tag{9.69}$$

Similarly, we have

$$\begin{aligned}
\hat{J}_+\hat{J}_-|a, b_{min}\rangle &= (\hat{\mathbf{J}}^2 - \hat{J}_z^2 + \hbar\hat{J}_z)|a, b_{min}\rangle \\
&= (a - b_{min}^2 + \hbar b_{min})|a, b_{min}\rangle \\
&= 0.
\end{aligned} \tag{9.70}$$

Then we obtain

$$a = b_{min}(b_{min} - \hbar). \tag{9.71}$$

Comparing Eq.(9.68) with Eq.(9.71) gives

$$b_{max} = -b_{min}. \tag{9.72}$$

The allowed values of b are limited within

$$-b_{max} \leq b \leq b_{max}. \tag{9.73}$$

Applying $\hat{J}_+$ successively to $|a, b_{min}\rangle$, we will be able to reach $|a, b_{max}\rangle$. Suppose that we obtain $|a, b_{max}\rangle$ after operating $\hat{J}_+$ n times, we have

$$b_{max} = b_{min} + n\hbar = -b_{max} + n\hbar, \tag{9.74}$$

which gives

$$b_{max} = \frac{n\hbar}{2}. \tag{9.75}$$

We introduce a quantum number j defined by

$$j \equiv \frac{n}{2}. \tag{9.76}$$

Since n is a nonnegative integer, j is either a nonnegative integer or a half-integer. Using Eq.(9.68), we have

$$a = \hbar^2 j(j + 1). \tag{9.77}$$

We also introduce $m \equiv b/\hbar$ as another quantum number. Thus

$$b = m\hbar. \tag{9.78}$$

m takes the following $2j + 1$ value for a given j.

$$m = 0, \pm 1, \pm 2, \cdots, \pm j. \tag{9.79}$$

We usually use $|j, m\rangle$ to denote the eigenstates of $\hat{\mathbf{J}}^2$ and $\hat{J}_z$ instead of $|a, b\rangle$. We have

$$\hat{\mathbf{J}}^2|j, m\rangle = j(j + 1)\hbar^2|j, m\rangle \tag{9.80}$$

and

$$\hat{J}_z|j, m\rangle = m\hbar|j, m\rangle. \tag{9.81}$$

Since the eigenvalues of $\hat{\mathbf{L}}^2$ is $l(l + 1)$, Eq.(9.49) is the usual form of the eigenvalue equation for $\hat{\mathbf{L}}^2$, which is the reason we use $l(l + 1)$ as the separation constant in Eq.(9.47).

9.2.5 *Matrix elements of angular momentum operators*

Now we evaluate the matrix elements of the angular momentum operators. From Eqs.(9.80) and (9.81), we have

$$\langle j', m'|\hat{\mathbf{J}}^2|j, m\rangle = j(j+1)\hbar^2 \delta_{j'j}\delta_{m'm} \tag{9.82}$$

and

$$\langle j', m'|\hat{J}_z|j, m\rangle = m\hbar\delta_{j'j}\delta_{m'm}. \tag{9.83}$$

The matrix elements of $\hat{J}_\pm$ can be determined using the following equation

$$\hat{J}_\pm|j, m\rangle = C^\pm_{jm}\hbar|j, m \pm 1\rangle. \tag{9.84}$$

Eq.(9.84) is just Eq.(9.58) rewritten in terms of j and m. Using the relation

$$\begin{aligned}\langle j, m|\hat{J}^\dagger_\pm \hat{J}_\pm|j, m\rangle &= \langle j, m|\hat{\mathbf{J}}^2 - \hat{J}^2_z \mp \hbar\hat{J}_z|j, m\rangle \\ &= [j(j+1) - m^2 \mp m]\hbar^2,\end{aligned} \tag{9.85}$$

we obtain

$$\begin{aligned}|C^\pm_{jm}|^2 &= [j(j+1) - m(m \pm 1)]\hbar^2 \\ &= (j \mp m)(j \pm m + 1)\hbar^2.\end{aligned} \tag{9.86}$$

Thus the matrix elements of $\hat{J}_\pm$ are given by

$$\langle j', m'|\hat{J}_\pm|j, m\rangle = \sqrt{(j \mp m)(j \pm m + 1)}\hbar\delta_{j'j}\delta_{m'm\pm1}. \tag{9.87}$$

9.2.6 *Spherical harmonics*

In Eq.(9.48), $Y(\theta, \varphi)$ is the wave function of the angular part in the spherical coordinate representation. We have shown that there are two quantum numbers l and m. For the spherical coordinate representation, we introduce the direction eigenstate $|\hat{\mathbf{n}}\rangle$. Then

$$\langle\hat{\mathbf{n}}|j, m\rangle = Y^m_l(\hat{\mathbf{n}}) = Y^m_l(\theta, \varphi). \tag{9.88}$$

Since $|j, m\rangle$ is the eigenstate of $\hat{J}_z$, we have

$$-i\hbar\frac{\partial}{\partial\varphi}Y^m_l(\theta, \varphi) = m\hbar Y^m_l(\theta, \varphi). \tag{9.89}$$

Thus

$$Y^m_l(\theta, \varphi) = e^{im\varphi}\phi^m_l(\theta). \tag{9.90}$$

In order to fulfill the requirement that the wave function is single valued, we impose

$$e^{im(\varphi+2\pi)} = e^{im\varphi}, \tag{9.91}$$

which demands m to be an integer. According to Eq.(9.79),

$$m = 0, \pm 1, \pm 2, \cdots, \pm l. \tag{9.92}$$

Thus l should be integer. To obtain the θ-dependence of $Y_l^m(\theta, \varphi)$, we start with the case of $m = l$. According to Eq.(9.63), we have

$$\hat{L}_+|l, l\rangle = 0 \tag{9.93}$$

or equivalently

$$
\begin{aligned}
-i\hbar e^{i\varphi} &\left(i\frac{\partial}{\partial\theta} - \cot\theta\frac{\partial}{\partial\varphi} \right) Y_l^l(\theta, \varphi) \\
&= -i\hbar e^{i\varphi} \left(i\frac{\partial}{\partial\theta} - \cot\theta\frac{\partial}{\partial\varphi} \right) e^{il\varphi}\phi_l^l(\theta) \\
&= 0.
\end{aligned}
\tag{9.94}
$$

The solution of Eq.(9.94) is given by

$$\phi_l^l(\theta) = c_l \sin^l \theta, \tag{9.95}$$

where c_l is the normalization constant. The normalization condition is

$$\int_0^{2\pi} d\varphi \int_{-1}^{1} d(\cos\theta) Y_l^{m*}(\theta, \varphi) Y_l^m(\theta, \varphi) = \delta_{ll'}\delta_{m'm}. \tag{9.96}$$

From the normalization condition, we can only determine the modula of c_l. There is an undetermined phase factor $e^{i\delta}$. Generally we take δ to be zero. The undetermined phase factor comes from the complex wave function of the Dirac fermions. We have shown in the chapter of quantum field theory that the Dirac fermions are composite in order to fulfill the causality and covariance principles. The doublet field operators are needed and thus the Dirac fermion field operators are complex. Although the doublet field operators are not independent, there is an constant phase factor undetermined because the composite can not be broken into two independent particles due to the causality and covariance principles. The phase factor can be important in some periodic systems and adiabatic evolution where the phase factor is called Berry phase or geometric phase.

Inserting Eq.(9.95) into Eq.(9.96), we have

$$c_l = \frac{(-1)^l}{2^l l!} \sqrt{\frac{(2l+1)(2l)!}{4\pi}}. \tag{9.97}$$

Starting from $|l, l\rangle$, we can apply $\hat{L}_-$ successively to $|l, l\rangle$ to obtain all other $|l, m\rangle$ with l fixed. We use the following formula

$$
e^{-i\varphi}\left(i\frac{\partial}{\partial\theta} + \cot\theta\frac{\partial}{\partial\varphi}\right)\left[f(\theta)e^{im\varphi}\right]
$$
$$
= -ie^{i(m-1)\varphi}\sin^{1-m}\theta\frac{d(f(\theta)\sin^m\theta)}{d(\cos\theta)}. \tag{9.98}
$$

We define the term on the right hand side of Eq.(9.98) as $f_1(\theta)e^{i(m-1)\varphi}$ and apply $e^{-i\varphi}(i\frac{\partial}{\partial\theta} + \cot\theta\frac{\partial}{\partial\varphi})$ repeatedly. Then we obtain

$$
\left[-i\hbar e^{-i\varphi}\left(i\frac{\partial}{\partial\theta} + \cot\theta\frac{\partial}{\partial\varphi}\right)\right]^{l-m}Y_l^l(\theta, \varphi)
$$
$$
= \left[-i\hbar e^{-i\varphi}\left(i\frac{\partial}{\partial\theta} + \cot\theta\frac{\partial}{\partial\varphi}\right)\right]^{l-m}(c_l e^{il\varphi}\sin^l\theta) \tag{9.99}
$$
$$
= c_l(-\hbar)^{l-m}e^{im\varphi}\sin^{-m}\theta\frac{d^{l-m}(\sin^{2l}\theta)}{(d\cos\theta)^{l-m}}.
$$

Using the relation

$$
\hat{L}_-|l, m+1\rangle = \sqrt{(l-m)(l+m+1)]}|l, m\rangle \tag{9.100}
$$

and Eq.(9.99), we obtain

$$
Y_l^m(\theta, \varphi) = e^{im\varphi}(-1)^l\sqrt{\frac{(2l+1)(l+m)!}{4\pi(l-m)!}}\frac{1}{2^l l!\sin^m\theta}\frac{d^{l-m}(\sin^{2l}\theta)}{(d\cos\theta)^{l-m}}
$$
$$
= (-1)^m i^l\sqrt{\frac{(2l+1)(l-m)!}{4\pi(l+m)!}}P_l^m(\cos\theta)e^{im\varphi}, \tag{9.101}
$$

where P_l^m is the *associate Legendre function* defined by

$$
P_l^m(x) \equiv (1-x^2)^{\frac{|m|}{2}}\left(\frac{d}{dx}\right)^{|m|}P_l(x). \tag{9.102}
$$

P_l is the lth *Legendre polynomial* defined by the *Rodriguez formula*

$$
P_l(x) \equiv \frac{1}{2^l l!}\left(\frac{d}{dx}\right)^l(x^2-1)^l. \tag{9.103}
$$

Eq.(9.101) is for $m \geq 0$. For $m < 0$, we use the definition

$$
Y_l^{-m}(\theta, \varphi) = (-1)^m\left[Y_l^m(\theta, \varphi)\right]^*. \tag{9.104}
$$

Then the complete expression of the spherical harmonics for all the values of m is

$$
Y_l^m(\theta, \varphi) = (-1)^{\frac{(m+|m|)}{2}}i^l\sqrt{\frac{(2l+1)(l-|m|)!}{4\pi(l+|m|)!}}P_l^{|m|}(\cos\theta)e^{im\varphi}. \tag{9.105}
$$

9.2.7   *Radial equation*

Now we consider the radial part of the wave function $R(r)$. R is determined by Eq.(9.47a), which can be rewritten as

$$\frac{d}{dr}\left(r^2\frac{dR}{dr}\right) - \frac{2mr^2}{\hbar^2}[V(r) - E]R = l(l+1)R. \qquad (9.106)$$

We define

$$u(r) \equiv rR(r). \qquad (9.107)$$

Then

$$\frac{dR}{dr} = \frac{1}{r^2}\left(r\frac{du}{dr} - u\right), \qquad (9.108a)$$

$$\frac{d}{dr}\left(r^2\frac{dR}{dr}\right) = r\frac{d^2u}{dr^2}. \qquad (9.108b)$$

Substituting u for R, Eq.(9.106) becomes

$$-\frac{\hbar^2}{2m}\frac{d^2u}{dr^2} + \left[V + \frac{\hbar^2}{2m}\frac{l(l+1)}{r^2}\right]u = Eu, \qquad (9.109)$$

which is called the *radial equation*. It can be considered as a one-dimensional Schrödinger equation with an effective potential

$$V_{eff} = V + \frac{\hbar^2}{2m}\frac{l(l+1)}{r^2}. \qquad (9.110)$$

The term $\frac{\hbar^2}{2m}\frac{l(l+1)}{r^2}$ is called the *centrifugal term*. The normalization condition for u is

$$\int_0^\infty |u|^2 dr = 1. \qquad (9.111)$$

9.2.8   *Hydrogen atom*

9.2.8.1   *Reduction to one-body problem*

An electron with negative charge $-e$ and a proton with positive charge e can form a composite particle, which is called *hydrogen atom*. The positive charge particle has a much larger mass than the electron and is localized in an atom. Thus we call it the *nucleus*. Since a hydrogen atom consists of two particles, the proton and electron, it is a two-body problem. However, it can be reduced to a one-body problem. The Schrödinger equation of the electron and nucleus is given by

$$i\hbar\frac{\partial}{\partial t}\Psi(\mathbf{x}_e, \mathbf{x}_p, t)$$
$$= \left(-\frac{\hbar^2}{2m_e}\nabla^2_{\mathbf{x}_e} - \frac{\hbar^2}{2m_p}\nabla^2_{\mathbf{x}_p} + U\right)\Psi(\mathbf{x}_e, \mathbf{x}_p, t), \qquad (9.112)$$

where $\mathbf{x}_e$ and $\mathbf{x}_p$ are the coordinates of the electron and nucleus, respectively. m_e and m_p are the masses of the electron and nucleus, respectively. U is the Coulomb potential between the nucleus and electron. According to Eq.(7.99)

$$U(\mathbf{x}_e - \mathbf{x}_p) = -\frac{e^2}{|\mathbf{x}_e - \mathbf{x}_p|}. \tag{9.113}$$

We introduce two coordinates, the relative coordinate and the mass center coordinate to replace the coordinates $\mathbf{x}_e$ and $\mathbf{x}_p$. The relative coordinate $\mathbf{x}$ is defined as

$$\mathbf{x} \equiv \mathbf{x}_e - \mathbf{x}_p \tag{9.114}$$

and the mass center coordinate $\mathbf{X}$ is defined as

$$\mathbf{X} \equiv \frac{m_e \mathbf{x}_e + m_p \mathbf{x}_p}{m_e + m_p}. \tag{9.115}$$

We can express the differentials in terms of the relative and mass center coordinates.

$$\frac{\partial}{\partial x_e^i} = \frac{\partial X^i}{\partial x_e^i}\frac{\partial}{\partial X^i} + \frac{\partial x^i}{\partial x_e^i}\frac{\partial}{\partial x^i} = \frac{m_e}{m_e + m_p}\frac{\partial}{\partial X^i} + \frac{\partial}{\partial x^i}, \tag{9.116}$$

$$\frac{\partial^2}{(\partial x_e^i)^2} = \left(\frac{m_e}{m_e + m_p}\frac{\partial}{\partial X^i} + \frac{\partial}{\partial x^i}\right)\left(\frac{m_e}{m_e + m_p}\frac{\partial}{\partial X^i} + \frac{\partial}{\partial x^i}\right) \tag{9.117}$$

and

$$\frac{\partial}{\partial x_p^i} = \frac{\partial X^i}{\partial x_p^i}\frac{\partial}{\partial X^i} + \frac{\partial x^i}{\partial x_p^i}\frac{\partial}{\partial x^i} = \frac{m_p}{m_e + m_p}\frac{\partial}{\partial X^i} - \frac{\partial}{\partial x^i}, \tag{9.118}$$

$$\frac{\partial^2}{(\partial x_p^i)^2} = \left(\frac{m_p}{m_e + m_p}\frac{\partial}{\partial X^i} - \frac{\partial}{\partial x^i}\right)\left(\frac{m_p}{m_e + m_p}\frac{\partial}{\partial X^i} - \frac{\partial}{\partial x^i}\right). \tag{9.119}$$

Inserting the above equations into the Schrödinger equation Eq.(9.112), we have

$$i\hbar\frac{\partial \Psi}{\partial t} = \left(-\frac{\hbar^2}{2M}\nabla_{\mathbf{X}}^2 - \frac{\hbar^2}{2m}\nabla_{\mathbf{x}}^2 + U(\mathbf{x})\right)\Psi, \tag{9.120}$$

where

$$M \equiv m_e + m_p \tag{9.121}$$

is the total mass of a hydrogen atom and

$$m \equiv \frac{m_e m_p}{m_e + m_p} \tag{9.122}$$

is called the reduced mass. We consider the solution which is separated
into the product

$$\Psi(\mathbf{x}, \mathbf{X}, t) = \psi(\mathbf{x})\Phi(\mathbf{X})e^{-\frac{i}{\hbar}E_t t}. \tag{9.123}$$

Then we obtain the stationary Schrödinger equation

$$-\frac{\hbar^2}{2M}\frac{1}{\Phi}\nabla_{\mathbf{X}}^2\Phi - \frac{\hbar^2}{2m}\frac{1}{\psi}\nabla_{\mathbf{x}}^2\psi + U(\mathbf{x}) = E_t. \tag{9.124}$$

The first term depends only on $\mathbf{X}$. The other two terms depend only on $\mathbf{x}$.
Therefore, each should be a constant. We have

$$-\frac{\hbar^2}{2m}\nabla_{\mathbf{x}}^2\psi + U(\mathbf{x})\psi = E\psi \tag{9.125}$$

and

$$-\frac{\hbar^2}{2M}\nabla_{\mathbf{X}}^2\Phi = (E_t - E)\Phi. \tag{9.126}$$

Eq.(9.126) is the Schrödinger equation for the wave function $\Phi(\mathbf{X})$ of mass
center. It is equivalent to that of a free particle with the energy $E_t - E$.
Eq.(9.125) is the Schrödinger equation of an electron in a relative coordinates. It is equivalent to that of a particle in a central potential

$$U(r) = -\frac{e^2}{r}, \tag{9.127}$$

where r is the distance of the electron to the nucleus. Thus we can consider
the nucleus as a localized charge. We introduce Z to denote the charge
number of the localized positive charge so that we can easily generalize
our results to more applications. For the hydrogen atom, $Z = 1$. The
potential energy for an electron in the Coulomb potential field generated
by a localized charge of Ze is

$$U(r) = -\frac{Ze^2}{r}. \tag{9.128}$$

9.2.8.2 *Solution of the radial equation in a central potential*

The radial equation reads

$$-\frac{\hbar^2}{2m}\frac{d^2 u}{dr^2} + \left[-\frac{Ze^2}{r} + \frac{\hbar^2}{2m}\frac{l(l+1)}{r^2}\right]u = Eu. \tag{9.129}$$

To simplify notation, we introduce

$$\kappa \equiv \frac{\sqrt{-2mE}}{\hbar}. \tag{9.130}$$

We consider the bound states for which E is negative. Thus κ is real. In terms of κ, we can rewritten Eq.(9.129) as

$$\frac{1}{\kappa^2}\frac{d^2u}{dr^2} = \left[1 - \frac{2mZe^2}{\hbar^2\kappa}\frac{1}{\kappa r} + \frac{l(l+1)}{(\kappa r)^2}\right]u. \qquad (9.131)$$

we introduce

$$\rho_0 \equiv \frac{2mZe^2}{\hbar^2\kappa} = \frac{2mZe^2}{\hbar\sqrt{-2mE}} \qquad (9.132)$$

and

$$\rho \equiv \kappa r. \qquad (9.133)$$

Then Eq.(9.131) becomes

$$\frac{d^2u}{d\rho^2} = \left[1 - \frac{\rho_0}{\rho} + \frac{l(l+1)}{\rho^2}\right]u. \qquad (9.134)$$

First we consider the asymptotic form of the solutions of Eq.(9.134). As $\rho \to \infty$, keeping the dominated terms in Eq.(9.134) gives

$$\frac{d^2u}{d\rho^2} = u. \qquad (9.135)$$

The solution of Eq.(9.135) is

$$u(\rho) = Ae^{-\rho} + Be^{\rho}. \qquad (9.136)$$

The term e^{ρ} goes to ∞ as $\rho \to \infty$. Thus it should be dropped and then Eq.(9.136) becomes

$$u(\rho) = Ae^{-\rho}. \qquad (9.137)$$

On the other hand, as $\rho \to 0$, the dominated terms in Eq.(9.134) gives

$$\frac{d^2u}{d\rho^2} = \frac{l(l+1)}{\rho^2}u. \qquad (9.138)$$

The solution of Eq.(9.138) is given by

$$u(\rho) = C\rho^{l+1} + D\rho^{-l}. \qquad (9.139)$$

Since $\rho^{-l} \to \infty$ as $\rho \to 0$, we have

$$u(\rho) = C\rho^{l+1}. \qquad (9.140)$$

Thus the solution of Eq.(9.134) should have the form

$$u(\rho) = C\rho^{l+1}e^{-\rho}v(\rho). \qquad (9.141)$$

In terms of $v(\rho)$, the radial equation becomes

$$\rho\frac{d^2v}{d\rho^2} + 2(l+1-\rho)\frac{dv}{d\rho} + [\rho_0 - 2(l+1)]v = 0. \qquad (9.142)$$

To find the solution, we expand $v(\rho)$ into a power series in ρ

$$v(\rho) = \sum_{k=0}^{\infty} C_k\rho^k. \qquad (9.143)$$

Inserting Eq.(9.143) into Eq.(9.142), we have

$$\sum_{k=0}^{\infty} k(k+1)C_{k+1}\rho^k + 2(l+1)\sum_{k=0}^{\infty}(k+1)C_{k+1}\rho^k$$
$$-2\sum_{k=0}^{\infty} kC_k\rho^k + [\rho_0 - 2(l+1)]\sum_{k=0}^{\infty} C_k\rho^k = 0 \qquad (9.144)$$

or

$$\sum_{k=0}^{\infty}\{k(k+1)C_{k+1} + 2(l+1)(k+1)C_{k+1}$$
$$- 2kC_k + [\rho_0 - 2(l+1)]C_k\}\rho^k = 0. \qquad (9.145)$$

The coefficient of ρ^k should be zero. We have

$$k(k+1)C_{k+1} + 2(l+1)(k+1)C_{k+1}$$
$$- 2kC_k + [\rho_0 - 2(l+1)]C_k = 0, \qquad (9.146)$$

which gives

$$C_{k+1} = \frac{2(k+l+1) - \rho_0}{(k+1)(k+2l+2)}C_k. \qquad (9.147)$$

Eq.(9.147) is a recursion formula which determines the coefficients of the expansion Eq.(9.143).

If $2(k+l+1)-\rho_0 \neq 0$, we have infinite terms in the expansion Eq.(9.143). As $k \to \infty$, we have

$$C_{k+1} \cong \frac{2}{k}C_k. \qquad (9.148)$$

Inserting Eq.(9.148) into Eq.(9.143), we obtain

$$v(\rho) \sim \mathcal{O}\left(\sum_{k=0}^{\infty}\frac{2^k}{k!}\rho^k\right) = \mathcal{O}(e^{2\rho}). \qquad (9.149)$$

Then $v(\rho) \to \infty$ as $\rho \to \infty$, which is not the solution we want. Therefore, there should be an integer $k = k_0$ fulfilling the relation

$$2(k+l+1) - \rho_0 = 0. \qquad (9.150)$$

We introduce

$$n \equiv k_0 + l + 1, \tag{9.151}$$

which is the so-called *principal quantum number*. From Eq.(9.150), we have

$$\rho_0 = 2n. \tag{9.152}$$

Inserting $\rho_0 = 2n$ into Eq.(9.132), we obtain the energy

$$E_n = -\frac{\hbar^2 \kappa^2}{2m} = -\frac{mZ^2 e^4}{2\hbar^2 n^2}. \tag{9.153}$$

Eq.(9.153) is the *Bohr formula*. The smallest n gives the energy of the ground state. Since the smallest k_0 and l are zero, the smallest n is 1. Thus n takes the values $n = 1, 2, \cdots$. For the ground state, $n = 1$. The energy of the ground state is

$$E_1 = -\frac{mZ^2 e^4}{2\hbar^2}. \tag{9.154}$$

When $Z = 1$, $E_1 = -13.3 eV$. $n = 1$ yields $l = 0$ and $m = 0$. The recursion formula Eq.(9.147) gives $c_1 = 0$. So $v(\rho) = c_0$ is a constant. Then we have

$$R_{10}(r) = \frac{c_0}{a} e^{-\frac{r}{a}} \tag{9.155}$$

with

$$a = \frac{\hbar^2}{mZe^2}. \tag{9.156}$$

When $Z = 1$, $a = a_0 \equiv \frac{\hbar^2}{me^2} = 0.529 \times 10^{-10} M$, which is called the *Bohr radius*. The radial part of the wave function for a hydrogen atom is given by

$$R_{nl}(r) = \frac{1}{r} \rho^{l+1} e^{-\rho} v(\rho), \tag{9.157}$$

where $v(\rho)$ is a polynomial of the order $k = n - l - 1$ in ρ. Inserting $\rho_0 = 2n$ into the recursion formula Eq.(9.147), we have

$$C_{k+1} = \frac{2(k + l + 1 - n)}{(k + 1)(k + 2l + 2)} C_k. \tag{9.158}$$

Using the recursion formula Eq.(9.158), we obtain $v(\rho)$ in an expansion form

$$
\begin{aligned}
v(\rho) &= \sum_{k=0}^{n-l-1} C_k \rho^k \\
&= c_0 \left[1 - \frac{n-l-1}{1!(2l+2)}(2\rho) + \frac{(n-l-1)(n-l-2)}{2!(2l+2)(2l+3)}(2\rho)^2 + \cdots \right. \\
&\qquad \left. + (-1)^{n-l-1} \frac{(n-l-1)(n-l-2)\cdots 1}{(n-l-1)!(2l+2)(2l+3)\cdots(n+l)}(2\rho)^{n-l-1} \right] \\
&= -c_0 \frac{(2l+1)!(n-l-1)!}{[(n+l)!]^2} L_{n+l}^{2l+1}(2\rho),
\end{aligned}
\tag{9.159}
$$

where $\mathcal{L}_{n+l}^{2l+1}(x)$ is the *associate Laguerre polynomial* defined by

$$\mathcal{L}_{n+l}^{2l+1}(x) = \sum_{k=0}^{n-l-1} (-1)^{n-l-1} \frac{[(n+l)!]^2 x^k}{(n-l-1-k)!(2l+1+k)!k!}. \tag{9.160}$$

From Eq.(9.130), we have

$$\kappa = \frac{mZe^2}{n\hbar^2} = \frac{Z}{na_0}. \tag{9.161}$$

Thus Eq.(9.157) becomes

$$R_{nl}(r) = N_{nl} e^{-\frac{Z}{na_0}r} \left(\frac{2Zr}{na_0}\right)^l \mathcal{L}_{n+l}^{2l+1}\left(\frac{2Zr}{na_0}\right), \tag{9.162}$$

where N_{nl} is the normalization constant. The normalization condition for R_{nl} is

$$\int_0^\infty R_{nl}^2(r) r^2 dr = 1, \tag{9.163}$$

which gives N_{nl} as

$$N_{nl} = \left\{ \left(\frac{2Z}{na_0}\right)^3 \frac{(n-l-1)!}{2n[(n+l)!]^3} \right\}^{\frac{1}{2}}. \tag{9.164}$$

Together with the angular part, the wave function for hydrogen is given by

$$\psi_{nlm}(r, \theta, \varphi) = R_{nl}(r) Y_l^m(\theta, \varphi), \tag{9.165}$$

which is labeled by the three quantum numbers n, l and m. Eigenvalue of energy E_n depends only on n. Thus the energy is degenerate. For a fixed n, l takes the values $l = 0, 1, 2, \cdots, n-1$. For a fixed l, m takes the values $m = 0, \pm 1, \pm 2, \cdots, \pm l$. Thus the degeneracy of energy is

$$\sum_{l=0}^{n-1} (2l+1) = n^2. \tag{9.166}$$

l is the quantum number for the orbital angular momentum. The states with $l = 0, 1, 2, 3 \cdots$ are often called the s, p, d, f, $\cdots$ states, respectively.

Chapter 10

Statistical mechanics

10.1 Equi-probability principle and statistical distributions

Now we consider systems containing large number of particles. For an isolated system which does not exchange energy and particles with external environment, the energy of the system is fixed. The different states are the degenerate states of energy. We call these states microscopic states in statistic mechanics. One can not tell the differences between these states statistically. We have the basic principle that all the microscopic states are equally probable to be occupied in an isolated system. This is called the *Boltzmann equi-probability principle*, which is the standard assumption in statistical physics. A set of identical isolated systems is called the *micro-canonical ensemble*. Thus the statistical distribution for a system in micro-canonical ensemble, which is called the *micro-canonical distribution*, is a constant.

Now we consider a system (denoted as system 1) contacted with a larger system(denoted as system 2). The larger system is called the thermal reservoir. The two systems contain large number of particles N (typically $N \sim 10^{23}$). We call the systems with large number of particles the *macroscopic systems*. When the two systems are isolated, the states in each system are equally probable to be occupied. After the two systems are contacted, the two systems will be in thermal equilibrium with each other when the thermal quantities of the two systems are not changed. The system and the thermal reservoir can be considered to form an isolated system. This is valid because we can always include the contacted surroundings into the thermal reservoir until the total system is isolated. For a macroscopic system, the boundary part can be neglected. Thus the two systems are sufficiently weakly interacting that we can write

$$E_t = E + E', \tag{10.1}$$

where E is the energy of the system 1 and E' the energy of the system 2. E_t is the total energy of the two systems. According to the equi-probability principle, the probability $p(E_n)$ (also denoted as p_n) that the system 1 in the state n with the energy E_n is directly proportional to the number of the possible states of the total isolated system for this situation. Since the system 1 is in a definite state n and the system 2 can be in any states with the energy E', we have

$$p(E_n) \propto 1 \cdot \Omega'(E'), \tag{10.2}$$

where $\Omega'(E)$ is the number of the states of the system 2 with the energy E'. In the derivation of Eq.(10.2), we have used the condition that the two systems are not correlated statistically because the two systems are macroscopic ones with large number of particles and the contact boundary between them is only a minor part of the total system.[1] Eq.(10.2) is not hold in the microscopic scale. Thus the two systems fulfilling Eq.(10.2) should be macroscopic systems. We introduce a macroscopic quantity

$$S \equiv k \ln \Omega(E), \tag{10.3}$$

where $\Omega(E)$ is the number of the states of the system with energy E. S is called the *entropy* of the system. k is a constant defining the unit of entropy. Eq.(10.3) is usually called the *Boltzmann entropy relation*. Using Eq.(10.1), we have

$$p(E_n) \propto \Omega'(E_t - E_n) \tag{10.4}$$

with

$$\Omega'(E_t - E_n) = \exp\left[\frac{1}{k}S'(E_t - E_n)\right]. \tag{10.5}$$

When the thermal reservoir is large, we have $E_n \ll E' \approx E_t$. We can expand S' in a Taylor series:[2]

$$S'(E_t - E_n) = S'(E_t) - E_n\frac{\partial S'}{\partial E} + \frac{1}{2}E_n^2\frac{\partial^2 S'}{\partial E^2} + \cdots. \tag{10.6}$$

---

[1]It also requires that time is long enough that the statistical correlation between the two systems is dissolved. This is always possible because massless photons that exist everywhere can interact with all the particles in any energy scale.

[2]One may have noted that the Taylor expansion is made for S' instead of Ω, which is the key to obtain Eq.(10.11). From Eqs.(10.52) and (10.53), we can see that the entropy S' is an extensive quantity while Ω is not an extensive quantity. Thus S' is selected for the Taylor expansion.

We define the *temperature* of a macroscopic system as

$$T \equiv \frac{\partial E}{\partial S}. \tag{10.7}$$

For any macroscopic system, we have the lowest energy state as the ground state. However, there is no limit to the highest energy generally because we have selected the positive sign in the kinetic term for the Hamiltonian. An opposite selection would result in an opposite sign of temperature but the physics is not changed. When we increase the energy, the increased energy allows the particles to occupy the states with higher energy. The number of possible microscopic states will increase. Ω represents the number of the microscopic states. Thus the increase of energy leads to the increase of entropy. According to Eq.(10.7), the temperature should be positive.

$$T > 0. \tag{10.8}$$

There are some artificial systems having only finite energy levels. In these systems, negative temperature can be achieved.

In terms of temperature, Eq.(10.5) becomes

$$\Omega'(E_t - E_n) \propto \exp\left(-\frac{E_n}{kT} + \frac{1}{2k}E_n^2\frac{\partial^2 S'}{\partial E^2} + \cdots\right). \tag{10.9}$$

T is the temperature of the thermal reservoir. We usually use β to denote $\frac{1}{kT}$,

$$\beta \equiv \frac{1}{kT}. \tag{10.10}$$

For a large thermal reservoir, the high order terms in Eq.(10.9) can be neglected. Then Eq.(10.4) can be rewritten as

$$p(E_n) = Z^{-1}e^{-\beta E_n} \tag{10.11}$$

with

$$Z = \sum_n e^{-\beta E_n}. \tag{10.12}$$

The summation is over all the states in the system 1. The normalization factor Z is called the *partition function*. We usually call a set of the identical systems (the identical systems are the systems with the same Hamiltonian) contacted with a thermal reservoir as *canonical ensemble*. Thus $p(E)$ in Eq.(10.11) is called the *canonical distribution*. In a narrow sense, a system in the canonical ensemble exchanges only energy with a thermal reservoir and there is no particle exchange. Since Eq.(10.11) is also applicable to the system with the particle number variable, which can be seen from the derivation procedure of Eq.(10.11), we usually do not restrict the particle exchange for a contact with a thermal reservoir, which is also more consistent with the concept that a quasi-particle is an energy excitation. When the particle number is conserved, the summation in Eq.(10.12) is over the states in the Hilbert space. When the particle number is variable, the summation is over the states in the Fock space.

Using the symbol of the trace Tr defined by

$$\text{Tr}\hat{A} \equiv \sum_n \langle n|\hat{A}|n\rangle, \tag{10.13}$$

where $\{|n\rangle\}$ is an arbitrary completely orthonormal basis states, we can express Z in terms of the Hamiltonian operator,

$$Z = \sum_n e^{-\beta E_n} = \text{Tr}e^{-\beta\hat{H}}. \tag{10.14}$$

10.2   Average of an observable $\hat{A}$

10.2.1   *Statistical average*

Now let us consider the mean value of an observable $\hat{A}$. When a system is in the state $|\psi\rangle$, the quantum mean value of the observable $\hat{A}$ is given by

$$\langle\hat{A}\rangle = \langle\psi|\hat{A}|\psi\rangle. \tag{10.15}$$

In order to distinguish with the statistical average, we have used the notation $\langle\hat{A}\rangle$, instead of $\overline{\hat{A}}$, to denote the quantum mean value. We use $|\psi_i\rangle$ to denote an orthonormal basis and p_i to represent the probability for the state $|\psi_i\rangle$ to be occupied. Then the statistical average of the observable $\hat{A}$ reads

$$\overline{\hat{A}} = \sum_i p_i \langle\psi_i|\hat{A}|\psi_i\rangle. \tag{10.16}$$

We introduce the density matrix ρ defined by

$$\hat{\rho} \equiv \sum_i p_i |\psi_i\rangle\langle\psi_i| \tag{10.17}$$

with

$$\text{Tr}\hat{\rho} = \sum_i p_i = 1. \tag{10.18}$$

Eq.(10.18) is the normalization condition for the probability p_i. Using Eq.(10.17), Eq.(10.16) becomes

$$\overline{A} = \sum_i p_i \langle \psi_i | \hat{A} | \sum_n |n\rangle\langle n|\psi_i\rangle$$
$$= \sum_n \langle n| \sum_i p_i |\psi_i\rangle\langle \psi_i | \hat{A} | n\rangle. \tag{10.19}$$
$$= \mathrm{Tr}\hat{\rho}\hat{A}.$$

10.2.2 *Average using canonical distribution*

The canonical density matrix is defined by

$$\hat{\rho}_c \equiv \sum_n p(E_n)|n\rangle\langle n|$$
$$= Z^{-1} \sum_n e^{-\frac{E_n}{kT}} |n\rangle\langle n|$$
$$= Z^{-1} \sum_n e^{-\frac{\hat{H}}{kT}} |n\rangle\langle n| \tag{10.20}$$
$$= Z^{-1} e^{-\frac{\hat{H}}{kT}} \sum_n |n\rangle\langle n|$$
$$= Z^{-1} e^{-\frac{\hat{H}}{kT}}.$$

The mean value of the observable $\hat{A}$ for a system is given by

$$\overline{A} = \mathrm{Tr}\hat{\rho}_c\hat{A} = Z^{-1}\mathrm{Tr}\left(e^{-\frac{\hat{H}}{kT}}\hat{A}\right). \tag{10.21}$$

10.2.3 *Average using grand canonical distribution*

We consider a system contacted with the particle reservoir. When particles can be exchanged, we write the particle number dependence of $\Omega'(E')$ in Eq.(10.2) explicitly. For the probability $p(E_n, N)$ that the system in a state n with the particle number N, we have

$$p(E_n, N) \propto 1 \cdot \Omega'(E', N'), \tag{10.22}$$

where N' is the particle number of the reservoir. Since the particle number is conserved, the particle number N_t of the total system (the system + reservoir) is a constant. We have

$$N_t = N + N'. \tag{10.23}$$

Using Eqs.(10.1) and (10.23), we obtain

$$p(E_n, N) \propto 1 \cdot \Omega'(E_t - E_n, N_t - N)$$

$$\propto \exp\left[\frac{1}{k}S'(E_t - E_n, N_t - N)\right]. \tag{10.24}$$

When the particle reservoir is large, we have

$$E_n \ll E_t \approx E', \tag{10.25a}$$

$$N \ll N_t \approx N'. \tag{10.25b}$$

We can expand S' in a Taylor series.

$$S'(E_t - E_n, N_t - N) = S'(E_t, N_t) - E_n\frac{\partial S'}{\partial E'} - N\frac{\partial S'}{\partial N'} + \cdots . \tag{10.26}$$

Inserting Eq.(10.26) into Eq.(10.24), we have

$$p(E_n, N) \propto e^{-\alpha - \beta E_n} \tag{10.27}$$

with

$$\alpha = \frac{1}{k}\frac{\partial S'}{\partial N'}, \tag{10.28a}$$

$$\beta = \frac{1}{k}\frac{\partial S'}{\partial E'}. \tag{10.28b}$$

We introduce chemical potential μ defined by

$$\mu \equiv -T\left(\frac{\partial S'}{\partial N'}\right)_{E,V} \tag{10.29}$$

or

$$\alpha = -\beta\mu. \tag{10.30}$$

Thus the probability that the system in a state n with the particle number N for a system that particle number is changeable is given by

$$p(E_n, N) = \Xi^{-1}e^{-\beta(E_n - \mu N)}. \tag{10.31}$$

where Ξ is the normalization constant. To simplify the notation, $p(E_n, N)$ is often denoted as $p_{n,N}$.

We usually call a set of identical systems contacted with particle reservoir as *grand canonical ensemble*. Systems contacted with particle reservoir are called the open systems. $p(E_n, N)$ is thus called the *grand canonical distribution*. Since Ξ is the normalization constant, we have

$$\Xi = \sum_N \sum_n e^{-\beta(E_n - \mu N)}$$

$$= \sum_N \mathrm{Tr} e^{-\beta(\hat{H} - \mu\hat{N})} \tag{10.32}$$

$$= \sum_N Z(N)e^{\beta\mu N}.$$

Ξ is called the *grand partition function*. We introduce the density matrix of the grand canonical ensemble defined by

$$\hat{\rho}_G \equiv \Xi^{-1} e^{-\beta(\hat{H} - \mu \hat{N})}. \tag{10.33}$$

Then the average value of an observable $\hat{A}$ is given by

$$\overline{\hat{A}} = \text{Tr}(\hat{\rho}_G \hat{A}). \tag{10.34}$$

The trace here is to be understood as a double summations $\sum_N \text{Tr}$. The first summation Tr is over the state for a fixed particle number N and the second is over all the particle numbers $N = 0, 1, 2, \cdots$.

10.3 Functional integral representation of partition function

According to the definition of the partition function Eq.(10.12), we have

$$Z = \sum_n e^{-\beta E_n} = \sum_n \langle n | e^{-\beta \hat{H}} | n \rangle. \tag{10.35}$$

Similar to the derivation of the functional integral representation of the transition amplitude $\langle \phi' | e^{-i \hat{H} t} | \phi \rangle$, we can derive the functional integral representation of the partition function as

$$Z = \sum_n \langle n | e^{-\beta \hat{H}} | n \rangle = \int_P D\phi \, e^{-\int_0^\beta d\tau \int d^3 x \mathcal{L}(\phi, \dot{\phi})}, \tag{10.36}$$

where the subscript P denotes the periodic boundary condition which demands that the functional integral should be done over all $\phi(\mathbf{x}, \tau)$ with the boundary condition $\phi'(\mathbf{x}, \beta) = \phi(\mathbf{x}, 0)$. In comparison with the transition amplitude, Eq.(10.36) can be obtained by simply replacing the time t by $-i\beta$ in the transition amplitude and summing over $|\phi, \tau\rangle$ with the condition $\phi(\beta) = \phi(0)$. Thus the functional integral formalism of the partition function is equivalent to the Euclidean quantum field formalism with $0 < \tau < \beta$ and the periodic boundary condition imposed.

In Eq.(10.36), we have used the Lagrangian of the boson field. The functional representation for the fermion field can be given similarly. One can obtain the functional representation for the fermion field by simply replacing the Lagrangian density of the boson field by that of the fermion field . Also, we can use the Hamiltonian of quantum mechanics and obtain the partition function when quantum mechanics is applicable.

$$Z = \text{Tr} e^{-\beta \hat{H}} = \int_P Dq \, e^{-\int_0^\beta d\tau L(q, \dot{q})}. \tag{10.37}$$

In statistical mechanics, we deal with the many-particle systems. We usually use the field expression Eq.(10.36).

When temperature approaches zero ($\beta \to \infty$), we recover the Wick rotated quantum field theory in the Euclidean representation, which gives the ground state properties as is expected.

10.4   First law of thermodynamics

Let us consider the properties of the macroscopic systems. The properties of the macroscopic systems are described by a set of macroscopic quantities such as temperature, entropy, etc. The relations of macroscopic quantities are determined by the laws of thermodynamics. We will derive these laws of thermodynamics from the principle of statistical mechanics.

We consider an equilibrium system. We denote the probability for the system in the state r as p_r. The average energy of the system is given by

$$\overline{E} = \sum_r E_r p_r. \tag{10.38}$$

The energy can be changed with a small external disturbance. The variation of the energy can be written as

$$d\overline{E} = \sum_r E_r dp_r + \sum_r p_r dE_r. \tag{10.39}$$

The second term comes from the change of E_r. E_r is the energy of a quantum state which can only be changed by applying an external field. This way of changing energy is called performing work. The change of energy from the first term is caused by the change of p_r. Since E_r is not changed, there is no work done on the system. In this case, the way of changing energy is called heat transfer. We define the heat transfer $đQ$ by

$$đQ \equiv \sum_r E_r dp_r. \tag{10.40}$$

The second term in Eq.(10.39) corresponds to the work performed on the system. We denote the work performed on the system by $đW$. Then

$$đW = \sum_r p_r dE_r = \overline{dE_r}. \tag{10.41}$$

The small bar is added in the symbols $đW$ and $đQ$ because the work W and heat Q are not state functions. $đW$ and $đQ$ depend on the process. In terms of $đW$ and $đQ$, Eq.(10.39) becomes

$$d\overline{E} = đQ + đW. \tag{10.42}$$

Eq.(10.42) is the so-called *first law of thermodynamics*. It states that for any process of a macroscopic system, the variation of the energy is equal to the sum of the adsorbed heat and the work performed by the external fields.

The energy E_r is a function of the parameters y_i related to the external fields.

$$E_r = E_r(y_1, \cdots, y_n).$$ (10.43)

y_i ($i = 1, 2, \cdots, n$) are often called the *generalized coordinates* in thermodynamics. According to Eq.(10.41), we have

$$\begin{aligned}
\dbar W &= \sum_r p_r dE_r \\
&= \sum_r p_r \sum_{i=1}^{n} \frac{\partial E_r}{\partial y_i} dy_i \\
&= \sum_r \sum_{i=1}^{n} p_r \frac{\partial E_r}{\partial y_i} dy_i \\
&= \sum_{i=1}^{n} \overline{Y}_i dy_i
\end{aligned}$$ (10.44)

with

$$\overline{Y}_i \equiv \sum_r p_r \frac{\partial E_r}{\partial y_i} = \overline{\frac{\partial E_r}{\partial y_i}}.$$ (10.45)

$\overline{Y}_i$ is called the *generalized force*. Eq.(10.44) shows that the work done on the system is equal to the generalized force times the variation of the generalized coordinate. For example, the general coordinate is the volume V for a hydrodynamic system. The work done on a system is equal to force timing displacement.[3] We define pressure P as the force on a unit area. We

---

[3]We show that the force timing displacement is equal to the energy variation caused by the external force for a particle with a Hamiltonian

$$H = \frac{\mathbf{p}^2}{2m} + U_i(\mathbf{x}) + U_e(\mathbf{x}),$$ (10.46)

where $U_e(\mathbf{x})$ is the potential due to the external field and $U_i(\mathbf{x})$ is the potential excluding the $U_e(\mathbf{x})$. From Eq.(8.67), we have

$$\begin{aligned}
\frac{d\mathbf{p}}{dt} &= -\boldsymbol{\nabla}_\mathbf{x} U_i(\mathbf{x}) - \boldsymbol{\nabla}_\mathbf{x} U_e(\mathbf{x}) \\
&= -\boldsymbol{\nabla}_\mathbf{x} U_i(\mathbf{x}) + \mathbf{F}_e,
\end{aligned}$$ (10.47)

denote the displacement of area $\mathbf{ds}$ by $\mathbf{dx}$. Then work done on the system is given by

$$\dbar W = \int P\mathbf{ds} \cdot \mathbf{dx} = -PdV. \tag{10.51}$$

In the hydrodynamic systems, the generalized force corresponds to the minus pressure.

For a non-equilibrium system, we can divide the system into subsystems. When subsystems are small enough, the subsystems can be considered as equilibrium systems. Another method to treat non-equilibrium systems is to use non-equilibrium distribution function. Thus Eq.(10.42) is also valid for the non-equilibrium systems.

10.5   Second law of thermodynamics

10.5.1   *Entropy increase principle*

First let us consider a system (subsystem A) with energy E_A contacted with a thermal reservoir (subsystem B) with energy E_B. we have shown that the total system consisting of the system and the thermal reservoir can be considered as an isolated system. For the total system consisting of two macroscopic subsystems, we have

$$\Omega_t(E_A, E_B) = \Omega_A(E_A)\Omega_B(E_B), \tag{10.52}$$

where $\mathbf{F}_e = -\boldsymbol{\nabla}_\mathbf{x} U_e(\mathbf{x})$ is the external force. Eq.(10.47) can be rewritten as

$$\frac{d\mathbf{p}}{dt} + \boldsymbol{\nabla}_\mathbf{x} U_i(\mathbf{x}) = \mathbf{F}_e. \tag{10.48}$$

Let $d\mathbf{x}$ be the displacement vector of the particle caused by the disturbance. Multiplying Eq.(10.48) by $d\mathbf{x}$, we have

$$\begin{aligned}
\mathbf{F}_e \cdot d\mathbf{x} &= \frac{d\mathbf{p}}{dt} \cdot d\mathbf{x} + d\mathbf{x} \cdot \boldsymbol{\nabla}_\mathbf{x} U_i(\mathbf{x}) \\
&= \frac{d\mathbf{p}}{dt} \cdot \frac{d\mathbf{x}}{dt} dt + dU_i(\mathbf{x}) \\
&= d\left(\frac{\mathbf{p}^2}{2m}\right) + dU_i(\mathbf{x}) \\
&= dE_i,
\end{aligned} \tag{10.49}$$

where $E_i = \frac{\mathbf{p}^2}{2m} + U_i(\mathbf{x})$ is the energy of the particle excluding the external potential. $dW \equiv \mathbf{F}_e \cdot d\mathbf{x}$ is called the work done by the force $\mathbf{F}_e$ on the particle in the time interval dt during which the displacement of the particle is $d\mathbf{x}$. Then we have

$$\dbar W = \mathbf{F}_e \cdot d\mathbf{x} = dE_i. \tag{10.50}$$

Thus force timing displacement is equal to the energy variation caused by the force.

where $\Omega_t(E_A, E_B)$ is the number of the states of the total system consisting a subsystem A with energy E_A and a subsystem B with energy E_B. Eq.(10.52) means that the two macroscopic systems are not correlated statistically. $S = k \ln \Omega(E)$ is a function of the energy of system. Then we obtain

$$\begin{aligned} S_t(E_A, E_B) &= k \ln \Omega_t(E_A, E_B) \\ &= k \ln \Omega_A(E_A) + k \ln \Omega_B(E_B) \\ &= k \ln \Omega_A(E_A) + k \ln \Omega_B(E_t - E_A), \end{aligned} \tag{10.53}$$

where S_t and E_t are the entropy and energy of the total system, respectively. For a thermodynamic process that a system with the initial energy $E_A = E_{Ai}$ approaches to an equilibrium state, the evolution path should be along a most probable one. Thus the evolved state of the total system should has larger Ω_t than the previous state because it is more probable to have such a state. Therefore the thermodynamic process is along a path with increasing entropy $S_t = k \ln \Omega_t(E_A, E_t - E_A)$ as the system exchanges energy with the reservoir. The equilibrium state is the most probable state with $\frac{\partial S_t}{\partial E_A} = 0$.

Next we consider a general isolated macroscopic system with the energy E and volume V. For a macroscopic system, we can divide the system into small parts (macroscopically small, but still microscopically large). We divide the system into N_s parts (we call them subsystems) with the same volume $v = V/N_s$. We use n_i to denote the number of the subsystems taking the energy ε_i. Each energy ε_i has a degeneracy Ω_i. Then the number of the microscopic states of the subsystem with the energy ε_i is Ω_i. For an isolated system, we have the following constrained relations:

$$N_s = \sum_i n_i, \tag{10.54a}$$

$$E = \sum_i n_i \varepsilon_i. \tag{10.54b}$$

We call the distribution $\{n_i\}$ a *macroscopic state* of the system with N subsystems.

Now we calculate the number of the microscopic states $\Omega_{\{n_i\}}$ for a system taking a macroscopic state $\{n_i\}$. $\Omega_{\{n_i\}}$ is just the number of different possible ways to select n_1 subsystems taking energy ε_1, n_2 subsystems taking energy $\varepsilon_2, \cdots$. We first select n_1 subsystems to take the energy ε_1. There are

$$C_{n_1}^{N_s} = \frac{N_s!}{n_1!(N_s - n_1)!} \tag{10.55}$$

ways of selecting n_1 subsystems. Then we select the remaining $N_s - n_1$ subsystems to take ε_2. There are

$$C_{n_2}^{N_s-n_1} = \frac{(N_s - n_1)!}{n_2!(N_s - n_1 - n_2)!} \tag{10.56}$$

selecting ways. We continue the selections in a similar way until all the subsystems are consumed. In total, we have W selecting ways.

$$
\begin{aligned}
W &= C_{n_1}^{N_s} C_{n_2}^{N_s-n_1} \cdots \\
&= \frac{N_s!}{n_1!(N_s - n_1)!} \frac{(N_s - n_1)!}{n_2!(N_s - n_1 - n_2)!} \frac{(N_s - n_1 - n_2)!}{n_3!(N_s - n_1 - n_2 - n_3)!} \cdots \\
&= \frac{N_s!}{\prod_i n_i!}.
\end{aligned} \tag{10.57}
$$

For each selection, the subsystems with energy ε_i can occupy any of the Ω_i degenerate states. We have an additional factor $\Omega_i^{n_i}$ to count the possible ways to arrange n_i subsystems with energy ε_i onto different degenerate states. Thus we have

$$\Omega_{\{n_i\}} = \frac{N_s!}{\prod_i n_i!} \prod_i \Omega_i^{n_i} = N_s! \prod_i \frac{\Omega_i^{n_i}}{n_i!}. \tag{10.58}$$

Different distribution $\{n_i\}$ has different number of microscopic states $\Omega_{\{n_i\}}$. There is a maximum value for $\Omega_{\{n_i\}}$. The macroscopic state $\{n_i\}$ with the maximum $\Omega_{\{n_i\}}$ is called the most probable state. In the probability interpretation, any initial macroscopic state $\{n_i\}$ has the largest probability to evolve into the most probable state. Thus the equilibrium state should be the most probable state $\{n_i\}_m$. Since there are two constraint conditions given by Eq.(10.54), we use the Lagrange multiplier method to determine the most probable state. We introduce two Lagrange multipliers α and β for the constraint conditions Eqs.(10.54a) and (10.54b), respectively. Then the most probable state $\{n_i\}_m$ is determined by the following equation

$$\frac{\partial \ln \Omega_{\{n_i\}}}{\partial n_i} + \alpha \frac{\partial(N_s - \sum_i n_i)}{\partial n_i} + \beta \frac{\partial(E - \sum_i n_i \varepsilon_i)}{\partial n_i} = 0. \tag{10.59}$$

Since $\ln \Omega_{\{n_i\}}$ has the same maximum position with $\Omega_{\{n_i\}}$, we have equivalently used $\ln \Omega_{\{n_i\}}$ instead of $\Omega_{\{n_i\}}$ in Eq.(10.59). To simplify the deduction, we consider that the subsystems are so small that they already in equilibrium and Ω_i remains unchanged. From Eq.(10.58), we have

$$\ln \Omega_{\{n_i\}} = \ln N_s! + \sum_i (n_i \ln \Omega_i - \ln n_i!). \tag{10.60}$$

Since $n_i \gg 1$, we can use the Stirling formula

$$\ln x! \cong x(\ln x - 1) \quad \text{for} \quad x \gg 1. \tag{10.61}$$

Then Eq.(10.61) becomes

$$\ln \Omega_{\{n_i\}} = \ln N_s! + \sum_i (n_i \ln \Omega_i - n_i \ln n_i + n_i)$$

$$= \ln N_s! + \sum_i n_i \left(\ln \frac{\Omega_i}{n_i} + 1 \right). \tag{10.62}$$

The derivative of $\Omega_{\{n_i\}}$ is given by

$$\frac{\partial \ln \Omega_{\{n_i\}}}{\partial n_i} = \ln \frac{\Omega_i}{n_i}. \tag{10.63}$$

Inserting Eq.(10.63) into Eq.(10.59), we have

$$n_{im} = \Omega_i e^{-\alpha - \beta \varepsilon_i}. \tag{10.64}$$

n_{im} is the number of the subsystems with the energy ε_i in the most probable state. Eq.(10.64) is called the *Boltzmann distribution*. Then

$$p_i = \frac{n_{im}}{N_s} = \frac{\Omega_i}{N_s} e^{-\alpha - \beta \varepsilon_i} \tag{10.65}$$

is the probability that a subsystem has the energy ε_i. Since a subsystem in an isolated system can be considered as an arbitrary macroscopic system contacted with a heat reservoir or an environment, p_i is also the probability for a system contacted with a reservoir to take the energy ε_i.

A thermodynamic process is an evolution process with time. Although a macroscopic isolated system can occupy any microscopic state with equal probability, it can not occupy all the microscopic states at the same time. It can only occupy one microscopic state at one time. An experimental measurement is an average process. Thus experiments give only averaged quantities. For a thermodynamic process from a non-equilibrium state to an equilibrium state, the equilibrium occurs always locally first. An inhomogeneous thermodynamic system can be divided into small subsystems which are relatively in equilibrium. These subsystems are large microscopically and small macroscopically. These subsystems can be characterized by their energies since the energy is a key measured quantity in a thermodynamic system and also a quantity which changes slowly due to the conservation of energy-momentum. Thus macroscopic measurement gives average quantities of the system in a macroscopic state that has a definite energy distribution. Since the entropy is given by $S = k \ln \Omega$ and Ω is proportional to the probability, the more probable macroscopic state is also the state with the larger entropy. Any macroscopic state will evolve to a more probable state. Thus we have for an isolated system with constant E and V

$$\delta S_{\mathrm{M}} \geq 0, \tag{10.66}$$

where S_{M} is the entropy of the macroscopic state. Usually we omit the subscript M although this is a bad custom conceptually. Eq.(10.66) is the so-called *principle of entropy increase* or Clausius principle, which states that the entropy of any isolated macroscopic system always increases. The principle of entropy increase is one of the formulations for the *second law of thermodynamics*. The second law of thermodynamics have many equivalent formulations. Any formulation predicting the evolving direction of an irreversible process can be used as one of the formulations of the second law of thermodynamics. We can show that all these formulations are equivalent.

Now we show that the entropy of the most probable macroscopic state $S_{\mathrm{Mm}} = k \ln \Omega_{\mathrm{Mm}} \equiv k \ln \Omega_{\{n_i\}_m}$ is equal to $k \ln \Omega$ in the thermodynamic limit (the system is infinite large). Here $\Omega_{\mathrm{Mm}} \equiv \Omega_{\{n_i\}_m}$ is the number of the microscopic states in the most probable state $\{n_i\}_m$. We expand $\ln \Omega_{\{n_i\}}$ around the most probable state $\{n_i\}_m$

$$\ln \Omega_{\{n_i\}} = \ln \Omega_{\mathrm{Mm}} - \sum_i \frac{n_{im}}{2} \left(\frac{\delta n_i}{n_{im}} \right)^2 + \cdots, \tag{10.67}$$

where $\delta n_i = n_i - n_{im}$ is the deviation from the most probable state n_{im}. The total number of states is given by

$$\Omega = \sum_{\{n_i\}} \Omega_{\{n_i\}} = \sum_{\{\delta n_i\}} \Omega_{\mathrm{Mm}} \exp\left[-\frac{1}{2} \sum_i n_{im} \left(\frac{\delta n_i}{n_{im}} \right)^2 + \cdots \right]. \tag{10.68}$$

Since the fluctuation is small in the thermodynamic limit, we can estimate the summation in Eq.(10.68) using Gaussian integration. We have

$$\begin{aligned} \Omega &= \sum_{\{\delta n_i\}} \Omega_{\mathrm{Mm}} \exp\left[-\frac{1}{2} \sum_i n_{im} \left(\frac{\delta n_i}{n_{im}} \right)^2 + \cdots \right] \\ &= \Omega_{\mathrm{Mm}} \prod_i \mathcal{O}(n_{im}^{\frac{1}{2}}). \end{aligned} \tag{10.69}$$

Since $\frac{\ln n_{im}}{n_{im}} \to 0$ in the thermodynamic limit, we have

$$\ln \Omega = \ln \Omega_{\mathrm{Mm}}. \tag{10.70}$$

According to Eq.(10.12),

$$Z = \sum_i \Omega_i e^{-\beta \varepsilon_i}. \tag{10.71}$$

is the partition function of the subsystem. Using the normalization condition

$$\sum_i p_i = 1,$$ (10.72)

we have

$$\sum_i \Omega_i e^{-\alpha - \beta \varepsilon_i} = N_s.$$ (10.73)

From Eq.(10.71), we have

$$e^\alpha = \frac{Z}{N_s}.$$ (10.74)

Then

$$p_i = \frac{\Omega_i}{Z} e^{-\beta \varepsilon_i}.$$ (10.75)

Each energy ε_i has Ω_i equivalent quantum states. Dividing p_i by Ω_i, we get the probability $p_{i\alpha}$ $(\alpha = 1, \cdots, \Omega_i)$ for a macroscopic system on a microscopic state

$$p_{i\alpha} = \frac{1}{Z} e^{-\beta \varepsilon_i}.$$ (10.76)

$p_{i\alpha}$ is the canonical distribution. p_i and $p_{i\alpha}$ in Eqs.(10.75) and (10.76) are also called the *macroscopic and microscopic distributions* for an equilibrium state, respectively.

10.5.2 *Extensiveness of* $\ln Z$

From Eq.(10.71), we can show that $\ln Z$ is an extensive quantity. For a system consisted of two subsystems (subsystem 1 and subsystem 2), we have

$$
\begin{aligned}
\ln Z &= \ln \left(\sum_k \Omega_k e^{-\beta \varepsilon_k} \right) \\
&= \ln \left[\sum_{ij} \Omega_i^{(1)} \Omega_j^{(2)} e^{-\beta(\varepsilon_i^{(1)} + \varepsilon_j^{(2)})} \right] \\
&= \ln \left[\left(\sum_i \Omega_i^{(1)} e^{-\beta \varepsilon_i^{(1)}} \right) \left(\sum_j \Omega_j^{(2)} e^{-\beta \varepsilon_j^{(2)}} \right) \right] \\
&= \ln Z^{(1)} + \ln Z^{(2)}.
\end{aligned}
$$ (10.77)

where we have used the relation

$$\Omega_k = \Omega_i^{(1)}\Omega_j^{(2)} \tag{10.78}$$

in the derivation of the above equation. Eq.(10.78) means that the two macroscopic systems are not correlated statistically. Eq.(10.77) shows that $\ln Z$ is an extensive quantity.

10.5.3   *Thermodynamic quantities in terms of partition function*

The entropy is defined as $S = k \ln \Omega$. Using Eqs.(10.62), (10.65)and (10.74), we have

$$
\begin{aligned}
S &= k \ln \Omega_{\{n_i\}_{\mathrm{m}}} \\
&= k \ln N! + k \sum_i (n_{\mathrm{m}i} \ln \Omega_i - n_{\mathrm{m}i} \ln n_{\mathrm{m}i} + n_{\mathrm{m}i}) \\
&= kN(\ln N - 1) + k \sum_i n_{\mathrm{m}i}(\alpha + \beta\varepsilon_i + 1) \\
&= kN(\ln Z_{sub} + \beta \overline{E}_{sub}).
\end{aligned}
\tag{10.79}
$$

In Eq.(10.79), Z_{sub} and $\overline{E}_{sub}$ are the partition function and average energy of one subsystem. N is the number of subsystems. We have shown that $\ln Z$ is an extensive quantity. Thus we have

$$S = k(\ln Z + \beta\overline{E}), \tag{10.80}$$

where Z is the partition function of the system and $\overline{E}$ is the total energy of the system. Eq.(10.80) shows that S is an extensive quantity.

If we calculate the average of p_i, we have

$$\overline{\ln p} = \overline{\ln (Z^{-1}e^{-\beta\varepsilon_i})} = -\ln Z - \beta\overline{E}. \tag{10.81}$$

Thus Eq.(10.80) becomes

$$S = -k\overline{\ln p} = -k\mathrm{Tr}(\hat{\rho}\ln\hat{\rho}). \tag{10.82}$$

Eq.(10.82) is the Gibbs formulation of entropy.

Now let us express $\overline{E}$ in terms of the partition function Z. Using Eq.(10.11), we have

$$
\begin{aligned}
\overline{E} &= \sum_n p_n E_n \\
&= Z^{-1} \sum_n E_n e^{-\beta E_n} \\
&= -Z^{-1}\frac{\partial Z}{\partial \beta} \\
&= -\frac{\partial \ln Z}{\partial \beta}.
\end{aligned}
\tag{10.83}
$$

Inserting Eq.(10.83) into Eq.(10.80), we have

$$S = k \left(\ln Z - \beta \frac{\partial \ln Z}{\partial \beta} \right). \tag{10.84}$$

Using Eqs.(10.80) and (10.83), we have

$$\begin{aligned} \frac{\partial S}{\partial \bar{E}} &= k \frac{\partial \ln Z}{\partial \beta} \frac{\partial \beta}{\partial \bar{E}} + k\beta + k \frac{\partial \beta}{\partial \bar{E}} \bar{E} \\ &= k\beta. \end{aligned} \tag{10.85}$$

According to Eq.(10.7), we have

$$\beta = \frac{1}{kT}. \tag{10.86}$$

Using Eq.(10.45), we obtain the formula to calculate the generalized force $\overline{Y}_i$

$$\begin{aligned} \overline{Y}_i &= \sum_n p_n \frac{\partial E_n}{\partial y_i} \\ &= \sum_n Z^{-1} \frac{\partial E_n}{\partial y_i} e^{-\beta E_n} \\ &= -\frac{1}{\beta} \frac{\partial \ln Z}{\partial y_i}. \end{aligned} \tag{10.87}$$

Eq.(10.87) is also called the *equation of state*. If y_i is the volume V of the system, the corresponding generalized force is $-P$. Then the equation of state for hydrodynamic system is

$$P = \frac{1}{\beta} \frac{\partial \ln Z}{\partial V}. \tag{10.88}$$

According to Eq.(10.42), we have

$$\begin{aligned} đQ &= d\bar{E} - đW \\ &= d\bar{E} - \sum_i \overline{Y}_i dy_i \\ &= -d \frac{\partial \ln Z}{\partial \beta} + \frac{1}{\beta} \sum_i \frac{\partial \ln Z}{\partial y_i} dy_i. \end{aligned} \tag{10.89}$$

Using

$$d(\ln Z) = \frac{\partial \ln Z}{\partial \beta} d\beta + \sum_i \frac{\partial \ln Z}{\partial y_i} dy_i, \tag{10.90}$$

we can eliminate the summation over i in Eq.(10.89) and obtain

$$
\begin{aligned}
\mathchar'26\mkern-12mu dQ &= -d\frac{\partial \ln Z}{\partial \beta} + \frac{1}{\beta}\sum_i \frac{\partial \ln Z}{\partial y_i} dy_i \\
&= -d\frac{\partial \ln Z}{\partial \beta} + \frac{1}{\beta}\left[d(\ln Z) - \frac{\partial \ln Z}{\partial \beta}d\beta\right] \\
&= \frac{1}{\beta}d\left(\ln Z - \beta\frac{\partial \ln Z}{\partial \beta}\right) \\
&= \frac{1}{\beta k}dS \\
&= TdS.
\end{aligned}
\tag{10.91}
$$

Eq.(10.91) is the *Clausius relation*. Thus Eq.(10.42) becomes

$$
d\overline{E} = TdS + \sum_i \overline{Y}_i dy_i.
\tag{10.92}
$$

Eq.(10.92) is called the *fundamental thermodynamic relation*. When y_i is the volume V, we have

$$
d\overline{E} = TdS - PdV.
\tag{10.93}
$$

We often denote $\overline{E}$ as E or U to simplify notation.

10.5.4   *Kelvin formulation of second law of thermodynamics*

The second law of thermodynamics determines the direction of irreversible processes. There are infinite kinds of irreversible processes. Thus we could have infinite kinds of formulations of the second law of thermodynamics. They are all equivalent. We have shown that $\Delta S \geq 0$ for a process in an isolated system. Now we will show another formulation of the second law of thermodynamics-the Kelvin formulation: There exists no thermodynamic transformation whose sole effect is to convert entirely a quantity of heat from a heat reservoir into work. We will prove the Kevin formulation from the entropy increase principle.

We consider a heat machine as shown in Fig.10.1. There are two processes supposed to be operated by the heat machine. One is the normal process. In this process, the work w is transferred into heat q by a machine and released into the heat reservoir. The conservation of energy demands $q = w$. The second process is the reverse process. In the reverse process, the heat in the heat reservoir is transferred into work and released to outside. The Kelvin formulation is equivalent to the statement that the reverse process is not possible. We will prove the Kelvin statement by *reductio ad absurdum*. Suppose that the Kelvin formulation is false. The reverse process is possible. We evaluate the change of entropy ΔS in the reverse process. There are three parts: the machine, the heat reservoir, the outside. (i) The machine returns to its starting position after a cycle of operation. We have the change of the entropy $\Delta S_m = 0$ (ii) Only work is released to the outside. We have the change of the entropy $\Delta S_o = 0$. (iii) The heat reservoir release a quantity of heat q. Thus the change of the entropy is given by $\Delta S_r = đ Q/T = -q/T$. Then the total change of the entropy

$$\Delta S = \Delta S_m + \Delta S_o + \Delta S_r = -\frac{q}{T} < 0. \tag{10.94}$$

This is in contradiction with the entropy increase principle. Therefore, the reverse process is not possible and the Kelvin formulation must be true.

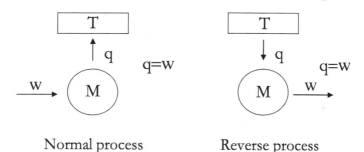

Fig. 10.1   Heat engine with one heat reservoir.

10.5.5   *Carnot theorem*

Now we consider a thermal engine as shown in Fig.10.2. A machine operates between two heat reservoirs. It absorbs heat q_1 from the high temperature reservoir and transfer the heat into work w. According to the Kelvin statement, the machine can not transfer the entire heat into work. It should release some heat q_2 into the low temperature reservoir. The conservation of energy demands $w = q_1 - q_2$. Using $\Delta S \geq 0$, we have

$$\Delta S = -\frac{q_1}{T_1} + \frac{q_2}{T_2} = -\frac{q_1}{T_1} + \frac{q_1 - w}{T_2} \geq 0. \tag{10.95}$$

Eq.(10.95) can be rewritten as

$$\eta \equiv \frac{w}{q_1} \leq 1 - \frac{T_2}{T_1} = \frac{T_1 - T_2}{T_1}. \tag{10.96}$$

where η is called the *efficiency of engine*. The equal sign holds when the process is the quasi static process, which is defined as an ideal process so slow that the system can be considered as in equilibrium during all the process. The quasi static process is reversible. A thermal engine that operates with the reversible process is called the *Carnot engine*. The efficiency of Carnot engine is given by

$$\eta_c = 1 - \frac{T_2}{T_1}. \tag{10.97}$$

All Carnot engines operating between two given temperatures have the same efficiency. According to Eq.(10.96), we have the Carnot theorem: No engine operating between two heat reservoirs is more efficient than a Carnot engine.

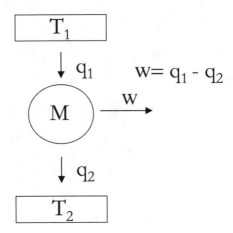

Fig. 10.2   Heat engine with two heat reservoirs.

10.5.6   *Clausius inequality*

We examine a system in contact with a heat reservoir. The temperature of the heat reservoir is T. The system absorbs an amount of heat $đQ$. When the heat reservoir is large, the heat releasing process of the heat reservoir can be considered as a quasi static process. However, the process in the system is not a quasi static process generally. The change of entropy in the heat reservoir is given by

$$dS_r = -\frac{đQ}{T}.$$ (10.98)

We denote dS as the change of entropy in the system. Then the total change of the entropy is given by

$$dS_t = dS + dS_r = dS - \frac{đQ}{T}.$$ (10.99)

The heat reservoir and the system together can be considered as an isolated system. We have

$$dS_t = dS - \frac{đQ}{T} \geq 0.$$ (10.100)

Thus

$$dS \geq \frac{đQ}{T}.$$ (10.101)

Eq.(10.101) is called the *Clausius inequality*, which can also be considered as a formulation of the second law of thermodynamics. For an adiabatic process, $đQ = 0$, we recover the entropy increase principle

$$dS \geq 0.$$ (10.102)

10.5.7 *Characteristic functions*

We define the *free energy F* by

$$F \equiv E - TS.$$ (10.103)

Using Eq.(10.80), we have

$$F = -kT \ln Z.$$ (10.104)

F is also called the Helmholtz free energy. The variation of the free energy is given by

$$dF = -SdT - PdV.$$ (10.105)

We call the replacement E by $F = E - TS$ as a *Legendre's transformation*. The term TdS in Eq.(10.93) is replaced by $-SdT$ in Eq.(10.105). Eq.(10.93) is often used when S and V are independent variables while Eq.(10.105) is used when T and V are independent variables. When a function with suitable variables (so-called *natural variables*) contains all thermodynamic information, we call this function as a *characteristic function*. $E(S, V)$ is such a function. We can obtain all other thermodynamic functions when we know the function $E(S, V)$. From Eq.(10.93), we have

$$T = \left(\frac{\partial E}{\partial S}\right)_V, \quad P = -\left(\frac{\partial E}{\partial V}\right)_S. \tag{10.106}$$

The equation of state $P = (V, T)$ can be obtained by eliminating the variable S in Eq.(10.106). Therefore, $E(S, V)$ is a characteristic function.

$F(T, V)$ is also a characteristic function. There are two other important characteristic functions which can be constructed through the Legendre's transformation. One is the *enthalpy* defined by

$$H \equiv E + PV. \tag{10.107}$$

$H(S, P)$ is a characteristic function with the variables S and P. The other is the *Gibbs free energy* defined by

$$G \equiv F + PV = E - TS + PV. \tag{10.108}$$

$G(T, P)$ is a characteristic function with the variables T and P.

10.5.8 *Maxwell relations*

The following fundamental relations are related through the Legendre's transformations:

$$dE = TdS - PdV, \tag{10.109a}$$
$$dH = TdS + VdP, \tag{10.109b}$$
$$dF = -SdT - PdV, \tag{10.109c}$$
$$dG = -SdT + VdP. \tag{10.109d}$$

From Eq.(10.109), we can easily obtain the following four relations between derivatives

$$\left(\frac{\partial T}{\partial V}\right)_S = -\left(\frac{\partial P}{\partial S}\right)_V, \tag{10.110a}$$

$$\left(\frac{\partial T}{\partial P}\right)_S = \left(\frac{\partial V}{\partial S}\right)_P, \tag{10.110b}$$

$$\left(\frac{\partial S}{\partial V}\right)_T = \left(\frac{\partial P}{\partial T}\right)_V, \tag{10.110c}$$

$$\left(\frac{\partial S}{\partial P}\right)_T = -\left(\frac{\partial V}{\partial T}\right)_P. \tag{10.110d}$$

For example,

$$\left(\frac{\partial T}{\partial V}\right)_S = \frac{\partial^2 E}{\partial V \partial S} = \frac{\partial^2 E}{\partial S \partial V} = -\left(\frac{\partial P}{\partial S}\right)_V. \qquad (10.111)$$

The four relations between derivatives are called the *Maxwell relations*. They are useful to express the thermodynamic variables in terms of measurable variables.

10.5.9 *Isothermal processes*

We have shown that in an isolated system, the entropy of the system always increases. It reaches its maximin when the system reaches its equilibrium state.

When the system is in contact with a heat reservoir and is not isolated, we have other criteria to determine the process direction. For a system with constant temperature T and volume V, we have

$$\begin{aligned}
\Delta F &= \Delta(E - TS) \\
&= \Delta E - T\Delta S \\
&= \Delta Q + \Delta W - T\Delta S \qquad (10.112) \\
&= \Delta Q - T\Delta S \\
&\leq 0.
\end{aligned}$$

In the derivation, we have used $\Delta W = 0$ because $V = const.$ Eq.(10.112) shows that an isothermal and isochoral process takes the direction that the free energy F decreases.

For a system with constant temperature T and pressure P, we have

$$\begin{aligned}
\Delta G &= \Delta(E - TS + PV) \\
&= \Delta Q + \Delta W - T\Delta S - P\Delta V \\
&= \Delta Q - T\Delta S \qquad (10.113) \\
&\leq 0.
\end{aligned}$$

Eq.(10.113) states that an isothermal and isobaric process takes the direction that the Gibbs free energy G decreases.

10.5.10 *Derivatives of thermodynamic quantities*

In this section, we will introduce several important thermodynamic derivatives. We define the *heat capacity* by

$$C_x \equiv \left.\frac{d\!\!^- Q}{dT}\right|_x. \qquad (10.114)$$

The specific heat capacity is the heat capacity per unit mass. Since dQ depends on the path of a process, the heat capacity also depends on the path of process. For an adiabatic process, $dQ = 0$. Thus the adiabatic heat capacity $C_S = 0$. For an isothermal process, $dT = 0$. Then the isothermal heat capacity $C_T = \infty$. The two most important heat capacities are the heat capacity C_V at constant V and the heat capacity C_P at constant P. They can be expressed as the derivatives of state functions

$$C_V = T \left(\frac{\partial S}{\partial T} \right)_{V,N} = \left(\frac{\partial E}{\partial T} \right)_{V,N} \tag{10.115}$$

and

$$C_P = T \left(\frac{\partial S}{\partial T} \right)_{P,N} = \left(\frac{\partial H}{\partial T} \right)_{P,N}. \tag{10.116}$$

Other important thermodynamic derivatives include the compressibility, the coefficient of thermal expansion, and the thermal pressure coefficient. The *compressibility* is defined by

$$\kappa \equiv -\frac{1}{V} \frac{dV}{dP}. \tag{10.117}$$

When there is no heat transfer, it is called the adiabatic (isentropic) compressibility

$$\kappa_S = -\frac{1}{V} \left(\frac{\partial V}{\partial P} \right)_{S,N}. \tag{10.118}$$

For a compression at a constant temperature, we have the isothermal compressibility

$$\kappa_T = -\frac{1}{V} \left(\frac{\partial V}{\partial P} \right)_{T,N}. \tag{10.119}$$

To describe the thermal expansion, we use the *coefficient of thermal expansion* defined by

$$\alpha \equiv \frac{1}{V} \left(\frac{\partial V}{\partial T} \right)_{P,N}. \tag{10.120}$$

The *thermal pressure coefficient* is defined by

$$\beta \equiv \frac{1}{P} \left(\frac{\partial P}{\partial T} \right)_{V,N}. \tag{10.121}$$

Using Eq.(G.7) in the appendix G, we have

$$\alpha = \kappa_T \beta P. \tag{10.122}$$

The quantities such as κ_T show how the extensive quantities vary with the change of the intensive quantities. We also called them *susceptibilities*.

10.6 Third law of thermodynamics

When $T = 0$, the value of the entropy depends on the degeneracy of the ground state. We denote the degeneracy of the ground state energy E_0 as Ω_0. The density matrix of the canonical ensemble can be written as

$$\begin{aligned}
\hat{\rho} &= \frac{e^{-\beta \hat{H}}}{\mathrm{Tr} e^{-\beta \hat{H}}} \\
&= \frac{\sum_n e^{-\beta E_n} |n\rangle\langle n|}{\sum_n e^{-\beta E_n}} \\
&= \frac{\sum_{i=1}^{\Omega_0} |0\rangle_{ii}\langle 0| + \sum_{n \neq 0} e^{-\beta(E_n - E_0)} |n\rangle\langle n|}{\Omega_0 + \sum_{n \neq 0} e^{-\beta(E_n - E_0)}},
\end{aligned} \tag{10.123}$$

where $|0\rangle_i$ is the ith degenerate ground state. Thus the entropy at $T = 0$ is given by

$$S(T = 0) = -k \mathrm{Tr}(\hat{\rho} \ln \hat{\rho}) = k \ln \Omega_0. \tag{10.124}$$

The ground states of all known systems are found to have degeneracy $\Omega_0 = \mathcal{O}(1)$. We have

$$\lim_{\substack{T \to 0 \\ N \to \infty}} \frac{S}{kN} = 0. \tag{10.125}$$

Even if $\Omega_0 = \mathcal{O}(N)$, Eq.(10.125) is still hold. Eq.(10.125) is not a proved results. It is only a summary of known properties of the ground states. Although Eq.(10.125) can not be proven strictly, it is expected to be hold generally because all the known results on the ground states suggest the validity of Eq.(10.125). Eq.(10.125) is called *Nernst's theorem* or the *third law of thermodynamics*. The Nernst's theorem has some restrictions on the specific heat capacity and other thermodynamic quantities.

Using Eq.(10.114), we have

$$S(T) - S(T = 0) = \int_0^T dT \frac{C_x}{T}. \tag{10.126}$$

Since $\frac{S(T=0)}{N} = 0$ in the thermodynamic limit $N \to \infty$, Eq.(10.126) becomes

$$S(T) = \int_0^T dT \frac{C_x}{T}. \tag{10.127}$$

In order to have convergent integration, we have

$$C_x(T) \to 0 \quad \text{for} \quad T \to 0. \tag{10.128}$$

Let us consider other thermodynamic derivatives. For the coefficient of thermal expansion α, we have

$$\alpha = \frac{1}{V}\left(\frac{\partial V}{\partial T}\right)_P = -\frac{1}{V}\left(\frac{\partial S}{\partial P}\right)_T \xrightarrow[T\to 0]{} 0. \tag{10.129}$$

The ratio of the thermal expansion coefficient α to the isothermal compressibility κ_T also approaches zero when $T \to 0$.

$$\begin{aligned}
\frac{\alpha}{\kappa_T} &= \frac{\dfrac{1}{V}\left(\dfrac{\partial V}{\partial T}\right)_P}{-\dfrac{1}{V}\left(\dfrac{\partial V}{\partial P}\right)_T} \\
&= \left(\frac{\partial P}{\partial T}\right)_V \\
&= \left(\frac{\partial S}{\partial V}\right)_T \to 0.
\end{aligned} \tag{10.130}$$

In the derivation of Eq.(10.130), we have used Eq.(G.7) in the appendix G and the Maxwell relations.

When Eq.(10.125) holds, we can show that one needs infinitively many steps to reach the temperature of zero, which is called the *Nernst principle*.

10.7   Thermodynamic quantities expressed in terms of grand partition function

Thermodynamic functions can also be evaluated using the grand partition function. The average value of an observable $\hat{A}$ is evaluated by Eq.(10.34). First we calculate the average number of particles for an open system in equilibrium with a heat and particle reservoir.

$$\begin{aligned}
\overline{N} &= \sum_N \sum_s N p_{s,N} \\
&= \Xi^{-1} \sum_N \sum_s N e^{-\alpha N - \beta E_s} \\
&= -\Xi^{-1} \frac{\partial}{\partial \alpha} \sum_N \sum_s e^{-\alpha N - \beta E_s} \\
&= -\frac{\partial \ln \Xi}{\partial \alpha},
\end{aligned} \tag{10.131}$$

where

$$\alpha = -\beta\mu. \tag{10.132}$$

The energy $\overline{E}$ of the system is given by

$$\overline{E} = \sum_N \sum_s E_s p_{s,N}$$

$$= \Xi^{-1} \sum_N \sum_s E_s e^{-\alpha N - \beta E_s}$$

$$= -\Xi^{-1} \frac{\partial}{\partial \beta} \sum_N \sum_s e^{-\alpha N - \beta E_s} \tag{10.133}$$

$$= -\frac{\partial \ln \Xi}{\partial \beta}.$$

The generalized force is the average of $\frac{\partial E_s}{\partial y}$.

$$\overline{Y}_i = \sum_N \sum_s \frac{\partial E_s}{\partial y} p_{s,N}$$

$$= \Xi^{-1} \sum_N \sum_s \frac{\partial E_s}{\partial y} e^{-\alpha N - \beta E_s}$$

$$= -\Xi^{-1} \frac{1}{\beta} \frac{\partial}{\partial y} \sum_N \sum_s e^{-\alpha N - \beta E_s} \tag{10.134}$$

$$= -\frac{1}{\beta} \frac{\partial \ln \Xi}{\partial y}.$$

If y_i is the volume V of the system, we have the equation of state for a hydrodynamic system

$$P = \frac{1}{\beta} \frac{\partial \ln \Xi}{\partial V}. \tag{10.135}$$

Using Eq.(10.82), the entropy is given by

$$S = -k \overline{\ln p}$$

$$= k\beta(\overline{E} - \mu \overline{N}) + k \ln \Xi$$

$$= k \left(\ln \Xi - \alpha \frac{\partial \ln \Xi}{\partial \alpha} - \beta \frac{\partial \ln \Xi}{\partial \beta} \right). \tag{10.136}$$

10.8 Relation between grand partition function and partition function

Eq.(10.32) shows that the grand partition function has the following relation with the partition function

$$\Xi(\alpha, \beta, y) = \sum_{N=0}^{\infty} e^{\beta \mu N} Z_N(\beta, y) \tag{10.137}$$

with

$$Z_N(\beta, y) = \sum_s e^{-\beta E_s}. \tag{10.138}$$

Z_N is the partition function for an N-particle system. When $N = 0$, we define $Z_0 \equiv 1$. Using the *fugacity* q defined by

$$q \equiv e^{-\alpha} = e^{\frac{\mu}{kT}}, \tag{10.139}$$

Eq.(10.137) can be rewritten as

$$\Xi(q, \beta, y) = \sum_{N=0}^{\infty} q^N Z_N(\beta, y). \tag{10.140}$$

Generally, for a classical system, it is more easy to calculate the partition function Z. The grand partition function can also be calculated using Eq.(10.137). For a quantum system, the grand partition is usually more easy to obtain. Then the partition functions can be calculated using Eq.(10.140) from the grand partition function as the expansion coefficients.

10.9   Systems with particle number changeable

10.9.1   *Thermodynamic relations for open systems*

Let us consider entropy $S(E, V, N)$ with variables E, V, N.

$$\begin{aligned} dS &= \left(\frac{\partial S}{\partial E}\right)_{V,N} dE + \left(\frac{\partial S}{\partial V}\right)_{E,N} dV + \left(\frac{\partial S}{\partial N}\right)_{E,V} dN \\ &= \frac{1}{T} dE + \frac{P}{T} dV - \frac{\mu}{T} dN \end{aligned} \tag{10.141}$$

or

$$dE = TdS - PdV + \mu dN. \tag{10.142}$$

For a system with multi-components, we denote μ_i as the chemical potential of the ith component ($i = 1, \cdots, n$). The chemical potential term in Eq.(10.141) should be replaced by a summation over all components for a multi-component system. Then we have

$$dE = TdS - PdV + \sum_i \mu_i dN_i. \tag{10.143}$$

Eq.(10.143) is the fundamental thermodynamic relation for the open systems.

The other derivative relations can be obtained through Legendre's transformations.

$$dH = TdS + VdP + \sum_i \mu_i dN_i, \tag{10.144a}$$

$$dF = -SdT - PdV + \sum_i \mu_i dN_i, \tag{10.144b}$$

$$dG = -SdT + VdP + \sum_i \mu_i dN_i. \tag{10.144c}$$

Thus the chemical potentials can be evaluated through a variety of relations.

$$
\begin{aligned}
\mu_i &= \left(\frac{\partial E}{\partial N_i} \right)_{S,V,N_j \neq N_i} \\
&= \left(\frac{\partial H}{\partial N_i} \right)_{S,P,N_j \neq N_i} \\
&= \left(\frac{\partial F}{\partial N_i} \right)_{T,V,N_j \neq N_i} \\
&= \left(\frac{\partial G}{\partial N_i} \right)_{T,P,N_j \neq N_i}.
\end{aligned}
\tag{10.145}
$$

Eq.(10.144a) also gives

$$\left(\frac{\partial S}{\partial N_i} \right)_{E,V,N_j \neq N_i} = -\frac{\mu_i}{T}. \tag{10.146}$$

10.9.2 Gibbs-Duhem relation

Chemical potential is one of the most important functions in statistical mechanics. From Eq.(10.145), we have

$$\mu = \left. \frac{\partial F}{\partial N} \right|_{V,T} = \left. \frac{\partial}{\partial N} \left(-\frac{1}{\beta} \ln Z \right) \right|_{V,\beta}. \tag{10.147}$$

According to Eq.(10.77), $F(T, V, N)$ is an *extensive quantity* with the property

$$F(T, \alpha V, \alpha N) = \alpha F(T, V, N), \tag{10.148}$$

where α is a factor by which the system is enlarged. Since $T = \partial \bar{E}/\partial S$, T is not changed by enlarging the system. We call T an *intensive quantity*.

Differentiating Eq.(10.148) with respect to α and then setting $\alpha = 1$, we find

$$
\begin{aligned}
F &= \left[V \frac{\partial}{\partial (\alpha V)} F(T, \alpha V, \alpha N) + N \frac{\partial}{\partial (\alpha N)} F(T, \alpha V, \alpha N) \right]_{\alpha=1} \\
&= -PV + \mu N,
\end{aligned}
\tag{10.149}
$$

which gives

$$E = TS - PV + \mu N. \tag{10.150}$$

Eq.(10.150) is called the *Gibbs-Duhem relation*. Using Eq.(10.150), we obtain

$$G = E - TS + PV = \mu N, \tag{10.151}$$

which shows that the chemical potential μ is the Gibbs free energy per particle.

The Gibbs-Duhem relation for multi-component systems is given by

$$G = \sum_i \mu_i N_i. \tag{10.152}$$

Similar to the free energy defined by Eq.(10.104), we introduce the grand potential J defined by

$$J \equiv -kT \ln \Xi. \tag{10.153}$$

From Eq.(10.136), we have

$$k \ln \Xi = S - k\beta(\overline{E} - \mu\overline{N}). \tag{10.154}$$

Replacing $k \ln \Xi$ in Eq.(10.153) by $S - k\beta(\overline{E} - \mu\overline{N})$, we have

$$J = \overline{E} - TS - \mu\overline{N}. \tag{10.155}$$

Since $G = \mu\overline{N} = \overline{E} - TS + PV$, we have

$$J = -PV, \tag{10.156}$$

which gives

$$PV = kT \ln \Xi. \tag{10.157}$$

For the grand potential J, we have

$$\begin{aligned}
dJ &= dF - dG \\
&= dF - d(\sum_i \mu_i N_i) \\
&= -SdT - PdV - \sum_i N_i d\mu_i.
\end{aligned} \tag{10.158}$$

Eq.(10.158) is the fundamental thermodynamic relation for the grand potential. Inserting $J = -PV$ into Eq.(10.158), we have

$$SdT - VdP + \sum_i N_i d\mu_i = 0. \tag{10.159}$$

This is the *differential Gibbs-Duhem relation*. In the case of one component, it shows that T, P, and μ can not be varied independently. We have only two independent intensive variables for a homogeneous system with one component.

10.9.3 *Equilibrium conditions of two systems*

Let us consider two macroscopic systems. In the following, we discuss what is the conditions that these two systems are in equilibrium with each other. We denote the two systems as A_1 and A_2. The two systems A_1 and A_2 can be considered to form an isolated system A_t with the total energy E_t and total volume V_t. We have two constraint conditions

$$E_t = E_1 + E_2, \tag{10.160a}$$

$$V_t = V_1 + V_2. \tag{10.160b}$$

If the particle number is conserved, we have another constraint condition

$$N_t = N_1 + N_2. \tag{10.161}$$

The number of microscopic states Ω_t of the total system is given by

$$\begin{aligned}
\Omega_t(E_t, V_t, N_t) &= \Omega_1(E_1, V_1, N_1)\Omega_2(E_2, V_2, N_2) \\
&= \Omega_1(E_1, V_1, N_1)\Omega_2(E_t - E_1, V_t - V_1, N_t - N_1).
\end{aligned} \tag{10.162}$$

The equilibrium state is the state with the maximum entropy. Thus the equilibrium condition is

$$dS = kd(\ln \Omega_t) = 0, \tag{10.163}$$

which gives

$$\begin{aligned}
&\left(\frac{\partial \ln \Omega_1}{\partial E_1} - \frac{\partial \ln \Omega_2}{\partial E_2}\right) dE_1 \\
&+ \left(\frac{\partial \ln \Omega_1}{\partial V_1} - \frac{\partial \ln \Omega_2}{\partial V_2}\right) dV_1 \\
&+ \left(\frac{\partial \ln \Omega_1}{\partial N_1} - \frac{\partial \ln \Omega_2}{\partial N_2}\right) dN_1 = 0.
\end{aligned} \tag{10.164}$$

Using Eq.(10.141), we have

$$\frac{\partial \ln \Omega}{\partial E} = \frac{1}{kT}, \tag{10.165a}$$

$$\frac{\partial \ln \Omega}{\partial V} = \frac{P}{kT}, \tag{10.165b}$$

$$\frac{\partial \ln \Omega}{\partial N} = -\frac{\mu}{kT}. \tag{10.165c}$$

Since dE_1, dV_1 and dN_1 in Eq.(10.164) are independent, we obtain

$$T_1 = T_2 \quad \text{(thermal equilibrium condition)}, \tag{10.166a}$$

$$P_1 = P_2 \quad \text{(mechanical equilibrium condition)}, \tag{10.166b}$$

$$\mu_1 = \mu_2 \quad \text{(chemical equilibrium condition)}. \tag{10.166c}$$

These are the three equilibrium conditions between two macroscopic systems. Eq.(10.166a) is also called the *zeroth law of thermodynamics*, which states that when two systems are in thermal equilibrium with a third system, then they are in equilibrium with one another.

10.9.4   *Phase equilibrium conditions*

In thermodynamics, a phase is defined as a homogeneous part of a macroscopic system. When the system is homogeneous, there is only one phase in the system. If the system can be divided into two homogeneous parts, then there are two phases in the system. For example, in the low temperature, the atoms arrange themselves orderly to form a solid state to achieve the lowest energy. With the increase of temperature, the entropy begins to have effect. The entropy makes the system become disordered. When the temperature is high enough, the solid phase begins to melt into a disordered state (called liquid phase). Then we have a system with two phases.

The equilibrium conditions for two phases are similar to the equilibrium conditions Eq.(10.166) for two systems. We can generalize Eq.(10.166) to a system of the φ phases with k components in each phase. The equilibrium conditions are given by

$$T_\alpha = T_\beta = \cdots = T_\varphi, \tag{10.167a}$$

$$P_\alpha = P_\beta = \cdots = P_\varphi, \tag{10.167b}$$

$$\mu_{\alpha,1} = \mu_{\beta,1} = \cdots = \mu_{\varphi,1}, \tag{10.167c}$$

$$\cdots \tag{10.167d}$$

$$\mu_{\alpha,k} = \mu_{\beta,k} = \cdots = \mu_{\varphi,k}. \tag{10.167e}$$

Together we have $(k+2)(\varphi-1)$ equations. Since the concentrations $x_{\alpha,i} = \frac{N_{\alpha,i}}{N_\alpha}$ satisfy the following normalization condition

$$x_{\sigma,1} + x_{\sigma,2} + \cdots + x_{\sigma,k} = 1, \tag{10.168}$$

we have $2\varphi + (k-1)\varphi = (k+1)\varphi$ variables. Then the number of independent variables (also called freedom number) f is given by

$$f = (k+1)\varphi - (k+2)(\varphi-1) = k+2 - \varphi \geq 0. \tag{10.169}$$

Eq.(10.169) is the *Gibbs phase rule*. For a system with one component, there are only pure phases. From Eq.(10.169), we have $\varphi \leq 3$. Thus the maximum number of the coexistent phases is three for a pure phase system. For example, liquid water, water vapor, and one type of solid ice can coexist at $T_t = 273.16\text{K}$ and $P_t = 4.58\text{Torr}$. The coexistent point is called the triple point. The absolute temperature scale is defined by the triple point of water. The constant k in $S = k \ln \Omega$ fixed by this temperature unit is called the Boltzmann constant. We usually denote the Boltzmann constant as k_B.

10.10 Equilibrium distributions of nearly independent particle systems

Now we discuss the calculations of the thermodynamic properties of the systems composed of nearly independent particles. For example, in a gas, the distance between particles are large. Thus the interaction of particles in such systems are weak and can be neglected. When the interaction can be neglected, the energy of the system can be described by the single-particle energy. We can then use the distribution functions of single particle to describe the statistical properties of the system.

10.10.1 *Derivations of distribution functions of single particle from macro-canonical distribution*

10.10.1.1 *Expressions in terms of single particle quantities*

We denote ε_i as the single-particle energy and g_i as the degeneracy of the energy ε_i. We define the distribution function n_i as the number of particles occupying the energy level ε_i. The distribution $\{n_i\}$ is a macroscopic state of the system. For a system on the state s with particle number N and energy E_s, we have

$$N = \sum_i n_i, \tag{10.170a}$$

$$E_s = \sum_i \varepsilon_i. \tag{10.170b}$$

The equilibrium state is described by the macro-canonical distribution

$$p_{s,N} = \Xi^{-1} e^{-\alpha N - \beta E_s}. \tag{10.171}$$

$p_{s,N}$ is the probability of the system occupying the state s with the particle number N and the energy E_s. If there are $\Omega_{\{n_i\}}$ microscopic states for the distribution $\{n_i\}$, the probability of the system occupying the macroscopic state $\{n_i\}$ is given by

$$p_{\{n_i\},N} = \Xi^{-1} \Omega_{\{n_i\}} e^{-\alpha N - \beta E_s}, \tag{10.172}$$

where N is the particle number of the system in the macroscopic state $\{n_i\}$ and E_s is the energy of the system in the macroscopic state $\{n_i\}$. They are given by Eq.(10.170). Using the normalization condition

$$\sum_N \sum_{\{n_i\}'} p_{\{n_i\},N} = 1, \tag{10.173}$$

we have

$$\Xi(\alpha, \beta, y) = \sum_N \sum_{\{n_i\}'} \Omega_{\{n_i\}} e^{-\alpha N - \beta E_s}. \tag{10.174}$$

where the prime in $\{n_i\}'$ represents that the summation is over all the distributions $\{n_i\}$ at fixed N.

For an independent particle system, $\Omega_{\{n_i\}}$ is the number of the microscopic states corresponding to the distribution $\{n_i\}$. We denote Ω_i as the number of distinct ways of assigning the n_i particles to g_i degenerate states of ε_i. Then $\Omega_{\{n_i\}}$ is equal to the multiplying of all Ω_i, which gives

$$\Omega_{\{n_i\}} = \prod_i \Omega_i. \tag{10.175}$$

Inserting Eq.(10.170) into Eq.(10.172), we have

$$p_{\{n_i\}, N} = \Xi^{-1} \Omega_{\{n_i\}} e^{-\alpha \sum_i n_i - \beta \sum_i n_i \varepsilon_i} \tag{10.176}$$

and

$$\Xi(\alpha, \beta, y) = \sum_{N=0}^{\infty} \sum_{\{n_i\}'} \Omega_{\{n_i\}} e^{-\alpha \sum_i n_i - \beta \sum_i n_i \varepsilon_i}. \tag{10.177}$$

The second summation is over all the distributions $\{n_i\}$ at fixed N. Since we have the summation over N together with that over $\{n_i\}'$, which releases the restriction on $\{n_i\}'$, we can change the summations over N and $\{n_i\}'$ to the summation over $\{n_i\}$ without restriction. Eq.(10.177) becomes

$$\begin{aligned}
\Xi(\alpha, \beta, y) &= \sum_{n_1=0}^{\infty} \sum_{n_2=0}^{\infty} \cdots \sum_{n_i=0}^{\infty} \cdots \prod_i \Omega_i e^{-(\alpha + \beta \varepsilon_i) n_i} \\
&= \sum_{n_1=0}^{\infty} \Omega_1 e^{-(\alpha + \beta \varepsilon_1) n_1} \sum_{n_2=0}^{\infty} \Omega_2 e^{-(\alpha + \beta \varepsilon_2) n_2} \cdots \\
&= \prod_i \sum_{n_i=0}^{\infty} \Omega_i e^{-(\alpha + \beta \varepsilon_i) n_i} \\
&= \prod_i \Xi_i(\alpha, \beta, y)
\end{aligned} \tag{10.178}$$

with

$$\Xi_i(\alpha, \beta, y) = \sum_{n_i=0}^{\infty} \Omega_i e^{-(\alpha + \beta \varepsilon_i) n_i}. \tag{10.179}$$

Now we calculate the average particle number $\bar{n}_i$ on the ε_i energy level.

$$\bar{n}_i = \sum_N \sum_{\{n_i\}'} n_i p_{\{n_i\},N}$$

$$= \Xi^{-1} \sum_N \sum_{\{n_i\}'} n_i \Omega_{\{n_i\}} e^{-\alpha N - \beta E_s}$$

$$= \Xi^{-1} \sum_N \sum_{\{n_i\}'} n_i \Omega_{\{n_i\}} e^{-\alpha \sum_i n_i - \beta \sum_i n_i \varepsilon_i} \qquad (10.180)$$

$$- \Xi^{-1} \sum_{n_i=0}^{\infty} n_i \Omega_i c^{-(\alpha+\beta\varepsilon_i)n_i} \prod_{n_j \neq n_i} \sum_{n_j=0}^{\infty} \Omega_j e^{-(\alpha+\beta\varepsilon_j)n_j}.$$

Using Eq.(10.178), we have

$$\bar{n}_i = \Xi^{-1} \sum_{n_i=0}^{\infty} n_i \Omega_i e^{-(\alpha+\beta\varepsilon_i)n_i} \frac{\Xi}{\sum_{n_i=0}^{\infty} \Omega_i e^{-(\alpha+\beta\varepsilon_i)n_i}}$$

$$= -\frac{1}{\Xi_i} \frac{\partial \Xi_i}{\partial \alpha} \qquad (10.181)$$

$$= -\frac{\partial \ln \Xi_i}{\partial \alpha}.$$

In order to calculate Ξ_i, we need to evaluate Ω_i. There are two types of identical particles: bosons and fermions.

Bosons

First we discuss the system consisting of bosons, which is called Bose system. To facilitate the calculations, we use the graph in Fig.10.3 to represent the configurations of the n_i particles occupying g_i quantum states of ε_i. We use circles (○) to represent the particles and squares (□) to represent the quantum states. There are g_i squares. In Fig.10.3, the circles on the right of a quantum state □ denote the particles occupying the quantum state on their left. Since every particle should at least occupy one quantum state, the first position on the left should be a quantum state represented by a square. As an example, the graph in Fig.10.3 represents a configuration that ten particles on an energy level with four quantum states. In this configuration, there are two particles on the state 1, zero particle on the state 2, five particles on the state 3 and three particles on the state 4. Let us count the number of the distinct configurations of graphs, which gives the number Ω_i of the different ways assigning the n_i particles to the g_i degenerate states of ε_i. Since the first position on the left has to be occupied by a quantum state, we have g_i selections. There are n_i circles and

$g_i - 1$ squares left. If the circles and squares were labeled, there would be $(n_i + g_i - 1)!$ different ways to arrange them. However, the circles represent identical particles and are all equivalent, we need divide the configuration number by the number of the ways permuting particles which is $n_i!$. Likewise, quantum states are all equivalent and a factor $g_i!$ should be divided. We have

$$\Omega_i = \frac{g_i(n_i + g_i - 1)!}{n_i! g_i!} = \frac{(n_i + g_i - 1)!}{n_i!(g_i - 1)!}. \tag{10.182}$$

Multiplying all the factors Ω_i of each energy level ε_i, we have the total number of the microscopic states for a boson system

$$\Omega^{(B)}_{\{n_i\}} = \prod_i \frac{(n_i + g_i - 1)!}{n_i!(g_i - 1)!}. \tag{10.183}$$

$\square$ quantum states, $\mathbf{O}$ particles

Fig. 10.3   Schematic configuration of a microscopic state for a boson system.

Fermions

The system consisting of fermions is called Fermi system. Since fermions obey the Pauli exclusion principle, no more than one particle can occupy each quantum state. The number of possible ways to select n_i states from the g_i quantum states of ε_i for n_i particles to occupy is

$$\Omega_i = \frac{g_i!}{n_i!(g_i - n_i)!}. \tag{10.184}$$

Multiplying all the factors Ω_i of each energy level ε_i, we obtain the total number of the microscopic states for a fermion system

$$\Omega^{(F)}_{\{n_i\}} = \prod_i \frac{g_i!}{n_i!(g_i - n_i)!}. \tag{10.185}$$

10.10.1.2   *Bose distribution*

Using Ω_i given by Eq.(10.182) for Bose systems, we can evaluate $\Xi_i(\alpha, \beta, y)$ in Eq.(10.179). For Bose systems, we have

$$\Xi_i = \sum_{n_i=0}^{\infty} \frac{(n_i + g_i - 1)!}{n_i!(g_i - 1)!} e^{-(\alpha + \beta \varepsilon_i)n_i} \tag{10.186}$$

$$= \left(1 - e^{-\alpha - \beta \varepsilon_i}\right)^{-g_i}.$$

In the derivation of Eq.(10.186), we have used the following summation formula

$$(1 - x)^{-m} = 1 + mx + \frac{m(m+1)}{2!}x^2 + \cdots$$
$$= \sum_{n=0}^{\infty} \frac{(m+n-1)!}{n!(m-1)!}x^n. \tag{10.187}$$

Inserting Eq.(10.186) into Eq.(10.181), we have

$$\bar{n}_i = -\frac{\partial \ln \Xi_i}{\partial \alpha}$$
$$= \frac{g_i \partial \ln(1 - e^{-\alpha - \beta \varepsilon_i})}{\partial \alpha} \tag{10.188}$$
$$= \frac{g_i}{e^{\alpha + \beta \varepsilon_i} - 1}.$$

Eq.(10.188) is called the *Bose-Einstein distribution* or *Bose distribution*. The grand partition function Ξ for the independent boson systems is given by

$$\Xi(\alpha, \beta, y) = \prod_i \left(1 - e^{-\alpha - \beta \varepsilon_i}\right)^{-g_i} \tag{10.189}$$

or

$$\ln \Xi(\alpha, \beta, y) = -\sum_i g_i \ln \left(1 - e^{-\alpha - \beta \varepsilon_i}\right). \tag{10.190}$$

10.10.1.3 *Fermi distribution*

For fermions, we insert Ω_i in Eq.(10.184) into Eq.(10.179) and obtain

$$\Xi_i = \sum_{n_i=0}^{g_i} \frac{g_i!}{n_i!(g_i - n_i)!} e^{-(\alpha + \beta \varepsilon_i)n_i}$$
$$= \left(1 + e^{-\alpha - \beta \varepsilon_i}\right)^{g_i}. \tag{10.191}$$

In the derivation of Eq.(10.191), we have used the following summation formula

$$(1 + x)^m = 1 + mx + \frac{m(m-1)}{2!}x^2 + \cdots$$
$$= \sum_{n=0}^{m} \frac{m!}{n!(m-n)!}x^n. \tag{10.192}$$

Then Eq.(10.181) becomes

$$
\begin{aligned}
\bar{n}_i &= -\frac{\partial \ln \Xi_i}{\partial \alpha} \\
&= \frac{-g_i \partial \ln(1 + e^{-\alpha - \beta \varepsilon_i})}{\partial \alpha} \\
&= \frac{g_i}{e^{\alpha + \beta \varepsilon_i} + 1}.
\end{aligned}
\tag{10.193}
$$

Eq.(10.193) is called the *Fermi-Dirac distribution* or *Fermi distribution*. The grand partition function Ξ for the independent fermion systems is given by

$$
\Xi(\alpha, \beta, y) = \prod_i \left(1 + e^{-\alpha - \beta \varepsilon_i}\right)^{g_i}
\tag{10.194}
$$

or

$$
\ln \Xi(\alpha, \beta, y) = \sum_i g_i \ln \left(1 + e^{-\alpha - \beta \varepsilon_i}\right).
\tag{10.195}
$$

10.10.1.4   *Semi-classical distribution*

The Bose and Fermi distributions can be approximated by the classical distribution when the quantum correlations can be neglected. When

$$
g_i \gg n_i,
\tag{10.196}
$$

which is called the *non-degenerate condition* (or *semi-classical condition*), Ω_i for both boson and fermion systems are approximated by

$$
\begin{aligned}
\Omega_i &= \begin{cases}
\dfrac{(n_i + g_i - 1)!}{n_i!(g_i - 1)!} & \text{for bosons} \\[2ex]
\dfrac{g_i!}{n_i!(g_i - n_i)!} & \text{for fermions}
\end{cases} \\[2ex]
&\doteq \frac{g_i^{n_i}}{n_i!}.
\end{aligned}
\tag{10.197}
$$

$g_i \gg n_i$ means that there are much less than one particle in one quantum state in average. Thus the quantum correlations, such as the correlation due to the Pauli exclusion principle and also the main effect of indistinguishability of identical particles can be neglected. Using Eq.(10.197), we have

$$
\begin{aligned}
\Xi_i &= \sum_{n_i=0}^{\infty} \frac{g_i^{n_i}}{n_i!} e^{-(\alpha + \beta \varepsilon_i)n_i} \\
&= \sum_{n_i=0}^{\infty} \frac{1}{n_i!} \left(g_i e^{-\alpha - \beta \varepsilon_i}\right)^{n_i} \\
&= \exp\left(g_i e^{-\alpha - \beta \varepsilon_i}\right).
\end{aligned}
\tag{10.198}
$$

Inserting Eq.(10.198) into Eq.(10.181), we have

$$\bar{n}_i = -\frac{\partial \ln \Xi_i}{\partial \alpha} = g_i e^{-\alpha - \beta \varepsilon_i}. \tag{10.199}$$

Eq.(10.199) is called the *semi-classical distribution* or *Boltzmann distribution* for identical particles. From Eq.(10.199), we have

$$e^\alpha = \frac{g_i}{\bar{n}_i} e^{-\beta \varepsilon_i}. \tag{10.200}$$

If $e^\alpha \gg 1$, then the non-degenerate condition $g_i \gg \bar{n}_i$ holds. The condition

$$e^\alpha \gg 1 \tag{10.201}$$

is also called the non-degenerate condition.

10.10.2 *Partition function of independent particle systems*

Now we discuss the calculations of the partition function for independent particle systems. For quantum boson and fermion systems, we have shown that the grand partition function Ξ can be calculated easily. For the classical cases, we will show that partition function Z can be calculated easily.

For the case of independent classical particles, whose positions can be designated, the particles can be labeled. The particles can occupy the energy level ε_s independently. The energy E_n of the system is given by

$$E_n = \varepsilon_{s_1} + \varepsilon_{s_2} + \cdots + \varepsilon_{s_N}, \tag{10.202}$$

where s_α is the quantum number and ε_{s_α} is the single-particle energy of the particle α. The partition function of the system is given by

$$\begin{aligned}
Z &= \sum_n e^{-\beta E_n} \\
&= \sum_{s_1} \sum_{s_2} \cdots \sum_{s_N} e^{-\beta(\varepsilon_{s_1} + \varepsilon_{s_2} + \cdots + \varepsilon_{s_N})}.
\end{aligned} \tag{10.203}$$

Since there are no quantum correlation and particles are distinguishable, the summations in Eq.(10.203) are independent. We have

$$\begin{aligned}
Z &= \sum_{s_1} \sum_{s_2} \cdots \sum_{s_N} e^{-\beta(\varepsilon_{s_1} + \varepsilon_{s_2} + \cdots + \varepsilon_{s_N})} \\
&= \left(\sum_s e^{-\beta \varepsilon_s} \right)^N \\
&\equiv z^N
\end{aligned} \tag{10.204}$$

with

$$z = \sum_s e^{-\beta\varepsilon_s}. \tag{10.205}$$

z is called the *partition function of single particle*. If we use i to denote different energy levels ε_i with the degeneracy of g_i, we have

$$z = \sum_i g_i e^{-\beta\varepsilon_i}. \tag{10.206}$$

Eq.(10.204) leads to the relation of the partition function Z of the system with the partition function z of single particle for the classical independent particle systems

$$Z = z^N. \tag{10.207}$$

When the particles are indistinguishable, exchanging two particles gives the same microscopic state. Since exchanging two particles on the same quantum state of ε_i also gives the same microscopic state for the classical system with distinguishable particles, we need to exclude this possibility to simplify the calculation. For the semi-classical case, $n_i \ll g_i$, we do not have the possibility that one quantum state is occupied by two particles. Since exchanging two particles gives the same microscopic state, we need divide a factor $N!$ which is the number of ways permuting particles in Eq.(10.204). Thus we have

$$Z = \frac{1}{N!} \sum_{s_1} \sum_{s_2} \cdots \sum_{s_N} e^{-\beta(\varepsilon_{s_1}+\varepsilon_{s_2}+\cdots+\varepsilon_{s_N})}$$
$$\equiv \frac{1}{N!} z^N. \tag{10.208}$$

Eq.(10.208) is the relation of the partition function Z of system with the partition function z of single particle for the semi-classical independent particle systems.

It should be noted that Eq.(10.208) can not be applied to the general quantum boson and fermion systems. For a boson system, one quantum state can be occupied by more than one particles. For a fermion system, there is a limitation that one quantum state can only be occupied by one particle, the summations over $\{s_1, s_2, \cdots, s_N\}$ in Eq.(10.208) are not independent for general boson and fermion systems.

10.10.3 *About summations in calculations of independent particle system*

For an independent particle system, the Hamiltonian of the system contains no interaction term. There is only kinetic term

$$\hat{H} = H(\hat{\mathbf{p}}). \tag{10.209}$$

The summations involved in the calculations of thermodynamic quantities often have the form

$$\sum_s F(\varepsilon_s) = \mathrm{Tr}F(\hat{H}) = \mathrm{Tr}F(H(\hat{\mathbf{p}})). \tag{10.210}$$

The trace in Eq.(10.210) can be transformed into integration in the Γ space that spanned by the momentum $\mathbf{p}$ and position $\mathbf{q}$.

$$\begin{aligned}
\mathrm{Tr}F(H(\hat{\mathbf{p}})) &= \int d^f p \langle \mathbf{p}|F(H(\hat{\mathbf{p}}))|\mathbf{p}\rangle \\
&= \int d^f p \int d^f q \langle \mathbf{p}|F(H(\hat{\mathbf{p}}))|\mathbf{q}\rangle\langle\mathbf{q}|\mathbf{p}\rangle \\
&= \int d^f p \int d^f q F(H(\mathbf{p}))\langle\mathbf{p}|\mathbf{q}\rangle\langle\mathbf{q}|\mathbf{p}\rangle \qquad (10.211) \\
&= \int d^f p \int d^f q F(H(\mathbf{p}))\frac{1}{(2\pi\hbar)^f} \\
&= \int \frac{d^f p d^f q}{(2\pi\hbar)^f} F(H(\mathbf{p})).
\end{aligned}$$

When the energy-momentum relation is $\varepsilon = \frac{p^2}{2m}$ and $f = 3$, Eq.(10.211) has the form

$$\begin{aligned}
\mathrm{Tr}F(H(\hat{\mathbf{p}})) &= \frac{V}{(2\pi\hbar)^3}\int d^3 p F(\varepsilon(\mathbf{p})) \\
&= \frac{4\pi V}{(2\pi\hbar)^3}\int p^2 dp F(\varepsilon) \\
&= \frac{4\pi V}{(2\pi\hbar)^3}\int 2m\varepsilon \frac{\sqrt{2m}}{2}\frac{d\varepsilon}{\sqrt{\varepsilon}} F(\varepsilon) \qquad (10.212) \\
&= \frac{2\pi V}{h^3}(2m)^{\frac{3}{2}}\int \sqrt{\varepsilon}F(\varepsilon)d\varepsilon \\
&\equiv \int g(\varepsilon)F(\varepsilon)d\varepsilon
\end{aligned}$$

with

$$g(\varepsilon) = \frac{2\pi V}{h^3}(2m)^{\frac{3}{2}}\sqrt{\varepsilon}. \tag{10.213}$$

$g(\varepsilon)$ is called the *density of states*.

10.11   Fluctuations

The thermodynamic properties of a macroscopic system are determined by the statistical average of observables. However, there are always fluctuations around the average for a finite system. Now we discuss the fluctuations.

10.11.1   *Absolute and relative fluctuations*

For a physical quantity u, its deviation from the average is $\Delta u = u - \bar{u}$. Since $\overline{u - \bar{u}} = 0$, we define

$$\Delta \bar{u} \equiv \left[\overline{(u - \bar{u})^2} \right]^{\frac{1}{2}} \tag{10.214}$$

as the fluctuation of u.

$$\Delta \bar{u} = \left[\overline{(\Delta u)^2} \right]^{\frac{1}{2}} = (\overline{u^2} - \bar{u}^2)^{\frac{1}{2}}. \tag{10.215}$$

$\Delta \bar{u}$ is called the *absolute fluctuation*. The *relative fluctuation* is defined as

$$\delta \bar{u} \equiv \frac{\Delta \bar{u}}{\bar{u}} = \frac{(\overline{u^2} - \bar{u}^2)^{\frac{1}{2}}}{\bar{u}}. \tag{10.216}$$

10.11.2   *Fluctuations in systems of canonical ensemble*

First we consider the closed systems in which there is only energy exchange and no particle exchange. They are the systems of canonical ensemble. The fluctuation of energy is defined as

$$\Delta \overline{E} \equiv (\overline{E^2} - \overline{E}^2)^{\frac{1}{2}}. \tag{10.217}$$

$\overline{E}$ is given by Eq.(10.83). $\overline{E^2}$ can be evaluated as follows:

$$\begin{aligned}
\overline{E^2} &= \sum_n p_n E_n^2 \\
&= Z^{-1} \sum_n E_n^2 e^{-\beta E_n} \\
&= Z^{-1} \frac{\partial^2}{\partial \beta^2} \sum_n e^{-\beta E_n} \\
&= Z^{-1} \frac{\partial^2 Z}{\partial \beta^2} \\
&= \frac{\partial^2 \ln Z}{\partial \beta^2} + \left(\frac{\partial \ln Z}{\partial \beta} \right)^2.
\end{aligned} \tag{10.218}$$

Thus we have

$$\Delta \overline{E} = (\overline{E^2} - \overline{E}^2)^{\frac{1}{2}}$$

$$= \left(\frac{\partial^2 \ln Z}{\partial \beta^2} \right)^{\frac{1}{2}}$$

$$= \left(-\frac{\partial \overline{E}}{\partial \beta} \right)^{\frac{1}{2}} \tag{10.219}$$

$$= \left(-\frac{\partial \overline{E}}{\partial T} \frac{\partial T}{\partial \beta} \right)^{\frac{1}{2}}$$

$$= (k_B T^2 C_V)^{\frac{1}{2}}.$$

The relative fluctuation is given by

$$\delta \overline{E} = \frac{1}{\overline{E}} \left(k_B T^2 C_V \right)^{\frac{1}{2}}. \tag{10.220}$$

Eq.(10.219) shows that the heat capacity C_V is always positive

$$C_V \geq 0. \tag{10.221}$$

Since $\overline{E} \propto N$ and $C_V \propto N$, we have

$$\delta \overline{E} \propto N^{-\frac{1}{2}}. \tag{10.222}$$

For a macroscopic system ($N \sim 10^{23}$), the relative fluctuation is very small.

10.11.3 *Fluctuations in systems of grand canonical ensemble*

For an open system, there are both energy and particle exchange with reservoir. We call such system as the system of grand canonical ensemble. There is fluctuation of particle number in an open system in addition to the fluctuation of energy. The fluctuation of particle number is given by

$$\Delta \overline{N} = (\overline{N^2} - \overline{N}^2)^{\frac{1}{2}}. \tag{10.223}$$

$\overline{N}$ is calculated using Eq.(10.131). $\overline{N^2}$ can be evaluated as follows:

$$\overline{N^2} = \sum_N \sum_s N^2 p_{s,N}$$

$$= \Xi^{-1} \sum_N \sum_s N^2 e^{-\alpha N - \beta E_s}$$

$$= \Xi^{-1} \frac{\partial^2 \Xi}{\partial \alpha^2} \tag{10.224}$$

$$= \frac{\partial^2 \ln \Xi}{\partial \alpha^2} + \left(\frac{\partial \ln \Xi}{\partial \alpha} \right)^2.$$

Thus we have

$$
\begin{aligned}
\Delta \overline{N} &= (\overline{N^2} - \overline{N}^2)^{\frac{1}{2}} \\
&= \left(\frac{\partial^2 \ln \Xi}{\partial \alpha^2} \right)^{\frac{1}{2}} \\
&= \left[-\left(\frac{\partial \overline{N}}{\partial \alpha} \right)_{\beta V} \right]^{\frac{1}{2}} \\
&= \left[k_B T \left(\frac{\partial \overline{N}}{\partial \mu} \right)_{TV} \right]^{\frac{1}{2}}.
\end{aligned}
\tag{10.225}
$$

Eq.(10.225) shows that $\left(\frac{\partial \overline{N}}{\partial \mu} \right)_{TV}$ is positive. The relative fluctuation is given by

$$
\delta \overline{N} = \left[\frac{k_B T}{\overline{N}^2} \left(\frac{\partial \overline{N}}{\partial \mu} \right)_{TV} \right]^{\frac{1}{2}}.
\tag{10.226}
$$

Next we consider the fluctuation of energy.

$$
\begin{aligned}
\Delta \overline{E} &= (\overline{E^2} - \overline{E}^2)^{\frac{1}{2}} \\
&= \left(\frac{\partial^2 \ln \Xi}{\partial \beta^2} \right)^{\frac{1}{2}} \\
&= \left[-\left(\frac{\partial \overline{E}}{\partial \beta} \right)_{\alpha V} \right]^{\frac{1}{2}} \\
&= \left[k_B T^2 \left(\frac{\partial \overline{E}}{\partial T} \right)_{\alpha V} \right]^{\frac{1}{2}}.
\end{aligned}
\tag{10.227}
$$

The relative fluctuation of energy is given by

$$
\delta \overline{E} = \left[\frac{k_B T^2}{\overline{E}^2} \left(\frac{\partial \overline{E}}{\partial T} \right)_{\alpha V} \right]^{\frac{1}{2}}.
\tag{10.228}
$$

10.12   Classic statistical mechanics and quantum corrections

10.12.1   *Classic limit of statistical distribution functions*

Now we consider the classical limit of the statistical distribution functions. The classical limit corresponds to the case of high temperature and low densities.

First we consider the simplest case, i.e. one particle systems, in which we do not need deal with quantum correlation. Then we consider realistic many-particle systems. The Hamiltonian operator for one particle systems is given by

$$\hat{H} = \frac{\hat{\mathbf{p}}^2}{2m} + V(\hat{\mathbf{q}}) \equiv K(\hat{\mathbf{p}}) + V(\hat{\mathbf{q}}), \qquad (10.229)$$

where $\hat{\mathbf{p}}$ and $\hat{\mathbf{q}}$ are the momentum and position operators respectively. They obey the commutation relation Eq.(6.116)

$$[\hat{q}_i, \hat{p}_i] = i\hbar. \qquad (10.230)$$

Their eigenstates are defined by the following equations:

$$\hat{q}_i|q_i\rangle = q_i|q_i\rangle, \qquad (10.231a)$$
$$\hat{p}_i|p_i\rangle = p_i|p_i\rangle. \qquad (10.231b)$$

The normalization conditions for the eigenstates are given by

$$\langle q_i'|q_i\rangle = \delta(q_i' - q_i), \qquad (10.232a)$$
$$\langle p_i'|p_i\rangle = \delta(p_i' - p_i). \qquad (10.232b)$$

We can derive following relation directly from the commutation relation Eq.(10.230).

$$\langle q_i|p_i\rangle = \frac{1}{\sqrt{2\pi\hbar}} e^{\frac{i}{\hbar}p_i q_i}. \qquad (10.233)$$

From commutation relation, we have

$$[\hat{q}_i, \hat{p}_i^{\,n}] = i\hbar n \hat{p}_i^{\,n-1}, \qquad (10.234a)$$
$$[\hat{p}_i, \hat{q}_i^{\,n}] = -i\hbar n \hat{q}_i^{\,n-1}. \qquad (10.234b)$$

Using the Taylor expansion, we have

$$\hat{q}_i \exp\left(-\frac{i}{\hbar}q_i\hat{p}_i\right)|0\rangle_{q_i}$$

$$= \left[\hat{q}_i, \exp\left(-\frac{i}{\hbar}q_i\hat{p}_i\right)\right]|0\rangle_{q_i} \qquad (10.235)$$

$$= q_i \exp\left(-\frac{i}{\hbar}q_i\hat{p}_i\right)|0\rangle_{q_i}.$$

Thus the eigenstates of $\hat{q}_i$ are given by

$$|q_i\rangle = \exp\left(-\frac{i}{\hbar}q_i\hat{p}_i\right)|0\rangle_{q_i}. \qquad (10.236)$$

Similarly, we can show that the eigenstates of $\hat{p}_i$ are given by

$$|p_i\rangle = \exp\left(\frac{i}{\hbar}p_i\hat{q}_i\right)|0\rangle_{p_i}. \tag{10.237}$$

Then we can calculate $\langle q_i|p_i\rangle$

$$\begin{aligned}
\langle q_i|p_i\rangle &= \langle q_i|\exp\left(\frac{i}{\hbar}p_i\hat{q}_i\right)|0\rangle_{p_i} \\
&= \exp\left(\frac{i}{\hbar}p_iq_i\right)\langle q_i|0\rangle_{p_i} \\
&= \exp\left(\frac{i}{\hbar}p_iq_i\right){}_{q_i}\langle 0|\exp\left(-\frac{i}{\hbar}q_i\hat{p}_i\right)|0\rangle_{p_i} \\
&= \exp\left(\frac{i}{\hbar}p_iq_i\right){}_{q_i}\langle 0|0\rangle_{p_i}.
\end{aligned} \tag{10.238}$$

${}_{q_i}\langle 0|0\rangle_{p_i}$ is just a constant for normalization, which we will take as $1/\sqrt{2\pi\hbar}$. Thus we get Eq.(10.233). The normalization condition is consistent with the following completeness relations of $|q_i\rangle$ and $|p_i\rangle$

$$\int dq_i|q_i\rangle\langle q_i| = 1, \tag{10.239a}$$

$$\int dp_i|p_i\rangle\langle p_i| = 1. \tag{10.239b}$$

We associate with the operator $\hat{A}(\hat{p}, \hat{q})$ a function $A_c(\mathbf{p}, \mathbf{q})$.

$$A_c(\mathbf{p}, \mathbf{q}) \equiv \frac{\langle\mathbf{p}|\hat{A}|\mathbf{q}\rangle}{\langle\mathbf{p}|\mathbf{q}\rangle}. \tag{10.240}$$

$A_c(\mathbf{p}, \mathbf{q})$ is the classical quantity corresponding to the operator $\hat{A}(\hat{p}, \hat{q})$. Then the classical Hamiltonian function is given by

$$\begin{aligned}
H_c(p, q) &= \langle\mathbf{p}|\left[\frac{\hat{\mathbf{p}}^2}{2m} + V(\hat{\mathbf{q}})\right]|\mathbf{q}\rangle\frac{1}{\langle\mathbf{p}|\mathbf{q}\rangle} \\
&= \left[\frac{\mathbf{p}^2}{2m} + V(\mathbf{q})\right]\langle\mathbf{p}|\mathbf{q}\rangle\frac{1}{\langle\mathbf{p}|\mathbf{q}\rangle} \\
&= \frac{\mathbf{p}^2}{2m} + V(\mathbf{q}) \\
&= H(p, q).
\end{aligned} \tag{10.241}$$

Now we calculate the partition function

$$Z = \mathrm{Tr}e^{-\beta\hat{H}}$$

$$= \int d^3p\langle\mathbf{p}|e^{-\beta\hat{H}(\hat{\mathbf{p}},\hat{\mathbf{q}})}|\mathbf{p}\rangle$$

$$= \int d^3p\int d^3q\langle\mathbf{p}|e^{-\beta\hat{H}(\hat{\mathbf{p}},\hat{\mathbf{q}})}|\mathbf{q}\rangle\langle\mathbf{q}|\mathbf{p}\rangle$$

$$= \int d^3p\int d^3q\langle\mathbf{p}|[e^{-\beta\hat{K}(\hat{\mathbf{p}})}e^{-\beta\hat{V}(\hat{\mathbf{q}})} + \mathcal{O}(\hbar)]|\mathbf{q}\rangle\frac{\langle\mathbf{p}|\mathbf{q}\rangle}{\langle\mathbf{p}|\mathbf{q}\rangle}\langle\mathbf{q}|\mathbf{p}\rangle \quad (10.242)$$

$$= \int d^3p\int d^3q[e^{-\beta H(\mathbf{p},\mathbf{q})} + \mathcal{O}(\hbar)]\frac{1}{(2\pi\hbar)^3}$$

$$= \int\frac{d^3p\,d^3q}{(2\pi\hbar)^3}[e^{-\beta H(\mathbf{p},\mathbf{q})} + \mathcal{O}(\hbar)].$$

$\mathcal{O}(\hbar)$ comes from the commutators between $\hat{K}(\hat{\mathbf{p}})$ and $\hat{V}(\hat{\mathbf{q}})$, which can be evaluated using the Campbell-Baker-Hausdorff formula

$$e^{\hat{A}}e^{\hat{B}} = e^{\hat{A}+\hat{B}+\frac{1}{2}[\hat{A},\hat{B}]+\frac{1}{12}\{[[\hat{A},\hat{B}],\hat{B}]+[[\hat{B},\hat{A}],\hat{A}]\}+\cdots}. \quad (10.243)$$

Thus we have the classical partition function

$$Z = \int\frac{d^3p\,d^3q}{(2\pi\hbar)^3}e^{-\beta H(\mathbf{p},\mathbf{q})}. \quad (10.244)$$

We can also define the *Wigner function* by

$$\rho(\mathbf{p},\mathbf{q}) \equiv \frac{\langle\mathbf{q}|\hat{\rho}|\mathbf{p}\rangle}{(2\pi\hbar)^3\langle\mathbf{q}|\mathbf{p}\rangle}. \quad (10.245)$$

The Wigner function satisfies the normalization condition and has the same average value with the density matrix $\hat{\rho}$. We have the normalization condition

$$\int d^3q\int d^3p\,\rho(\mathbf{p},\mathbf{q}) = \int d^3q\int d^3p\frac{1}{(2\pi\hbar)^3}\frac{\langle\mathbf{q}|\hat{\rho}|\mathbf{p}\rangle}{\langle\mathbf{q}|\mathbf{p}\rangle}$$

$$= \int d^3q\int d^3p\frac{1}{(2\pi\hbar)^3}\frac{\langle\mathbf{q}|\hat{\rho}|\mathbf{p}\rangle\langle\mathbf{p}|\mathbf{q}\rangle}{\langle\mathbf{q}|\mathbf{p}\rangle\langle\mathbf{p}|\mathbf{q}\rangle} \quad (10.246)$$

$$= \mathrm{Tr}\hat{\rho} = 1$$

and the average values

$$\int d^3q\int d^3p\,\rho(\mathbf{p},\mathbf{q})A(\mathbf{p},\mathbf{q}) = \int d^3q\int d^3p\frac{1}{(2\pi\hbar)^3}\frac{\langle\mathbf{q}|\hat{\rho}|\mathbf{p}\rangle\langle\mathbf{p}|\hat{A}|\mathbf{q}\rangle}{\langle\mathbf{q}|\mathbf{p}\rangle\langle\mathbf{p}|\mathbf{q}\rangle}$$

$$= \mathrm{Tr}(\hat{\rho}\hat{A}).$$

$$(10.247)$$

For the canonical distribution, we have

$$\rho(\mathbf{p}, \mathbf{q}) = \frac{1}{(2\pi\hbar)^3} \frac{\langle \mathbf{q}|Z^{-1}e^{-\beta\hat{H}}|\mathbf{p}\rangle}{\langle \mathbf{q}|\mathbf{p}\rangle}$$

$$= \frac{1}{(2\pi\hbar)^3 Z} \frac{\langle \mathbf{q}|e^{-\beta\hat{K}}e^{-\beta\hat{V}} + \mathcal{O}(\hbar)|\mathbf{p}\rangle}{\langle \mathbf{q}|\mathbf{p}\rangle} \qquad (10.248)$$

$$= \frac{1}{(2\pi\hbar)^3 Z} e^{-\beta H(\mathbf{p}, \mathbf{q})} + \mathcal{O}(\hbar).$$

Thus we have the classical limit of the distribution function

$$\rho(\mathbf{p}, \mathbf{q}) = \frac{1}{(2\pi\hbar)^3 Z} e^{-\beta H(\mathbf{p}, \mathbf{q})}. \qquad (10.249)$$

The average value of an observable $\hat{A}$ is given by

$$\overline{\hat{A}} = \frac{\iint d^3q d^3p A(\mathbf{p}, \mathbf{q}) e^{-\beta H(\mathbf{p}, \mathbf{q})}}{\iint d^3q d^3p e^{H(\mathbf{p}, \mathbf{q})}}. \qquad (10.250)$$

We can easily generalize the above formalism to the N-particle system in three-dimensions with the Hamiltonian operator

$$\hat{H} = \sum_{i=1}^{N} \left[\frac{\hat{\mathbf{p}}^2}{2m} + V(\hat{\mathbf{q}}_1, \cdots, \hat{\mathbf{q}}_N) \right]. \qquad (10.251)$$

We introduce the following eigenstates of the position operators $\hat{\mathbf{q}}_i$ and momentum operators $\hat{\mathbf{p}}_i$ for many-particle states.

$$|\mathbf{q}\rangle \equiv |\mathbf{q}_1\rangle \cdots |\mathbf{q}_N\rangle \qquad (10.252a)$$

$$|\mathbf{p}\rangle \equiv |\mathbf{p}_1\rangle \cdots |\mathbf{p}_N\rangle \qquad (10.252b)$$

They are orthonormal

$$\langle \mathbf{q}_i|\mathbf{q}'_i\rangle = \delta^3(\mathbf{q}_i - \mathbf{q}'_i) \qquad (10.253a)$$

$$\langle \mathbf{p}_i|\mathbf{p}'_i\rangle = \delta^3(\mathbf{p}_i - \mathbf{p}'_i) \qquad (10.253b)$$

and satisfy the completeness condition

$$\int d^3p_i |\mathbf{p}_i\rangle\langle \mathbf{p}_i| = 1. \qquad (10.254)$$

We have also

$$\langle \mathbf{q}_i|\mathbf{p}_i\rangle = \frac{1}{(2\pi\hbar)^{\frac{3}{2}}} e^{\frac{i}{\hbar}\mathbf{p}_i \cdot \mathbf{q}_i}. \qquad (10.255)$$

The many-body states are either symmetric (bosons) or antisymmetric (fermions). According to Eq.(6.78), we have

$$|\mathbf{p}\rangle_S = \frac{1}{\sqrt{N!}} \sum_P \xi^{S_P} P|\mathbf{p}\rangle. \qquad (10.256)$$

The subscript S is used to denote that the state has been symmetrized according to the features of identical particles. The symmetrical states ($\xi = 1$) are for bosons and antisymmetrical states ($\xi = -1$) are for fermions. The sum runs over all the permutations P of $\{\mathbf{p}_1, \mathbf{p}_2 \cdots, \mathbf{p}_N\}$. According to Eq.(6.81), the normalized state is given by

$$|\mathbf{p}\rangle_{SN} = \frac{1}{\sqrt{n_1! n_2! \cdots}} |\mathbf{p}\rangle_S, \tag{10.257}$$

where n_i is the number of particles with momentum $\mathbf{p}_i$. The trace of an operator $\hat{A}$ is then given by

$$
\begin{aligned}
\mathrm{Tr}\hat{A} &= \sideset{}{'}\sum_{\mathbf{p}_1 \cdots \mathbf{p}_N} {}_{SN}\langle \mathbf{p}|\hat{A}|\mathbf{p}\rangle_{SN} \\
&= \sum_{\mathbf{p}_1 \cdots \mathbf{p}_N} \frac{n_1! n_2! \cdots}{N!} {}_{SN}\langle \mathbf{p}|\hat{A}|\mathbf{p}\rangle_{SN} \\
&= \sum_{\mathbf{p}_1 \cdots \mathbf{p}_N} \frac{1}{N!} {}_{S}\langle \mathbf{p}|\hat{A}|\mathbf{p}\rangle_{S}.
\end{aligned}
\tag{10.258}
$$

The prime in Eq.(10.258) indicates that the sum is limited to different states. Similar to Eq.(10.242), we can deduce the classical form of the partition function for N-particle systems.

$$
\begin{aligned}
Z &= \mathrm{Tr} e^{-\beta \hat{H}} \\
&= \frac{1}{N!} \int d^{3N}p \, {}_{S}\langle \mathbf{p}|e^{-\beta \hat{H}}|\mathbf{p}\rangle_{S} \\
&= \frac{1}{N!} \int d^{3N}q \int d^{3N}p \, {}_{S}\langle \mathbf{p}|e^{-\beta \hat{H}}\mathbf{q}\rangle\langle \mathbf{q}|\mathbf{p}\rangle_{S} \\
&= \frac{1}{N!} \int d^{3N}p \int d^{3N}q \, e^{-\beta H(\mathbf{p},\mathbf{q})} |\langle \mathbf{q}|\mathbf{p}\rangle_S|^2 + \mathcal{O}(\hbar).
\end{aligned}
\tag{10.259}
$$

The factor $|\langle \mathbf{q}|\mathbf{p}\rangle_S|^2$ can be expressed as

$$(2\pi\hbar)^{3N} |\langle \mathbf{q}|\mathbf{p}\rangle_S| = 1 + f(\mathbf{p}, \mathbf{q}), \tag{10.260}$$

where $f(\mathbf{p}, \mathbf{q})$ comes from the symmetrization contribution of identical particles, which is the quantum correlation effect. The leading term 1 corresponds to the pure classical limit. Then we have the classical partition function

$$Z = \frac{1}{N!(2\pi\hbar)^{3N}} \int d^{3N}p \int d^{3N}q \, e^{-\beta H(\mathbf{p},\mathbf{q})}. \tag{10.261}$$

10.12.2   *Quantum corrections*

Now we discuss the quantum correlation correction to the classical partition function. There are two sources for the corrections: i) The symmetrization of wave function; ii) the non-commutativity of $\hat{K}$ and $\hat{V}$. Since the first contribution is more important when the interaction is weak, we consider the first contribution. The first contribution comes from the factor $|\langle \mathbf{q}|\mathbf{p}\rangle_S|^2$. Using Eq.(10.256), we have

$$
\begin{aligned}
|\langle \mathbf{q}|\mathbf{p}\rangle_S|^2 &= \frac{1}{N!} \sum_P \sum_{P'} (\pm 1)^{S_P}(\pm 1)^{S_{P''}} \langle \mathbf{q}|P'|\mathbf{p}\rangle \langle \mathbf{q}|P|\mathbf{p}\rangle^* \\
&= \frac{1}{N!} \sum_P \sum_{P'} (\pm 1)^{S_P}(\pm 1)^{S_{P''}} \langle P'\mathbf{q}|\mathbf{p}\rangle \langle P\mathbf{q}|\mathbf{p}\rangle^* \\
&= \frac{1}{N!} \sum_P \sum_{P'} (\pm 1)^{S_P}(\pm 1)^{S_{P''}} \langle \mathbf{q}|\mathbf{p}\rangle \langle PP'^{-1}\mathbf{q}|\mathbf{p}\rangle^* \qquad (10.262) \\
&= \sum_P (\pm 1)^{S_P} \langle \mathbf{q}|\mathbf{p}\rangle \langle P\mathbf{q}|\mathbf{p}\rangle^* \\
&= \frac{1}{(2\pi\hbar)^{3N}} \sum_P (\pm 1)^{S_P} e^{\frac{i}{\hbar}(\mathbf{p}_1 \cdot (\mathbf{q}_1 - \hat{P}\mathbf{q}_1) + \cdots + \mathbf{p}_N \cdot (\mathbf{q}_N - \hat{P}\mathbf{q}_N))} .
\end{aligned}
$$

In the derivation of Eq.(10.262), we have used the fact that the permutations of the particles in the configuration space are equivalent to the permutations of the space coordinates. We can rewrite Eq.(10.259) using the following formula

$$
\begin{aligned}
\int d^3p \, e^{-\frac{\beta \mathbf{p}^2}{2m} + \frac{i}{\hbar}\mathbf{p}\cdot\mathbf{x}} &= \int d^3p \, e^{-\frac{\beta}{2m}\left(\mathbf{p} - \frac{im\mathbf{x}}{\hbar\beta}\right)^2 - \frac{m\mathbf{x}^2}{2\beta\hbar^2}} \\
&= \int d^3p \, e^{-\frac{\beta \mathbf{p}^2}{2m}} e^{-\frac{\pi \mathbf{x}^2}{\lambda_T^2}} ,
\end{aligned} \qquad (10.263)
$$

where

$$
\lambda_T = \frac{2\pi\hbar}{\sqrt{2\pi m k_B T}} . \qquad (10.264)
$$

λ_T is called the thermal de Broglie wave length of particle. It represents the de Broglie wave length of a particle with an energy of $\pi k_B T$, which can be seen easily if we evaluate the energy of a particle with the wave length of λ_T,

$$
E_{\lambda_T} = \frac{\mathbf{p}^2}{2m} = \frac{(\hbar\mathbf{k})^2}{2m} = \frac{1}{2m}\frac{h^2}{(\lambda_T)^2} = \pi k_B T . \qquad (10.265)
$$

We define

$$
f(\mathbf{x}) \equiv e^{-\frac{\pi \mathbf{x}^2}{\lambda_T^2}} . \qquad (10.266)
$$

Then the partition function Z with the quantum corrections due to the symmetrization of wave function becomes

$$Z = \int \frac{d^{3N}p\, d^{3N}q}{N!(2\pi\hbar)^{3N}} e^{-\beta H(\mathbf{p},\mathbf{q})}$$
$$\times \sum_P (\pm 1)^{S_P} f(\mathbf{q}_1 - P\mathbf{q}_1) \cdots f(\mathbf{q}_N - P\mathbf{q}_N). \tag{10.267}$$

Arranging the terms in the summation according to the number of permutation exchanges, we have

$$\sum_P (\pm 1)^{S_P} f(\mathbf{q}_1 - P\mathbf{q}_1) \cdots f(\mathbf{q}_N - P\mathbf{q}_N)$$
$$= 1 \pm \sum_{i<j} (f(\mathbf{q}_i - \mathbf{q}_j))^2$$
$$+ \sum_{ijk} f(\mathbf{q}_i - \mathbf{q}_j) f(\mathbf{q}_j - \mathbf{q}_k) f(\mathbf{q}_k - \mathbf{q}_i) \pm \cdots, \tag{10.268}$$

where the upper sign corresponds to bosons and the lower sign to fermions. The first term of the expansion in Eq.(10.268) corresponds to the unit element of P. The second term corresponds to the P with one transposition in which only one pair of particles is exchanged, and so on. With the increase of temperature, λ_T decreases and $f(\mathbf{x}) = e^{-\frac{\pi x^2}{\lambda_T^2}}$ decreases rapidly. When $(\frac{V}{N})^{\frac{1}{3}} \gg \lambda_T$, $f(\mathbf{q}_i - \mathbf{q}_j)$ becomes exponentially small in most configuration space. $f(\mathbf{q}_i - \mathbf{q}_j)$ is significant nonzero only for a very small configuration space with $|\mathbf{q}_i - \mathbf{q}_j| \leq \lambda_T$. We consider only the leading quantum corrections in the expansion Eq.(10.268). When we consider up to the second terms, we have

$$1 \pm \sum_{i<j} f^2(\mathbf{q}_i - \mathbf{q}_j) = \prod_{i<j} (1 \pm f^2(\mathbf{q}_i - \mathbf{q}_j))$$
$$= e^{-\beta \sum_{i<j} \tilde{v}_i(\mathbf{q}_i - \mathbf{q}_j)} \tag{10.269}$$

with

$$\tilde{v}_i(\mathbf{q}_i - \mathbf{q}_j) \equiv -kT \ln(1 \pm e^{-\frac{2\pi|\mathbf{q}_i - \mathbf{q}_j|}{\lambda_T^2}}). \tag{10.270}$$

$\tilde{v}_i(\mathbf{q}_i - \mathbf{q}_j)$ is the effective potential. For bosons, the potential $\tilde{v}_i(\mathbf{q}_i - \mathbf{q}_j)$ is negative and equivalently attractive. For fermions, $\tilde{v}_i(\mathbf{q}_i - \mathbf{q}_j)$ is positive and equivalently repulsive. Using the effective potential $\tilde{v}_i(\mathbf{q}_i - \mathbf{q}_j)$, Eq.(10.267) is approximated as

$$Z = \int \frac{d^{3N}p\, d^{3N}q}{N!(2\pi\hbar)^{3N}} e^{-\beta H(\mathbf{p},\mathbf{q})} e^{-\beta \sum_{i<j} \tilde{v}_i(\mathbf{q}_i - \mathbf{q}_j)}$$
$$= \int \frac{d^{3N}p\, d^{3N}q}{N!(2\pi\hbar)^{3N}} e^{-\beta H'(\mathbf{p},\mathbf{q})} \tag{10.271}$$

with

$$H'(\mathbf{p}, \mathbf{q}) \equiv H(\mathbf{p}, \mathbf{q}) + \tilde{v}_i(\mathbf{q}_i - \mathbf{q}_j). \tag{10.272}$$

$H'(\mathbf{p}, \mathbf{q})$ is the effective Hamiltonian with the potential added with $\tilde{v}_i(\mathbf{q}_i - \mathbf{q}_j)$.

10.12.3   *Equipartition theorem*

For a classical system, the average energy of the system can be calculated using a very simple method based on the theorem of equipartition of energy. We will prove this theorem in the following. The energy for a classical system $E(\mathbf{p}, \mathbf{q}, y)$ is a function of momentums $\mathbf{p}$ and positions $\mathbf{q}$. If $E(\mathbf{p}, \mathbf{q}, y) \to \infty$ when p_i and $q_i \to \pm\infty$, we can prove the following relations

$$\overline{p_i \frac{\partial E}{\partial p_i}} = k_B T, \tag{10.273a}$$

$$\overline{q_i \frac{\partial E}{\partial q_i}} = k_B T. \tag{10.273b}$$

First we prove Eq.(10.273a)

$$\begin{aligned}
\overline{p_i \frac{\partial E}{\partial p_i}} &= \frac{1}{N!Z} \int \cdots \int \frac{d^f p d^f q}{(2\pi\hbar)^f} p_i \frac{\partial E}{\partial p_i} e^{-\beta E} \\
&= \frac{1}{N!h^f Z} \left(-\frac{1}{\beta} \right) \int \cdots \int dp_i d^{f-1} p d^f q \, p_i \frac{\partial e^{-\beta E}}{\partial p_i},
\end{aligned} \tag{10.274}$$

where $dp_i d^{f-1} p d^f q = d^f p d^f q$. Integrating by parts, we have

$$\begin{aligned}
\overline{p_i \frac{\partial E}{\partial p_i}} &= \frac{-1}{\beta N!h^f Z} \left[\int \cdots \int d^{f-1} p d^f q \left(p_i e^{-\beta E} \right) \Big|_{p_i=-\infty}^{\infty} \right. \\
&\quad \left. - \int \cdots \int e^{-\beta E} d^f p d^f q \right] \\
&= \frac{1}{\beta N!h^f Z} \int \cdots \int e^{-\beta E} d^f p d^f q \\
&= k_B T.
\end{aligned} \tag{10.275}$$

Similarly we can prove Eq.(10.273b).

Suppose that the energy has the following form

$$E = \sum_{i=1}^{f_1} c_{1i} p_i^{l_1} + \sum_{j=1}^{f_2} c_{2j} q_j^{l_2}. \tag{10.276}$$

Then

$$\frac{\partial E}{\partial p_i} = l_1 c_{1i} p_i^{l_1-1}, \tag{10.277a}$$

$$\frac{\partial E}{\partial q_i} = l_2 c_{2i} q_i^{l_2-1}. \tag{10.277b}$$

We have

$$
\begin{aligned}
\bar{E} &= \overbrace{\sum_{i=1}^{f_1} c_{1i} p_i^{l_1}}^{f_1} + \overbrace{\sum_{j=1}^{f_2} c_{2j} q_j^{l_2}}^{f_2} \\
&= \frac{1}{l_1} \overbrace{\sum_{i=1}^{f_1} p_i \frac{\partial E}{\partial p_i}}^{f_1} + \frac{1}{l_2} \overbrace{\sum_{j=1}^{f_2} q_i \frac{\partial E}{\partial q_i}}^{f_2} \\
&= \frac{f_1}{l_1} k_B T + \frac{f_2}{l_2} k_B T \\
&= \left(\frac{f_1}{l_1} + \frac{f_2}{l_2} \right) k_B T.
\end{aligned}
\tag{10.278}
$$

Eq.(10.278) is the *generalized theorem of equipartition of energy*.

For an independent particle system, the particle energy $\varepsilon \propto p^2$ in the non-relativistic case. Then the average energy of the system is given by

$$\bar{E} = \frac{3N}{2} k_B T, \tag{10.279}$$

where N is the particle number. Eq.(10.279) is the *Boltzmann theorem of equipartition of energy* . It shows that each degree of freedom contributes an energy of $\frac{1}{2} k_B T$. In the extreme relativistic case, the particle energy is given by

$$\varepsilon = cp = c(p_x^2 + p_y^2 + p_z^2)^{\frac{1}{2}}. \tag{10.280}$$

The energy of the system reads

$$E = \sum_{i=1}^{N} c(p_{ix}^2 + p_{iy}^2 + p_{iz}^2)^{\frac{1}{2}}. \tag{10.281}$$

We have

$$\frac{\partial E}{\partial p_{i\alpha}} = c p_{i\alpha} (p_{ix}^2 + p_{iy}^2 + p_{iz}^2)^{-\frac{1}{2}} \quad (\alpha = x, y, z). \tag{10.282}$$

Thus we obtain

$$
\begin{aligned}
\bar{E} &= \sum_{i=1}^{N} \overline{c(p_{ix}^2 + p_{iy}^2 + p_{iz}^2)^{\frac{1}{2}}} \\
&= \sum_{i=1}^{N} \sum_{\alpha} \overline{p_{i\alpha} \frac{\partial E}{\partial p_{i\alpha}}} \\
&= \sum_{i=1}^{N} 3k_B T \\
&= 3Nk_B T.
\end{aligned}
\tag{10.283}
$$

In the relativistic case, each degree of freedom contributes an energy of $k_B T$, instead of $\frac{1}{2}k_B T$ in the non-relativistic case.

Chapter 11

Applications of statistical mechanics

11.1   Ideal gas

In the high temperature and low density, the state of matter is usually a gas due to the entropy effect which prefers a disordered state. For a gas state, the distance of molecules are much larger than the molecular size in average and thus the interactions of molecules are small. Now we consider the properties of the ideal gas. An ideal gas is a gas in which the interactions of molecules can be neglected and the condition

$$e^\alpha \gg 1 \qquad (11.1)$$

holds. $e^\alpha \gg 1$ is the non-degenerate condition. If Eq.(11.1) is not satisfied, we call the gas quantum gas. When $e^\alpha \gg 1$, the system obeys the semi-classical distribution Eq.(10.199). The partition function z of single particle for the ideal gas is given by

$$z = \sum_{\varepsilon_a} g_a e^{-\beta \varepsilon_a}, \qquad (11.2)$$

where ε_a is the energy eigenvalues determined from the Schrödinger equation of single particle and g_a is the degeneracy of ε_a. For a gas, the independent particles are molecules which consist of atoms. We can divide the energy of the molecules into the part of mass center and the part of internal degrees of freedom, which is similar to what we have done when we treat the Schrödinger equation of the hydrogen atom in quantum mechanics. When the Hamiltonian $\hat{H}_c$ of mass center and the Hamiltonian $\hat{H}_i$ of internal degrees of freedom commute, they share the same eigenstates described by the same set of quantum numbers. We denote the energy of mass center as ε_c with the degeneracy g_c and the energy of internal degrees of freedom as ε_i with the degeneracy g_i. We have

$$\varepsilon_a = \varepsilon_c + \varepsilon_i \quad \text{and} \quad g_a = g_c + g_i. \qquad (11.3)$$

Inserting Eq.(11.3) into Eq.(11.2), we have

$$z(\beta, V) = \sum_{\varepsilon_a} g_a e^{-\beta \varepsilon_a}$$

$$= \sum_{\varepsilon_c} g_c e^{-\beta \varepsilon_c} \sum_{\varepsilon_i} g_i e^{-\beta \varepsilon_i} \qquad (11.4)$$

$$= z_c(\beta, V) z_i(\beta)$$

with

$$z_c(\beta, V) \equiv \sum_{\varepsilon_c} g_c e^{-\beta \varepsilon_c}, \qquad (11.5a)$$

$$z_i(\beta) \equiv \sum_{\varepsilon_i} g_i e^{-\beta \varepsilon_i}. \qquad (11.5b)$$

Eq.(11.4) shows that the partition function $z(\beta, V)$ for a semi-classical system can be divided into two parts, $z_c(\beta, V)$ and $z_i(\beta)$, which can be evaluated independently.

The particle number N is related to the partition function z through the following relation

$$N = \sum_{\varepsilon_a} \bar{n}_a$$

$$= \sum_{\varepsilon_a} g_a e^{-\alpha - \beta \varepsilon_a} \qquad (11.6)$$

$$= e^{-\alpha} z(\beta, V).$$

When N is known, Eq.(11.6) can be used to determine α by

$$\alpha = \ln \frac{z}{N}. \qquad (11.7)$$

The average energy E (also called the internal energy) is given by

$$E = -\frac{\partial \ln Z}{\partial \beta} = -N \frac{\partial \ln z}{\partial \beta} = E_c + E_i \qquad (11.8)$$

with

$$E_c = -N \frac{\partial \ln z_c}{\partial \beta}, \qquad (11.9a)$$

$$E_i = -N \frac{\partial \ln z_i}{\partial \beta}. \qquad (11.9b)$$

The total average energy is the sum of the energy of mass center and the energy of internal degrees of freedom.

The equation of state is determined by

$$P = \frac{1}{\beta}\frac{\partial \ln Z}{\partial V} = \frac{N}{\beta}\frac{\partial \ln z}{\partial V} = \frac{N}{\beta}\frac{\partial \ln z_c}{\partial V}. \tag{11.10}$$

Eq.(11.10) shows that the equation of state of an ideal gas is independent of the internal degrees of freedom. Thus all the equations of state of ideal gases are same, independent of the structures of molecules.

The entropy of the ideal gas is given by

$$\begin{aligned} S &= k_B \left(\ln Z - \beta \frac{\partial \ln Z}{\partial \beta} \right) \\ &= k_B N \left(\ln z - \beta \frac{\partial \ln z}{\partial \beta} \right) - k_B \ln N! \\ &= S_c + S_i \end{aligned} \tag{11.11}$$

with

$$S_c = k_B N \left(\ln z_c - \beta \frac{\partial \ln z_c}{\partial \beta} \right) - N k_B (1 - \ln N), \tag{11.12a}$$

$$S_i = k_B N \left(\ln z_i - \beta \frac{\partial \ln z_i}{\partial \beta} \right). \tag{11.12b}$$

We have included the term $\ln N!$ into the entropy of mass center because the factor $N!$ comes from the identical properties of molecules. When the molecules do not have the internal degrees of freedom, we still have the factor $N!$.

11.1.1 *Partition function for mass center motion*

The Hamiltonian for the mass center motion has the form

$$H_c = \frac{1}{2m}(p_x^2 + p_y^2 + p_z^2) + U \tag{11.13}$$

with

$$U = \begin{cases} \infty & \texttt{outside container} \\ 0 & \texttt{inside container} \end{cases}. \tag{11.14}$$

Using Eq.(10.211), we have

$$
\begin{aligned}
z_c &= \sum_{\varepsilon_c} g_c e^{-\beta \varepsilon_c} \\
&= \mathrm{Tr} e^{-\beta \hat{H}_c} \\
&= \int \frac{d^3 q d^3 p}{(2\pi\hbar)^3} e^{-\beta H_c} \\
&= h^{-3} \int dx dy dz e^{-\beta U} \int dp_x dp_y dp_z \exp\left(-\frac{\beta}{2m}(p_x^2 + p_y^2 + p_z^2)\right) \\
&= h^{-3} V \int 4\pi p^2 dp \exp\left(-\frac{\beta}{2m}p^2\right) \\
&= h^{-3} V (2\pi m k_B T)^{\frac{3}{2}}.
\end{aligned}
\tag{11.15}
$$

11.1.2   Ideal gas of single-atom molecules

The simplest molecule is the single-atom molecule in which there is only one atom. Since the excitation energy of electrons is in order of eV or equivalent to 10^4 K, We can neglect the excitation of electrons and consider the atom as one single particle. Thus there is no internal degree of freedom for the single-atom molecules. The partition function z of single particle for the ideal gas of single-atom molecules is equal to the partition function for mass center. We have

$$
z = z_c = h^{-3} V (2\pi m k_B T)^{\frac{3}{2}}.
\tag{11.16}
$$

Then

$$
e^\alpha = \frac{z}{N} = \frac{V}{N}\left(\frac{2\pi m k_B T}{h^2}\right)^{\frac{3}{2}}.
\tag{11.17}
$$

We can estimate e^α using Eq.(11.17). In the condition of room temperature and one atmospheric pressure, $e^\alpha \approx 10^5 \gg 1$. Therefore, the ordinary gas can be approximated as an ideal gas.

Inserting Eq.(11.16) into Eq.(11.8), we have

$$
E = -N\frac{\partial \ln z_c}{\partial \beta} = \frac{3}{2}N k_B T,
\tag{11.18}
$$

which is the same as that given by the equipartition theorem. Then the heat capacity is given by

$$
C_V = \left(\frac{\partial E}{\partial T}\right)_V = \frac{3}{2}N k_B.
\tag{11.19}
$$

The equation of state of the ideal gas has the form

$$P = \frac{N}{\beta} \frac{\partial \ln z_c}{\partial V} = \frac{N k_B T}{V}. \tag{11.20}$$

It should be noted that Eq.(11.20) is also applicable for the ideal gases of multi-atom molecules.

The entropy is given by

$$S = N k_B \left(\ln z_c - \beta \frac{\partial \ln z_c}{\partial \beta} \right) + N k_B (1 - \ln N)$$

$$= N k_B \left[\frac{V}{N} \left(\frac{2\pi m k_B T}{h^2} \right)^{\frac{3}{2}} \right] + \frac{5}{2} N k_B. \tag{11.21}$$

Eq.(11.21) is called the *Sackur-Tetrode equation*. Eq.(11.21) shows that S is an extensive quantity. It should be noted that the term $-\ln N!$ is crucial for the entropy S to be an extensive quantity. The factor $N!$ results from the indistinguishability of the particles. Before quantum mechanics was established, the particles were not considered as indistinguishable, which gives an entropy formula without the term $-\ln N!$. Without the term $-\ln N!$, the entropy is not an extensive quantity. This is so-called Gibbs's paradox.

The free energy can be calculated by

$$F = E - TS = -N k_B T \ln \left[\frac{V}{N} \left(\frac{2\pi m k_B T}{h^2} \right)^{\frac{3}{2}} \right] - N k_B T, \quad (11.22)$$

which gives the chemical potential

$$\mu = \left(\frac{\partial F}{\partial N} \right)_{T,V} = -k_B T \ln \left[\frac{V}{N} \left(\frac{2\pi m k_B T}{h^2} \right)^{\frac{3}{2}} \right]. \tag{11.23}$$

11.1.3 *Internal degrees of freedom*

Now we consider the contribution from the internal degrees of freedom. The partition function for the internal degrees of freedom is given by

$$z_i(\beta) = \sum_{\varepsilon_i} g_i e^{-\beta \varepsilon_i}. \tag{11.24}$$

We can conceive a macroscopic system as a huge molecule and then consider that there are $M (M \gg 1)$ huge molecules to form an ideal gas. The thermodynamic properties of the system is determined by z_i in Eq.(11.24).

Physically, this is equivalent to the ensemble method. The ideal gas consisting of the macroscopic systems can be considered as a canonical ensemble. For these systems, z_i is then the canonical partition function of the system.

z_i can only be evaluated analytically in a few simple cases. In the following, we will deal with an ideal gas consisting of two-atom molecules as an example. If we consider atoms as particles without internal degrees of freedom, each two-atom molecule has $3+3 = 6$ degrees of freedom with three for mass center motion, two for rotations and one for vibration. Since the Hamiltonian operator for rotation commutes with the Hamiltonian operator for vibration, we have

$$\varepsilon_i = \varepsilon_r + \varepsilon_v, \tag{11.25}$$

where ε_r is the eigenvalue of energy for rotation and ε_v is the eigenvalue of energy for vibration. The total degeneracy g_i for the internal degrees of freedom is given by

$$g_i = g_r g_v, \tag{11.26}$$

where g_r is the degeneracy of ε_r and g_v is the degeneracy of ε_v. Then Eq.(11.24) becomes

$$\begin{aligned} z_i &= \sum_{\varepsilon_i} g_i e^{-\beta \varepsilon_i} \\ &= \sum_{\varepsilon_r} g_r e^{-\beta \varepsilon_r} \sum_{\varepsilon_v} g_v e^{-\beta \varepsilon_v} \\ &= z_r(\beta) z_v(\beta) \end{aligned} \tag{11.27}$$

with

$$z_r(\beta) \equiv \sum_{\varepsilon_r} g_r e^{-\beta \varepsilon_r}, \tag{11.28a}$$

$$z_v(\beta) \equiv \sum_{\varepsilon_v} g_v e^{-\beta \varepsilon_v}. \tag{11.28b}$$

11.1.3.1   *Vibration*

The Hamiltonian operator for the vibration of the molecules is given by Eq.(9.7)

$$H = -\frac{\hbar^2}{2m}\nabla^2 + \frac{1}{2}m\omega^2 x^2, \tag{11.29}$$

where $m = \frac{m_1 m_2}{m_1 + m_2}$ is the reduced mass with $m_i(i = 1, 2)$ the mass of the atoms. ω is the vibration frequency. The eigenvalues of the Hamiltonian operator for vibration are given by Eq.(9.16). We have

$$\varepsilon_v(n) = \hbar\omega(n + \frac{1}{2}), \quad n = 0, 1, 2, \cdots \tag{11.30}$$

with

$$g_n = 1. \tag{11.31}$$

The partition function for vibration reads

$$z_v = \sum_n g_n e^{-\beta \varepsilon_v(n)}$$

$$= e^{-\frac{1}{2}\beta\hbar\omega} \sum_{n=0}^{\infty} e^{-\beta\hbar\omega n} \tag{11.32}$$

$$= \exp\left(-\frac{\hbar\omega}{2k_BT}\right) \frac{1}{1 - \exp\left(-\frac{\hbar\omega}{k_BT}\right)}.$$

Using the partition function given by Eq.(11.32), the average energy of vibration can be evaluated as follows

$$E_v = -N\frac{\partial \ln z_v}{\partial \beta}$$

$$= N\hbar\omega \left[\frac{1}{2} + \frac{\exp\left(-\frac{\hbar\omega}{k_BT}\right)}{1 - \exp\left(-\frac{\hbar\omega}{k_BT}\right)}\right]. \tag{11.33}$$

In Eq.(11.33), the first term is the zero-point energy of vibration and the second term is the excitation energy of vibration. Using Eq.(11.33), we obtain the heat capacity

$$C_V = \left(\frac{\partial E_v}{\partial T}\right)_V = Nk_B\varepsilon\left(\frac{\hbar\omega}{k_BT}\right), \tag{11.34}$$

where $\varepsilon(x)$ is the *Einstein function* defined by

$$\varepsilon(x) \equiv \frac{x^2 \exp(x)}{[1 - \exp(x)]^2}. \tag{11.35}$$

In the low temperature limit $\frac{\hbar\omega}{k_BT} \gg 1$, Eq.(11.34) becomes

$$C_V \approx Nk_B\left(\frac{\hbar\omega}{k_BT}\right)^2 \exp\left(-\frac{\hbar\omega}{k_BT}\right). \tag{11.36}$$

Eq.(11.36) shows that the heat capacity of vibration approaches to zero when $T \to 0$. This is due to the quantum effect. We introduce a characteristic temperature θ_v defined by

$$\frac{\hbar\omega}{k_B\theta_v} = 1, \tag{11.37}$$

which gives

$$\theta_v = \frac{\hbar\omega}{k_B}. \tag{11.38}$$

When $T \ll \theta_v$, which gives the low temperature region, the vibration degree of freedom is frozen and the heat capacity is small. When $T \gg \theta_v$, which gives the high temperature region, the energy becomes

$$E_v \approx N\hbar\omega \left(\frac{1}{2} + \frac{1}{\beta\hbar\omega} \right) = Nk_BT. \tag{11.39}$$

This result agrees with the equipartition theorem. The heat capacity C_V is then equal to Nk_B.

11.1.3.2 *Rotation*

For a two-atom molecule, the bonding length of the two atoms is approximately a constant. The small variation of the distance between the two atoms is described by vibration. The Hamiltonian of the two atoms is given by

$$H = -\frac{\hbar^2}{2m_1}\nabla^2 - \frac{\hbar^2}{2m_2}\nabla^2 + U(r), \tag{11.40}$$

where $m_i(i = 1, 2)$ are the masses of the two atoms. $U(r)$ is the interaction potential of the two atoms, which depends on the distance between the atoms. We introduce the mass center coordinates $\mathbf{X}$ and relative coordinates $\mathbf{x}$

$$\mathbf{X} = \frac{m_1\mathbf{x}_1 + m_2\mathbf{x}_2}{m_1 + m_2}, \quad \mathbf{r} = \mathbf{x}_2 - \mathbf{x}_1. \tag{11.41}$$

In terms of $\mathbf{X}$ and $\mathbf{x}$, the Hamiltonian operator Eq.(11.41) is expressed as

$$H = -\frac{\hbar^2}{2M}\nabla_{\mathbf{X}}^2 - \frac{\hbar^2}{2m}\nabla_{\mathbf{x}}^2 + U(r), \tag{11.42}$$

where $m = \frac{m_1 m_2}{m_1 + m_2}$ is the reduced mass and $M = m_1 + m_2$ is the total mass. Expressing the solution in the separable product

$$\Psi(\mathbf{x}, \mathbf{X}) = \psi(\mathbf{x})\Phi(\mathbf{X}). \tag{11.43}$$

We have

$$-\frac{\hbar^2}{2m}\nabla_{\mathbf{x}}^2\psi + U(r)\psi = E\psi \tag{11.44}$$

and

$$-\frac{\hbar^2}{2M}\nabla_{\mathbf{X}}^2\Phi = (E_t - E)\Phi. \tag{11.45}$$

Thus $H_c = -\frac{\hbar^2}{2M}\nabla_{\mathbf{X}}^2$ is the Hamiltonian operator of mass center, while

$$H_i = -\frac{\hbar^2}{2m}\nabla_{\mathbf{x}}^2 + U(r) \qquad (11.46)$$

is the Hamiltonian operator of internal degrees of freedom. In the spherical coordinates, H_i has the form

$$H_i = -\frac{\hbar^2}{2m}\left[\frac{1}{r^2}\frac{d}{dr}\left(r^2\frac{d}{dr}\right)\right.$$
$$\left. + \frac{1}{r^2\sin\theta}\frac{\partial}{\partial\theta}\left(\sin\theta\frac{\partial}{\partial\theta}\right) + \frac{1}{r^2\sin^2\theta}\frac{\partial^2}{\partial\varphi^2}\right] + U(r) \qquad (11.47)$$
$$= \frac{\hat{L}^2}{2I} - \frac{\hbar^2}{2m}\nabla_r^2 + U(r)$$

with

$$L^2 = \frac{1}{\sin\theta}\frac{\partial}{\partial\theta}\left(\sin\theta\frac{\partial}{\partial\theta}\right) + \frac{1}{\sin^2\theta}\frac{\partial^2}{\partial\varphi^2} \qquad (11.48)$$

and

$$I = mr^2, \qquad (11.49)$$

where I is the rotation inertia.[1] The first term in H_i is the Hamiltonian operator for rotation and the last two terms form the Hamiltonian operator for vibration. From Eq.(9.80), the eigenvalue of energy of rotation is given by

$$\varepsilon_r = \frac{\hbar^2}{2I}l(l+1), \quad l = 0, 1, 2, \cdots \qquad (11.53)$$

---

[1]In the following, we give a note to show $I = mr^2$. We introduce the position vectors $\mathbf{r}_1$ and $\mathbf{r}_2$ of the two atoms relative to the mass center.

$$\mathbf{r}_1 = \mathbf{x}_1 - \frac{m_1\mathbf{x}_1 + m_2\mathbf{x}_2}{m_1 + m_2} = \frac{m_2(\mathbf{x}_1 - \mathbf{x}_2)}{m_1 + m_2} = -\frac{m_2\mathbf{r}}{m_1 + m_2} \qquad (11.50)$$

and

$$\mathbf{r}_2 = \mathbf{x}_2 - \frac{m_1\mathbf{x}_1 + m_2\mathbf{x}_2}{m_1 + m_2} = \frac{m_1(\mathbf{x}_2 - \mathbf{x}_1)}{m_1 + m_2} = \frac{m_1\mathbf{r}}{m_1 + m_2}. \qquad (11.51)$$

The rotation inertial I is defined by $I \equiv \sum_i m_i r_i^2$. Using the above relations, we have

$$I = \sum_i m_i r_i^2$$
$$= \frac{m_1 m_2^2}{(m_1 + m_2)^2}r^2 + \frac{m_1^2 m_2}{(m_1 + m_2)^2}r^2 \qquad (11.52)$$
$$= \frac{m_1 m_2}{m_1 + m_2}r^2$$
$$= mr^2.$$

with the degeneracy

$$g_r = 2l + 1. \tag{11.54}$$

When the two atoms in the molecules are the same kind, we need to consider the effect of identical particles, which leads to the limitation on the values of l. We will consider the simple case that the two atoms in the molecules are different. The partition function for rotation has the form

$$z_r = \sum_{l=0}^{\infty} (2l+1) \exp\left[-\frac{\beta\hbar^2}{2I} l(l+1)\right]. \tag{11.55}$$

We define a characteristic temperature for rotation by

$$\theta_r \equiv \frac{1}{k_B} \frac{\hbar^2}{2I}. \tag{11.56}$$

In terms of θ_r, Eq.(11.55) becomes

$$\begin{aligned} z_r &= \sum_{l=0}^{\infty} (2l+1) \exp\left[-\frac{\beta\hbar^2}{2I} l(l+1)\right] \\ &= \sum_{l=0}^{\infty} (2l+1) \exp\left[-\frac{\theta_r}{T} l(l+1)\right]. \end{aligned} \tag{11.57}$$

At the high temperature $T \gg \theta_r$, we use the Euler-Maclaurin summation formula

$$\sum_{l=0}^{\infty} f(l) = \int_0^{\infty} dl f(l) + \frac{1}{2} f(0) + \sum_{k=1}^{\infty} \frac{-B_{2k}}{(2k)!} f^{(2k-1)}(0). \tag{11.58}$$

to evaluate the summation in Eq.(11.57). For the case of $f(\infty) = f'(\infty) = \cdots = 0$, B_n is the first Bernoulli numbers. Thus we have

$$\begin{aligned} z_r &= \int_0^{\infty} dl(2l+1) \exp\left[-\frac{\beta\hbar^2}{2I} l(l+1)\right] + \frac{1}{2} + \mathcal{O}\left(\frac{\theta_r}{T}\right) \\ &= \int_0^{\infty} dx \exp\left(-\frac{\beta\hbar^2}{2I} x\right) + \frac{1}{2} + \mathcal{O}\left(\frac{\theta_r}{T}\right) \\ &= \frac{T}{\theta_r} + \frac{1}{2} + \mathcal{O}\left(\frac{\theta_r}{T}\right) \\ &\approx \frac{T}{\theta_r}. \end{aligned} \tag{11.59}$$

Using $z_r = \frac{T}{\theta_r}$, the energy contributed by the rotational degrees of freedom can be evaluated. We have

$$E_r = -N \frac{\partial \ln z_r}{\partial \beta} = N k_B T. \tag{11.60}$$

Eq.(11.60) can also be obtained by the equipartition theorem. According to the energy equipartition theorem, there are two rotational degrees of freedom for one molecule and each contributes $\frac{1}{2}k_BT$ to the internal energy. The total energy is then Nk_BT. The heat capacity at constant volume reads

$$C_V = \left(\frac{\partial E_v}{\partial T}\right)_V = Nk_B. \tag{11.61}$$

At the low temperature $T \ll \theta_r$, only the small values of l contribute in Eq.(11.57). Then we have

$$z_r = 1 + 3\exp\left(-\frac{2\theta_r}{T}\right) + \mathcal{O}\left(\exp\left(-\frac{\theta_r}{T}\right)\right). \tag{11.62}$$

The energy and the heat capacity are given by

$$E_r = 6Nk_BT\frac{\theta_r}{T}\exp\left(-\frac{2\theta_r}{T}\right) + \cdots, \tag{11.63a}$$

$$C_V = 12Nk_B\left(\frac{\theta_r}{T}\right)^2\exp\left(-\frac{2\theta_r}{T}\right) + \cdots. \tag{11.63b}$$

Eq.(11.63) shows that the rotational degrees of freedom are not thermally excited in the low temperature region and the rotational contribution to the internal energy is exponentially small.

11.2 Weakly degenerate quantum gas

The ideal gases satisfy the non-degenerate condition

$$e^\alpha = \frac{V}{N}\left(\frac{2\pi mk_BT}{h^2}\right)^{\frac{3}{2}} \gg 1. \tag{11.64}$$

If the condition Eq.(11.64) is not satisfied, the gas is called the degenerate gas or quantum gas. We use the thermal de Broglie wave length (Eq.(10.264))

$$\lambda_T = \frac{h}{(2\pi mk_BT)^{\frac{1}{2}}} \tag{11.65}$$

to characterize the degenerate level of a gas. Using λ_T, Eq.(11.64) can be rewritten as

$$\left(\frac{1}{n}\right)^{\frac{1}{3}} \gg \lambda_T = \frac{h}{(2\pi mk_BT)^{\frac{1}{2}}}. \tag{11.66}$$

Eq.(11.66) shows that when the average distance of particles is larger than the thermal de Broglie wave length of particles, the particles can be considered as the classical particles. Otherwise, the waves of particles interweaves and the quantum correlation plays role. When $e^{-\alpha} = n\lambda_T^3 < 1$, the quantum correlation is weak. We call this case the weakly degenerate quantum gas. When $e^{-\alpha} = n\lambda_T^3 \geq 1$, it is the strong degenerate case. In the weakly degenerate case, $e^{-\alpha} = n\lambda_T^3 < 1$ is a small parameter and we can use the expansion method to calculate the thermodynamic quantities.

We consider a gas without internal degrees of freedom in the weakly degenerate quantum case. The grand partition function is given by

$$\ln \Xi(\alpha, \beta, V) = \pm \sum_i g_i \ln(1 \pm e^{-\alpha - \beta \varepsilon_i})$$
$$= \pm \int_0^\infty d\varepsilon g_s g(\epsilon) \ln(1 \pm e^{-\alpha - \beta \varepsilon}) \qquad (11.67)$$
$$= \pm CV \int_0^\infty d\varepsilon \sqrt{\varepsilon} \ln(1 \pm e^{-\alpha - \beta \varepsilon})$$

with $C = g_s 2\pi (2m)^{\frac{3}{2}} h^{-3}$, where $g_s = 2s + 1$ is the spin degeneracy factor for particles with spin s. The upper sign '+' corresponds to the Fermi gas and the lower sign '−' is for the Bose gas. Since $e^{-\alpha} < 1$, we use the Taylor expansion formula

$$\ln(1 \pm x) = \pm \sum_{n=1}^\infty (\mp)^{n-1} \frac{x^n}{n}. \qquad (11.68)$$

Then we have

$$\int_0^\infty d\varepsilon \sqrt{\varepsilon} \ln(1 \pm e^{-\alpha - \beta \varepsilon}) = \pm \sum_{n=1}^\infty (\mp)^{n-1} \frac{1}{n} \int_0^\infty d\varepsilon \sqrt{\varepsilon} e^{-n(\alpha + \beta \varepsilon)}$$
$$= \pm \sum_{n=1}^\infty (\mp)^{n-1} \frac{1}{n} e^{-n\alpha} \int_0^\infty dx \, 2x^2 e^{-n\beta x^2}$$
$$= \pm \sum_{n=1}^\infty (\mp)^{n-1} \frac{1}{n} e^{-n\alpha} \sqrt{\frac{\pi}{n\beta}} \frac{1}{2n\beta} \qquad (11.69)$$
$$= \pm \sqrt{\frac{\pi}{\beta}} \frac{1}{2\beta} f(\alpha)$$

with

$$f(\alpha) = \sum_{n=1}^\infty (\mp)^{n-1} n^{-\frac{5}{2}} e^{-n\alpha}. \qquad (11.70)$$

Thus Eq.(11.67) becomes

$$\ln \Xi(\alpha, \beta, V) = CV \frac{\sqrt{\pi}}{2} \beta^{-\frac{3}{2}} f(\alpha) = V \lambda_T^{-3} g_s f(\alpha). \qquad (11.71)$$

α can be determined by solving the following equation.

$$N = -\frac{\partial \ln \Xi}{\partial \alpha} = -V \lambda_T^{-3} g_s f'(\alpha). \qquad (11.72)$$

Eq.(11.72) can be rewritten as

$$\frac{1}{g_s} n \lambda_T^3 = \sum_{n=1}^{\infty} (\mp)^{n-1} n^{-\frac{3}{2}} e^{-n\alpha} = e^{-\alpha} (1 \mp 2^{-\frac{3}{2}} e^{-\alpha} + \cdots), \quad (11.73)$$

where $n = \frac{N}{V}$. Eq.(11.73) can be solved by the iteration method, which gives

$$e^{-\alpha} = \frac{1}{g_s} n \lambda_T^3 (1 \pm 2^{-\frac{3}{2}} \frac{1}{g_s} n \lambda_T^3 + \cdots) \qquad (11.74)$$

or

$$\alpha = -\ln \frac{n \lambda_T^3}{g_s} \mp 2^{-\frac{3}{2}} \frac{1}{g_s} n \lambda_T^3 + \cdots. \qquad (11.75)$$

Inserting Eq.(11.74) into Eq.(11.70), we have

$$f(\alpha) = e^{-\alpha} (1 \mp 2^{-\frac{5}{2}} e^{-\alpha} + \cdots)$$
$$= \frac{1}{g_s} n \lambda_T^3 (1 \pm 2^{-\frac{5}{2}} \frac{1}{g_s} n \lambda_T^3 + \cdots). \qquad (11.76)$$

Now we can evaluate the thermodynamic quantities using the grand partition function given by Eq.(11.71). The average energy has the form

$$E = -\frac{\partial \ln \Xi}{\partial \beta} = -\ln \Xi \frac{\partial \ln \ln \Xi}{\partial \beta} = \frac{3}{2} \frac{\ln \Xi}{\beta}$$
$$= \frac{3}{2} N k_B T (1 \pm 2^{-\frac{5}{2}} \frac{1}{g_s} n \lambda_T^3 + \cdots). \qquad (11.77)$$

The first term corresponds to the semi-classical approximation, which gives the energy of the ideal gas. The second term is contributed by the quantum effect. The quantum effect adds a positive value to the internal energy for the Fermi gas due to the Pauli exclusion principle, while the contribution of quantum effect is negative for the Bose gas. Using Eq.(11.77), the heat capacity can be obtained.

$$C_V = \left(\frac{\partial E}{\partial T} \right)_V = \frac{3}{2} N k_B (1 \mp 2^{-\frac{7}{2}} \frac{1}{g_s} n \lambda_T^3 + \cdots). \qquad (11.78)$$

Other thermodynamic quantities can be evaluated similarly. The equation of state has the form

$$P = \frac{1}{\beta} \frac{\partial \ln \Xi}{\partial V} = \frac{\ln \Xi}{\beta V} = \frac{2}{3} \frac{E}{V}$$
$$= n k_B T \left(1 \pm 2^{-\frac{5}{2}} \frac{1}{g_s} n \lambda_T^3 + \cdots \right). \tag{11.79}$$

The effect of the Pauli exclusion principle for the Fermi gas is equivalent to an repulsive force and contributes a positive pressure modification. The effect of quantum correlation for the Bose gas is equivalent to an attractive force and contributes a negative pressure modification. The entropy is given by

$$S = k_B \left(\ln \Xi - \alpha \frac{\partial \ln \Xi}{\partial \alpha} - \beta \frac{\partial \ln \Xi}{\partial \beta} \right)$$
$$= k_B \left(\frac{5}{3} \beta E + N \alpha \right) \tag{11.80}$$
$$= N k_B T \left[\left(\frac{g_s}{n \lambda_T^3} + \frac{5}{2} \right) \pm 2^{-\frac{7}{2}} \frac{1}{g_s} n \lambda_T^3 + \cdots \right].$$

The lower temperature T, the larger mass m and the higher density n lead to larger $n \lambda_T^3$ and thus stronger quantum effect. For a Bose gas, Eq.(11.78) shows that the heat capacity C_V increases with the decrease of temperature T. Since C_V should become zero as temperature approaches zero, we would expect that there is other mechanism making C_V decrease. Thus there should be a peak in the curve of C_V as a function of T. We will show that there is a phase transition for the Bose gas in the following section. The peak of C_V corresponds to a transition point. The heat capacity will be shown to decrease with the decrease of temperature in the strong degenerate region below the transition temperature.

11.3   Bose gas

11.3.1   *Bose-Einstein condensation*

Now we consider the strongly degenerate case $n \lambda_T^3 \geq 1$ for the Bose gas. In this case, the quantum correlation effect is strong. According to Eq.(10.188), we have

$$N = \sum_i \bar{n}_i = \sum_i \frac{g_i}{e^{\alpha + \beta \varepsilon_i} - 1}. \tag{11.81}$$

Using Eq.(10.132), Eq.(11.81) becomes

$$N = \sum_i \frac{g_i}{e^{\beta(\varepsilon_i - \mu)} - 1}. \tag{11.82}$$

We can set $\varepsilon_0 = 0$, which means that ε_0 is the reference energy. Since $n_i \geq 0$, we have

$$\varepsilon_i - \mu \geq 0, \tag{11.83}$$

which gives

$$\mu \leq 0. \tag{11.84}$$

With the decrease of temperature T, the chemical potential μ has to increase in order to maintain a constant N. Thus

$$\frac{\partial \mu}{\partial T} < 0. \tag{11.85}$$

According to Eq.(11.84), μ has an upper limit $\mu_0 \leq 0$. For simplicity, we consider a Bose gas composed of single-atom molecules. Then

$$N = \frac{2\pi V}{h^3} (2m)^{\frac{3}{2}} g_s \int_0^\infty d\varepsilon \frac{\sqrt{\varepsilon}}{e^{\beta(\varepsilon - \mu)} - 1}. \tag{11.86}$$

Using Eq.(11.86), we have

$$
\begin{aligned}
\left(\frac{\partial \mu}{\partial T} \right)_N &= \frac{\dfrac{\partial}{\partial T} \displaystyle\int_0^\infty d\varepsilon \frac{\sqrt{\varepsilon}}{e^{\beta(\varepsilon - \mu)} - 1}}{\dfrac{\partial}{\partial \mu} \displaystyle\int_0^\infty d\varepsilon \frac{\sqrt{\varepsilon}}{e^{\beta(\varepsilon - \mu)} - 1}} \\[2ex]
&= \frac{-\displaystyle\int_0^\infty d\varepsilon \frac{\varepsilon - \mu}{k_B T^2} e^{\beta(\varepsilon - \mu)} \frac{\sqrt{\varepsilon}}{(e^{\beta(\varepsilon - \mu)} - 1)^2}}{\displaystyle\int_0^\infty d\varepsilon \frac{1}{k_B T} e^{\beta(\varepsilon - \mu)} \frac{\sqrt{\varepsilon}}{(e^{\beta(\varepsilon - \mu)} - 1)^2}} \\[2ex]
&\leq 0.
\end{aligned}
\tag{11.87}
$$

The grand partition function is given by

$$
\begin{aligned}
\ln \Xi &= -\sum_i g_i \ln(1 - e^{-\beta(\varepsilon_i - \mu)}) \\
&= -CV \int_0^\infty d\varepsilon \sqrt{\varepsilon} \ln(1 - e^{-\beta(\varepsilon - \mu)}),
\end{aligned}
\tag{11.88}
$$

where $C = g_s 2\pi (2m)^{\frac{3}{2}} h^{-3}$. We introduce the fugacity

$$q \equiv e^{\beta \mu}. \tag{11.89}$$

When $\mu = 0$, $q = 1$. Using Eq.(11.89), Eq.(11.88) becomes

$$
\begin{aligned}
\ln \Xi &= -CV \int_0^\infty d\varepsilon \varepsilon^{\frac{1}{2}} \ln(1 - qe^{-\beta\varepsilon}) \\
&= \frac{2}{3} CV\beta \int_0^\infty d\varepsilon \frac{\varepsilon^{\frac{3}{2}}}{q^{-1}e^{\beta\varepsilon} - 1} \\
&= \frac{2}{3} CV\beta \mathcal{G}_{\frac{5}{2}}(q)\Gamma(\frac{5}{2})
\end{aligned}
\tag{11.90}
$$

with

$$
\mathcal{G}_n(q) \equiv \frac{1}{\Gamma(n)} \int_0^\infty \frac{x^{n-1}dx}{q^{-1}e^x - 1},
\tag{11.91}
$$

where $\Gamma(n)$ is the Gamma function. Expanding the integrand in Eq.(11.91) gives

$$
\begin{aligned}
\mathcal{G}_n(q) &= \frac{1}{\Gamma(n)} \int_0^\infty dx \frac{x^{n-1}qe^{-x}}{1 - qe^{-x}} \\
&= \frac{1}{\Gamma(n)} \sum_{k=1}^\infty q^k \int_0^\infty dx x^{n-1}e^{-kx} \\
&= \frac{1}{\Gamma(n)} \sum_{k=1}^\infty \frac{q^k}{k^n} \int_0^\infty dy y^{n-1}e^{-y} \\
&= \sum_{k=1}^\infty \frac{q^k}{k^n}.
\end{aligned}
\tag{11.92}
$$

When $q = 1$, $\mathcal{G}_n(q)$ becomes the Riemann Zeta function $\zeta(n)$,

$$
\mathcal{G}_n(1) = \sum_{k=1}^\infty \frac{1}{k^n} = \zeta(n) \quad (n > 1).
\tag{11.93}
$$

Using Eq.(11.92), Eq.(11.86) becomes

$$
N = V g_s \left(\frac{2\pi m k_B T}{h^2} \right)^{\frac{3}{2}} \mathcal{G}_{\frac{3}{2}}(q).
\tag{11.94}
$$

According to Eq.(11.92), $\mathcal{G}_n(q)$ increases with the increase of q, which has an upper limit at $q_m = 1$ or correspondingly $\mu = 0$. Thus the right hand of Eq.(11.86) is smaller than N_{max} given by

$$
N_{max} = V g_s \left(\frac{2\pi m k_B T}{h^2} \right)^{\frac{3}{2}} \zeta\left(\frac{3}{2}\right) = \frac{V g_s}{\lambda_T^3} \zeta\left(\frac{3}{2}\right) \propto VT^{\frac{3}{2}}.
\tag{11.95}
$$

With the decrease of temperature T, N_{max} becomes smaller than N below a certain temperature T_c. This situation is caused by the using of Eq.(10.212) when we replace the summation with integration. The contribution from the ground state $\varepsilon = 0$ does not appear in the integration since $g(0) = 0$. For bosons, there is no limitation on the number of particles in a quantum state. The lowest state $\varepsilon = 0$ can be occupied by $\mathcal{O}(N)$ particles. In this case, the contribution from the ground state can not be neglected. We should express explicitly the term that accounts for the contribution from the state $\varepsilon = 0(k = 0)$. Thus Eq.(11.86) should be replaced by

$$
\begin{aligned}
N &= \frac{g_0}{e^{-\beta\mu} - 1} + \frac{2\pi V}{h^3}(2m)^{\frac{3}{2}} g_s \int_0^\infty d\varepsilon \frac{\sqrt{\varepsilon}}{e^{\beta(\varepsilon - \mu)} - 1} \\
&= \frac{q g_0}{1 - q} + \frac{V g_s}{\lambda_T^3} g_{\frac{3}{2}}(q) \\
&= N_0 + N_{\varepsilon > 0}
\end{aligned}
\tag{11.96}
$$

with

$$
N_{\varepsilon > 0} \equiv \frac{V g_s}{\lambda_T^3} g_{\frac{3}{2}}(q)
\tag{11.97}
$$

and

$$
N_0 \equiv \frac{q g_0}{1 - q}.
\tag{11.98}
$$

$N_{\varepsilon > 0}$ is the number of particle in the excited states and N_0 is the number of particles in the ground state. When $\mu \to 0$, the first term turns to be important. The number of particles in the ground state becomes $\mathcal{O}(N)$. Therefore, with the decrease of temperature T, μ becomes zero at a certain temperature T_c. This temperature T_c is determined by

$$
\begin{aligned}
N &= \frac{2\pi V}{h^3}(2m)^{\frac{3}{2}} g_s \int_0^\infty d\varepsilon \sqrt{\varepsilon} \left[\exp\left(\frac{\varepsilon}{k_B T_c} \right) - 1 \right]^{-1} \\
&= \frac{2\pi V}{h^3}(2m)^{\frac{3}{2}} g_s (k_B T_c)^{\frac{3}{2}} \int_0^\infty dx \frac{\sqrt{x}}{e^x - 1} \\
&= \frac{2\pi V}{h^3}(2m)^{\frac{3}{2}} g_s (k_B T_c)^{\frac{3}{2}} \Gamma\left(\frac{3}{2} \right) \zeta\left(\frac{3}{2} \right) \\
&= \frac{2\pi V}{h^3}(2m)^{\frac{3}{2}} g_s (k_B T_c)^{\frac{3}{2}} \frac{\sqrt{\pi}}{2} \times 2.612,
\end{aligned}
\tag{11.99}
$$

which gives

$$
T_c = \frac{h^2}{2\pi m k_B} \left(\frac{N}{2.612 g_s V} \right)^{\frac{2}{3}}.
\tag{11.100}
$$

As $T \to T_c$, $\mu \to 0$. The particle number in the ground state $\varepsilon_0 = 0$ increases significantly. At T_c, μ reaches its upper limit of zero and remains zero for the temperature below T_c. Therefore, when $T < T_c$, we have

$$
\begin{aligned}
N_{\varepsilon>0} &= \frac{2\pi V}{h^3}(2m)^{\frac{3}{2}} g_s \int_0^\infty d\varepsilon \sqrt{\varepsilon} \left[\exp\left(\frac{\varepsilon}{k_B T}\right) - 1\right]^{-1} \\
&= \frac{2\pi V}{h^3}(2m)^{\frac{3}{2}} g_s (k_B T)^{\frac{3}{2}} \int_0^\infty dx \frac{\sqrt{x}}{e^x - 1} \qquad (11.101) \\
&= N\left(\frac{T}{T_c}\right)^{\frac{3}{2}}
\end{aligned}
$$

and

$$
N_0 = N - N_{\varepsilon>0} = N\left[1 - \left(\frac{T}{T_c}\right)^{\frac{3}{2}}\right]. \qquad (11.102)
$$

T_c is called the critical temperature. Above T_c, $\mu > 0$. Below T_c, the chemical potential $\mu = 0$ and all orders of its derivatives are zero. Thus $T = T_c$ is a point with singularity in the thermodynamic limit, which means that there is a phase transition at $T = T_c$. This phase transition is called the *Bose-Einstein condensation* (BEC). It can be seen that the phase transition is caused by the condensation on the ground state or $k = 0$ state. This transition is important for the properties of macroscopic systems in low temperature. It transforms a classical phase of a macroscopic system into a quantum phase which we call the macroscopic quantum state. This phase transition mechanism is also responsible for the superfluidity of liquid ^{4}He and ^{3}He, and superconductivity in solids.

11.3.2   *Thermodynamic properties of BEC*

After the Bose-Einstein condensation, $\mu = 0$, which gives

$$
G = \mu N = E + PV - ST = 0. \qquad (11.103)
$$

First we consider the contribution of the ground state $\varepsilon = 0$. $E = 0$ and $S = 0$ for the ground state. According to Eq.(11.103), we have

$$
P = \frac{ST - E}{V} = 0. \qquad (11.104)
$$

Thus we can neglect the contributions of the ground state in the calculations of E, P and S. The $\varepsilon = 0$ state only plays the role as a particle source.

Next we consider the contribution of the excited states.

$$
\begin{aligned}
\ln \Xi|_{\varepsilon>0} &= -CV \int_0^\infty d\varepsilon \sqrt{\varepsilon} \ln(1 - e^{-\beta\varepsilon}) \\
&= \frac{2}{3} CV\beta \int_0^\infty d\varepsilon \frac{\varepsilon^{\frac{3}{2}}}{e^{-\beta\varepsilon} - 1} \\
&= \frac{2}{3} CV\beta^{-\frac{3}{2}} \int_0^\infty dx \frac{x^{\frac{3}{2}}}{e^{-x} - 1} \\
&= \frac{2}{3} CV\beta^{-\frac{3}{2}} \frac{3\sqrt{\pi}}{4} \times 1.341.
\end{aligned}
\tag{11.105}
$$

Thus we have

$$
E = -\frac{\partial \ln \Xi|_{\varepsilon>0}}{\partial \beta} = CV\beta^{-\frac{5}{2}} \frac{3\sqrt{\pi}}{4} \times 1.341 \propto m^{\frac{3}{2}} g_s V T^{\frac{5}{2}} \tag{11.106}
$$

and

$$
\begin{aligned}
C_V &= \left(\frac{\partial E}{\partial T}\right)_V \\
&= \frac{5}{2} C_V k_B \beta^{-\frac{3}{2}} \frac{3\sqrt{\pi}}{4} \times 1.341 \\
&= 1.926 N k_B \left(\frac{T}{T_c}\right)^{\frac{3}{2}}.
\end{aligned}
\tag{11.107}
$$

Eq.(11.107) shows that $C_V \to 0$ as $T \to 0$.

Other physical quantities can be calculated similarly. We find

$$
P = \frac{1}{\beta} \frac{\partial \ln \Xi|_{\varepsilon>0}}{\partial V} = \frac{2}{3} C\beta^{-\frac{3}{2}} \frac{3\sqrt{\pi}}{4} \times 1.341 \propto m^{\frac{3}{2}} g_s T^{\frac{3}{2}} \tag{11.108}
$$

and

$$
\begin{aligned}
S &= k(\ln \Xi|_{\varepsilon>0} + \alpha N + \beta E) \\
&= \frac{5}{3} CV k_B \beta^{-\frac{3}{2}} \frac{3\sqrt{\pi}}{4} \times 1.341 \\
&\propto m^{\frac{3}{2}} g_s V T^{\frac{3}{2}}.
\end{aligned}
\tag{11.109}
$$

It can be seen that P is independent of V due to the existence of the $\varepsilon = 0$ state as a particle source.

11.4 Photon gas

Now we turn to the photon gas. An equilibrium photon gas is also called the black body. The radiation emitted from a small opening on a cavity can be approximately considered as a black body radiation. Photons are massless spin-1 vector bosons. There are almost no interaction between photons. Thus we can use the Bose distribution for independent particles to calculate the properties of the photon gas. Since photons do not have mass, the particle number of a photon gas is not a constant. According to Eqs.(10.147) and (10.112), the equilibrium condition for a photon gas with constant temperature T and volume V is

$$\mu = \left.\frac{\partial F}{\partial N}\right|_{V,T} = 0. \tag{11.110}$$

Thus the chemical potential of an equilibrium photon gas is zero, which gives $\alpha = -\beta\mu = 0$. The Bose distribution for photon gas becomes

$$n_i = \frac{g_i}{e^{\beta\varepsilon_i} - 1}, \tag{11.111}$$

where $g_i = 2$ because photons have two spin components as shown by Eq.(2.578). From Eq.(2.565), the energy-momentum relation for photons is given by

$$\varepsilon = cp \tag{11.112}$$

with

$$p = \hbar k. \tag{11.113}$$

Then we have

$$\varepsilon = \hbar\omega = \hbar kc. \tag{11.114}$$

We introduce the wave length

$$\lambda \equiv \frac{2\pi}{k}. \tag{11.115}$$

Then we have

$$\lambda = \frac{2\pi c}{\omega} = \frac{c}{\nu} \tag{11.116}$$

with

$$\nu \equiv \frac{\omega}{2\pi}. \tag{11.117}$$

Both ν and ω are called frequency. ω is also called angular frequency. ε can be expressed in terms of ν as

$$\varepsilon = h\nu. \tag{11.118}$$

The photon number in the momentum interval $p \to p + dp$ is

$$n(p)dp = \frac{g_s V}{(2\pi\hbar)^3} \frac{1}{e^{\beta cp} - 1} 4\pi p^2 dp, \tag{11.119}$$

where $g_s = 2$ is the spin degeneracy of photons. We can express Eq.(11.119) in terms of frequency ν. The photon number in the interval $\nu \to \nu + d\nu$ is given by

$$n(\nu)d\nu = \frac{4\pi g_s V}{c^3} \nu^2 \frac{1}{e^{\beta h\nu} - 1} d\nu. \tag{11.120}$$

Thus the energy in the interval $\nu \to \nu + d\nu$ is given by

$$U(\nu)d\nu = n(\nu)h\nu d\nu = 8\pi V h\nu^3 c^{-3} \frac{1}{e^{\beta h\nu} - 1} d\nu. \tag{11.121}$$

Eq.(11.121) is called the *Planck law for black body radiation*.

When $\frac{h\nu}{k_B T} \ll 1$, Eq.(11.121) becomes

$$U(\nu)d\nu = 8\pi V \nu^2 k_B T c^{-3} d\nu. \tag{11.122}$$

Eq.(11.122) is called the *Rayleigh-Jeans law*, which is the classical version of the Planck law. When $\frac{h\nu}{k_B T} \gg 1$, Eq.(11.121) becomes

$$U(\nu)d\nu = 8\pi V h\nu^3 c^{-3} e^{-\beta h\nu} d\nu. \tag{11.123}$$

Eq.(11.123) is called the *Wien law*, which is the quantum version of the Planck law.

The total energy of the radiation field of a photon gas is calculated by the integration over frequency ν

$$U = \int_0^\infty U(\nu)d\nu = \int_0^\infty 8\pi V h\nu^3 c^{-3} \frac{1}{e^{\beta h\nu} - 1} d\nu = bVT^4 \tag{11.124}$$

with

$$b = \frac{8\pi k_B^4}{h^3 c^3} \int_0^\infty dx \frac{x^3}{e^x - 1} = \frac{8\pi^5 k_B^4}{15 h^3 c^3}. \tag{11.125}$$

Then the energy density u has the form

$$u = \frac{U}{V} = bT^4, \tag{11.126}$$

which gives the specific heat capacity per unit volume

$$C_v = \frac{32\pi^5 k_B^4}{15 h^3 c^3} T^3. \tag{11.127}$$

The specific heat capacity does not approaches to a constant as $T \to \infty$ because the photon number increases with the increase of temperature.

Using Eq.(11.121), we can also calculate the radiation escaped through an opening of a unit area on a black body cavity per unit time. For an opening with a unit area oriented in the **n** direction, the radiation flux density of the photons with frequency ν through the opening is given by

$$
\begin{aligned}
j &= \int \frac{U(\nu)}{V} c \frac{\mathbf{k}}{|\mathbf{k}|} \cdot \mathbf{n} \frac{d\Omega}{4\pi} \\
&= \int \frac{U(\nu)}{V} c \cos\theta \frac{d\Omega}{4\pi} \\
&= \int_0^{\frac{\pi}{2}} \frac{U(\nu)}{V} c \cos\theta \frac{2\pi \sin\theta d\theta}{4\pi} \\
&= \frac{1}{4} c \frac{U(\nu)}{V}.
\end{aligned}
\tag{11.128}
$$

In Eq.(11.128), the angular integration in the first line extends only over a hemisphere. Integrating over the frequency gives the total radiation flux J.

$$
J = \int j d\nu = \frac{1}{4} cu = \sigma T^4 \tag{11.129}
$$

with

$$
\sigma = \frac{2\pi^5 k_B^4}{15 h^3 c^3}. \tag{11.130}
$$

Eq.(11.129) is called the *Stefan-Boltzmann law* and σ is the *Stefan constant*.

In order to evaluate other thermodynamic quantities, we calculate the grand partition function of photon gas

$$
\begin{aligned}
\ln \Xi &= -\frac{V}{h^3} \int d^3 p g_s \ln(1 - e^{-\beta\varepsilon}) \\
&= -\int_0^\infty \frac{8\pi V}{h^3} p^2 dp \ln(1 - e^{-\beta\varepsilon}) \\
&= -\frac{8\pi V}{h^3 c^3 \beta^3} \int_0^\infty dx x^2 \ln(1 - e^{-x}) \\
&= \frac{8\pi V}{3 h^3 c^3 \beta^3} \int_0^\infty dx \frac{x^3}{e^x - 1} \\
&= \frac{8\pi^5 V}{45 h^3 c^3 \beta^3}.
\end{aligned}
\tag{11.131}
$$

The equation of state is given by

$$
P = \frac{1}{\beta} \frac{\partial \ln \Xi}{\partial V} = \frac{8\pi^5}{45 h^3 c^3 \beta^4} = \frac{b}{3} T^4 \tag{11.132}
$$

and the entropy has the form

$$
S = k_B (\ln \Xi - \beta \frac{\partial \ln \Xi}{\partial \beta}) = 4 k_B \ln \Xi = \frac{4}{3} b V T^3. \tag{11.133}
$$

Then we obtain the Helmholtz free energy

$$F = U - TS = -\frac{1}{3}U = -\frac{8\pi^5 V}{45h^3c^3\beta^4} = -\frac{1}{3}bVT^4, \qquad (11.134)$$

which gives

$$G = F + PV = 0. \qquad (11.135)$$

Since $G = \mu N$, we have $\mu = 0$. This is consistent with Eq.(11.110).

11.5 Fermi gas

Now we discuss the degenerate Fermi gas such as the electron gas in metals. We neglect the interaction between fermions and treat the gas as an independent particle system. The grand partition function for the Fermi gas is given by

$$\ln \Xi = \int_0^\infty d\varepsilon g(\varepsilon) \ln(1 + e^{-\alpha-\beta\varepsilon}) \qquad (11.136)$$

with

$$g(\varepsilon) = CVg_s \varepsilon^{\frac{1}{2}}. \qquad (11.137)$$

Integrating by parts, Eq.(11.136) becomes

$$\ln \Xi = \frac{2}{3}CVg_s \int_0^\infty d\varepsilon^{\frac{3}{2}} \ln(1 + e^{-\alpha-\beta\varepsilon})$$
$$= \frac{2}{3}CVg_s\beta \int_0^\infty d\varepsilon \frac{\varepsilon^{\frac{3}{2}}}{e^{\alpha+\beta\varepsilon} + 1}. \qquad (11.138)$$

α in Eq.(11.138) can be determined by

$$N = \int_0^\infty d\varepsilon g(\varepsilon) \frac{1}{e^{\alpha+\beta\varepsilon} + 1}$$
$$= CVg_s \int_0^\infty d\varepsilon \frac{\varepsilon^{\frac{1}{2}}}{e^{\alpha+\beta\varepsilon} + 1}. \qquad (11.139)$$

The integrals in Eqs.(11.138) and (11.139) have the similar form. We write them as

$$Q_l = \int_0^\infty dx x^l f(x). \qquad (11.140)$$

with

$$f(x) = \frac{1}{e^{\alpha+\beta x} + 1}. \qquad (11.141)$$

We use $q = e^{-\alpha}$ to replace $e^{-\alpha}$ in Eq.(11.141). q is the fugacity. In the case of the weakly degenerate quantum gas, we have used the following expansion

$$
\begin{aligned}
Q_l(q) &= \int_0^\infty dx \frac{x^l q e^{-\beta x}}{1 + q e^{-\beta x}} \\
&= \frac{1}{\beta^{l+1}} \int_0^\infty dx \frac{x^l q e^{-x}}{1 + q e^{-x}} \\
&= \frac{1}{\beta^{l+1}} \int_0^\infty dx\, x^l q e^{-x} \sum_{k=0}^\infty (-q e^{-x})^k \\
&= \frac{1}{\beta^{l+1}} \int_0^\infty dx\, x^l \sum_{k=1}^\infty (-1)^{k-1} q^k e^{-kx} \\
&= \frac{1}{\beta^{l+1}} \sum_{k=1}^\infty (-1)^{k-1} \frac{q^k}{k^{l+1}} \int_0^\infty dx\, x^l e^{-x} \\
&= \frac{1}{\beta^{l+1}} \Gamma(l+1) \sum_{k=1}^\infty (-1)^{k-1} \frac{q^k}{k^{l+1}}.
\end{aligned}
\tag{11.142}
$$

In the derivation of the last line of Eq.(11.142), the definition of the Γ-function is used. The expansion Eq.(11.142) can only be used when $q = e^{\beta\mu}$ is small. In low temperature, $\beta\mu$ is large, we consider another kind of expansion. We first integrate by parts, which gives

$$
\begin{aligned}
Q_l &= \frac{1}{l+1} \varepsilon^{l+1} f(\varepsilon) \Big|_0^\infty - \frac{1}{l+1} \int_0^\infty d\varepsilon\, \varepsilon^{l+1} \frac{df}{d\varepsilon} \\
&= -\frac{1}{l+1} \int_0^\infty d\varepsilon\, \varepsilon^{l+1} \frac{df}{d\varepsilon} \\
&= \int_0^\infty d\varepsilon\, v(\varepsilon) f'(\varepsilon)
\end{aligned}
\tag{11.143}
$$

with

$$
v(\varepsilon) \equiv -\frac{1}{l+1} \varepsilon^{l+1}.
\tag{11.144}
$$

We expand the function $v(\varepsilon)$ at $\varepsilon = \mu$

$$
v(\varepsilon) = \sum_{n=0}^\infty \frac{v^{(n)}(\mu)}{n!} (\varepsilon - \mu)^n.
\tag{11.145}
$$

Inserting the expansion Eq.(11.145) into Eq.(11.143), we have

$$
Q_l = \sum_{n=0}^\infty \frac{v^{(n)}(\mu)}{n!} \int_0^\infty d\varepsilon\, f'(\varepsilon)(\varepsilon - \mu)^n.
\tag{11.146}
$$

We introduce $\eta = \beta(\varepsilon - \mu)$. Then

$$f(\varepsilon) = \frac{1}{e^\eta + 1} \tag{11.147}$$

and

$$f'(\varepsilon) = -\beta \frac{e^\eta}{(e^\eta + 1)^2}. \tag{11.148}$$

Thus Eq.(11.146) becomes

$$Q_l = -\sum_{n=0}^{\infty} \frac{v^{(n)}(\mu)}{n!} \beta^{-n} \int_{-\beta\mu}^{\infty} d\eta \frac{e^\eta}{(e^\eta + 1)^2} \eta^n. \tag{11.149}$$

In low temperature, $\beta\mu \gg 1$. We neglect the exponentially small terms and obtain

$$\begin{aligned}
Q_l &= -\sum_{n=0}^{\infty} \frac{v^{(n)}(\mu)}{n!} \beta^{-n} \int_{-\beta\mu}^{\infty} d\eta \frac{e^\eta}{(e^\eta + 1)^2} \eta^n \\
&\doteq -\sum_{n=0}^{\infty} \frac{v^{(n)}(\mu)}{n!} \beta^{-n} \int_{-\infty}^{\infty} d\eta \frac{e^\eta}{(e^\eta + 1)^2} \eta^n \\
&= -v(\mu) - \frac{v^{(2)}(\mu)}{2!} \beta^{-2} \frac{\pi^2}{3} + \mathcal{O}\left[\left(\frac{k_B T}{\mu}\right)^4\right].
\end{aligned} \tag{11.150}$$

The expansion in Eq.(11.150) is called the *Sommerfeld expansion*.

The chemical potential μ in Eq.(11.150) is determined by

$$\begin{aligned}
N &= CV g_s Q_{\frac{1}{2}} \\
&= \frac{2}{3} CV g_s \mu^{\frac{3}{2}} \left\{ 1 + \frac{\pi^2}{8}\left(\frac{k_B T}{\mu}\right)^2 + \mathcal{O}\left[\left(\frac{k_B T}{\mu}\right)^4\right] \right\}. \tag{11.151}
\end{aligned}$$

$\mu(T)$ can be evaluated using the iteration method. The zero order term is $\mu(0)$ which is the chemical potential at $T = 0$. Let $T = 0$ in Eq.(11.151), we have

$$N = \frac{2}{3} CV g_s \mu(0)^{\frac{3}{2}}. \tag{11.152}$$

Solving $\mu(0)$ in Eq.(11.152) gives

$$\mu(0) = \left(\frac{3N}{2CV g_s}\right)^{\frac{2}{3}} \equiv \varepsilon_F. \tag{11.153}$$

We have introduced ε_F to denote $\mu(0)$ because the chemical potential at zero temperature is also called the *Fermi energy*. At zero temperature, the energy levels are filled in a way that one fermion occupies one state until all particles are exhausted. According to Eq.(10.193), the boundary between the occupied energy and unoccupied energy at zero temperature is the Fermi energy. In terms of $\mu(0)$, Eq.(11.151) is expressed as

$$\varepsilon_F^{\frac{3}{2}} = \mu^{\frac{3}{2}} \left\{ 1 + \frac{\pi^2}{8} \left(\frac{k_B T}{\mu} \right)^2 + \mathcal{O} \left[\left(\frac{k_B T}{\mu} \right)^4 \right] \right\}. \tag{11.154}$$

Solving μ gives

$$\mu = \varepsilon_F \left\{ 1 - \frac{\pi^2}{12} \left(\frac{k_B T}{\varepsilon_F} \right)^2 + \mathcal{O} \left[\left(\frac{k_B T}{\mu} \right)^4 \right] \right\}. \tag{11.155}$$

Using the Sommerfeld expansion, we can evaluate the thermodynamic quantities of the Fermi gas. The energy U of the Fermi gas is given by

$$
\begin{aligned}
U &= -\frac{\partial \ln \Xi}{\partial \beta} \\
&= CV g_s \int d\varepsilon \frac{\varepsilon^{\frac{3}{2}}}{e^{\alpha + \beta \varepsilon} + 1} \\
&= \frac{3}{2\beta} \ln \Xi \\
&= \frac{2}{5} CV g_s \mu^{\frac{5}{2}} \left\{ 1 + \frac{5\pi^2}{8} \left(\frac{k_B T}{\mu} \right)^2 + \mathcal{O} \left[\left(\frac{k_B T}{\mu} \right)^4 \right] \right\}.
\end{aligned}
\tag{11.156}
$$

Inserting Eq.(11.155) into Eq.(11.156), we have

$$
\begin{aligned}
U &= \frac{2}{5} CV g_s \varepsilon_F^{\frac{5}{2}} \left\{ 1 + \frac{5\pi^2}{12} \left(\frac{k_B T}{\varepsilon_F} \right)^2 + \mathcal{O} \left[\left(\frac{k_B T}{\varepsilon_F} \right)^4 \right] \right\} \\
&= U_0 \left\{ 1 + \frac{5\pi^2}{12} \left(\frac{k_B T}{\varepsilon_F} \right)^2 + \mathcal{O} \left[\left(\frac{k_B T}{\varepsilon_F} \right)^4 \right] \right\}
\end{aligned}
\tag{11.157}
$$

with $U_0 = \frac{3}{5} N \varepsilon_F$. U_0 is the ground state energy of the Fermi gas. The equation of state for the Fermi gas is given by

$$
\begin{aligned}
P &= \frac{1}{\beta} \frac{\partial \ln \Xi}{\partial V} \\
&= \frac{4}{15} C g_s \mu^{\frac{5}{2}} \left\{ 1 + \frac{5\pi^2}{8} \left(\frac{k_B T}{\mu} \right)^2 + \mathcal{O} \left[\left(\frac{k_B T}{\mu} \right)^4 \right] \right\} \\
&= \frac{2}{5} \frac{N}{V} \varepsilon_F \left\{ 1 + \frac{5\pi^2}{12} \left(\frac{k_B T}{\varepsilon_F} \right)^2 + \mathcal{O} \left[\left(\frac{k_B T}{\varepsilon_F} \right)^4 \right] \right\} \\
&= \frac{2U}{3V}.
\end{aligned}
\tag{11.158}
$$

Eq.(11.158) shows that at zero temperature, there is a nonzero pressure

$$P_0 = \frac{2U_0}{3V}. \tag{11.159}$$

The entropy is given by

$$
\begin{aligned}
S &= k_B(\ln \Xi + \alpha N + \beta U) \\
&= \frac{4}{15} k_B C V g_s \beta \mu^{\frac{5}{2}} \left\{ 0 + \frac{5\pi^2}{4} \left(\frac{k_B T}{\mu} \right)^2 + \mathcal{O} \left[\left(\frac{k_B T}{\mu} \right)^4 \right] \right\} \\
&= \frac{\pi^2}{3} C V g_s \varepsilon_F^{\frac{1}{2}} k_B^2 T \left\{ 1 + \mathcal{O} \left[\left(\frac{k_B T}{\varepsilon_F} \right)^2 \right] \right\}.
\end{aligned}
\tag{11.160}
$$

Eq.(11.160) shows that $S \to 0$ as $T \to 0$. The specific heat capacity is given by

$$
\begin{aligned}
C_V &= \left(\frac{\partial U}{\partial T} \right)_V \\
&= U_0 \left\{ \frac{5\pi^2}{6} \frac{k_B}{\varepsilon_F^2} T + \mathcal{O} \left[\left(\frac{k_B T}{\varepsilon_F} \right)^2 \right] \right\} \\
&= N k_B \frac{\pi^2}{2} \frac{k_B T}{\varepsilon_F} \left\{ 1 + \mathcal{O} \left[\left(\frac{k_B T}{\varepsilon_F} \right)^2 \right] \right\}.
\end{aligned}
\tag{11.161}
$$

Eq.(11.161) shows that the specific heat capacity of the Fermi gas in low temperature is a linear function of temperature T. We introduce a characteristic temperature T_F defined by

$$T_F \equiv \frac{\varepsilon_F}{k_B}. \tag{11.162}$$

T_F is called the *Fermi temperature*. When $T \gg T_F$, the Fermi gas becomes the ideal gas. Otherwise, it is a quantum gas. In terms of T_F, Eq.(11.161) can be expressed as

$$C_V = \frac{\pi^2}{2} N k_B \frac{T}{T_F} \left\{ 1 + \mathcal{O} \left[\left(\frac{T}{T_F} \right)^2 \right] \right\}. \tag{11.163}$$

For the electrons in metals, $T_F \sim 10^4 K$. Thus C_V of the electron gas in metals is small as compared with that contributed from the vibration of atoms in the room temperature.

Chapter 12

Relativity theory

12.1 Classical energy-momentum tensor

In the Einstein field equations Eq.(3.23), the metric tensor $g_{\mu\nu}$ is determined by the energy-momentum tensor T^μ_ν. Now we consider the energy-momentum tensor T^μ_ν of a classical system. The conservation of energy-momentum in the local flat metric gives

$$\frac{\partial T^\mu_\nu}{\partial x^\mu} = 0. \tag{12.1}$$

T^{00} is the energy density and we denote it as $W \equiv T^{00}$. We can separate the conservation equations into the space and time parts.

$$\frac{1}{c}\frac{\partial T^{00}}{\partial t} + \frac{\partial T^{0i}}{\partial x^i} = 0, \tag{12.2a}$$

$$\frac{1}{c}\frac{\partial T^{i0}}{\partial t} + \frac{\partial T^{ij}}{\partial x^j} = 0. \tag{12.2b}$$

Integrating the first equation over a volume V in space, we have

$$\frac{1}{c}\frac{\partial}{\partial t}\int T^{00}dV + \int \frac{\partial T^{0i}}{\partial x^i}dV = 0. \tag{12.3}$$

Using Gauss's theorem, we obtain

$$\frac{\partial}{\partial t}\int T^{00}dV = -c\oint T^{0i}ds_i, \tag{12.4}$$

where the integral on the right is taken over the surface surrounding the volume V. Since the expression on the left is the change rate of the energy contained in the volume V, the expression on the right is the amount of energy transferred across the boundary of the volume V. We define a vector $\mathbf{S}$,

$$\mathbf{S} \equiv \{cT^{01}, cT^{02}, cT^{03}\}. \tag{12.5}$$

From Eq.(12.4), we can see that $\mathbf{S}$ is the amount of energy passing through unit surface in unit time. Thus $\mathbf{S}$ is called the *flux density of energy*.

Eq.(2.202) shows that $\frac{\mathbf{S}}{c^2}$ is the momentum density. Thus the flux density of energy is equal to the momentum density multiplied by c^2. Now we consider the second equation in Eq.(12.2). We have

$$\frac{\partial}{\partial t}\int \frac{1}{c}T^{i0}dV = -\int \frac{\partial T^{ij}}{\partial x^j}dV$$
$$= -\oint T^{ij}ds_i.$$

(12.6)

The term on the left is the change of the momentum of the system in V per unit time. Therefore, $\oint T^{ij}ds_i$ is the momentum leaving the system in V per unit time. The components T^{ij} of the energy-momentum tensor constitute the three-dimensional tensor of momentum flux density, which we denote as $-\sigma_{ij}$. σ_{ij} is also called the *stress tensor*, which has the meaning that the component σ_{ij} is the amount of i-component of the momentum passing though unit surface perpendicular to the x^j axis per unit time (the direction entering the system is taken as positive) according to Eq.(12.6). Thus $T^{\mu\nu}$ can be expressed as the following matrix

$$T^{\mu\nu} = \begin{pmatrix} W & \dfrac{S_x}{c} & \dfrac{S_y}{c} & \dfrac{S_z}{c} \\ \dfrac{S_x}{c} & -\sigma_{xx} & -\sigma_{xy} & -\sigma_{xz} \\ \dfrac{S_y}{c} & -\sigma_{yx} & -\sigma_{yy} & -\sigma_{yz} \\ \dfrac{S_z}{c} & -\sigma_{zx} & -\sigma_{zy} & -\sigma_{zz} \end{pmatrix}.$$

(12.7)

The flux of momentum through the surface element $d\mathbf{s}$ is the force acting on the surface element according to Newton's law (Eq.(8.69)). Thus $\sigma_{ij}ds_j$ is the i-component of the force acting on the surface element. Now we use the reference system in which the elements of volume of the system are at rest. For an equilibrium hydrodynamic system, the pressure P is equal everywhere. The pressure P has the meaning of being the force acting on unit surface element. We have

$$\sigma_{ij}ds_j = -Pds_i.$$

(12.8)

Thus,

$$\sigma_{ij} = -P\delta_{ij}.$$

(12.9)

We denote T^{00} in the local rest frame by ε. $\rho = \frac{\varepsilon}{c^2}$ is defined as the mass density of the system, i.e. the mass per unit volume. It should be noted that the volume element here is the one in the reference frame in which the corresponding portion of body is at rest. We call such volume as *proper volume*. Thus in the reference frame that the system is at rest, the energy-momentum tensor has the form

$$T^{\mu\nu} = \begin{pmatrix} \varepsilon & 0 & 0 & 0 \\ 0 & P & 0 & 0 \\ 0 & 0 & P & 0 \\ 0 & 0 & 0 & P \end{pmatrix}. \tag{12.10}$$

We can use tensor transformation to obtain the expression for the energy-momentum tensor in an arbitrary reference frame. We introduce the four-velocity defined by Eq.(A.34) in Appendix A to describe the macroscopic motion of a body element. In the rest frame, $u^\alpha = (c, \mathbf{0})$. Generally,

$$u^\mu = \frac{dx^\mu}{d\tau}. \tag{12.11}$$

Since $dx_\mu dx^\mu = ds^2 = -c^2 (d\tau)^2$, we have

$$u^\mu u_\mu = -c^2. \tag{12.12}$$

When the velocity is v, we have

$$v^2 = \frac{dx^2 + dy^2 + dz^2}{dt^2}. \tag{12.13}$$

In a local rest frame, $dx' = dy' = dz' = 0$. We have

$$ds^2 = -c^2 dt^2 + dx^2 + dy^2 + dz^2 = -c^2 dt'^2. \tag{12.14}$$

Then

$$dt' = d\tau = dt \sqrt{1 - \frac{v^2}{c^2}}. \tag{12.15}$$

Thus the four-velocity has the form

$$u^\mu = \left(\frac{c}{\sqrt{1 - \dfrac{v^2}{c^2}}}, \frac{\mathbf{v}}{\sqrt{1 - \dfrac{v^2}{c^2}}} \right). \tag{12.16}$$

Using the four-velocity, we can write the right-hand side of Eq.(12.10) in a tensor form.

$$T^{\mu\nu} = (P + \varepsilon) \frac{1}{c^2} u^\mu u^\nu + P g^{\mu\nu}. \tag{12.17}$$

Since a tensor equation is hold in any frame, Eq.(12.17) is also valid for a general reference frame.

Thus the energy density W, energy flow vector $\mathbf{S}$ and stress tensor σ_{ij} are given by

$$W = \frac{\varepsilon + P\dfrac{v^2}{c^2}}{1 - \dfrac{v^2}{c^2}}, \tag{12.18a}$$

$$\mathbf{S} = \frac{(\varepsilon + P)\mathbf{v}}{c\left(1 - \dfrac{v^2}{c^2}\right)}, \tag{12.18b}$$

$$\sigma_{ij} = \frac{(\varepsilon + P)v_i v_j}{c^2\left(1 - \dfrac{v^2}{c^2}\right)} - P\delta_{ij}. \tag{12.18c}$$

12.2   Equation of motion in Riemann spacetime

The curvature of metric has the similar effect as an interaction. This effect is called the gravity. Now we derive the equation of motion for a classical particle in the curved metric.

The energy-momentum tensor is given by Eq.(12.17). For a classical particle, there is no pressure term, The energy-momentum tensor becomes

$$T^{\mu\nu} = \frac{\varepsilon}{c^2} u^\mu u^\nu = \rho u^\mu u^\nu. \tag{12.19}$$

The conservation of energy-momentum reads

$$T^{\mu\nu}{}_{;\nu} \equiv D_\nu T^{\mu\nu} = 0. \tag{12.20}$$

Using Eq.(A.78a) in Appendix A, we have

$$T^{\mu\nu}{}_{;\nu} = T^{\mu\nu}{}_{,\nu} + \Gamma^\mu_{\sigma\nu} T^{\sigma\nu} + \Gamma^\nu_{\sigma\nu} T^{\mu\sigma}. \tag{12.21}$$

The connection obeys the following relation

$$\Gamma^\mu_{\alpha\mu} = \frac{1}{2} g^{\mu\nu} g_{\mu\nu,\alpha} = -\frac{1}{2} g_{\mu\nu} g^{\mu\nu}{}_{,\alpha} = \frac{1}{2g} \frac{\partial g}{\partial x^\alpha} = \frac{\partial}{\partial x^\alpha}(\ln \sqrt{-g}). \tag{12.22}$$

Inserting Eq.(12.22) into Eq.(12.21), we have

$$T^{\mu\nu}{}_{;\nu} = \frac{1}{\sqrt{-g}} \frac{\partial}{\partial x^\nu}(T^{\mu\nu}\sqrt{-g}) + \Gamma^\mu_{\nu\sigma} T^{\nu\sigma} = 0. \tag{12.23}$$

We can change this equation into the following form

$$\frac{\partial}{\partial x^0}(\sqrt{-g}T^{0\mu}) + \frac{\partial}{\partial x^i}(\sqrt{-g}T^{i\mu}) + \sqrt{-g}\Gamma^\mu_{\alpha\beta} T^{\alpha\beta} = 0, \quad i = 1,2,3. \tag{12.24}$$

Integrating Eq.(12.24) over the volume of the particle, we have

$$\int \frac{\partial}{\partial x^0}(\sqrt{-g}T^{0\mu})d^3x + \int \frac{\partial}{\partial x^i}(\sqrt{-g}T^{i\mu})d^3x + \int \sqrt{-g}\Gamma^\mu_{\alpha\beta}T^{\alpha\beta}d^3x$$
$$= 0. \tag{12.25}$$

Using Gauss's theorem, the second term can be transformed into the surface integration and be dropped away because $\rho = 0$ at surface.

$$\int \frac{\partial}{\partial x^i}(\sqrt{-g}T^{i\mu})d^3x = \int \sqrt{-g}\rho u^i u^\mu ds_i = 0. \tag{12.26}$$

Inserting the expression of the energy momentum tensor given by Eq.(12.19), we have

$$\frac{d}{dx^0}\int \sqrt{-g}\rho u^0 u^\mu d^3x + \int \sqrt{-g}\Gamma^\mu_{\alpha\beta}\rho u^\alpha u^\beta d^3x = 0. \tag{12.27}$$

We have changed $\frac{\partial}{\partial x^0}$ to $\frac{d}{dx^0}$ in Eq.(12.27) because the spatial variables have been integrated over.

For a point-like particle, we can take u^μ and $\Gamma^\mu_{\alpha\beta}$ out of the integral. Then Eq.(12.27) becomes

$$\frac{d}{dx^0}\left(u^\mu \int \sqrt{-g}\rho u^0 d^3x\right) + \Gamma^\mu_{\alpha\beta}u^\alpha u^\beta \frac{1}{u^0}\int \sqrt{-g}\rho u^0 d^3x = 0. \tag{12.28}$$

We define the mass m of the particle in gravity as

$$m \equiv \frac{1}{c}\int \sqrt{-g}\rho u^0 d^3x. \tag{12.29}$$

In the local flat rest frame, $u^0 = c$, $\sqrt{-g} = 1$. Then

$$m = \int \rho d^3x. \tag{12.30}$$

m is just the rest mass of particle. In the local flat rest frame, ρ is a constant due to the conservation of energy. We have

$$(\rho u^\nu)_{;\nu} = 0. \tag{12.31}$$

Eq.(12.31) is a tensor equation, which should be valid in any reference frame. Eq.(12.31) is also called the *continuity equation of mass conservation*. In an arbitrary reference frame, we have

$$(\rho u^\nu)_{;\nu} = \frac{1}{\sqrt{-g}}(\sqrt{-g}\rho u^\nu)_{;\nu} = 0. \tag{12.32}$$

Multiplying Eq.(12.32) with $\sqrt{-g}$ and integrating, we obtain

$$\frac{d}{dx^0}\int \sqrt{-g}\rho u^0 d^3x + \int \frac{\partial}{\partial x^i}(\sqrt{-g}\rho u^i)d^3x = 0. \tag{12.33}$$

Using Gauss's theorem, the second term turns to be the surface integration and vanishes. We have

$$\frac{d}{dx^0} \int \sqrt{-g}\rho u^0 d^3 x = \frac{d}{dx^0} m = 0. \tag{12.34}$$

Thus, m is not dependent on x^0. Using Eq.(12.34), Eq.(12.28) becomes

$$m\frac{dx^0}{d\tau}\frac{du^\mu}{dx^0} + m\Gamma^\mu_{\alpha\beta}u^\alpha u^\beta = 0. \tag{12.35}$$

Thus we obtain the *geodesic equation* for the motion of a classical particle

$$\frac{du^\mu}{d\tau} + \Gamma^\mu_{\alpha\beta}u^\alpha u^\beta = 0 \tag{12.36}$$

or

$$\frac{d^2 x^\mu}{d\tau^2} + \Gamma^\mu_{\alpha\beta}\frac{dx^\alpha}{d\tau}\frac{dx^\beta}{d\tau} = 0. \tag{12.37}$$

Eq.(12.37) is called the geodesic equation because it is also the equation describing the shortest route connecting two points in the Riemann spacetime. The detailed derivation is shown in Appendix H. Thus particles move along the shortest route in the Riemann spacetime.

12.3   Weak field limit

12.3.1   *Static weak field limit-Newtonian gravitation*

The metric tensor is determined by the Einstein field equations Eq.(3.23), which reads in the classical form

$$R_{\mu\nu} - \frac{1}{2}g_{\mu\nu}R = \kappa T_{\mu\nu}. \tag{12.38}$$

where $\kappa = \frac{8\pi G}{c^4}$. We have neglected the cosmological constant. Using the Ricci scalar defined as

$$R \equiv g^{\mu\nu}R_{\mu\nu} = g^{\mu\nu}g^{\alpha\beta}R_{\alpha\mu\beta\nu}, \tag{12.39}$$

we have

$$R = -\kappa T. \tag{12.40}$$

with $T \equiv T^\mu_\mu$. Then Einstein field equations can be rewritten as

$$R_{\mu\nu} = \kappa\left(T_{\mu\nu} - \frac{1}{2}g_{\mu\nu}T\right). \tag{12.41}$$

In the normal temperature and weak field, the thermal velocity of particles is much smaller than the speed of light. Thus the pressure P is much smaller than the density ε. Using Eq.(12.17), we have

$$T^{\mu\nu} = \rho u^\mu u^\nu. \tag{12.42}$$

The four-velocity u^μ in the proper (local rest) frame has the form

$$u^\mu \equiv \frac{dx^\mu}{d\tau} = \frac{1}{\sqrt{-g_{00}}}(c, 0, 0, 0), \tag{12.43}$$

where $d\tau = \frac{ids}{c} = \frac{1}{c}\sqrt{-g_{\mu\nu}dx^\mu dx^\nu} = \frac{1}{c}\sqrt{-g_{00}}dx^0 = \sqrt{-g_{00}}dt$.

In the local rest frame, the energy-momentum tensor has only one nonzero component T^{00}

$$T^{00} = \frac{\rho c^2}{\sqrt{-g_{00}}}. \tag{12.44}$$

The trace of the energy-momentum tensor T is given by

$$T = g_{\mu\nu}T^{\mu\nu} = \rho u^\mu u_\mu = -\rho c^2. \tag{12.45}$$

The energy of a point-like particle is defined by

$$E = -mu^\mu u_\mu = mc^2, \tag{12.46}$$

which corresponds to T^{00} in the local flat rest frame. Since E is a scalar and conserved in the local flat rest frame, E is conserved in any frame.

When the curvature effect is weak, we can write the metric tensor as follows

$$g_{\mu\nu} = \eta_{\mu\nu} + h_{\mu\nu}, \tag{12.47}$$

where $h_{\mu\nu}$ is the term describing the deviation of the curved metric from the Minkowski metric. We have

$$|h_{\mu\nu}| \ll 1. \tag{12.48}$$

Its derivative is also small.

$$|g_{\mu\nu,i}| = |h_{\mu\nu,i}| \ll 1 \quad i = 1, 2, 3. \tag{12.49}$$

We consider the static case, in which ρ is not time dependent and correspondingly

$$|g_{\mu\nu,0}| = |h_{\mu\nu,0}| = 0. \tag{12.50}$$

The Ricci tensor is defined as

$$R_{\mu\nu} = R^\lambda_{\mu\lambda\nu} = \Gamma^\lambda_{\mu\nu,\lambda} - \Gamma^\lambda_{\mu\lambda,\nu} + \Gamma^\lambda_{\mu\nu}\Gamma^\rho_{\lambda\rho} - \Gamma^\rho_{\lambda\nu}\Gamma^\lambda_{\rho\mu}, \tag{12.51}$$

where $\Gamma^\lambda_{\mu\nu}$ is the Levi-Civita connection of the Riemann metric given by

$$\Gamma^\lambda_{\mu\nu} = \frac{1}{2}g^{\lambda\rho}(g_{\rho\mu,\nu} + g_{\rho\nu,\mu} - g_{\mu\nu,\rho}).$$ (12.52)

We keep only the linear terms of $h_{\mu\nu}$ and $h_{\mu\nu,\rho}$ in the expansion of $\Gamma^\lambda_{\mu\nu}$. We have

$$\Gamma^\lambda_{\mu\nu} = \frac{1}{2}\eta^{\lambda\rho}(h_{\rho\mu,\nu} + h_{\rho\nu,\mu} - h_{\mu\nu,\rho}).$$ (12.53)

We can neglect the quadratic terms of $\Gamma^\lambda_{\mu\nu}$ in the Ricci tensor and obtain

$$R_{\mu\nu} = \Gamma^\lambda_{\mu\nu,\lambda} - \Gamma^\lambda_{\mu\lambda,\nu}.$$ (12.54)

Inserting Eq.(12.53) into Eq.(12.54), we have

$$R_{00} = -\frac{1}{2}h_{00,i,i},$$ (12.55a)

$$R_{0i} = \frac{1}{2}(h_{k0,i,k} - h_{0i,k,k}),$$ (12.55b)

$$R_{ij} = -\frac{1}{2}(-h_{00,i,j} + h_{kk,i,j} - h_{ki,j,k} - h_{kj,i,k} + h_{ij,k,k}).$$ (12.55c)

The most important term is the (00) component, which obeys the following equation

$$R_{00} = \kappa(T_{00} - \frac{1}{2}g_{00}T).$$ (12.56)

In the weak field limit, Eq.(12.56) becomes

$$h_{00,i,i} = -c^2\kappa\rho.$$ (12.57)

We define

$$\varphi \equiv -\frac{c^2}{2}h_{00},$$ (12.58)

which is called the *gravitational potential function*. Then Eq.(12.57) reads

$$\Delta\varphi = \frac{c^4}{2}\kappa\rho = 4\pi G\rho.$$ (12.59)

Eq.(12.59) is the *Poisson equation for Newtonian gravitation*, which has the solution

$$\varphi(\mathbf{x}) = -G\int_V \frac{\rho(\mathbf{x}')d^3x'}{|\mathbf{x} - \mathbf{x}'|}.$$ (12.60)

For point-like particles with mass of M, Eq.(12.60) becomes

$$\varphi(r) = -\frac{GM}{r}.$$ (12.61)

12.3.2 *Equation of motion in Newtonian approximation*

Now we discuss the equation of motion in the static weak field and non-relativistic limit, which is called the Newtonian gravitational equation. We approximate the connection by keeping up to one order terms

$$\Gamma^\lambda_{\mu\nu} = \frac{1}{2}\eta^{\lambda\rho}(h_{\rho\mu,\nu} + h_{\rho\nu,\mu} - h_{\mu\nu,\rho}). \tag{12.62}$$

In the nonrelativistic limit, we have

$$\left|\frac{dx^i}{d\tau}\right| \ll \left|\frac{dx^0}{d\tau}\right|. \tag{12.63}$$

The geodesic equation Eq.(12.37) becomes

$$\frac{d^2x^0}{d\tau^2} = 0, \tag{12.64a}$$

$$\frac{d^2x^i}{d\tau^2} + \Gamma^i_{00}\left(\frac{dx^0}{d\tau}\right)^2 = 0. \tag{12.64b}$$

Solving Eq.(12.64a), we have

$$x^0 = a\tau + b, \tag{12.65}$$

where a and b are constants. Inserting Eq.(12.65) into Eq.(12.64b), we obtain

$$\frac{d^2x^i}{dx^{0^2}} = -\Gamma^i_{00} = \frac{1}{2}h_{00,i}. \tag{12.66}$$

Using $x^0 = ct$ and Eq.(12.58), we have

$$\frac{d^2x^i}{dt^2} = -\frac{\partial\varphi}{\partial x^i}. \tag{12.67}$$

Multiplying the equation by mass m, we obtain the *Newton's equation of motion* for a particle in a gravitational potential

$$m\frac{d^2x^i}{dt^2} = -\frac{\partial}{\partial x^i}(m\varphi). \tag{12.68}$$

The mass on the left hand side of Eq.(12.68) is usually called the inertial mass and the mass on the right hand side is called the gravitational mass due to the historical reason. Eq.(12.68) shows that the inertial mass is the same as the gravitational mass. We define the gravitational force $\mathbf{F}_g$ as

$$F_{g_i} \equiv -\frac{\partial}{\partial x^i}(m\varphi). \tag{12.69}$$

Then Eq.(12.68) becomes

$$m\frac{d^2x^i}{dt^2} = F_{g_i}. \tag{12.70}$$

Thus

$$V(\mathbf{x}) \equiv m\varphi(\mathbf{x}) \tag{12.71}$$

can be considered as the potential energy.

When the curvature source is a particle, from Eq.(12.61), the potential energy reads

$$V(r) = -\frac{GMm}{r}. \tag{12.72}$$

Eq.(12.72) is the *Newtonian gravitational law*. Using Eq.(12.58), we have

$$g_{00} = -\left(1 - \frac{2GM}{c^2 r}\right). \tag{12.73}$$

The weak field limit demands

$$\frac{2GM}{c^2 r} \ll 1 \tag{12.74}$$

or

$$r \gg r_g \equiv \frac{2GM}{c^2}. \tag{12.75}$$

r_g is called the gravitational radius of star. It is also called the radius of black hole. Although Newtonian gravitational potential could lead to gravitational collapse to black hole, Eq.(12.75) shows that it is invalid to use the Newtonian theory to treat the collapse to black hole.

12.3.3   *Harmonic coordinate*

In the calculations of the particle motion, we have the freedom to select a coordinate frame. The most convenient selection is the frame determined by the harmonic coordinate condition, which is also called harmonic gauge. With this condition, when the curvature source disappears, we recover the inertial frame in the flat spacetime. The *harmonic gauge* is defined by

$$\Gamma^\lambda \equiv g^{\mu\nu}\Gamma^\lambda_{\mu\nu} = 0. \tag{12.76}$$

Using the relation

$$\delta^\lambda_{\mu,\nu} = (g^{\lambda\rho}g_{\rho\mu})_{,\nu} = g^{\lambda\rho}g_{\rho\mu,\nu} + g_{\rho\mu}g^{\lambda\rho}{}_{,\nu} = 0, \tag{12.77}$$

we have

$$\begin{aligned}
\Gamma^\lambda &= \frac{1}{2}g^{\mu\nu}g^{\lambda\rho}(g_{\rho\mu,\nu} + g_{\rho\nu,\mu} - g_{\mu\nu,\rho}) \\
&= \frac{1}{2}g^{\mu\nu}(-g_{\rho\mu}g^{\lambda\rho}{}_{,\nu}) + \frac{1}{2}g^{\mu\nu}(-g_{\rho\nu}g^{\lambda\rho}{}_{,\mu}) - \frac{1}{2}g^{\mu\nu}g^{\lambda\rho}g_{\mu\nu,\rho} \\
&= -\frac{1}{2}g^{\mu\nu}(g_{\rho\mu}g^{\lambda\rho}{}_{,\nu}) - \frac{1}{2}g^{\nu\mu}(g_{\rho\mu}g^{\lambda\rho}{}_{,\nu}) - g^{\lambda\rho}\frac{\partial}{\partial x^\rho}(\ln\sqrt{-g}) \\
&= -\delta^\nu_\rho g^{\lambda\rho}{}_{,\nu} - g^{\lambda\rho}\frac{1}{\sqrt{-g}}\frac{\partial}{\partial x^\rho}(\sqrt{-g}) \\
&= -\frac{1}{\sqrt{-g}}\frac{\partial}{\partial x^\rho}(\sqrt{-g}g^{\lambda\rho}).
\end{aligned} \tag{12.78}$$

In the derivation of Eq.(12.78), we have used Eq.(A.94) in Appendix A.

Using Eq.(A.96) in Appendix A, we have

$$\Box(x^\mu) = \frac{1}{\sqrt{-g}} \frac{\partial}{\partial x^\rho} \left(\sqrt{-g} g^{\lambda\rho} \frac{\partial x^\mu}{\partial x^\lambda} \right)$$
$$= \frac{1}{\sqrt{-g}} \frac{\partial}{\partial x^\rho} (\sqrt{-g} g^{\mu\rho}), \tag{12.79}$$

where the symbol $\Box$ is the four-dimensional Laplacian operator defined as

$$\Box f = f^{;\mu}{}_{;\mu} = g^{\mu\nu} f_{;\mu;\nu} = f^{,\mu}{}_{;\mu} = \left(-\frac{1}{c^2} \frac{\partial^2}{\partial t^2} + \nabla^2 \right) f, \tag{12.80}$$

where f is a scalar. The symbol $\Box$ is also called the d'Alembert or wave operator. Eq.(12.78) becomes

$$\Gamma^\mu = -\Box x^\mu = 0. \tag{12.81}$$

Eq.(12.81) is the harmonic gauge for the coordinates. Since it was considered to be similar to the gauge in the electromagnetic field, it is also called the Lorentz gauge in gravity.

For the Minkowski metric,

$$g_{\mu\nu} = \eta_{\mu\nu}. \tag{12.82}$$

Since $\Gamma^\alpha_{\mu\nu} = 0$, the harmonic gauge Eq.(12.76) is satisfied for the Minkowski metric. Thus the harmonic gauge is a generalization of the inertial frame in the flat spacetime to the Riemann spacetime.

12.3.4 Weak field approximation in harmonic gauge

12.3.4.1 Radiation of gravitational waves

When we use the harmonic gauge, the weak field formulas become much more simpler. In the weak field limit, the metric can be written as

$$g_{\mu\nu} = \eta_{\mu\nu} + h_{\mu\nu} \tag{12.83}$$

with

$$|h_{\mu\nu}| \ll 1. \tag{12.84}$$

The Christoffel symbol becomes

$$\Gamma^\mu_{\alpha\beta} = \frac{1}{2} \eta^{\mu\nu} (h_{\nu\alpha,\beta} + h_{\nu\beta,\alpha} - h_{\alpha\beta,\nu})$$
$$= \frac{1}{2} (h^\mu{}_{\alpha,\beta} + h^\mu{}_{\beta,\alpha} - h_{\alpha\beta}{}^{,\mu}). \tag{12.85}$$

The Ricci tensor has the form

$$R_{\mu\nu} = \Gamma^{\lambda}_{\mu\nu,\lambda} - \Gamma^{\lambda}_{\mu\lambda,\nu} = -\frac{1}{2}(h_{\mu\nu}{}^{,\alpha}{}_{,\alpha} + h_{,\mu,\nu} - h^{\alpha}{}_{\mu,\nu,\alpha} - h^{\alpha}{}_{\nu,\mu,\alpha}), \quad (12.86)$$

where h is defined as

$$h \equiv h^{\alpha}_{\alpha} = \eta^{\alpha\beta}h_{\alpha\beta} = -h_{00} + h_{11} + h_{22} + h_{33}. \quad (12.87)$$

We introduce a tensor $\bar{h}_{\mu\nu}$ defined as

$$\bar{h}_{\mu\nu} \equiv h_{\mu\nu} - \frac{1}{2}\eta_{\mu\nu}h. \quad (12.88)$$

Then the Einstein equation Eq.(12.38) becomes

$$\bar{h}_{\mu\nu}{}^{,\alpha}{}_{,\alpha} + \eta_{\mu\nu}\bar{h}_{\alpha\beta}{}^{,\alpha,\beta} - \bar{h}_{\mu\alpha}{}^{,\alpha}{}_{,\nu} - \bar{h}_{\nu\alpha}{}^{,\alpha}{}_{,\mu} = -16\pi\frac{G}{c^4}T_{\mu\nu}. \quad (12.89)$$

The harmonic gauge Eq.(12.76) in the weak field has the form

$$\begin{aligned} \Gamma^{\lambda} &= \frac{1}{2}g^{\mu\nu}g^{\lambda\rho}(g_{\rho\mu,\nu} + g_{\rho\nu,\mu} - g_{\mu\nu,\rho}) \\ &= \bar{h}_{\lambda\alpha}{}^{,\alpha} \\ &= 0. \end{aligned} \quad (12.90)$$

The harmonic gauge Eq.(12.90) makes the last three terms in Eq.(12.89) vanish. Then the Einstein equations become

$$\bar{h}_{\mu\nu}{}^{,\alpha}{}_{,\alpha} = -16\pi\frac{G}{c^4}T_{\mu\nu}. \quad (12.91)$$

Eq.(12.91) is a wave equation with a source. The source could emit the gravitational wave.

The solution of Eq.(12.91) is given by

$$\bar{h}_{\mu\nu}(\mathbf{x}, t) = \frac{\kappa}{2\pi}\int_V \frac{T_{\mu\nu}(\mathbf{x}', t - \frac{r}{c})}{|\mathbf{x} - \mathbf{x}'|}d^3x'. \quad (12.92)$$

$T_{\mu\nu}$ is the source and $\bar{h}_{\mu\nu}$ can be considered as the potential induced by the source $T_{\mu\nu}$.

In the region outside the source, Eq.(12.91) becomes

$$\left(\nabla^2 - \frac{1}{c^2}\frac{\partial^2}{\partial t^2}\right)\bar{h}_{\mu\nu}(\mathbf{x}, t) = 0. \quad (12.93)$$

The solutions of Eq.(12.93) are the superpositions of plane waves

$$\bar{h}_{\mu\nu}(\mathbf{x}, t) = \bar{h}_{\mu\nu 0}e^{i(\mathbf{k}\cdot\mathbf{x} - \omega t)} \quad (12.94)$$

with

$$k = \frac{\omega}{c}. \quad (12.95)$$

Thus the gravitational waves propagate at the speed of light.

12.3.4.2 *Newtonian gravitation*

In the nonrelativistic case, we have

$$\Box \bar{h}_{00} = -16\pi \frac{G}{c^2} \rho, \tag{12.96}$$

where we have used the approximation $T_{00} \approx \rho c^2$. Since the time variation is caused by the source moving with the velocity $\mathbf{v}$, $\frac{\partial}{\partial t}$ is of the same order as $\mathbf{v} \cdot \boldsymbol{\nabla}$, we have

$$\Box = \boldsymbol{\nabla}^2 + \mathcal{O}((\mathbf{v} \cdot \boldsymbol{\nabla})^2). \tag{12.97}$$

To the lowest order,

$$\boldsymbol{\nabla}^2 \bar{h}_{00} = -16\pi \frac{G}{c^2} \rho. \tag{12.98}$$

Since all other components of $\bar{h}_{\alpha\beta}$ ($\alpha, \beta \neq 0$) are negligible at this order, we have

$$h = h_\alpha^\alpha = -\bar{h}_\alpha^\alpha = -\bar{h}_{00}. \tag{12.99}$$

Using the relation

$$h_{\alpha\beta} = \bar{h}_{\alpha\beta} + \frac{1}{2}\eta_{\alpha\beta}h, \tag{12.100}$$

we have

$$h_{00} = \frac{1}{2}\bar{h}_{00}, \tag{12.101a}$$

$$h_{xx} = h_{yy} = h_{zz} = -\frac{1}{2}\bar{h}_{00}. \tag{12.101b}$$

Using Eq.(12.58) for the definition of φ, we have

$$\bar{h}_{00} = -\frac{4}{c^2}\varphi. \tag{12.102}$$

Inserting Eq.(12.102) into Eq.(12.98), we obtain the Poisson equation for Newtonian gravitation

$$\nabla^2 \varphi = 4\pi G\rho. \tag{12.103}$$

The metric in the weak field limit is given by

$$ds^2 = -c^2\left(1 + \frac{2\varphi}{c^2}\right)dt^2 + \left(1 - \frac{2\varphi}{c^2}\right)(dx^2 + dy^2 + dz^2). \tag{12.104}$$

We define the four-momentum p as

$$p^\mu \equiv m\frac{dx^\mu}{d\tau}. \tag{12.105}$$

Then

$$p \cdot p = -m^2 c^2 = g_{\alpha\beta} p^\alpha p^\beta. \tag{12.106}$$

Using Eq.(12.104), we have

$$-m^2 c^2 = -\left(1 + \frac{2\varphi}{c^2}\right)(p^0)^2 + \left(1 - \frac{2\varphi}{c^2}\right)\mathbf{p}^2. \tag{12.107}$$

We solve p^0 in Eq.(12.107) and get

$$(p^0)^2 = \frac{1}{\left(1 + \frac{2\varphi}{c^2}\right)}\left[m^2 c^2 + \left(1 - \frac{2\varphi}{c^2}\right)\mathbf{p}^2\right]. \tag{12.108}$$

Since $\frac{\varphi}{c^2} \ll 1$ and $p \ll mc$, Eq.(12.108) can be rewritten as

$$(p^0)^2 \approx m^2 c^2 \left(1 - \frac{2\varphi}{c^2} + \frac{\mathbf{p}^2}{m^2 c^2}\right) \tag{12.109}$$

or

$$p^0 \approx mc\left(1 - \frac{\varphi}{c^2} + \frac{\mathbf{p}^2}{2m^2 c^2}\right). \tag{12.110}$$

Lowering the index gives

$$p_0 = g_{0\alpha} p^\alpha = g_{00} p^0 = -\left(1 + \frac{2\varphi}{c^2}\right) p^0 = -\frac{1}{c}\left(mc^2 + m\varphi + \frac{\mathbf{p}^2}{2m}\right). \tag{12.111}$$

Now we consider the geodesic equation Eq.(H.3) in Appendix H, which can be expressed as an equation for the lowered components of p as follows

$$p^\alpha p_{\beta;\alpha} = 0. \tag{12.112}$$

Using

$$p_{\beta;\alpha} = p_{\beta,\alpha} - \Gamma^\gamma_{\beta\alpha} p_\gamma \tag{12.113}$$

and

$$m\frac{dp_\beta}{d\tau} = p^\alpha p_{\beta,\alpha}, \tag{12.114}$$

we have

$$\begin{aligned}
m\frac{dp_\beta}{d\tau} &= \Gamma^\gamma_{\beta\alpha} p^\alpha p_\gamma \\
&= \frac{1}{2} g^{\gamma\nu}(g_{\nu\beta,\alpha} + g_{\nu\alpha,\beta} - g_{\alpha\beta,\nu}) p^\alpha p_\gamma \\
&= \frac{1}{2}(g_{\nu\beta,\alpha} + g_{\nu\alpha,\beta} - g_{\alpha\beta,\nu}) g^{\gamma\nu} p_\gamma p^\alpha \\
&= \frac{1}{2}(g_{\nu\beta,\alpha} + g_{\nu\alpha,\beta} - g_{\alpha\beta,\nu}) p^\nu p^\alpha \\
&= \frac{1}{2} g_{\nu\alpha,\beta} p^\nu p^\alpha.
\end{aligned} \tag{12.115}$$

Thus the geodesic equation becomes

$$m\frac{dp_\beta}{d\tau} = \frac{1}{2}g_{\nu\alpha,\beta}p^\nu p^\alpha. \tag{12.116}$$

In the case of stationary (time-independent) field, $g_{\nu\alpha,0} = 0$. Thus p_0 is time-independent and thus conserved. We can call $-p_0 c$ as the energy of the particles in the gravitational field and denote it by $E_0 \equiv -p_0 c$. As we can see from Eq.(12.111) that E_0 consists of three terms. mc^2 is the rest energy. $m\varphi$ is the gravitational potential energy. $\frac{p^2}{2m}$ is the kinetic energy. It should be noted that this conserved law is only applicable to the stationary case.

12.4 Spherical solutions for stars

12.4.1 *Spherically symmetric spacetime*

Spherically symmetric systems are the most important gravitational systems because point-like particles and spherical stars are described by such systems.

12.4.1.1 *Minkowski spacetime in spherical coordinates*

Minkowski spacetime is a flat spacetime with the spherical symmetry. In the spherical coordinates, the line element of the Minkowski spacetime is given by

$$ds^2 = -dt^2 + dr^2 + r^2(d\theta^2 + \sin^2\theta d\phi^2). \tag{12.117}$$

The surface of constant t and r is a two dimensional spherical surface, which is often called two-sphere in a simple notation. Distances dl along curves on the two-sphere are given by Eq.(12.117) with the constraint $dt = dr = 0$.

$$dl^2 = r^2(d\theta^2 + \sin^2\theta d\phi^2) = r^2 d\Omega^2, \tag{12.118}$$

where the symbol $d\Omega^2$ defines the element of solid angle. A two-sphere has circumference $2\pi r$ and area $4\pi r^2$.

12.4.1.2 *Spherically symmetric metric*

For a Riemann spacetime with the spherical symmetry, every point of spacetime should be on a two-sphere, whose line element is given by

$$dl^2 = f(r',t)(d\theta^2 + \sin^2\theta d\phi^2), \tag{12.119}$$

where $f(r', t)$ is a function of two other coordinates r' and t. The area of each two-sphere is $4\pi f(r', t)$. We can make a coordinate transformation from (r', t) to (r, t) in such a way $f(r', t) = r^2$. Then in the new coordinates r and t, the area of a two-sphere is $4\pi r^2$ and circumference $2\pi r$. This coordinate r is called the *curvature coordinate* or *area coordinate*. It should be noted that r is generally not the distance from the center of the sphere to its surface in the Riemann spacetime.

Now we consider the spheres at r and $r + dr$. Each sphere has a coordinate system (θ, ϕ). We demand that a line with $\theta =$const and $\phi =$const is orthogonal to the two-spheres, which requires $\mathbf{e}_r \cdot \mathbf{e}_\theta = \mathbf{e}_r \cdot \mathbf{e}_\phi = 0$. Thus we have $g_{r\theta} = g_{r\phi} = 0$. Then the metric with spherical symmetry has the form

$$ds^2 = g_{00}dt^2 + 2g_{0r}dtdr + 2g_{0\theta}dtd\theta$$
$$+ 2g_{0\phi}dtd\phi + g_{rr}dr^2 + r^2 d\Omega^2. \tag{12.120}$$

Similarly, we consider the spheres at t and $t+dt$. The line with $r =$const, $\theta =$const and $\phi =$const should also be orthogonal to the two-spheres, which requires $\mathbf{e}_t \cdot \mathbf{e}_\theta = \mathbf{e}_t \cdot \mathbf{e}_\phi = 0$ or $g_{t\theta} = g_{t\phi} = 0$. Then Eq.(12.120) becomes

$$ds^2 = g_{00}(r, t)dt^2 + 2g_{0r}(r, t)dtdr + g_{rr}(r, t)dr^2 + r^2 d\Omega^2. \tag{12.121}$$

This is the general form of a spherically symmetric metric.

12.4.1.3 *Spherically symmetric metric for static systems*

Now we consider the static systems. For a static system, the energy-momentum tensor is independent of time t. Thus the metric can be chosen to be static, i.e. a metric whose components are independent of time t. Since the energy-momentum tensor has the time reversal symmetry, the geometry is not changed by time reversal, $t \to -t$. The metric should be unchanged by the coordinate transformation $(t, r, \theta, \phi) \to (-t, r, \theta, \phi)$. We have $g_{0r} = 0$. The causality principle demands that $g_{00} < 0$ and $g_{rr} > 0$. Then the metric of a static spacetime with the spherical symmetry is given by

$$ds^2 = -e^{2\Phi}dt^2 + e^{2\Lambda}dr^2 + r^2 d\Omega^2 \tag{12.122}$$

with

$$g_{00} \equiv -e^{2\Phi} \quad \text{and} \quad g_{rr} \equiv e^{2\Lambda}. \tag{12.123}$$

For a star, which is a bound system, the spacetime far from the star is flat. We have the boundary conditions on the Einstein field equations.

$$\lim_{r \to \infty} \Phi(r) = \lim_{r \to \infty} \Lambda(r) = 0. \tag{12.124}$$

This condition is called the *asymptotic flat condition of spacetime*.

12.4.1.4 *Einstein tensor in spherically symmetric metric for static systems*

Using the metric Eq.(12.122), we can calculate the Einstein tensor

$$G_{\mu\nu} \equiv R_{\mu\nu} - \frac{1}{2}g_{\mu\nu}R. \tag{12.125}$$

The components of the Einstein tensor are given by

$$G_{00} = \frac{1}{r^2}e^{2\Phi}\frac{d}{dr}\left[r(1 - e^{-2\Lambda})\right], \tag{12.126a}$$

$$G_{rr} = -\frac{1}{r^2}e^{2\Lambda}(1 - e^{-2\Lambda}) + \frac{2}{r}\dot{\Phi}, \tag{12.126b}$$

$$G_{\theta\theta} = r^2 e^{-2\Lambda}\left[\ddot{\Phi} + (\dot{\Phi})^2 + \frac{\dot{\Phi}}{r} - \dot{\Phi}\dot{\Lambda} - \frac{\dot{\Lambda}}{r}\right], \tag{12.126c}$$

$$G_{\phi\phi} = \sin^2\theta G_{\theta\theta}. \tag{12.126d}$$

All other components are zero.

12.4.1.5 *Gravitational redshift*

We have shown that any particle moving along a geodesic has a constant energy $E = -p_0 c$. However, a local inertia observer at rest measures a different energy. When one is at rest, $u^i = \frac{dx^i}{d\tau} = 0$. From the condition $u^\mu u_\mu = -c^2$, we have $u^0 = ce^{-\Phi}$. According to Eq.(12.46), the energy measured by the local observer at rest is

$$E' = -u^\mu p_\mu = e^{-\Phi}E. \tag{12.127}$$

Considering the asymptotic flat condition Eq.(12.124), $e^{-\Phi} = 1$ as $r \to \infty$. It can be seen that E is the energy that a distant observer would measure when the particle gets there. For a star, in the weak field limit, $e^{2\Phi} = 1 + \frac{2\varphi}{c^2}$ according to Eq.(12.104). Thus $\Phi \approx \frac{\varphi}{c^2} < 0$. We have $e^{-\Phi} > 1$. Then Eq.(12.127) shows that the particle has larger energy from the view point of local inertial observer. This extra energy is the kinetic energy gained by falling in a gravitational field.

When Eq.(12.127) is applied to photons, we get an important physical phenomenon called gravitational redshift. We consider a photon emitted at radius r_1 and received at r_2. We denote ν_{r_1} the frequency of the photon at r_1 in the local inertial frame, then its local energy is $h\nu_{r_1}$ and its conserved constant E is $h\nu_{r_1}\exp\left(\Phi(r_1)\right)$. When the photon reaches the radius r_2, it is measured to have energy

$$h\nu_{r_2} = E\exp\left(-\Phi(r_2)\right) = h\nu_{r_1}\exp\left(\Phi(r_1) - \Phi(r_2)\right). \tag{12.128}$$

The redshift of the photon is defined by

$$Z \equiv \frac{\lambda_{r_2} - \lambda_{r_1}}{\lambda_{r_1}} = \frac{\nu_{r_1}}{\nu_{r_2}} - 1. \tag{12.129}$$

Inserting Eq.(12.128), we have

$$Z = \exp\left(\Phi(r_2) - \Phi(r_1)\right) - 1. \tag{12.130}$$

When $\Delta r = r_2 - r_1 = h$ is small, we have

$$\frac{h\nu_{r_2}}{h\nu_{r_1}} = 1 - \frac{gh}{c^2}, \tag{12.131}$$

where g is gravitational acceleration, which is equal to GM/R^2 with M being the mass of the earth and R the radius of the earth according to Eq.(12.61). The effect of Eq.(12.131) is significant in the precision measurement.

12.4.2 *Einstein equations for static fluid*

12.4.2.1 *Energy-momentum tensor*

We consider the static stars, in which the fluid is at rest. u has only one nonzero component u^0. Using the formula $u^\mu u_\mu = -c^2$, we have

$$u^0 = ce^{-\Phi}, \quad u_0 = -ce^{\Phi}. \tag{12.132}$$

Inserting Eq.(12.132) into Eq.(12.17), we have

$$T_{00} = \rho c^2 e^{2\Phi}, \tag{12.133a}$$

$$T_{rr} = P e^{2\Lambda}, \tag{12.133b}$$

$$T_{\theta\theta} = r^2 P, \tag{12.133c}$$

$$T_{\phi\phi} = \sin^2\theta T_{\theta\theta}. \tag{12.133d}$$

All other components are zero.

12.4.2.2 *Equation of state*

In the energy-momentum tensor, which is often called the stress-energy tensor for a fluid, there are two thermodynamic variables P and ρ. From statistical mechanics, we can obtain a relation between them

$$P = P(\rho, T). \tag{12.134}$$

Eq.(12.134) is the equation of state. When the temperature T is low, we have

$$P = P(\rho). \tag{12.135}$$

The form of this relation depends on the constituents of star.

12.4.2.3   *Equation of motion*

The conservation of energy-momentum gives

$$T^{\alpha\beta}{}_{;\beta} = 0. \tag{12.136}$$

Using Eq.(12.133), we have

$$(\rho c^2 + P)\frac{d\Phi}{dr} = -\frac{dP}{dr}. \tag{12.137}$$

Due to the symmetries, the tensor equation Eq.(12.136) becomes a scalar equation.

12.4.2.4   *Einstein equations*

Using Eqs.(12.126) and (12.133), we obtain the Einstein equations for a fluid.

For the (0,0) component, we have

$$\frac{du(r)}{dr} = 4\pi r^2 \rho \tag{12.138}$$

with

$$u(r) \equiv \frac{1}{2}\frac{c^2}{G}r(1 - e^{-2\Lambda}). \tag{12.139}$$

Eq.(12.138) shows that $u(r)$ has the meaning of the mass apart from a constant.

$$u(r) = \int_0^r 4\pi r^2 \rho + u_0. \tag{12.140}$$

u_0 can be nonzero and is determined by Eq.(12.139) using the boundary condition. u_0 has only geometric meaning. In the Newtonian approximation, it can be shown $u_0 = 0$. Then $u(r)$ is the mass. In the case of the strong field, we will show that u_0 can be nonzero.

For the (r, r) component, we have

$$\frac{d\Phi(r)}{dr} = \frac{Gc^2 u(r) + 4\pi Gr^3 P(r)}{c^2 r[c^2 r - 2Gu(r)]}. \tag{12.141}$$

Due to the symmetry, (θ, θ) and (ϕ, ϕ) components can be derived from Eqs.(12.138) and (12.141) by the Bianchi identity. Now we have four equations(Eqs.(12.135), (12.137), (12.138) and (12.141)) with four functions $(P(r), \rho(r), \Phi(r)$ and $u(r))$. We can solve the equations to obtain the four functions $P(r), \rho(r), \Phi(r)$ and $u(r)$.

Generally, we use the boundary conditions at the boundary of the star, which reads

$$P|_{r=r_b} = 0 \quad \rho|_{r=r_b} = 0, \tag{12.142}$$

where r_b is the radius of the star.

12.4.3   *Metric outside a star*

Outside the star ($r > R \equiv r_b$) is the vacuum. We have $\rho = 0$ and $P = 0$. The four equations reduce to the two effective equations with the two functions u and Φ.

$$\frac{du(r)}{dr} = 0, \tag{12.143a}$$

$$\frac{d\Phi(r)}{dr} = \frac{Gu(r)}{r[c^2 r - 2Gu(r)]}. \tag{12.143b}$$

The solution of Eq.(12.143) has the form

$$u(r) = M = const, \tag{12.144a}$$

$$e^{2\Phi} = 1 - \frac{2GM}{c^2 r}. \tag{12.144b}$$

We have used the asymptotic flat boundary condition $\Phi \to 0$ as $r \to \infty$ for the solution. For the vacuum region outside the star, we have the following metric

$$ds^2 = -\left(1 - \frac{2GM}{c^2 r}\right) dt^2 + \left(1 - \frac{2GM}{c^2 r}\right)^{-1} dr^2 + r^2 d\Omega^2. \tag{12.145}$$

This metric is called the *Schwarzschild metric*. At large r, Eq.(12.145) becomes

$$ds^2 = -\left(1 - \frac{2GM}{c^2 r}\right) dt^2 + \left(1 + \frac{2GM}{c^2 r}\right) dr^2 + r^2 d\Omega^2. \tag{12.146}$$

We can see that this far field metric of a star is equivalent to the metric of point-like particles with mass M given by Eq.(12.104).

The Schwarzschild metric is the vacuum solution outside stars. The Minkowski metric is also the vacuum metric. When the whole space is vacuum, the only physical solution is the Minkowski metric. When the space contains a star with the spherical symmetry, the physical solution is the Schwarzschild metric. Therefore, Schwarzschild metric should be used for $r > R$ outside the star with the radius of $R = r_b$. It can not be used for $r < R$. Until now, all solutions for the fluid stars have $R > r_g \equiv \frac{2GM}{c^2}$.

12.4.4   *Interior structure of a star*

Inside the star, since $\rho \neq 0$ and $P \neq 0$, we can divide Eq.(12.137) by $(\rho c^2 + P)$, and eliminate $\frac{d\Phi}{dr}$ using Eq.(12.141). Then we have

$$\frac{dP}{dr} = -\frac{(c^2 \rho + P)(Gc^2 u + 4\pi G r^3 P)}{c^2 r[c^2 r - 2Gu(r)]}. \tag{12.147}$$

This equation is called the *Tolman-Oppenheimer-Volkov (TOV) equation.* Combined with Eq.(12.138) and the equation of state, we have three equations for u, ρ and P. Eqs.(12.138) and (12.147) are two first order differential equations. We have two constants of integration. There are two ways to determine the constants of integration: One is to use the boundary conditions for the integration from the center of the star; The other is to use the boundary conditions for the integration from the boundary of the star.

There are two types of solutions. One is that of the stars without void and the other gives the stars with void.

i) The solution of stars without void

In the first case, we use $u(r = 0)$ and $P(r = 0)$ as the initial values of integration. Solving $e^{-2\Lambda}$ from the equation $u(r) = \frac{1}{2}\frac{c^2}{G}r(1 - e^{-2\Lambda})$, we have

$$e^{-2\Lambda} = 1 - \frac{2Gu(r)}{c^2 r}. \tag{12.148}$$

Since $g_{rr} = e^{2\Lambda}$ is positive, we have $u(0) \leq 0$. If $u(0) \neq 0$, $e^{-2\Lambda}$ will approach infinite at the origin $r = 0$. When $u(0) \neq 0$, from Eq.(12.147), around the origin $r = 0$, we have

$$\frac{dP}{dr} = \frac{c^2\rho + P}{2r} = \frac{c^2\alpha P^s + P}{2r}, \tag{12.149}$$

where we have expressed the equation of state as $\rho = \alpha P^s$. When $s < 1$, the solution of Eq.(12.149) is

$$P^{1-s} \approx \frac{1}{2}c^2\alpha(1 - s)\ln r + c, \tag{12.150}$$

which will result in a negative p when $r \to 0$. Thus $u(0)$ can only be zero for a star without a void. When $s \geq 1$, we have $P \sim cr$. Then $P|_{r=0} = 0$. It is possible that $u(0)$ can be nonzero in this case. Since for most kinds of cold stars, $s < 1$. We will mainly focus on the case of $s < 1$, which is applicable for most star matters.

ii) The solution of stars with void

We can rewrite the solution (Eq.(12.150)) of Eq.(12.149) as follows,

$$P^{1-s} = \frac{1}{2}c^2\alpha(1 - s)\ln\left(\frac{r}{r_i}\right). \tag{12.151}$$

We find that $P = 0$ at $r = r_i$. If we consider $r = r_i$ as an inner boundary, P will remain zero when $r \leq r_i$ and we could avoid a negative p. Thus, we have another type of solutions with $P_i = 0$ at $r = r_i$. Since $\rho = 0$ for

$r \leq r_i$, there is a void around $r = 0$ for this type of solutions. From $r = 0$ to $r = r_i$, $\rho = 0$ and $P = 0$. There are no particles and thus pressure is zero in this void region. In the void region, we have the Minkowski-type metric

$$ds^2 = -Adt^2 + Bdr^2 + r^2 d\theta^2 + r^2 \sin^2 \theta d\phi^2, \qquad (12.152)$$

where $A = e^{2\Phi_i}$ and $B = (1 - 2Gu(r_i)/c^2 r_i)^{-1}$ are constants. Eq.(12.140) shows that $u(r_i) = u(0) = u_0$. The parameter Φ_i is the value of Φ at $r = r_i$, which can be obtained by integrating the equation Eq.(12.141). The initial condition for this type of solutions should take the values at the inner radius $r = r_i$ instead of $r = 0$. The differential equations can be solved by integrating from the initial values. The outer radius r_o is reached when $P = 0$.

Outside the star, the metric is the Schwarzschild metric. The metric functions must be continuous at $r = r_o$. Inside the star, the metric is

$$g_{rr} = \left(1 - \frac{2Gu(r)}{c^2 r}\right)^{-1}. \qquad (12.153)$$

Outside the star, we have

$$g_{rr} = \left(1 - \frac{2GM}{c^2 r}\right)^{-1}. \qquad (12.154)$$

The continuity of the metric demands

$$M = u(r_o). \qquad (12.155)$$

Thus the gravitational mass of the star is determined by

$$M = \int_{r_i}^{r_o} dr 4\pi r^2 \rho + u(r_i). \qquad (12.156)$$

For the stars without void, $r_i = 0$ and $u(r_i) = 0$. The gravitational mass in Eq.(12.156) is just the mass of the star.

It should be noted that we always have $\frac{2G}{c^2} u(r) < r$ for stars. If it ever happened that $r - \frac{2G}{c^2} u(r) = \varepsilon$ is small near some radius r_1, from the TOV equation Eq.(12.147), the pressure gradient $\frac{dP}{dr}$ is of order $\frac{1}{\varepsilon}$ and negative. This would leads to the rapid decrease of the pressure P and drop to zero before ε reaches zero. At $P = 0$, we reach the surface of the star. Outside the star, u is constant and r increases. Thus $u(r)$ of a star can not reach $\frac{c^2}{2G} r$ and M can not be larger than $\frac{c^2}{2G} r_o$, which means that $\frac{2MG}{c^2}$ can only be smaller than r. Thus it is impossible to have a black hole solution.

We can also solve the differential equation using the boundary conditions at the outer surface of the star. We use the initial values of $u(r_o)$ and $P = 0$

at $r = r_o$. Integrating from the surface of the star to the inner center, we can solve the TOV equations. We could obtain two types of solutions: solutions without void and those with void, without assuming that there is a void inside the star *a priori*. The solutions with void can only occur in strong field. In the weak field limit , there are only the solutions without void.

12.4.5 *Structure of a Newtonian star*

In the weak field and nonrelativistic limit, $P \ll \rho c^2$. We have $4\pi r^3 P \ll uc^2$ and $\frac{2Gu}{c^2 r} \ll 1$. Thus the TOV equation becomes

$$\frac{dP}{dr} = -\frac{G\rho u}{r^2}.$$
(12.157)

This equation is equivalent to the Newtonian gravitational equation. Since $P \to \infty$ as $r \to 0$, Eq.(12.157) does not have the solutions with void. Thus $u_0 = 0$, which gives

$$u = \int_0^r dr 4\pi r^2 \rho = m(r).$$
(12.158)

We consider a volume element as shown in Fig.12.1. The inward gravitational force by Newton theory is given by Eq.(12.69).

$$F = \Delta V \rho \frac{Gm(r)}{r^2}.$$
(12.159)

The outward force is given by

$$-P(r + \Delta r)\Delta A + P(r)\Delta A = -\frac{dP}{dr}\Delta V.$$
(12.160)

The balance of the force is Eq.(12.157). Thus the Newtonian gravitational equation is equivalent to Eq.(12.157).

12.4.6 *Simple model for interior structure of stars*

The TOV equation is hard to solve analytically for a given equation of state. We will show a simplified solution for the spherical stars.

To simplify the problem, we consider the approximation

$$\rho = \texttt{const}.$$
(12.161)

inside the star with a radius of R. This approximation is proposed by Schwarzschild. It should be noted that the speed of sound v_s which is

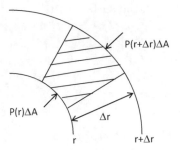

Fig. 12.1   The pressure force on a small volume element $\Delta V = \Delta A \Delta r$ of a spherical star.

proportional to $\left(\frac{dP}{d\rho}\right)^{\frac{1}{2}}$ [1] is infinite for this approximated equation of state. Thus it has the problem of causality. We consider the case of $u_0 = 0$. From

---

[1] According to Eq.(12.160) and Newton's law, we have

$$\rho\frac{d\mathbf{v}}{dt} = -\nabla_{\mathbf{r}}P. \tag{12.162}$$

In order to derive the relation of the time derivative of velocity with that of pressure, we use the equation of continuity

$$\frac{\partial\rho}{\partial t} + \nabla_{\mathbf{r}}\cdot\mathbf{j} = 0. \tag{12.163}$$

Neglecting the high order term of velocity, Eq.(12.163) can be rewritten as

$$\frac{\partial\rho}{\partial t} + \rho\nabla_{\mathbf{r}}\cdot\mathbf{v} = 0. \tag{12.164}$$

For a gas, we can approximately use the expansion relation

$$\rho = \left(\frac{\partial\rho}{\partial P}\right)_S P. \tag{12.165}$$

We have neglected the damping effect in Eq.(12.165). Combining Eq.(12.164) with Eq.(12.162), we have

$$\rho\frac{\partial}{\partial t}\left(\frac{1}{\rho}\frac{\partial\rho}{\partial t}\right) = -\rho\frac{\partial\nabla_{\mathbf{r}}\cdot\mathbf{v}}{\partial t} = \nabla_{\mathbf{r}}\cdot\nabla_{\mathbf{r}}P. \tag{12.166}$$

Inserting Eq.(12.165) into Eq.(12.166) and neglecting the high order terms, we have

$$\left(\frac{\partial\rho}{\partial P}\right)_S\frac{\partial^2 P}{\partial t^2} = \Delta P, \tag{12.167}$$

which gives

$$v_s \propto \sqrt{\left(\frac{\partial P}{\partial\rho}\right)_S}. \tag{12.168}$$

Eq.(12.140), we have

$$u(r) = \frac{4\pi}{3}\rho r^3 \quad r \leq R. \tag{12.169}$$

Outside the star, $\rho = 0$. $u(r)$ is constant and is denoted by M.

$$M = u(r)|_R = \frac{4\pi}{3}\rho R^3 \quad r \geq R. \tag{12.170}$$

This M is often called the Schwarzschild mass. The TOV equation now has the form

$$\frac{dP}{dr} = -\frac{4\pi G}{3c^4}\frac{r(\rho c^2 + P)(\rho c^2 + 3P)}{\left(1 - \dfrac{8\pi G}{3c^2}r^2\rho\right)}. \tag{12.171}$$

We denote the pressure P at $r = 0$ as P_0. Eq.(12.171) can be integrated from $P = P_0$ at $r = 0$, which gives

$$\frac{\rho c^2 + 3P}{\rho c^2 + P} = \frac{\rho c^2 + 3P_0}{\rho c^2 + P_0}\left(1 - \frac{2GM}{c^2 r}\right)^{\frac{1}{2}}. \tag{12.172}$$

At $r = R$, $P = 0$. We obtain the relation between P_0 and R

$$P_0 = \rho c^2 \left[1 - \left(1 - \frac{2GM}{c^2 R}\right)^{\frac{1}{2}}\right]\frac{1}{3\left(1 - \dfrac{2GM}{c^2 R}\right)^{\frac{1}{2}} - 1}. \tag{12.173}$$

Inserting Eq.(12.173) into Eq.(12.172), we find

$$P = \rho c^2 \frac{\left(1 - \dfrac{2GMr^2}{c^2 R^3}\right)^{\frac{1}{2}} - \left(1 - \dfrac{2GM}{c^2 R}\right)^{\frac{1}{2}}}{3\left(1 - \dfrac{2GM}{c^2 R}\right)^{\frac{1}{2}} - \left(1 - \dfrac{2GMr^2}{c^2 R^3}\right)^{\frac{1}{2}}}. \tag{12.174}$$

Eq.(12.174) is called the Schwarzschild constant-density interior solution. From Eq.(12.174), we can see that $P_0 = P|_{r=0} \to \infty$ as $\frac{GM}{c^2 R} \to \frac{4}{9}$. Thus the radii of an uniform-density star can not be smaller than $\frac{9}{4}\frac{GM}{c^2}$. For a star with $R = \frac{9}{4}\frac{GM}{c^2}$, the pressure at the center of the star is infinite.

12.4.7 *Pressure of relativistic Fermi gas*

12.4.7.1 *Thermal properties*

Now we show how to calculate the pressure that supports the compact stars such as white dwarfs and neutron stars. We start from the Hamiltonian for N non-interacting relativistic fermions given by Eq.(2.410)

$$\hat{H} = \sum_s \int d^3 p\, \hbar\omega_\mathbf{p}\hat{a}^\dagger_\mathbf{p}\hat{a}_\mathbf{p}, \tag{12.175}$$

where $\omega_{\mathbf{p}} = \frac{1}{\hbar}\sqrt{p^2c^2 + m^2c^4}$. The Hamiltonian operator is diagonal in the momentum space $|\mathbf{p}\rangle$. We can rewrite the Hamiltonian Eq.(12.175) as

$$\hat{H} = \sum_s \int d^3p\, \hbar\omega_{\mathbf{p}} |\mathbf{p}\rangle\langle\mathbf{p}|. \qquad (12.176)$$

$\hat{H}$ is the one-body operator. The N-particle basis is given by

$$|\mathbf{p}_1\mathbf{p}_2\cdots\mathbf{p}_N\rangle_{SN} = \frac{1}{\sqrt{N!}}\sum_P (-1)^{S_P} P|\mathbf{p}_1\rangle\cdots|\mathbf{p}_N\rangle. \qquad (12.177)$$

The sum runs over all the permutation P of $\{1, 2, \cdots, N\}$.

Since the total number of the occupied states equals the number of particles, we have

$$N = \sum_s\sum_{\mathbf{p}} n_{\mathbf{p}}. \qquad (12.178)$$

$|\mathbf{p}_1\mathbf{p}_2\cdots\mathbf{p}_N\rangle_{SN}$ is the eigenstate of $\hat{H}$. Thus the energy eigenvalue of N-particle state is given by

$$E(\{n_{\mathbf{p}}\}) = \sum_s\sum_{\mathbf{p}} n_{\mathbf{p}}\hbar\omega_{\mathbf{p}}. \qquad (12.179)$$

We can calculate the grand partition function Ξ.

$$\begin{aligned}
\Xi &= \sum_{N=0}^{\infty} \sum_{\substack{\{n_{\mathbf{p}}\} \\ \sum_s\sum_{\mathbf{p}} n_{\mathbf{p}}=N}} e^{-\beta(E(\{n_{\mathbf{p}}\})-\mu N)} \\
&= \sum_{\{n_{\mathbf{p}}\}} e^{-\beta\sum_s\sum_{\mathbf{p}}(\hbar\omega_{\mathbf{p}}-\mu)n_{\mathbf{p}}} \\
&= \prod_s\prod_{\mathbf{p}}\sum_{n_{\mathbf{p}}} e^{-\beta\sum_{\mathbf{p}}(\hbar\omega_{\mathbf{p}}-\mu)n_{\mathbf{p}}} \\
&= \prod_s\prod_{\mathbf{p}}\left[1 + e^{-\beta(\hbar\omega_{\mathbf{p}}-\mu)}\right],
\end{aligned} \qquad (12.180)$$

where μ is the chemical potential. The grand potential has the form

$$J = -\beta^{-1}\ln\Xi = -\beta^{-1}\sum_s\sum_{\mathbf{p}}\ln\left(1 + e^{-\beta(\hbar\omega_{\mathbf{p}}-\mu)}\right). \qquad (12.181)$$

Then we can evaluate the average particle number using Eq.(10.131), which has the form

$$N = -\frac{\partial\ln\Xi}{\partial\alpha} = -\left(\frac{\partial J}{\partial\mu}\right)_\beta = \sum_s\sum_{\mathbf{p}}\frac{1}{e^{\beta(\hbar\omega_{\mathbf{p}}-\mu)}+1}. \qquad (12.182)$$

The internal energy is given by

$$E = -\frac{\partial \ln \Xi}{\partial \beta}$$

$$= \sum_s \sum_{\mathbf{p}} \frac{\hbar\omega_{\mathbf{p}}}{e^{\beta(\hbar\omega_{\mathbf{p}}-\mu)}+1} \tag{12.183}$$

$$= \frac{g_s V}{(2\pi\hbar)^3} \int d^3p \frac{\hbar\omega_{\mathbf{p}}}{e^{\beta(\hbar\omega_{\mathbf{p}}-\mu)}+1},$$

where g_s is the spin degeneracy factor given by

$$g_s = 2s + 1. \tag{12.184}$$

$s = \frac{1}{2}$ for electrons or neutrons. V is the volume of the system.

12.4.7.2   Ground state ($T = 0$)

Now we deal with the ground state of noninteracting Fermi gas. In the ground state, the N lowest single-particle states $|\mathbf{p}\rangle$ are occupied. All the states within an energy surface (called Fermi surface) are thus occupied. The radius of the Fermi surface is called the Fermi momentum p_F. The particle number is related to the Fermi momentum p_F.

$$N = g_s \sum_{p \leq p_F} 1$$

$$= g_s \frac{V}{(2\pi\hbar)^3} \int d^3p \Theta(p_F - p)$$

$$= g_s \frac{V}{(2\pi\hbar)^3} \int_0^{p_F} 4\pi p^2 dp \tag{12.185}$$

$$= \frac{8\pi V p_F^3}{3h^3}.$$

We can solve p_F as a function of the particle density $n = \frac{N}{V}$ from Eq.(12.185).

$$p_F = \left(\frac{3h^3}{8\pi}\right)^{\frac{1}{3}} n^{\frac{1}{3}}. \tag{12.186}$$

Each fermion has an energy $\omega_{\mathbf{p}} = \frac{1}{\hbar}\sqrt{m^2c^4 + p^2c^2}$. Therefore the energy density of the relativistic Fermi gas is

$$\rho c^2 = \frac{E}{V}$$

$$= \int_0^{p_F} \frac{8\pi p^2}{h^3} (m^2c^4 + p^2c^2)^{\frac{1}{2}} dp \tag{12.187}$$

$$= \frac{c}{8\pi^2\hbar^3} \left\{ p_F(2p_F^2 + m^2c^2)\sqrt{p_F^2 + m^2c^2} - (mc)^4 \sinh^{-1}\left(\frac{p_F}{mc}\right) \right\}.$$

The pressure is given by

$$
\begin{aligned}
P &= -\frac{\partial E}{\partial V} \\
&= \frac{c}{8\pi^2 \hbar^2} \left\{ p_F \left(\frac{2}{3} p_F^2 - m^2 c^2 \right) \sqrt{p_F^2 + m^2 c^2} \right. \\
&\quad \left. + (mc)^4 \sinh^{-1} \left(\frac{p_F}{mc} \right) \right\}.
\end{aligned}
\tag{12.188}
$$

We introduce a parameter

$$
\xi = 4 \sinh^{-1} \left(\frac{p_F}{mc} \right).
\tag{12.189}
$$

Then the formulas can be rewritten in the following parameter form.

$$
n = \left(\frac{mc}{\hbar} \right)^3 \frac{1}{3\pi^2} \sinh^3 \frac{\xi}{4},
\tag{12.190a}
$$

$$
P = \frac{m^4 c^5}{32\pi^2 \hbar^3} \left(\frac{1}{3} \sinh \xi - \frac{8}{3} \sinh \frac{\xi}{2} + \xi \right),
\tag{12.190b}
$$

$$
\rho c^2 = \frac{m^4 c^5}{32\pi^2 \hbar^3} \left(\sinh \xi - \xi \right).
\tag{12.190c}
$$

We can also use another parameter defined by

$$
x_F \equiv \frac{p_F}{mc}.
\tag{12.191}
$$

Then

$$
\xi = 4 \sinh^{-1} (x_F).
\tag{12.192}
$$

12.5 White dwarfs

When a star with about the mass of the sun runs out of the reaction energy, the pressure in the star resulted from the thermal effect becomes small. Then the pressure resulted from the quantum effect of fermions due to the Pauli exclusion principle dominates. White dwarfs are stars that the outwards pressure is delivered by the cold electron gas. Since the mass of electrons is much smaller than the mass of nuclei (mass of four protons for helium), the pressure of electrons is larger than that of nuclei. This can be seen from the following derivations.

The pressure increases with the mass of star. We consider the limit case that the pressure of electron gas can resist the gravitational potential. We denote the total mass of the star by M and the radius of the star by R. We have

$$M = (m_e + 2m_p)N \approx 2m_pN, \tag{12.193a}$$

$$R = \left(\frac{3V}{4\pi}\right)^{\frac{1}{3}}, \tag{12.193b}$$

where m_e is the mass of an electron and m_p the mass of a proton. The density is given by

$$\rho \equiv \frac{1}{v} = \frac{3}{8\pi}\frac{M}{m_pR^3} \tag{12.194}$$

with $v = \frac{V}{N}$.

$$x_F = \frac{p_F}{m_ec} = \frac{\hbar}{m_ec}\frac{1}{R}\left(\frac{9\pi M}{8m_p}\right)^{\frac{1}{3}}. \tag{12.195}$$

We introduce two parameters $\bar{M}$ and $\bar{R}$

$$\bar{M} = \frac{9\pi M}{8m_p}, \tag{12.196a}$$

$$\bar{R} = \frac{m_ec}{\hbar}R. \tag{12.196b}$$

In terms of $\bar{M}$ and $\bar{R}$, we have

$$x_F = \frac{\bar{M}^{\frac{1}{3}}}{\bar{R}}. \tag{12.197}$$

In the nonrelativistic limit where $x_F \ll 1$, we have

$$P \cong \frac{m_e^4c^5}{15\pi^2\hbar^3}x_F^5 = K\frac{\bar{M}^{\frac{5}{3}}}{\bar{R}^5}, \tag{12.198}$$

where

$$K = \frac{m_e^4c^5}{15\pi^2\hbar^3}. \tag{12.199}$$

In the extreme relativistic limit where $x_F \gg 1$, we have

$$P \cong \frac{m_e^4c^5}{12\pi^2\hbar^3}(x_F^4 - x_F^2) = \frac{5}{4}K\left(\frac{\bar{M}^{\frac{4}{3}}}{\bar{R}^4} - \frac{\bar{M}^{\frac{2}{3}}}{\bar{R}^2}\right). \tag{12.200}$$

According to Eq.(12.157), The Newtonian equations for the star are given by

$$\frac{dm(r)}{dr} = 4\pi r^2\rho(r), \tag{12.201a}$$

$$\frac{dP}{dr} = -G\rho(r)\frac{m(r)}{r^2}. \tag{12.201b}$$

In the following, we make the evaluations in order of magnitude. We introduce the typical density $\bar{\rho}$ and the typical pressure $\bar{P}$. $\bar{\rho}$ and $\bar{P}$ can be considered approximately as the average density and average pressure, respectively. Eqs.(12.201) are equivalent to

$$M = R^3 \bar{\rho}, \qquad (12.202\text{a})$$

$$\frac{\bar{P}}{R} = \bar{\rho}\frac{GM}{R^2}. \qquad (12.202\text{b})$$

Eliminating $\bar{\rho}$, we have

$$\bar{P} = \frac{GM^2}{R^4}. \qquad (12.203)$$

We will discuss two approximate cases: $x_F \ll 1$ and $x_F \gg 1$.

(i) When the mass of the star is not very large, the nonrelativistic limit ($x_F \ll 1$) can be used. Then we can use the relation Eq.(12.198) to eliminate $\bar{P}$ in Eq.(12.203)

$$K\frac{\bar{M}^{\frac{5}{3}}}{\bar{R}^5} = K'\frac{\bar{M}^2}{\bar{R}^4}, \qquad (12.204)$$

where

$$K' = G\left(\frac{8m_p}{9\pi}\right)^2 \left(\frac{m_e c}{\hbar}\right)^4. \qquad (12.205)$$

Eq.(12.204) can be rewritten as

$$\bar{M}^{\frac{1}{3}}\bar{R} = \frac{K}{K'}. \qquad (12.206)$$

Thus the radius of the star decreases with the increase of the mass of the star. Eq.(12.196b) shows that the radius of the star would be smaller if we replace electron mass with proton mass. The effect of pressure of electrons is stronger than that of nuclei and thus it is reasonable that we consider only the pressure of electrons.

(ii) When the mass of the star is large enough, the extreme relativistic limit ($x_F \gg 1$) should be used. The equilibrium condition is given by

$$\frac{5}{4}K\left(\frac{\bar{M}^{\frac{4}{3}}}{\bar{R}^4} - \frac{\bar{M}^{\frac{2}{3}}}{\bar{R}^2}\right) = K'\frac{\bar{M}^2}{\bar{R}^4} \qquad (12.207)$$

or

$$\bar{R} = \bar{M}^{\frac{1}{3}}\sqrt{1 - \left(\frac{\bar{M}}{\bar{M}_0}\right)^{\frac{2}{3}}}, \qquad (12.208)$$

where

$$\bar{M}_0 = \left(\frac{5K}{4K'}\right)^{\frac{3}{2}} = \left(\frac{27\pi}{256}\right)^{\frac{3}{2}} \left(\frac{\hbar c}{Gm_p^2}\right)^{\frac{3}{2}}. \tag{12.209}$$

Eq.(12.208) shows that no white dwarf can have a mass larger than M_0 which is given by

$$M_0 = \frac{8}{9\pi} m_p \bar{M}_0 \approx 10^{33} g \approx M_\odot, \tag{12.210}$$

where $\odot$ is the mass of the sun. According to Eq.(12.208), we can see that $R \to 0$ as $M \to M_0$. Thus Newtonian gravitational theory can have gravitational collapse. In contrast, if one uses the TOV equation, as mass increases, R would not approach zero. Instead, R approaches a finite value. Therefore, the general relativity does not have the similar gravitational collapse as predicted by the Newtonian gravitational theory. The underlying physics is that the general relativity allows only positive energy while the energy in the Newtonian gravitational theory can be negative and without a lowest limit.

More refined calculations give the result $M_0 = 1.4 M_\odot$. This value of mass is called the *Chandrasekhar limit*. When the mass of the star is larger than the Chandrasekhar limit, it will collapse until other repulsive mechanism is effective or the Newtonian gravitational theory is no more applicable.

12.6 Neutron stars

When a white dwarf is further compressed, the electrons could combine with the protons to release the energy. The final equilibrium stars are the neutron stars. For a neutron star, the gravitation effect is so large that the Newtonian gravitational theory is no more applicable. We will use the TOV equation to calculate the interior structure of neutron stars.

When we apply the TOV equation to the neutron stars, there are two types of solutions. The solutions without void and the ones with void. First we consider the solutions without void, which we call the normal solutions.

12.6.1   *Normal solutions*

Eqs.(12.138) and (12.147) are the two first-order differential equations for solving $u(r)$ and $P(r)$. They are

$$\frac{du(r)}{dr} = 4\pi r^2 \rho, \qquad (12.211a)$$

$$\frac{dP}{dr} = -\frac{(c^2\rho + P)(Gc^2u + 4\pi Gr^3 P)}{c^2 r[c^2 r - 2Gu(r)]}. \qquad (12.211b)$$

We denote the radius of the neutron star as R. One can integrate the two equations simultaneously from some initial values $u = u_0$ and $P = P_0$ at $r = 0$ to the values at $r = R$ where $P = 0$. The value of u at the boundary $r = R$ is connected with the value of the Schwarzschild metric outside the star. We have

$$u(r)|_{r=R} = \frac{c^2 R}{2G}\left(1 - e^{-2\Lambda}\right) = \frac{c^2 R}{2G}\left[1 - \left(1 - \frac{2GM}{c^2 R}\right)\right] = M. \qquad (12.212)$$

Thus $u(R)$ is the gravitational mass of the neutron star as measured by a distant observer.

For a neutron star consisting of the fermions with the rest mass of m_n, it is more convenient to use the parametric form of ρ and P with the parameter ξ related to the Fermi momentum p_F by Eq.(12.189).

$$\xi = 4\ln\left\{\frac{p_F}{m_n c} + \left[1 + \left(\frac{p_F}{m_n c}\right)^2\right]^{\frac{1}{2}}\right\}. \qquad (12.213)$$

Then the mass density and pressure are given by

$$\rho = K(\sinh\xi - \xi), \qquad (12.214)$$

$$P = \frac{c^2}{3}K(\sinh\xi - 8\sinh\frac{\xi}{2} + 3\xi), \qquad (12.215)$$

where $K = \pi m_n^4 c^3/(4h^3)$. The Fermi momentum p_F is related to the density of the particle number $n = N/V$ by $n = 8\pi p_F^3/(3h^3)$.

There are some restrictions on the choice of P_0 and u_0. First only positive pressure is meaningful, which gives $P \geq 0$. Since $g_{rr} = e^{2\Lambda}$ is positive, we have $u_0 \leq 0$. Eq.(12.139) shows that $u_0 = 0$ if $e^{-2\Lambda}$ takes finite value. If $u_0 \neq 0$, we express the equation of state by $\rho = \alpha P^s$ at $r \approx 0$. If $P_0 = 0$, $s = \frac{3}{5}$ (Expanding Eq.(12.190) at $\xi = 0$ gives $P \sim \rho^{\frac{5}{3}}$). According to Eq.(12.149), $p^{1-s} \approx 1/2c^2\alpha(1-s)\ln r + c$, which will result in a negative p when $r \to 0$. Thus u_0 can only be zero.

We can use ξ as the parameter in solving the differential equations. Then Eq.(12.211) becomes

$$\frac{du}{dr} = 4\pi K r^2 (\sinh \xi - \xi), \tag{12.216}$$

$$\frac{d\xi}{dr} = -\frac{4\left(\sinh \xi - 2\sinh \frac{\xi}{2}\right)}{c^2 r(c^2 r - 2Gu)\left(\cosh \xi - 4\cosh \frac{\xi}{2} + 3\right)} \tag{12.217}$$

$$\times \left[\frac{4\pi G K c^2 r^3}{3}\left(\sinh \xi - 8\sinh \frac{\xi}{2} + 3\xi\right) + Gc^2 u\right].$$

For a neutron star, the equations can only be solved numerically. Numerical results show that there is a maximum limit of mass which is about $0.7\odot$.

12.6.2 *Solutions with void*

Now we consider the solution with void. We have shown that the initial value $u_0 \le 0$. When $u_0 < 0$, near $r = 0$, the TOV equation becomes Eq.(12.149). The solution Eq.(12.150) of Eq.(12.149) can be rewritten as

$$P^{\frac{2}{5}} = \frac{1}{5}c^2 \alpha \ln\left(\frac{r}{r_i}\right). \tag{12.218}$$

We can see that $P = 0$ ar $r = r_i$. If we consider $r = r_i$ as an inner boundary, P will remain zero when $r \le r_i$ and we would avoid a negative P. Thus, we have another type of solutions with $P = 0$ at $r = r_i$. Since $\rho = 0$ for $r \le r_i$, there is a void around $r = 0$ in this type of solutions. In the void region from $r = 0$ to $r = r_i$, $\rho = 0$ and $P = 0$. There are no particles and thus pressure is zero in this void region. In the void region, we have the Minkowski-type metric described by Eq.(12.152).

From $r = r_i$ to $r = r_o = R$, $\rho \ge 0$ and $P \ge 0$. P and ρ increase first from zero at $r = r_i$. After reaching a maximum, P and ρ then decrease. At $r = r_o$, P and ρ decrease to zero, where we have the outer boundary. At the outer radius $r_o = R$, $u_o = c^2 r_o[1 - e^{-2\Lambda(r_o)}]/(2G) \equiv M$. M is the apparent mass of the star as measured by a distant observer.

Similar to the case of the normal solutions, we can use the parameter form of the TOV equation Eqs.(12.216) and (12.217) to obtain the numerical solutions. We can also calculate the particle number in the star by

$$N = \int_{r_i}^{r_o} 4\pi r^2 g_{rr}^{\frac{1}{2}} n dr$$

$$= \frac{4(m_n c)^3}{3\pi \hbar^3}\int_{r_i}^{r_o} r^2 \left(1 - \frac{2Gu}{c^2 r}\right)^{-\frac{1}{2}} \sinh^3\left(\frac{\xi}{4}\right) dr. \tag{12.219}$$

Now we discuss the case of the solutions with initial value $u_i < 0$ at $r_i \neq 0$. Numerical calculations show that u increases from the negative value to a positive value at the outer radius r_o. $u = u_o$ at $r = r_o$ corresponds to the mass of the star as measured by a distant observer. The structure parameter ξ increases from zero at the inner radius r_i to a maximum and then decreases to zero at the outer radius r_o. ρ and P, as functions of ξ, show the similar change tendency. P increases from 0 according to Eq.(12.211b). After reaching a maximum, P decreases to zero at the outer radius r_o. Figure 12.2 show the particle number N, mass M and outer radius r_o as functions of u_i at $r_i = a$. From Fig. 12.2, we can see that the particle number N increases with the increase of $|u_i|$, exhibiting a power law dependence. The mass M also increases with the increase of $|u_i|$ according to the power law with a crossover. The crossover value corresponds to the minimum in the curve of r_o. The outer radius r_o decreases first with the increase of $|u_i|$ to a minimum at $|u_{im}|$ and then increases with the increase of $|u_i|$. When $|u_i| < |u_{im}|$, although r_o decreases with the increase of $|u_i|$, the peak values of ρ and P are increased. The increase of N and M is mainly due to the increase of the peak value of ρ. When $|u_i| > |u_{im}|$, the increase of N and M is mainly due to the increase of r_o. When $|u_i|$ increases, $g_o = (1 - 2Gu_o/c^2r_o)$ decreases and approaches to zero.

One can also solve the differential equations of Eqs.(12.216) and (12.217) by integrating from the outer radius r_o. Then the solutions with the void inside the center emerge naturally. When we keep the outer radius r_o in constant and make the parameter $g_o = 1 - 2Gu_o/(c^2r_o)$ decrease and approach to zero, the particle number approaches to infinite and the void radius r_i approaches to zero.

The solutions without maximum mass limit do not depend on a special property of the equation of state for the star matter. Similar solutions can also be obtained for other equations of state $P = P(\rho)$. It shall be noted that the Newtonian gravitational theory does not give this type of solutions. From Eq.(12.157), $dP/dr = -\rho Gm(r)/r^2 < 0$. Thus P always decreases monotonously. The pressure in the solution with a void is zero at both inner radius r_i and outer radius r_o of the two boundaries, which is not compatible with the above Newtonian gravitational equation for pressure P. The solutions with void show that the Einstein general relativistic theory is significantly different with the Newtonian gravitational theory on the equilibrium mass distribution.

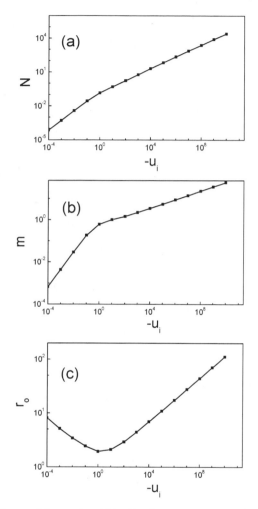

Fig. 12.2 (a) The particle number N inside the neutron star, (b) mass, and (c) outer radius r_o as functions of u_i at the inner radius $r_i = 1$. The unit of the length is taken to be $a = h^{3/2}/(\pi m_n^2 c^{1/2} G^{1/2}) = 1.36 \times 10^6$cm and the unit of the mass is $m_0 = ac^2/G = 1.83 \times 10^{34}$g. The unit of the particle number is $N_0 = 32\pi^2 (m_n ca)^3/(3h^3) = 1.174 \times 10^{59}$.

12.7   Relativistic particles in electromagnetic field

We have shown that a quasi-particle is a state which can be described as a wave. A free particle state is a plane wave

$$\varphi = a e^{i(\mathbf{k} \cdot r - \omega t)}. \tag{12.220}$$

Generally, for a classical particle, we can use the following expression of wave packet

$$\varphi = \sum_n a_n e^{i\Psi_n}.\tag{12.221}$$

The amplitude a_n is generally a function of time t and spatial coordinates x_i. The phase Ψ_n, which is also called eikonal, is a large quantity since $\lambda \to 0$ for a classical particle and Ψ_n changes by 2π when one moves through one wave length.

In a small space and time interval when the energy and momentum do not change much, Ψ_n can be expanded in series. We take the first order terms of Ψ_n and get

$$\Psi_n = \Psi_{n0} + \mathbf{r} \cdot \frac{\partial \Psi_n}{\partial \mathbf{r}} + t \frac{\partial \Psi_n}{\partial t}.\tag{12.222}$$

The wave packet in Eq.(12.221) becomes a plane wave packet described by Eq.(8.175). Comparing with Eq.(12.220), we can write

$$\mathbf{k}_n = \frac{\partial \Psi_n}{\partial \mathbf{r}}, \quad \omega_n = -\frac{\partial \Psi_n}{\partial t}.\tag{12.223}$$

Since $\Delta k \ll k$ and $\Delta \omega \ll \omega$ for a wave packet, we can omit the subscript n and write

$$\mathbf{k} = \frac{\partial \Psi}{\partial \mathbf{r}}, \quad \omega = -\frac{\partial \Psi}{\partial t}.\tag{12.224}$$

Eq.(12.224) can be expressed in a four-dimensional form as

$$k_\mu = -\frac{\partial \Psi}{\partial x^\mu}.\tag{12.225}$$

For a relativistic particle in a electromagnetic field, we have

$$\left(\frac{E - e\phi}{c}\right)^2 = m^2 c^2 + \left(\mathbf{P} - \frac{e}{c}\mathbf{A}\right)^2,\tag{12.226}$$

where E is the energy of the particle. Eq.(12.226) can be derived in a similar way as Eq.(2.374). Eq.(2.374) is the dispersion relation for a free particle. When we put a particle in an electromagnetic field, we need to replace ∂_μ by the covariant derivative $D_\mu = \partial_\mu + \frac{ie}{c} A_\mu$ defined by Eq.(2.582), which leads to the replacement of $\hbar\omega$ by $\hbar\omega - e\phi$ and $\mathbf{p}$ by $\mathbf{P} - \frac{e}{c}\mathbf{A}$. We have used $\mathbf{P}$ in $\mathbf{P} - \frac{e}{c}\mathbf{A}$ in accordance with the custom. $\mathbf{p}$ is called the ordinary momentum of a classical particle or simply momentum. $\mathbf{P}$ is usually called the generalized momentum of a classical particle. We have used E to represent the energy $\hbar\omega$ of the particle.

It should be noted that Eq.(12.226) can be applied to a composite particle. For a composite particle, the constituent particles have the same velocity and are in static relative to each other. The total internal force vanishes for each constituent particle. When we consider only the motion of composite particle, the internal interaction terms are constant approximately and can be neglected. The rest mass of the composite particle is the addition of all the rest mass of the constituent particles.

Solving Eq.(12.226) for E, we find

$$E = \sqrt{m^2c^4 + \left(\mathbf{P} - \frac{e}{c}\mathbf{A}\right)^2 c^2} + e\phi. \tag{12.227}$$

The wave frequency ω is given by

$$\omega = \frac{E}{\hbar} = \frac{1}{\hbar}\sqrt{m^2c^4 + \left(\mathbf{P} - \frac{e}{c}\mathbf{A}\right)^2 c^2} + \frac{1}{\hbar}e\phi. \tag{12.228}$$

The momentum of the particle is given by $\mathbf{P} = \hbar\mathbf{k}$.

A particle moves with a velocity that is equal to the group velocity of the wave packet given by Eq.(8.181)

$$\frac{d\mathbf{x}}{dt} = \mathbf{v} = \frac{\partial\omega}{\partial\mathbf{k}}. \tag{12.229}$$

Using Eq.(12.227), we find

$$\mathbf{v} = \frac{\partial\omega}{\partial\mathbf{k}} = \frac{\partial E}{\partial\mathbf{P}} = \frac{(\mathbf{P} - \frac{e}{c}\mathbf{A})c^2}{\sqrt{m^2c^4 + \left(\mathbf{P} - \frac{e}{c}\mathbf{A}\right)^2 c^2}}. \tag{12.230}$$

Solving $\mathbf{P}$ from Eq.(12.230), we have

$$\mathbf{P} = \frac{m\mathbf{v}}{\sqrt{1 - \frac{v^2}{c^2}}} + \frac{e}{c}\mathbf{A}. \tag{12.231}$$

For a classical particle, the Hamiltonian is an ordinary function. We will not distinguish the Hamiltonian H with the energy E. Using Eq.(12.230), we obtain

$$\frac{\partial H}{\partial\mathbf{P}} = \frac{(\mathbf{P} - \frac{e}{c}\mathbf{A})c^2}{\sqrt{m^2c^4 + \left(\mathbf{P} - \frac{e}{c}\mathbf{A}\right)^2 c^2}} = \mathbf{v}. \tag{12.232}$$

Using Eq.(12.224), we have

$$\dot{\mathbf{P}} = \hbar\dot{\mathbf{k}} = \frac{\partial\hbar\dot{\Psi}}{\partial\mathbf{r}} = -\frac{\partial H}{\partial\mathbf{r}}. \tag{12.233}$$

Eq.(12.232) and Eq.(12.233) form the Hamilton's equations

$$\dot{\mathbf{P}} = -\frac{\partial H}{\partial\mathbf{r}}, \quad \mathbf{v} = \dot{\mathbf{r}} = \frac{\partial H}{\partial\mathbf{P}}. \tag{12.234}$$

We introduce the Lagrangian L of a classical particle through a Legendre's transformation.

$$L(\mathbf{r}, \dot{\mathbf{r}}) \equiv \mathbf{P} \cdot \dot{\mathbf{r}} - H. \tag{12.235}$$

The total differential of the Lagrangian as a function of coordinate $\mathbf{r}$ and velocity $\mathbf{v}$ is

$$dL = \sum_i \frac{\partial L}{\partial r_i} dr_i + \sum_i \frac{\partial L}{\partial \dot{r}_i} d\dot{r}_i. \tag{12.236}$$

Using Eq.(12.235), we have

$$\begin{aligned}
dL &= \sum_i P_i d\dot{r}_i + \sum_i \dot{r}_i dP_i - dH \\
&= \sum_i P_i d\dot{r}_i + \sum_i \dot{r}_i dP_i - \sum_i \frac{\partial H}{\partial r_i} dr_i - \sum_i \frac{\partial H}{\partial P_i} dP_i \\
&= \sum_i P_i d\dot{r}_i + \sum_i \dot{r}_i dP_i + \sum_i \dot{P}_i dr_i - \sum_i \dot{r}_i dP_i \\
&= \sum_i P_i d\dot{r}_i + \sum_i \dot{P}_i dr_i.
\end{aligned} \tag{12.237}$$

Comparing Eq.(12.237) with Eq.(12.236), we find

$$\dot{P}_i = \frac{\partial L}{\partial r_i}, \quad P_i = \frac{\partial L}{\partial \dot{r}_i}, \tag{12.238}$$

which leads to the Euler-Lagrange equation.

$$\frac{d}{dt} \frac{\partial L}{\partial \dot{r}_i} - \frac{\partial L}{\partial r_i} = 0. \tag{12.239}$$

Inserting Eq.(12.227) and Eq.(12.230) into Eq.(12.235), we obtain

$$\begin{aligned}
L(\mathbf{r}, \dot{\mathbf{r}}) &= \mathbf{P} \cdot \mathbf{v} - H \\
&= \frac{\mathbf{P}(\mathbf{P} - \frac{e}{c}\mathbf{A})c^2}{\sqrt{m^2 c^4 + \left(\mathbf{P} - \frac{e}{c}\mathbf{A}\right)^2 c^2}} - \sqrt{m^2 c^4 + \left(\mathbf{P} - \frac{e}{c}\mathbf{A}\right)^2 c^2} - e\phi \\
&= -\frac{m^2 c^4}{\sqrt{m^2 c^4 + \left(\mathbf{p} - \frac{e}{c}\mathbf{A}\right)^2 c^2}} + \frac{e}{c}\mathbf{A} \cdot \mathbf{v} - e\phi \\
&= -mc^2 \sqrt{1 - \frac{v^2}{c^2}} + \frac{e}{c}\mathbf{A} \cdot \mathbf{v} - e\phi.
\end{aligned} \tag{12.240}$$

Eq.(12.240) is the Lagrangian of a relativistic particle in an electromagnetic field. When ϕ and A are zero, we obtain the L for a free particle. Integrating the Lagrangian over time t, we get the action S in the signature $[1,-1,-1,-1]$.

$$S \equiv \int_{t_a}^{t_b} L dt$$
$$= \int_{t_a}^{t_b} \left(-mc^2 \sqrt{1 - \frac{v^2}{c^2}} + \frac{e}{c} \mathbf{A} \cdot \mathbf{v} - e\phi \right) dt, \tag{12.241}$$

where a and b are the two boundary points. The action S in Eq.(12.241) can be written in a covariant form.

$$S = \int_{t_a}^{t_b} \left(-mcds + \frac{e}{c} \mathbf{A} \cdot d\mathbf{r} - e\phi dt \right)$$
$$= \int_{t_a}^{t_b} \left(-mcds - \frac{e}{c} A_\mu dx^\mu \right), \tag{12.242}$$

where

$$ds = \sqrt{dx_\mu dx^\mu}. \tag{12.243}$$

In the Minkowski spacetime,

$$ds^2 = c^2 dt^2 - dx_1^2 - dx_2^2 - dx_3^2. \tag{12.244}$$

The Euler-Lagrange equation Eq.(12.239) is equivalent to the variation equation

$$\delta S = \delta \int_{t_a}^{t_b} L dt = 0. \tag{12.245}$$

In the variation, the boundary points are fixed. Eq.(12.245) is the principle of least action.

In low velocity, omitting a constant, the Hamiltonian Eq.(12.227) becomes

$$H = \frac{1}{2m} \left(\mathbf{P} - \frac{e}{c} \mathbf{A} \right)^2 + e\phi, \tag{12.246}$$

which is the Hamiltonian of non-relativistic particles. The Lagrangian Eq.(12.240) turns into

$$L = \frac{mv^2}{2} + \frac{e}{c} \mathbf{A} \cdot \mathbf{v} - e\phi. \tag{12.247}$$

12.8   Equations of motion of a relativistic classical particle in an electromagnetic field

The equations of motion of a classical charge particle in an electromagnetic field is given by the Euler-Lagrange equation Eq.(12.239)

$$\frac{d}{dt}\frac{\partial L}{\partial \mathbf{v}} - \frac{\partial L}{\partial \mathbf{r}} = 0. \tag{12.248}$$

Using Eq.(12.238), we have

$$\frac{\partial L}{\partial \mathbf{v}} = \mathbf{P} = \frac{m\mathbf{v}}{\sqrt{1 - \dfrac{v^2}{c^2}}} + \frac{e}{c}\mathbf{A} = \mathbf{p} + \frac{e}{c}\mathbf{A}, \tag{12.249}$$

where

$$\mathbf{p} = \frac{m}{\sqrt{1 - \dfrac{v^2}{c^2}}}\mathbf{v} = m'\mathbf{v} \tag{12.250}$$

with

$$m' = \frac{m}{\sqrt{1 - \dfrac{v^2}{c^2}}}. \tag{12.251}$$

m' is called the relativistic mass of a particle. Then the left hand side of Eq.(12.248) becomes

$$\frac{d}{dt}\frac{\partial L}{\partial \mathbf{v}} = \frac{d}{dt}\left(\mathbf{p} + \frac{e}{c}\mathbf{A}\right). \tag{12.252}$$

The derivative $\frac{\partial L}{\partial \mathbf{r}}$ on the right hand side of Eq.(12.248) is given by

$$\frac{\partial L}{\partial \mathbf{r}} = \frac{e}{c}\nabla(\mathbf{A} \cdot \mathbf{v}) - e\nabla\phi. \tag{12.253}$$

Using the mathematical identity

$$\nabla(\mathbf{a} \cdot \mathbf{b}) = (\mathbf{a} \cdot \nabla)\mathbf{b} + (\mathbf{b} \cdot \nabla)\mathbf{a} + \mathbf{b} \times \nabla \times \mathbf{a} + \mathbf{a} \times \nabla \times \mathbf{b} \tag{12.254}$$

and taking into account that the partial differentiating of $\mathbf{v}$ with respect to $\mathbf{r}$ is zero, we find

$$\frac{\partial L}{\partial \mathbf{r}} = \frac{e}{c}(\mathbf{v} \cdot \nabla)\mathbf{A} + \frac{e}{c}\mathbf{v} \times \nabla \times \mathbf{A} - e\nabla\phi. \tag{12.255}$$

Thus the Euler-Lagrange equation becomes

$$\frac{d}{dt}\left(\mathbf{p} + \frac{e}{c}\mathbf{A}\right) = \frac{e}{c}(\mathbf{v} \cdot \nabla)\mathbf{A} + \frac{e}{c}\mathbf{v} \times \nabla \times \mathbf{A} - e\nabla\phi. \tag{12.256}$$

Inserting the following relation

$$\frac{d\mathbf{A}}{dt} = \frac{\partial \mathbf{A}}{\partial t} + (\mathbf{v} \cdot \nabla)\mathbf{A} \tag{12.257}$$

into Eq.(12.256), we find

$$\frac{d\mathbf{p}}{dt} = -\frac{e}{c}\frac{\partial \mathbf{A}}{\partial t} - e\nabla\phi + \frac{e}{c}\mathbf{v} \times \nabla \times \mathbf{A}. \tag{12.258}$$

Eq.(12.258) is the equations of motion for a charge particle in an electromagnetic field. In terms of electric field

$$\mathbf{E} = -\frac{1}{c}\frac{\partial \mathbf{A}}{\partial t} - \nabla\phi \tag{12.259}$$

and magnetic field

$$\mathbf{B} = \nabla \times \mathbf{A}, \tag{12.260}$$

Eq.(12.258) becomes

$$\frac{d\mathbf{p}}{dt} = e\mathbf{E} + \frac{e}{c}\mathbf{v} \times \mathbf{B}. \tag{12.261}$$

Eq.(12.261) is the relativistic form of Eq.(8.82). The right hand side of Eq.(12.261) is the Lorentz force.

12.9 Covariant form of equations of motion of a relativistic classical particle

We use the covariant form of the action in Eq.(12.242) to derive the covariant form of the equations of motion. The principle of least action given by Eq.(12.245) leads to

$$\delta S = \delta \int_{t_a}^{t_b} \left(-mcds - \frac{e}{c}A_\mu dx^\mu\right) = 0. \tag{12.262}$$

Using Eq.(12.243), we have

$$\delta S = -\int_{t_a}^{t_b} \left(mc\frac{dx_\mu d\delta x^\mu}{ds} + \frac{e}{c}A_\mu d\delta x^\mu + \frac{e}{c}\delta A_\mu dx^\mu\right) = 0. \tag{12.263}$$

Integrating by parts and using the four-velocity $u_\mu = \frac{dx_\mu}{d\tau}$, we find

$$\int_{t_a}^{t_b} \left(mdu_\mu\delta x^\mu + \frac{e}{c}\delta x^\mu dA_\mu - \frac{e}{c}\delta A_\mu dx^\mu\right)$$
$$- \left[\left(mu_\mu + \frac{e}{c}A_\mu\right)\delta x^\mu\right]\Big|_{x^\mu=a,b} = 0. \tag{12.264}$$

Since x^μ at the boundary points is fixed,

$$\delta x^\mu \Big|_{x^\mu = a, b} = 0. \tag{12.265}$$

The second term of Eq.(12.264) is zero. Using

$$\delta A_\mu = \frac{\partial A_\mu}{\partial x^\nu} \delta x^\nu, \quad dA_\mu = \frac{\partial A_\mu}{\partial x^\nu} dx^\nu, \tag{12.266}$$

We have

$$\int_{t_a}^{t_b} \left(m du_\mu \delta x^\mu - \frac{e}{c} \frac{\partial A_\mu}{\partial x^\nu} \delta x^\nu dx^\mu + \frac{e}{c} \frac{\partial A_\mu}{\partial x^\nu} dx^\nu \delta x^\mu \right) = 0. \tag{12.267}$$

Interchanging the indices μ and ν in the second term, we obtain

$$\int_{t_a}^{t_b} \left[mc \frac{du_\mu}{ds} - \frac{e}{c} \left(\frac{\partial A_\nu}{\partial x^\mu} - \frac{\partial A_\mu}{\partial x^\nu} \right) u^\nu \right] \delta x^\mu ds = 0. \tag{12.268}$$

Since δx^μ is an arbitrary variable, the integrand should be zero in order to make the integration zero. We find

$$\begin{aligned} mc \frac{du_\mu}{ds} &= \frac{e}{c} \left(\frac{\partial A_\nu}{\partial x^\mu} - \frac{\partial A_\mu}{\partial x^\nu} \right) u^\nu \\ &= \frac{e}{c} F_{\mu\nu} u^\nu. \end{aligned} \tag{12.269}$$

Eq.(12.269) is the equation of motion in the covariant form for a relativistic particle in an electromagnetic field. $F_{\mu\nu}$ satisfies the Maxwell equations.

$$\frac{\partial F^{\mu\nu}}{\partial x^\nu} = -\frac{4\pi}{c} j^\mu. \tag{12.270}$$

In the Riemann spacetime, the derivatives should be replaced by the covariant derivatives. Then Eq.(12.269) becomes

$$\begin{aligned} mc \frac{Du^\mu}{ds} &= mc \left(\frac{du^\mu}{ds} + \Gamma^\mu_{\alpha\beta} u^\alpha u^\beta \right) \\ &= \frac{e}{c} F^{\mu\nu} u_\nu. \end{aligned} \tag{12.271}$$

Using Eq.(A.96), the Maxwell equations has the form

$$F^{\mu\nu}{}_{;\nu} = \frac{1}{\sqrt{-g}} \frac{\partial}{\partial x^\nu} (\sqrt{-g} F^{\mu\nu}) = -\frac{4\pi}{c} j^\mu \tag{12.272}$$

with

$$j^\mu = \rho_e u^\mu, \tag{12.273}$$

where ρ_e is the charge density. When j^μ is generated by the point charges, we have

$$j^\mu = \sum_i \frac{e_i c}{\sqrt{-g}} \delta^3(\mathbf{r} - \mathbf{r}_i) \frac{dx^\mu}{dx^0}, \tag{12.274}$$

where e_i is the charge of ith particle at position $\mathbf{r}_i$.

Appendix A

Tensors

A.1 Vectors

A position in the space can be described by a three dimensional vector **x**. In the four-dimensional spacetime, it is a four-dimensional vector x. A vector in a certain coordinate system can be expressed by the components x^i ($i = 1, 2, 3$) in a three-dimensional space or x^α ($\alpha = 0, 1, 2, 3$) in a four-dimensional spacetime. $\alpha = 0$ is usually used to denote the time component. It is custom to use the Greek alphabet ($\alpha, \beta, \gamma, \cdots$) to denote the components in spacetime and Latin alphabet ($i, j, k, \cdots$) to denote the components in space only.

We have used superscripts for the components x^i of an ordinary vector, which is often called a *contravariant vector*. A vector **x** can be expressed in any coordinate system. We use x'^i to denote the components of the position vector in another coordinate system. Then the relation between x'^i and x^i can be written as

$$x'^i = \frac{\partial x'^i}{\partial x^j} x^j, \qquad (A.1)$$

where we have used the Einstein summation convention that a summation is carried over doubly repeated indices . Eq.(A.1) is the definition of a vector. A vector is an object whose components transform according to Eq.(A.1). x'^i is often written as $x^{i'}$ with the prime attached to the superscript rather than the main symbol, which can be used more conveniently. Thus Eq.(A.1) is often written as

$$x^{i'} = \frac{\partial x^{i'}}{\partial x^j} x^j. \qquad (A.2)$$

An infinitesimal dx^i can be expressed by a differential formula

$$dx^{i'} = \frac{\partial x^{i'}}{\partial x^j} dx^j. \qquad (A.3)$$

Thus dx^i forms an ordinary or contravariant vector.

Let us now consider $\frac{\partial}{\partial x^{i'}}$. For a function $f(x)$, we have from calculus

$$\frac{\partial f}{\partial x^{i'}} = \frac{\partial f}{\partial x^j} \frac{\partial x^j}{\partial x^{i'}}. \tag{A.4}$$

Thus the derivative transforms as

$$\frac{\partial}{\partial x^{i'}} = \frac{\partial x^j}{\partial x^{i'}} \frac{\partial}{\partial x^j}, \tag{A.5}$$

which shows that $\frac{\partial}{\partial x^{i'}}$ is also a vector. It is called a *covariant vector*. Thus we can define two types of vectors. A contravariant vector A^μ is defined as an object whose components transform as

$$A^{\mu'} = \frac{\partial x^{\mu'}}{\partial x^\nu} A^\nu. \tag{A.6}$$

A covariant vector A_μ (also called a one-form or covector) is defined by the transformation

$$A_{\mu'} = \frac{\partial x^\nu}{\partial x^{\mu'}} A_\nu. \tag{A.7}$$

A.2   Higher rank tensors

A vector has one index for its components. A scalar has zero indices. We can generalize them to the tensors with two or more indices. A contravariant tensor of rank two is of form $B^{\mu\nu}$ which obeys the following transformation relation

$$B^{\mu'\nu'} = \frac{\partial x^{\mu'}}{\partial x^\alpha} \frac{\partial x^{\nu'}}{\partial x^\beta} B^{\alpha\beta}. \tag{A.8}$$

A mixed tensor $B^\mu{}_\nu$ is partly covariant and partly contravariant with the transformation

$$B^{\mu'}{}_{\nu'} = \frac{\partial x^{\mu'}}{\partial x^\alpha} \frac{\partial x^\beta}{\partial x^{\nu'}} B^\alpha{}_\beta. \tag{A.9}$$

A covariant tensor is defined by

$$B_{\mu'\nu'} = \frac{\partial x^\alpha}{\partial x^{\mu'}} \frac{\partial x^\beta}{\partial x^{\nu'}} B_{\alpha\beta}. \tag{A.10}$$

More generally,

$$B^{\mu'\nu'\cdots}{}_{\rho'\cdots} = \frac{\partial x^{\mu'}}{\partial x^\alpha} \frac{\partial x^{\nu'}}{\partial x^\beta} \frac{\partial x^\gamma}{\partial x^{\rho'}} \cdots B^{\alpha\beta\cdots}{}_{\gamma\cdots}. \tag{A.11}$$

We introduce $\Lambda^\alpha_{\mu'} = \frac{\partial x^\alpha}{\partial x^{\mu'}}$, Eq.(A.11) becomes

$$B^{\mu'\nu'\cdots}_{\rho'\cdots} = \Lambda^{\mu'}_\alpha \Lambda^{\nu'}_\beta \Lambda^\gamma_{\rho'} \cdots B^{\alpha\beta\cdots}_{\gamma\cdots}. \tag{A.12}$$

Since

$$\frac{\partial x^\alpha}{\partial x^\beta} = \delta^\alpha_\beta, \tag{A.13}$$

where δ^μ_ν is the Kronecker delta, the chain rule gives

$$\frac{\partial x^\alpha}{\partial x^\beta} = \frac{\partial x^\alpha}{\partial x^{\mu'}} \frac{\partial x^{\mu'}}{\partial x^\beta} = \Lambda^\alpha_{\mu'} \Lambda^{\mu'}_\beta = \delta^\alpha_\beta. \tag{A.14}$$

The tensor product of two tensors produces a higher rank tensor. For example, A^α_γ and B^β_δ can form a higher rank tensor $C^\alpha_\gamma{}^\beta_\delta$ by

$$C^\alpha_\gamma{}^\beta_\delta = A^\alpha_\gamma B^\beta_\delta. \tag{A.15}$$

The tensor product (also called *outer product*) is often written as

$$C = A \otimes B. \tag{A.16}$$

When we set a covariant and contravariant index equal and sum over the index, we make a high rank tensor to a lower rank tensor. For example,

$$T^\alpha_\gamma{}^\beta_\beta \equiv T^\alpha_\gamma. \tag{A.17}$$

This process is called the *contraction*. The contraction over a pair of indices reduces the rank of a tensor by two.

We define an *inner product* of two tensors by forming the outer product and then contracting over a pair of indices. For example,

$$C^\alpha_\beta \equiv A^\alpha_\gamma B^\gamma_\beta. \tag{A.18}$$

The inner product of two vectors $\mathbf{A}$ and $\mathbf{B}$ produces a scalar C

$$C = A^\mu B_\mu \equiv \mathbf{A} \cdot \mathbf{B}. \tag{A.19}$$

A scalar is a tensor of rank zero. It is invariant under transformation

$$\begin{aligned} C' &= A^{\mu'} B_{\mu'} \\ &= \Lambda^{\mu'}_\mu A^\mu \Lambda^\nu_{\mu'} B_\nu \\ &= \delta^\nu_\mu A^\mu B_\nu \\ &= A^\mu B_\mu \\ &= C. \end{aligned} \tag{A.20}$$

A.3   Metric tensor

We define *metric tensor* as a tensor that realizes the mapping between contravariant and covariant tensors. We use the relation $\mathbf{A} \cdot \mathbf{B} = A_\mu B^\mu = g_{\mu\nu} A^\mu B^\nu$ for the definition of the metric tensor $g_{\mu\nu}$. Thus we have

$$A_\mu = g_{\mu\nu} A^\nu. \tag{A.21}$$

In this equation, the metric tensor plays the role of lowering the indices. The mapping should be invertible. We have the definition of $g^{\mu\nu}$:

$$A^\mu = g^{\mu\nu} A_\nu. \tag{A.22}$$

Thus $g^{\mu\nu}$ raises the indices. From the definition of metric tensor, we can see that $g_{\mu\nu}$ should be symmetric if we want $\mathbf{A} \cdot \mathbf{B} = A_\mu B^\mu = \mathbf{B} \cdot \mathbf{A} = B_\mu A^\mu$. Thus

$$g_{\mu\nu} = g_{\nu\mu}. \tag{A.23}$$

Using the relation

$$A^\mu = g^{\mu\alpha} A_\alpha = g^{\mu\alpha} g_{\alpha\nu} A^\nu, \tag{A.24}$$

we have

$$g^{\mu\alpha} g_{\alpha\nu} = \delta^\mu_\nu. \tag{A.25}$$

When we use the metric tensor $g_{\mu\nu}$ to lower the metric $g^{\mu\nu}$, we have

$$g^\mu_\nu = g_{\nu\alpha} g^{\alpha\mu} = \delta^\mu_\nu. \tag{A.26}$$

Thus g^μ_ν is also a Kronecker delta.

A.4   Flat spacetime

The simplest metric is the metric of flat space. The three-dimensional flat space characterized by the metric $g_{ij} = \delta_{ij}$ is an Euclidean space. The Minkowski metric of four-dimensional flat spacetime is given by $g_{\mu\nu} = \eta_{\mu\nu}$ or $g_{\mu\nu} = \eta'_{\mu\nu}$ with

$$\eta_{\mu\nu} = \begin{pmatrix} 1 & 0 & 0 & 0 \\ 0 & -1 & 0 & 0 \\ 0 & 0 & -1 & 0 \\ 0 & 0 & 0 & -1 \end{pmatrix} \quad \text{and} \quad \eta'_{\mu\nu} = \begin{pmatrix} -1 & 0 & 0 & 0 \\ 0 & 1 & 0 & 0 \\ 0 & 0 & 1 & 0 \\ 0 & 0 & 0 & 1 \end{pmatrix}. \tag{A.27}$$

The Minkowski metric is the only flat spacetime metric guaranteeing the causality principle. $\eta_{\mu\nu}$ is said to have the signature $[1, -1, -1, -1]$ and $\eta'_{\mu\nu}$

has the signature $[-1, 1, 1, 1]$. one can use either $\eta_{\mu\nu}$ or $\eta'_{\mu\nu}$ to describe the Minkowski spacetime. We use $\eta_{\mu\nu}$ in the quantum field theory and $\eta'_{\mu\nu}$ in the other parts of the book as a custom. Since $\eta_{\mu\nu} = -\eta'_{\mu\nu}$, one can easily make transformation. To simplify the notation, we often omit the prime in $\eta'_{\mu\nu}$.

When the signature $[-1, 1, 1, 1]$ is used, for a contravariant vector specified by

$$A^\mu = (A^0, A^i) = (A^0, \mathbf{A}), \tag{A.28}$$

the covariant vector $A_\mu = \eta_{\mu\nu} A^\nu$ has the form

$$A_\mu = (A_0, A_i) = (-A^0, A^i) = (-A^0, \mathbf{A}). \tag{A.29}$$

The distance of the spacetime is defined by

$$(ds)^2 = dx^\mu dx_\mu = g_{\mu\nu} dx^\mu dx^\nu. \tag{A.30}$$

In the Minkowski metric,

$$(ds)^2 = -(dt)^2 + (dx)^2 + (dy)^2 + (dz)^2 \tag{A.31}$$

when the signature $[-1, 1, 1, 1]$ is used. The *proper time* $d\tau$ is defined by

$$(d\tau)^2 = -(ds)^2 = (dt)^2 - (dx)^2 - (dy)^2 - (dz)^2. \tag{A.32}$$

Thus the signature $[1, -1, -1, -1]$ describes the proper time. In the unit $c \neq 1$, we have

$$(cd\tau)^2 = -(ds)^2 = c^2(dt)^2 - (dx)^2 - (dy)^2 - (dz)^2. \tag{A.33}$$

The proper time is the time measured in the local rest frame of observer. In terms of proper time, we can introduce an useful vector - the four-velocity u^μ, which is defined as

$$u^\mu \equiv \frac{dx^\mu}{d\tau}. \tag{A.34}$$

$(ds)^2$ can be positive or negative. When the signature $[-1, 1, 1, 1]$ is used, we call the distance $(ds)^2$ space-like if $(ds)^2 > 0$. Otherwise $((ds)^2 < 0)$, we call the distance time-like. $(ds)^2 = 0$ defines the light cone. Since $(ds)^2$ is a scalar, space-like (time-like) points remain space-like (time-like) under any coordinate transformation in the Minkowski spacetime (i.e. Lorentz transformation). Thus the causality principle is obeyed in the Minkowski spacetime.

A.5   Lorentz transformation

A.5.1   *Infinitesimal Lorentz transformation*

The coordinate transformation Eq.(A.1) in the Minkowski spacetime is called the Lorentz transformation. For an infinitesimal proper Lorentz transformation

$$(x')^\mu = \Lambda^\mu{}_\nu x^\nu, \tag{A.35}$$

$\Lambda^\mu{}_\nu$ can be expressed as

$$\Lambda^\mu{}_\nu = \delta^\mu{}_\nu + \Delta\omega^\mu{}_\nu, \tag{A.36}$$

where $\Delta\omega^\mu{}_\nu$ are infinitesimal parameters. Using Eq.(A.14), we have

$$\begin{aligned}
\Lambda^\lambda{}_\mu \Lambda_\lambda{}^\nu &= (\delta^\lambda{}_\mu + \Delta\omega^\lambda{}_\mu)(\delta_\lambda{}^\nu + \Delta\omega_\lambda{}^\nu)\\
&= \delta^\lambda{}_\mu \delta_\lambda{}^\nu + \delta^\lambda{}_\mu \Delta\omega_\lambda{}^\nu + \delta_\lambda{}^\nu \Delta\omega^\lambda{}_\mu\\
&= \delta^\nu_\mu + \Delta\omega_\mu{}^\nu + \Delta\omega^\nu{}_\mu\\
&= \delta^\nu_\mu.
\end{aligned} \tag{A.37}$$

We have omitted the second order terms of the infinitesimal $\Delta\omega^\mu{}_\nu$ in Eq.(A.37). Thus we have

$$\Delta\omega^\mu{}_\nu + \Delta\omega_\nu{}^\mu = 0 \tag{A.38}$$

or

$$\Delta\omega^{\mu\nu} = -\Delta\omega^{\nu\mu}, \tag{A.39}$$

which shows that $\Delta\omega^{\mu\nu}$ is antisymmetric.

There are six non-vanishing parameters in the antisymmetric $\Delta\omega^{\mu\nu}$. The following are two typical examples of $\Delta\omega^{\mu\nu}$.

1) Lorentz boost

We consider the case that $\Delta\omega^{10} = -\Delta\omega^{01} \equiv -\Delta\beta \neq 0$ and all other $\Delta\omega^{\mu\nu} = 0$. The components in mixed indices are $\Delta\omega^0{}_1 = -\Delta\omega_1{}^0 = -\eta_{1\mu}\Delta\omega^{\mu 0} = -\eta_{11}\Delta\omega^{10} = -\Delta\omega^{10} = -\Delta\beta$. Other components are zero. Thus we have the transformation

$$\begin{aligned}
x'^\nu &= (\delta^\nu_\mu + \Delta\omega^1{}_0 \delta^\nu_1 \delta^0_\mu + \Delta\omega^0{}_1 \delta^\nu_0 \delta^1_\mu)x^\mu\\
&= (\delta^\nu_\mu - \Delta\beta \delta^\nu_1 \delta^0_\mu - \Delta\beta \delta^\nu_0 \delta^1_\mu)x^\mu.
\end{aligned} \tag{A.40}$$

The explicit form of Eq.(A.40) reads

$$x'^0 = x^0 - \Delta\beta x^1, \tag{A.41a}$$

$$x'^1 = -\Delta\beta x^0 + x^1, \tag{A.41b}$$

$$x'^2 = x^2, \tag{A.41c}$$

$$x'^3 = x^3. \tag{A.41d}$$

This Lorentz transformation is a transformation relating the x frame to an inertial frame moving along x^1 with an infinitesimal velocity $\Delta\beta$ relative to the x frame. The Lorentz transformation Eq.(A.41) is called the *Lorentz boost*.

2) Spatial rotation

When $\Delta\omega^{12} = -\Delta\omega^{21} \equiv -\Delta\varphi \neq 0$ and all other $\Delta\omega^{\mu\nu} = 0$. The transformation relation is given by

$$x'^{\nu} = (\delta_{\mu}^{\nu} + \Delta\varphi\delta_1^{\nu}\delta_{\mu}^2 - \Delta\varphi\delta_2^{\nu}\delta_{\mu}^1)x^{\mu}. \tag{A.42}$$

The explicit form of Eq.(A.42) reads

$$x'^0 = x^0, \tag{A.43a}$$

$$x'^1 = x^1 + \Delta\varphi x^2, \tag{A.43b}$$

$$x'^2 = -\Delta\varphi x^1 + x^2, \tag{A.43c}$$

$$x'^3 = x^3. \tag{A.43d}$$

Eq.(A.43) is the transformation generated by an infinitesimal rotation about the z axis with the rotation angle of $\Delta\varphi$.

A.5.2 *Finite Lorentz transformation*

The finite Lorentz transformation can be generated by successive applications of the infinitesimal Lorentz transformations. We write

$$\Delta\omega^{\mu}{}_{\nu} = \Delta\omega(I_{\mathbf{n}})^{\mu}{}_{\nu}, \tag{A.44}$$

where $(I_{\mathbf{n}})^{\mu}{}_{\nu}$ is the 4×4 matrix for a unit rotation around the axis in the $\mathbf{n}$ direction. $\Delta\omega$ is the infinitesimal rotation angle around the $\mathbf{n}$ axis. Under a transformation of Lorentz boost described by Eq.(A.40), the $\mathbf{n}$ axis is perpendicular to the x^0 and x^1 axes. We use $\mathbf{n}(01)$ to denote the $\mathbf{n}$ axis. According to Eq.(A.40),

$$(I_{\mathbf{n}(01)}) = -(\delta_1^{\nu}\delta_{\mu}^0 + \delta_0^{\nu}\delta_{\mu}^1)$$

$$= \begin{pmatrix} 0 & -1 & 0 & 0 \\ -1 & 0 & 0 & 0 \\ 0 & 0 & 0 & 0 \\ 0 & 0 & 0 & 0 \end{pmatrix}. \tag{A.45}$$

Straightforward calculations can give the following relations

$$(I_{\mathbf{n}(01)})^2 = \begin{pmatrix} 0 & -1 & 0 & 0 \\ -1 & 0 & 0 & 0 \\ 0 & 0 & 0 & 0 \\ 0 & 0 & 0 & 0 \end{pmatrix} \begin{pmatrix} 0 & -1 & 0 & 0 \\ -1 & 0 & 0 & 0 \\ 0 & 0 & 0 & 0 \\ 0 & 0 & 0 & 0 \end{pmatrix}$$

$$= \begin{pmatrix} 1 & 0 & 0 & 0 \\ 0 & 1 & 0 & 0 \\ 0 & 0 & 0 & 0 \\ 0 & 0 & 0 & 0 \end{pmatrix}$$

(A.46)

and

$$(I_{\mathbf{n}(01)})^3 = \begin{pmatrix} 1 & 0 & 0 & 0 \\ 0 & 1 & 0 & 0 \\ 0 & 0 & 0 & 0 \\ 0 & 0 & 0 & 0 \end{pmatrix} \begin{pmatrix} 0 & -1 & 0 & 0 \\ -1 & 0 & 0 & 0 \\ 0 & 0 & 0 & 0 \\ 0 & 0 & 0 & 0 \end{pmatrix}$$

$$= \begin{pmatrix} 0 & -1 & 0 & 0 \\ -1 & 0 & 0 & 0 \\ 0 & 0 & 0 & 0 \\ 0 & 0 & 0 & 0 \end{pmatrix}$$

(A.47)

$$= (I_{\mathbf{n}(01)}).$$

For a spatial rotation around the z axis, according to Eq.(A.42), we have

$$(I_{\mathbf{n}(z)}) = (\delta^\nu_1 \delta^2_\mu - \delta^\nu_2 \delta^1_\mu)$$

$$= \begin{pmatrix} 0 & 0 & 0 & 0 \\ 0 & 0 & 1 & 0 \\ 0 & -1 & 0 & 0 \\ 0 & 0 & 0 & 0 \end{pmatrix}.$$

(A.48)

Straightforward calculations give the following relations

$$(I_{\mathbf{n}(z)})^2 = \begin{pmatrix} 0 & 0 & 0 & 0 \\ 0 & 0 & 1 & 0 \\ 0 & -1 & 0 & 0 \\ 0 & 0 & 0 & 0 \end{pmatrix} \begin{pmatrix} 0 & 0 & 0 & 0 \\ 0 & 0 & 1 & 0 \\ 0 & -1 & 0 & 0 \\ 0 & 0 & 0 & 0 \end{pmatrix}$$

$$= \begin{pmatrix} 0 & 0 & 0 & 0 \\ 0 & -1 & 0 & 0 \\ 0 & 0 & -1 & 0 \\ 0 & 0 & 0 & 0 \end{pmatrix}$$

(A.49)

and

$$(I_{\mathbf{n}(z)})^3 = \begin{pmatrix} 0 & 0 & 0 & 0 \\ 0 & -1 & 0 & 0 \\ 0 & 0 & -1 & 0 \\ 0 & 0 & 0 & 0 \end{pmatrix} \begin{pmatrix} 0 & 0 & 0 & 0 \\ 0 & 0 & 1 & 0 \\ 0 & -1 & 0 & 0 \\ 0 & 0 & 0 & 0 \end{pmatrix}$$

$$= \begin{pmatrix} 0 & 0 & 0 & 0 \\ 0 & 0 & -1 & 0 \\ 0 & 1 & 0 & 0 \\ 0 & 0 & 0 & 0 \end{pmatrix} \tag{A.50}$$

$$= -(I_{\mathbf{n}(z)}).$$

A finite rotation of ω can be divided into N successive infinitesimal rotations of $\Delta\omega = \frac{\omega}{N}$ with $N \to \infty$. Thus the finite Lorentz transformation can be written as

$$
\begin{aligned}
x'^{\nu} &= \lim_{N\to\infty} \left(1 + \frac{\omega}{N} I_{\mathbf{n}}\right)^{\nu}_{\ \mu_1} \left(1 + \frac{\omega}{N} I_{\mathbf{n}}\right)^{\mu_1}_{\ \mu_2} \cdots x^{\mu} \\
&= \lim_{N\to\infty} \left(\left(1 + \frac{\omega}{N} I_{\mathbf{n}}\right)^{N}\right)^{\nu}_{\ \mu} x^{\mu} \\
&= \left(e^{\omega I_{\mathbf{n}}}\right)^{\nu}_{\ \mu} x^{\mu}.
\end{aligned}
\tag{A.51}
$$

1) Lorentz boost
Under a finite pure Lorentz transformation (Lorentz boost), using Eqs.(A.45), (A.46) and (A.47), we have

$$
\begin{aligned}
x'^{\nu} &= \left(e^{\omega I_{\mathbf{n}(01)}}\right)^{\nu}_{\ \mu} x^{\mu} \\
&= \left(\cosh\left(\omega I_{\mathbf{n}(01)}\right) + \sinh\left(\omega I_{\mathbf{n}(01)}\right)\right)^{\nu}_{\ \mu} x^{\mu} \\
&= \left(\left[1 + \frac{1}{2!}(\omega I_{\mathbf{n}(01)})^2 + \frac{1}{4!}(\omega I_{\mathbf{n}(01)})^4 + \cdots\right]\right. \\
&\quad \left.+ \left[\omega I_{\mathbf{n}(01)} + \frac{1}{3!}(\omega I_{\mathbf{n}(01)})^3 + \cdots\right]\right)^{\nu}_{\ \mu} x^{\mu} \\
&= \left(\left[1 + \frac{\omega^2}{2!}(I_{\mathbf{n}(01)})^2 + \frac{\omega^4}{4!}(I_{\mathbf{n}(01)})^2 + \cdots\right]\right. \\
&\quad \left.+ \left[\omega + \frac{\omega^3}{3!} + \cdots\right] I_{\mathbf{n}(01)}\right)^{\nu}_{\ \mu} x^{\mu} \\
&= \left(1 - (I_{\mathbf{n}(01)})^2 + \cosh(\omega)(I_{\mathbf{n}(01)})^2 + \sinh(\omega) I_{\mathbf{n}(01)}\right)^{\nu}_{\ \mu} x^{\mu}.
\end{aligned}
\tag{A.52}
$$

The explicit matrix form of Eq.(A.52) is

$$
\begin{pmatrix} x'^0 \\ x'^1 \\ x'^2 \\ x'^3 \end{pmatrix} = \begin{pmatrix} \cosh(\omega) & -\sinh(\omega) & 0 & 0 \\ -\sinh(\omega) & \cosh(\omega) & 0 & 0 \\ 0 & 0 & 1 & 0 \\ 0 & 0 & 0 & 1 \end{pmatrix} \begin{pmatrix} x^0 \\ x^1 \\ x^2 \\ x^3 \end{pmatrix}. \tag{A.53}
$$

We introduce

$$
\beta \equiv \tanh(\omega). \tag{A.54}
$$

Then we have

$$
\cosh(\omega) = \frac{\cosh(\omega)}{\sqrt{\cosh^2(\omega) - \sinh^2(\omega)}} = \frac{1}{\sqrt{1 - \tanh(\omega)}} = \frac{1}{\sqrt{1 - \beta^2}}, \tag{A.55a}
$$

$$
\sinh(\omega) = \frac{\sinh(\omega)}{\sqrt{\cosh^2(\omega) - \sinh^2(\omega)}} = \frac{1}{\sqrt{\frac{1}{\beta^2} - 1}} = \frac{\beta}{\sqrt{1 - \beta^2}}. \tag{A.55b}
$$

Thus Eq.(A.53) becomes

$$
\begin{pmatrix} x'^0 \\ x'^1 \\ x'^2 \\ x'^3 \end{pmatrix} = \begin{pmatrix} \frac{1}{\sqrt{1-\beta^2}} & -\frac{\beta}{\sqrt{1-\beta^2}} & 0 & 0 \\ -\frac{\beta}{\sqrt{1-\beta^2}} & \frac{1}{\sqrt{1-\beta^2}} & 0 & 0 \\ 0 & 0 & 1 & 0 \\ 0 & 0 & 0 & 1 \end{pmatrix} \begin{pmatrix} x^0 \\ x^1 \\ x^2 \\ x^3 \end{pmatrix} \tag{A.56}
$$

or

$$
x'^0 = \frac{x^0 - \beta x^1}{\sqrt{1 - \beta^2}}, \tag{A.57a}
$$

$$
x'^1 = \frac{x^1 - \beta x^0}{\sqrt{1 - \beta^2}}, \tag{A.57b}
$$

$$
x'^2 = x^2, \tag{A.57c}
$$

$$
x'^3 = x^3. \tag{A.57d}
$$

Eq.(A.57) is called the *Lorentz transformation in the special relativity*.
When an observer is in rest in the x' frame,

$$
dx'^0 = \frac{dx^0 - \beta dx^1}{\sqrt{1 - \beta^2}} = 0. \tag{A.58}
$$

We have

$$
\frac{dx^1}{dx^0} = \beta \quad \text{or} \quad \frac{dx^1}{dt} = \beta. \tag{A.59}
$$

We can see that β is the velocity v of the frame x in the frame x'. β is also the velocity v of the frame x' in the frame x.

2) Spatial rotation

Under a finite spatial rotation in the z direction, we have

$$
\begin{aligned}
x'^{\nu} &= \left(e^{\omega I_{\mathbf{n}(z)}}\right)^{\nu}{}_{\mu} x^{\mu} \\
&= \left(\cosh\left(\omega I_{\mathbf{n}(z)}\right) + \sinh\left(\omega I_{\mathbf{n}(z)}\right)\right)^{\nu}{}_{\mu} x^{\mu} \\
&= \left(\left[1 + \frac{1}{2!}(\omega I_{\mathbf{n}(z)})^2 + \frac{1}{4!}(\omega I_{\mathbf{n}(z)})^4 + \cdots\right]\right. \\
&\qquad \left. + \left[\omega I_{\mathbf{n}(z)} + \frac{1}{3!}(\omega I_{\mathbf{n}(z)})^3 + \cdots\right]\right)^{\nu}{}_{\mu} r^{\mu} \\
&= \left(\left[1 + \frac{\omega^2}{2!}(I_{\mathbf{n}(z)})^2 - \frac{\omega^4}{4!}(I_{\mathbf{n}(z)})^2 + \cdots\right]\right. \\
&\qquad \left. + \left[\omega - \frac{\omega^3}{3!} + \cdots\right] I_{\mathbf{n}(z)}\right)^{\nu}{}_{\mu} x^{\mu} \\
&= \left(1 + (I_{\mathbf{n}(z)})^2 - \cos(\omega)(I_{\mathbf{n}(z)})^2 + \sin(\omega)I_{\mathbf{n}(z)}\right)^{\nu}{}_{\mu} x^{\mu}.
\end{aligned} \tag{A.60}
$$

The explicit matrix form of Eq.(A.60) is

$$
\begin{pmatrix} x'^0 \\ x'^1 \\ x'^2 \\ x'^3 \end{pmatrix} = \begin{pmatrix} 1 & 0 & 0 & 0 \\ 0 & \cos(\omega) & \sin(\omega) & 0 \\ 0 & -\sin(\omega) & \cos(\omega) & 0 \\ 0 & 0 & 0 & 1 \end{pmatrix} \begin{pmatrix} x^0 \\ x^1 \\ x^2 \\ x^3 \end{pmatrix}. \tag{A.61}
$$

A.6 Christoffel symbols

Now we consider a contravariant vector field $A^{\mu}(x^{\mu})$ as a function of contravariant coordinates. The direct derivative $\frac{\partial A^{\mu}}{\partial x^{\nu}}$ is often denoted as

$$
A^{\mu}{}_{,\nu} \equiv \frac{\partial A^{\mu}}{\partial x^{\nu}}. \tag{A.62}
$$

However, there is a problem here that the derivative $A^{\mu}{}_{,\nu}$ is not a tensor. This can be seen from the transformation of $A^{\mu}{}_{,\nu}$

$$
\begin{aligned}
A^{\mu'}{}_{,\nu'} &= \frac{\partial A^{\mu'}}{\partial x^{\nu'}} \\
&= \frac{\partial}{\partial x^{\nu'}} \left(\frac{\partial x^{\mu'}}{\partial x^{\alpha}} A^{\alpha}\right) \\
&= \frac{\partial x^{\mu'}}{\partial x^{\alpha}} \frac{\partial A^{\alpha}}{\partial x^{\nu'}} + \frac{\partial^2 x^{\mu'}}{\partial x^{\nu'} \partial x^{\alpha}} A^{\alpha}.
\end{aligned} \tag{A.63}
$$

Since A^α is a function of x^μ, we need express $\frac{\partial A^\alpha}{\partial x^{\nu\prime}}$ in terms of $\frac{\partial A^\alpha}{\partial x^\nu}$. Inserting $\frac{\partial A^\alpha}{\partial x^{\nu\prime}} = \frac{\partial A^\alpha}{\partial x^\gamma}\frac{\partial x^\gamma}{\partial x^{\nu\prime}}$, we have

$$
\begin{aligned}
A^{\mu\prime}{}_{,\nu\prime} &= \frac{\partial x^{\mu\prime}}{\partial x^\alpha}\frac{\partial x^\gamma}{\partial x^{\nu\prime}}\frac{\partial A^\alpha}{\partial x^\gamma} + \frac{\partial^2 x^{\mu\prime}}{\partial x^{\nu\prime}\partial x^\alpha}A^\alpha \\
&= \frac{\partial x^{\mu\prime}}{\partial x^\alpha}\frac{\partial x^\gamma}{\partial x^{\nu\prime}}A^\alpha{}_{,\gamma} + \frac{\partial^2 x^{\mu\prime}}{\partial x^{\nu\prime}\partial x^\alpha}A^\alpha.
\end{aligned}
\tag{A.64}
$$

Thus $A^\mu{}_{,\nu}$ does not follow the tensor transformation due to the existence of the second term in Eq.(A.64). This problem is caused by the definition of the derivative

$$
A^\mu{}_{,\nu} = \frac{\partial A^\mu}{\partial x^\nu} = \lim_{\delta x^\nu \to 0}\frac{A^\mu(x^\nu + \delta x^\nu) - A^\mu(x^\nu)}{\delta x^\nu}.
\tag{A.65}
$$

A vector has a direction. The direction also changes with the location in a curved spacetime. Therefore, the difference between two vectors is not a simple difference of components if they are located at different positions. In order to compare the direction of a vector, we need first put them at the same point in spacetime. This procedure is called the *parallel transport*.

We denote δA^μ as the change produced in the vector $A^\mu(x^\nu)$ at x^ν by an infinitesimal parallel transport of distance dx^ν. It should be directly proportional to dx^ν.

$$
\delta A^\mu \propto dx^\nu.
\tag{A.66}
$$

δA^μ should also be directly proportional to A^μ because larger A^μ would produces larger change. Thus

$$
\delta A^\mu = -\Gamma^\mu_{\alpha\nu}A^\alpha dx^\nu,
\tag{A.67}
$$

where $\Gamma^\mu_{\alpha\nu}$ is the constant of proportionality and is called the *Christoffel symbol* or *Levi-Civita connection*. It defines the parallel transport of a vector. The vector $A^\mu(x^\nu)$ at x^ν transported in parallel an infinitesimal distance dx^ν to the position at $x^\nu + dx^\nu$ has the components

$$
C^\mu = A^\mu + \delta A^\mu.
\tag{A.68}
$$

The vector $A^\mu(x^\nu)$ at $x^\nu + dx^\nu$ has the components $A^\mu(x^\nu + dx^\nu)$. The difference between them gives

$$
dA^\mu = A^\mu(x^\nu + dx^\nu) - [A^\mu(x^\nu) + \delta A^\mu].
\tag{A.69}
$$

This is the difference between two vector located at the same point, which should be also a vector. Thus we have a definition of derivative for a vector

$$
A^\mu{}_{;\nu} \equiv \frac{dA^\mu}{dx^\nu} = \lim_{\delta x^\nu \to 0}\frac{A^\mu(x^\nu + \delta x^\nu) - [A^\mu(x^\nu) + \delta A^\mu]}{\delta x^\nu}.
\tag{A.70}
$$

The derivative $A^{\mu}{}_{;\nu}$ is called the *covariant derivative*. Since dA^{μ} is a contravariant vector and $\frac{d}{dx^{\nu}}$ is a covariant vector, $A^{\mu}{}_{;\nu}$ should be a two rank tensor. Using Eq.(A.67), we have

$$dA^{\mu} = \frac{\partial A^{\mu}}{\partial x^{\nu}} dx^{\nu} - \delta A^{\mu}$$

$$= A^{\mu}{}_{,\nu} dx^{\nu} + \Gamma^{\mu}_{\alpha\nu} A^{\alpha} dx^{\nu}. \tag{A.71}$$

Inserting Eq.(A.71), Eq.(A.70) becomes

$$A^{\mu}{}_{;\nu} = A^{\mu}{}_{,\nu} + \Gamma^{\mu}_{\alpha\nu} A^{\alpha}. \tag{A.72}$$

Now we consider the derivative of a covariant vector B_{μ}. For any contravariant vector A^{μ},

$$\phi = B_{\alpha} A^{\alpha} \tag{A.73}$$

is a scalar. $\nabla_{\nu}\phi$ is thus a vector, which has the form

$$\nabla_{\nu}\phi = \phi_{,\nu} = \frac{\partial B_{\alpha}}{\partial x^{\nu}} A^{\alpha} + B_{\alpha} \frac{\partial A^{\alpha}}{\partial x^{\nu}}. \tag{A.74}$$

Expressing $A^{\alpha}{}_{,\nu}$ in terms of $A^{\alpha}{}_{;\nu}$, we have

$$\nabla_{\nu}\phi = \frac{\partial B_{\alpha}}{\partial x^{\nu}} A^{\alpha} + B_{\alpha} A^{\alpha}{}_{;\nu} - B_{\alpha} A^{\mu} \Gamma^{\alpha}_{\mu\nu}$$

$$= \left(\frac{\partial B_{\alpha}}{\partial x^{\nu}} - B_{\beta} \Gamma^{\beta}_{\alpha\nu} \right) A^{\alpha} + B_{\alpha} A^{\alpha}{}_{;\nu}. \tag{A.75}$$

This equation shows that $\frac{\partial B_{\alpha}}{\partial x^{\nu}} - B_{\beta}\Gamma^{\beta}_{\alpha\nu}$ should be a tensor because A^{μ} is an arbitrary vector. We define the covariant derivative of B_{α} as

$$B_{\mu;\nu} = B_{\mu;\nu} - B_{\alpha} \Gamma^{\alpha}_{\mu\nu}, \tag{A.76}$$

which is a two rank covariant tensor. Then Eq.(A.75) becomes

$$\nabla_{\nu}(B_{\alpha} A^{\alpha}) = B_{\alpha;\nu} A^{\alpha} + B_{\alpha} A^{\alpha}{}_{;\nu}. \tag{A.77}$$

Thus covariant differentiation obeys the product rule similar to the ordinary differentiation. Similarly we can obtain

$$A^{\mu\nu}{}_{;\alpha} = A^{\mu\nu}{}_{,\alpha} + A^{\beta\nu} \Gamma^{\mu}_{\beta\alpha} + A^{\mu\beta} \Gamma^{\nu}_{\beta\alpha}, \tag{A.78a}$$

$$B^{\mu}{}_{\nu;\alpha} = B^{\mu}{}_{\nu,\alpha} + B^{\beta}{}_{\nu} \Gamma^{\mu}_{\beta\alpha} - B^{\mu}{}_{\beta} \Gamma^{\beta}_{\nu\alpha}, \tag{A.78b}$$

$$C_{\mu\nu;\alpha} = C_{\mu\nu,\alpha} - C_{\beta\nu} \Gamma^{\beta}_{\mu\alpha} - C_{\mu\beta} \Gamma^{\beta}_{\nu\alpha}. \tag{A.78c}$$

Generally, $\Gamma^{\mu}_{\alpha\beta}$ is not symmetric. We can divide $\Gamma^{\mu}_{\alpha\beta}$ into the symmetric part $\Gamma^{\mu}_{(\alpha\beta)}$ and the asymmetric part $\Gamma^{\mu}_{[\alpha\beta]}$ by

$$\Gamma^{\mu}_{\alpha\beta} = \Gamma^{\mu}_{(\alpha\beta)} + \Gamma^{\mu}_{[\alpha\beta]} \tag{A.79}$$

with

$$\Gamma^{\mu}_{(\alpha\beta)} \equiv \frac{1}{2}(\Gamma^{\mu}_{\alpha\beta} + \Gamma^{\mu}_{\beta\alpha}), \tag{A.80a}$$

$$\Gamma^{\mu}_{[\alpha\beta]} \equiv \frac{1}{2}(\Gamma^{\mu}_{\alpha\beta} - \Gamma^{\mu}_{\beta\alpha}). \tag{A.80b}$$

A physical curved spacetime should have a local Minkowski metric to guarantee the local causality. In this local Minkowski metric, $\Gamma^{\mu}_{\alpha\beta}$ is zero and thus $\Gamma^{\mu}_{[\alpha\beta]}$ is zero. We can prove that $\Gamma^{\mu}_{[\alpha\beta]}$ is a tensor. If the tensor $\Gamma^{\mu}_{[\alpha\beta]}$ vanishes in one coordinate system, it must vanish in any coordinate system.

Now let us prove that $\Gamma^{\mu}_{[\alpha\beta]}$ is a tensor. The transformation relation for $\Gamma^{\tau}_{\mu\nu}$ is given by

$$\Gamma^{\tau'}_{\mu'\nu'} = \Gamma^{\rho}_{\alpha\beta}\frac{\partial x^{\alpha}}{\partial x^{\mu'}}\frac{\partial x^{\beta}}{\partial x^{\nu'}}\frac{\partial x^{\tau'}}{\partial x^{\rho}} + \frac{\partial^2 x^{\rho}}{\partial x^{\mu'}\partial x^{\nu'}}\frac{\partial x^{\tau'}}{\partial x^{\rho}}. \tag{A.81}$$

The second term is symmetric in μ' and ν'. Thus it cancels out in the transformation for $\Gamma^{\tau}_{[\mu\nu]}$. Therefore $\Gamma^{\tau}_{[\mu\nu]}$ transforms as a tensor

$$\Gamma^{\tau'}_{[\mu'\nu']} = \Gamma^{\rho}_{[\alpha\beta]}\frac{\partial x^{\alpha}}{\partial x^{\mu'}}\frac{\partial x^{\beta}}{\partial x^{\nu'}}\frac{\partial x^{\tau'}}{\partial x^{\beta}}. \tag{A.82}$$

$\Gamma^{\tau}_{[\mu\nu]}$ is often called the *torsion tensor*.

A.7   Riemann spacetime

Mathematically, at any position P, we can find a local flat space 'tangent' to any curved space if we do not restrict the transformation. Physically the local flat metric should be a local Minkowski metric to fulfill the causality. We call the spacetime with this property as the *Riemann spacetime* or strictly *pseudo-Riemann spacetime* (The metric of the Riemann space is positive-definite by definition). In this local flat spacetime at P, straight line is meaningful locally. Thus $\Gamma^{\tau}_{\mu\nu}$ is zero at P locally in this local Minkowski metric. In the local flat spacetime at P, the covariant derivative of a vector A^{α} is given by the ordinary derivative of the vector.

$$A^{\alpha}{}_{;\beta} = A^{\alpha}{}_{,\beta} \quad \text{at } P \text{ in the local flat spacetime.} \tag{A.83}$$

In this local flat metric, the metric is constant locally. We have

$$g_{\mu\nu;\lambda} = g_{\mu\nu,\lambda} = 0 \quad \text{at } P. \tag{A.84}$$

$g_{\mu\nu;\lambda}$ is a tensor. If the equation $g_{\mu\nu;\lambda} = 0$ is true in one frame, it will be valid in any frame. Therefore, we have

$$g_{\mu\nu;\lambda} = 0. \tag{A.85}$$

Since $\Gamma^\tau_{[\mu\nu]} = 0$ for a Riemann spacetime, we have

$$\Gamma^\tau_{\mu\nu} = \Gamma^\tau_{\nu\mu}. \tag{A.86}$$

The connection should be symmetric.

Eqs.(A.85) and (A.86) can lead to an important formula in which the Christoffel symbol is expressed in terms of the metric tensor and its derivatives. From Eq.(A.85), we have

$$g_{\mu\nu;\lambda} = g_{\mu\nu,\lambda} - \Gamma^\alpha_{\mu\lambda}g_{\alpha\nu} - \Gamma^\alpha_{\nu\lambda}g_{\mu\alpha} = 0. \tag{A.87}$$

This gives

$$g_{\mu\nu,\lambda} = \Gamma^\alpha_{\mu\lambda}g_{\alpha\nu} + \Gamma^\alpha_{\nu\lambda}g_{\mu\alpha}. \tag{A.88}$$

Permuting the $\mu\nu\lambda$ indices cyclically, we have

$$g_{\lambda\mu,\nu} = \Gamma^\alpha_{\lambda\nu}g_{\alpha\mu} + \Gamma^\alpha_{\mu\nu}g_{\lambda\alpha}, \tag{A.89a}$$

$$g_{\nu\lambda,\mu} = \Gamma^\alpha_{\nu\mu}g_{\alpha\lambda} + \Gamma^\alpha_{\lambda\mu}g_{\nu\alpha}. \tag{A.89b}$$

Adding the above two equations and subtracting Eq.(A.88), we obtain

$$g_{\lambda\mu,\nu} + g_{\nu\lambda,\mu} - g_{\mu\nu,\lambda} = 2\Gamma^\alpha_{\mu\nu}g_{\lambda\alpha}, \tag{A.90}$$

where we have used the symmetries $\Gamma^\alpha_{\mu\nu} = \Gamma^\alpha_{\nu\mu}$ and $g_{\mu\nu} = g_{\nu\mu}$. Multiplying Eq.(A.90) by $g^{\lambda\tau}$ and using Eq.(A.25), Eq.(A.90) becomes

$$\Gamma^\tau_{\mu\nu} = \frac{1}{2}g^{\lambda\tau}(g_{\lambda\mu,\nu} + g_{\nu\lambda,\mu} - g_{\mu\nu,\lambda}). \tag{A.91}$$

Eq.(A.91) is the relation between the metric tensor with the Christoffel symbol (or Levi-Civita connection) of the Riemann metric.

Using

$$g^{\alpha\tau}g_{\tau\beta,\alpha} = g^{\alpha\tau}g_{\beta\tau,\alpha} = g^{\tau\alpha}g_{\beta\alpha,\tau} = g^{\alpha\tau}g_{\beta\alpha,\tau} \tag{A.92}$$

and contracting over $\tau\nu$ of $\Gamma^\tau_{\mu\nu}$, Eq.(A.91) becomes

$$\begin{aligned}
\Gamma^\alpha_{\beta\alpha} &= \frac{1}{2}g^{\alpha\tau}(g_{\tau\beta,\alpha} + g_{\tau\alpha,\beta} - g_{\beta\alpha,\tau}) \\
&= \frac{1}{2}g^{\alpha\tau}g_{\tau\alpha,\beta}.
\end{aligned} \tag{A.93}$$

We use g to denote the determinant $|g_{\alpha\beta}|$ and calculate the differential dg of the determinant g. dg can be evaluated by taking the differential of each component of the tensor $g_{\alpha\beta}$ and multiplying it by its coefficient in the determinant which is the corresponding minor. Since the tensor $g^{\alpha\beta}$ is reciprocal to $g_{\alpha\beta}$, the components of $g^{\alpha\beta}$ are equal to the minors of the determinant of $g_{\alpha\beta}$ divided by the determinant. Thus, we have

$$dg = gg^{\alpha\beta}dg_{\alpha\beta} = -gg_{\alpha\beta}dg^{\alpha\beta}. \tag{A.94}$$

Using Eq.(A.94), Eq.(A.93) becomes

$$
\begin{aligned}
\Gamma^{\alpha}_{\beta\alpha} &= \frac{1}{2g}\frac{\partial g}{\partial g_{\tau\alpha}}\frac{\partial g_{\tau\alpha}}{\partial x^{\beta}} \\
&= \frac{1}{2g}\frac{\partial g}{\partial x^{\beta}} \\
&= \frac{1}{2}\frac{\partial \ln(-g)}{\partial x^{\beta}} = \frac{1}{2}[\ln(-g)]_{,\beta} \\
&= \frac{\partial \ln\sqrt{-g}}{\partial x^{\beta}} = (\ln\sqrt{-g})_{,\beta}.
\end{aligned}
\tag{A.95}
$$

Using Eq.(A.95), the divergence $A^{\alpha}_{;\alpha}$ of a vector A^{α} can be expressed as

$$
\begin{aligned}
A^{\alpha}_{;\alpha} &= A^{\alpha}_{,\alpha} + \frac{1}{\sqrt{-g}}A^{\alpha}(\sqrt{-g})_{,\alpha} \\
&= \frac{1}{\sqrt{-g}}(\sqrt{-g}A^{\alpha})_{,\alpha}.
\end{aligned}
\tag{A.96}
$$

A.8   Volume

Now we discuss the calculation of volumes for integrations in spacetime. In the local Minkowski metric, we have the volume element $dV \equiv d^4x = dx^0 dx^1 dx^2 dx^3$. In any other coordinate system $\{x^{\alpha'}\}$, we have

$$
dV = dx^0 dx^1 dx^2 dx^3 = \frac{\partial(x^0, x^1, x^2, x^3)}{\partial(x^{0'}, x^{1'}, x^{2'}, x^{3'})}dx^{0'} dx^{1'} dx^{2'} dx^{3'}, \tag{A.97}
$$

where the factor $\partial(\ \ )/\partial(\ \ )$ is the Jacobian determinant defined by

$$
\frac{\partial(x^0, x^1, x^2, x^3)}{\partial(x^{0'}, x^{1'}, x^{2'}, x^{3'})} = \det\begin{pmatrix} \dfrac{\partial x^0}{\partial x^{0'}} & \dfrac{\partial x^0}{\partial x^{1'}} & \cdots \\ \dfrac{\partial x^1}{\partial x^{0'}} & \cdots & \cdots \\ \vdots & \vdots & \ddots \end{pmatrix} \tag{A.98}
$$

$$
= \det(\Lambda^{\alpha}_{\beta'}).
$$

Meanwhile, we have

$$
g_{\alpha'\beta'} = \Lambda^{\mu}_{\alpha'}\Lambda^{\nu}_{\beta'}\eta_{\mu\nu}. \tag{A.99}
$$

To simplify the notation, we denote the matrix of $c_{\alpha\beta}$ as (c). Then Eq.(A.99) can be expressed in a matrix form

$$
(g) = (\Lambda)(\eta)(\Lambda)^{T}, \tag{A.100}
$$

where T denotes transpose. Evaluating the determinant of Eq.(A.100), we have

$$g \equiv \det(g) = \det(\Lambda)\det(\eta)\det(\Lambda^T) = -[\det(\Lambda)]^2. \tag{A.101}$$

Thus, Eq.(A.98) becomes

$$dV = \sqrt{-g}d^4x. \tag{A.102}$$

The factor $\sqrt{-g}d^4x$ is also called the *proper volume element*.

It should be noted that Gauss's theorem also applies on the Riemann spacetime. We integrate the divergence over a volume,

$$\int A^\alpha_{;\alpha}\sqrt{-g}d^4x = \int (\sqrt{-g}A^\alpha)_{,\alpha}d^4x. \tag{A.103}$$

We have used Eq.(A.96) in the derivation of Eq.(A.103). Using Gauss's theorem, we have

$$\int A^\alpha_{;\alpha}\sqrt{-g}d^4x = \oint A^\alpha n_\alpha\sqrt{-g}ds, \tag{A.104}$$

where n_α is the unit direction vector of the surface element ds. This is the version of Gauss's theorem in the Riemann spacetime. $n_\alpha\sqrt{-g}d^3S$ is also called the proper surface element.

A.9 Riemann curvature tensor

The connection $\Gamma^\alpha_{\mu\nu}$ is not a tensor. However, we can construct tensors using $\Gamma^\alpha_{\mu\nu}$. One of them is the Riemann curvature tensor, which is the most important tensor in describing the properties of the Riemann spacetime.

Let us consider a covariant vector field $A_\lambda(x)$. The second covariant derivative of A_λ reads

$$\begin{aligned}
A_{\lambda;\mu;\nu} =& A_{\lambda;\mu,\nu} - \Gamma^\rho_{\lambda\nu}A_{\rho;\mu} - \Gamma^\rho_{\mu\nu}A_{\lambda;\rho}\\
=& A_{\lambda,\mu,\nu} - \Gamma^\rho_{\lambda\mu,\nu}A_\rho - \Gamma^\rho_{\lambda\mu}A_{\rho,\nu} - \Gamma^\rho_{\lambda\nu}A_{\rho,\mu}\\
&+ \Gamma^\rho_{\lambda\nu}\Gamma^\sigma_{\rho\mu}A_\sigma - \Gamma^\rho_{\mu\nu}A_{\lambda;\rho}.
\end{aligned} \tag{A.105}$$

If we exchange the order of the differentials, we have

$$\begin{aligned}
A_{\lambda;\nu;\mu} =& A_{\lambda,\nu,\mu} - \Gamma^\rho_{\lambda\nu,\mu}A_\rho - \Gamma^\rho_{\lambda\nu}A_{\rho,\mu} - \Gamma^\rho_{\lambda\mu}A_{\rho,\nu}\\
&+ \Gamma^\rho_{\lambda\mu}\Gamma^\sigma_{\rho\nu}A_\sigma - \Gamma^\rho_{\nu\mu}A_{\lambda;\rho}.
\end{aligned} \tag{A.106}$$

The difference of them has the form

$$A_{\lambda;\mu;\nu} - A_{\lambda;\nu;\mu} = R^\rho_{\lambda\mu\nu}A_\rho - 2\Gamma^\rho_{[\mu\nu]}A_{\lambda;\rho} \tag{A.107}$$

with

$$R^{\rho}_{\lambda\mu\nu} \equiv \Gamma^{\rho}_{\lambda\nu,\mu} - \Gamma^{\rho}_{\lambda\mu,\nu} + \Gamma^{\rho}_{\sigma\mu}\Gamma^{\sigma}_{\lambda\nu} - \Gamma^{\rho}_{\sigma\nu}\Gamma^{\sigma}_{\lambda\mu}. \tag{A.108}$$

We call $R^{\rho}_{\lambda\mu\nu}$ the *Riemann curvature tensor*. $R^{\rho}_{\lambda\mu\nu}$ is a tensor because all other terms in Eq.(A.107) are tensors.

We have shown that the torsion tensor $\Gamma^{\rho}_{[\mu\nu]}$ is zero. Thus Eq.(A.107) becomes

$$A_{\lambda;\mu;\nu} - A_{\lambda;\nu;\mu} = R^{\rho}_{\lambda\mu\nu}A_{\rho}. \tag{A.109}$$

The Riemann curvature tensor can also be expressed as

$$R_{\rho\lambda\mu\nu} = g_{\rho\sigma}R^{\sigma}_{\lambda\mu\nu}. \tag{A.110}$$

For a flat spacetime with the Minkowski metric, $R^{\rho}_{\lambda\mu\nu} = 0$. If the Riemann curvature tensor is not zero, we have a curved spacetime. The Riemann curvature tensor has the following symmetry properties:

$$R^{\rho}_{\lambda\mu\nu} = -R^{\rho}_{\lambda\nu\mu}, \quad R_{\rho\lambda\mu\nu} = -R_{\rho\lambda\nu\mu}, \tag{A.111a}$$

$$R_{\rho\lambda\mu\nu} = -R_{\lambda\rho\nu\mu}, \tag{A.111b}$$

$$R_{\rho\lambda\mu\nu} = R_{\mu\nu\rho\lambda}. \tag{A.111c}$$

The symmetry relation Eq.(A.111a) can be obtained directly by exchanging the subscripts μ and ν in Eq.(A.108). The symmetry relations Eqs.(A.111b) and (A.111c) can be easily proved in the local flat frame. Since Eqs.(A.111b) and (A.111c) are tensor relations, they will be valid in any frame. In the local flat frame, $\Gamma^{\alpha}_{\mu\nu} = 0$. We have

$$\Gamma^{\alpha}_{\mu\nu,\sigma} = \frac{1}{2}g^{\alpha\beta}(g_{\beta\mu,\nu,\sigma} + g_{\beta\nu,\mu,\sigma} - g_{\mu\nu,\beta,\sigma}). \tag{A.112}$$

Inserting Eq.(A.112) into Eq.(A.108), we have

$$\begin{aligned} R^{\alpha}_{\beta\mu\nu} = &\frac{1}{2}g^{\alpha\sigma}(g_{\sigma\beta,\nu,\mu} + g_{\sigma\nu,\beta,\mu} - g_{\beta\nu,\sigma,\mu} - g_{\sigma\beta,\mu,\nu} \\ &- g_{\sigma\mu,\beta,\nu} + g_{\beta\mu,\sigma,\nu}). \end{aligned} \tag{A.113}$$

Since

$$g_{\alpha\beta,\mu,\nu} = g_{\alpha\beta,\nu,\mu}, \tag{A.114}$$

we have

$$R^{\alpha}_{\beta\mu\nu} = \frac{1}{2}g^{\alpha\sigma}(g_{\sigma\nu,\beta,\mu} - g_{\sigma\mu,\beta,\nu} + g_{\beta\mu,\sigma,\nu} - g_{\beta\nu,\sigma,\mu}) \tag{A.115}$$

or

$$\begin{aligned} R_{\alpha\beta\mu\nu} &= g_{\alpha\lambda}R^{\lambda}_{\beta\mu\nu} \\ &= \frac{1}{2}(g_{\alpha\nu,\beta,\mu} - g_{\alpha\mu,\beta,\nu} + g_{\beta\mu,\alpha,\nu} - g_{\beta\nu,\alpha,\mu}). \end{aligned} \tag{A.116}$$

Thus the symmetry relations Eqs.(A.111b) and (A.111c) can be easily obtained using Eq.(A.116).

There is another relation for the Riemann curvature tensor

$$R^\rho_{\lambda\mu\nu} + R^\rho_{\mu\nu\lambda} + R^\rho_{\nu\lambda\mu} = 0, \tag{A.117}$$

which can be easily verified using Eq.(A.108). Eq.(A.117) is called the *Ricci identity*.

A.10 Bianchi identities

In the following, we will prove an important derivative identity of the Riemann curvature tensor.

$$R^\rho_{\lambda\mu\nu;\sigma} + R^\rho_{\lambda\nu\sigma;\mu} + R^\rho_{\lambda\sigma\mu;\nu} = 0. \tag{A.118}$$

Eq.(A.118) is called the *Bianchi identity*. If a tensor equation is hold in one frame, it would be valid in any frame. Thus we only need to prove Bianchi identity in the local Minkowski metric. In the local Minkowski metric, $\Gamma^\rho_{\mu\nu} = 0$. We have

$$R^\rho_{\lambda\mu\nu;\sigma} = R^\rho_{\lambda\mu\nu,\sigma}. \tag{A.119}$$

Using Eq.(A.108), Eq.(A.119) becomes

$$\begin{aligned}
R^\rho_{\lambda\mu\nu;\sigma} &= (\Gamma^\rho_{\lambda\nu,\mu} - \Gamma^\rho_{\lambda\mu,\nu})_{,\sigma} + (\Gamma^\rho_{\kappa\mu}\Gamma^\kappa_{\lambda\nu} - \Gamma^\rho_{\kappa\nu}\Gamma^\kappa_{\lambda\mu})_{,\sigma} \\
&= (\Gamma^\rho_{\lambda\nu,\mu} - \Gamma^\rho_{\lambda\mu,\nu})_{,\sigma} \\
&= \Gamma^\rho_{\lambda\nu,\mu,\sigma} - \Gamma^\rho_{\lambda\mu,\nu,\sigma}.
\end{aligned} \tag{A.120}$$

Similarly, we can obtain

$$R^\rho_{\lambda\nu\sigma;\mu} = \Gamma^\rho_{\lambda\sigma,\nu,\mu} - \Gamma^\rho_{\lambda\nu,\sigma,\mu}, \tag{A.121a}$$

$$R^\rho_{\lambda\sigma\mu;\nu} = \Gamma^\rho_{\lambda\mu,\sigma,\nu} - \Gamma^\rho_{\lambda\sigma,\mu,\nu}. \tag{A.121b}$$

Adding Eqs.(A.121a), (A.121b) and (A.120), we obtain the Bianchi identity

$$R^\rho_{\lambda\mu\nu;\sigma} + R^\rho_{\lambda\nu\sigma;\mu} + R^\rho_{\lambda\sigma\mu;\nu} = 0. \tag{A.122}$$

It is a tensor equation and should be valid in any frame.

A.11   Ricci tensor

When we make contraction of the Riemann curvature tensor $R^{\mu}_{\ \alpha\nu\beta}$ on the first and third indices, we obtain a two rank tensor

$$R_{\alpha\beta} \equiv R^{\mu}_{\ \alpha\mu\beta} = R_{\beta\alpha}. \tag{A.123}$$

$R_{\alpha\beta}$ is called the *Ricci tensor*. Contractions on other indices either give zero or $\pm R_{\alpha\beta}$. Using $R_{\alpha\beta}$, we can define the *Ricci scalar*

$$R \equiv g^{\mu\nu} R_{\mu\nu} = g^{\mu\nu} g^{\alpha\beta} R_{\alpha\mu\beta\nu}, \tag{A.124}$$

R is also called the Ricci scalar curvature.

A.12   Einstein tensor

We contract the indices $\rho\sigma$ in the Bianchi identity Eq.(A.122) and obtain

$$R^{\sigma}_{\lambda\mu\nu;\sigma} - R_{\lambda\nu;\mu} + R_{\lambda\mu;\nu} = 0. \tag{A.125}$$

Multiplying $g^{\nu\lambda}$ and contracting, we have

$$R^{\sigma}_{\ \mu;\sigma} - R_{;\mu} + R^{\nu}_{\ \mu;\nu} = 0 \tag{A.126}$$

or

$$R^{\nu}_{\ \mu;\nu} - \frac{1}{2} R_{;\mu} = 0. \tag{A.127}$$

Eq.(A.127) can be rewritten as

$$(R^{\nu}_{\ \mu} - \frac{1}{2} \delta^{\nu}_{\mu} R)_{;\nu} = 0 \tag{A.128}$$

or

$$(R_{\mu\nu} - \frac{1}{2} g_{\mu\nu} R)^{;\nu} = 0, \tag{A.129a}$$

$$(R^{\mu\nu} - \frac{1}{2} g^{\mu\nu} R)_{;\nu} = 0. \tag{A.129b}$$

We define *Einstein tensor* as

$$G_{\mu\nu} \equiv R_{\mu\nu} - \frac{1}{2} g_{\mu\nu} R. \tag{A.130}$$

Eqs.(A.128) and (A.129) become

$$G^{\nu}_{\ \mu;\nu} = G_{\mu\nu}^{\ \ ;\nu} = G^{\mu\nu}_{\ \ ;\nu} = 0. \tag{A.131}$$

This property is used to show that the conservation of energy-momentum is fulfilled in the Einstein field equations.

Appendix B

Functional formula

A function of multi-variables can be expressed as an expansion form

$$F(\varphi_0, \varphi_1 \cdots, \varphi_n) = \sum_{m=0}^{\infty} \frac{1}{m!} \sum_{i_1} \cdots \sum_{i_m}$$
$$\left(\frac{\partial^m F}{\partial \varphi_{i_1} \cdots \partial \varphi_{i_m}} \right) \varphi_{i_1} \cdots \varphi_{i_m}. \tag{B.1}$$

Eq.(B.1) can also be considered as the definition of a function of multi-variables. Using the expansion form of Eq.(B.1), one can generalize the function of multi-variables to functionals. Let $\varphi(x)$ be a function of x. We can generalize Eq.(B.1) and express the functional $F[\varphi]$ in the expansion form

$$F[\varphi] = \sum_{m=0}^{\infty} \frac{1}{m!} \int dx_1 \cdots dx_m F^{(m)}(x_1, \cdots, x_m) \varphi(x_1) \cdots \varphi(x_m), \tag{B.2}$$

where $F^{(m)}$ is a symmetric function of its arguments.

We define the functional derivative by

$$\frac{\delta F[\varphi]}{\delta \varphi(x)} = \lim_{\epsilon \to 0} \frac{1}{\epsilon} \{ F[\varphi(y) + \epsilon \delta(x-y)] - F[\varphi(y)] \}. \tag{B.3}$$

The functional derivatives have the similar properties with the ordinary derivatives

$$\frac{\delta \varphi(y)}{\delta \varphi(x)} = \delta(x-y), \tag{B.4a}$$

$$\frac{\delta}{\delta \varphi(x)} (F_1[\varphi] + F_2[\varphi]) = \frac{\delta}{\delta \varphi(x)} F_1[\varphi] + \frac{\delta}{\delta \varphi(x)} F_2[\varphi], \tag{B.4b}$$

$$\frac{\delta}{\delta \varphi(x)} (F_1[\varphi] F_2[\varphi]) = F_1[\varphi] \frac{\delta}{\delta \varphi(x)} F_2[\varphi] + F_2[\varphi] \frac{\delta}{\delta \varphi(x)} F_1[\varphi]. \tag{B.4c}$$

According to Eq.(B.4), we have

$$\frac{\delta}{\delta\varphi(x)}F[\varphi] = \sum_{m=0}^{\infty} \frac{1}{m!} \int dx_1 \cdots dx_m F^{(m+1)}(x, x_1, \cdots, x_m)$$
$$\times \varphi(x_1) \cdots \varphi(x_m). \tag{B.5}$$

We can also define the functional integration as a limit of multi-variable integration.

$$\int D\varphi \Phi[\varphi] = \lim_{n\to\infty} \int d\varphi_1 \cdots d\varphi_n \Phi(\varphi_1, \cdots, \varphi_n). \tag{B.6}$$

Appendix C

Gaussian integrals

C.1 Gaussian integrals

The integral

$$I_0(a) = \int_0^\infty e^{-ax^2} dx \quad (a > 0) \tag{C.1}$$

is called the *Gaussian integral*. We can obtain the integral result in the following way. We consider $I_0^2(a)$,

$$
\begin{aligned}
I_0^2(a) &= \left(\frac{1}{2} \int_{-\infty}^\infty e^{-ax^2} dx \right) \left(\frac{1}{2} \int_{-\infty}^\infty e^{-ay^2} dy \right) \\
&= \frac{1}{4} \int\int dx dy e^{-a(x^2+y^2)}.
\end{aligned}
\tag{C.2}
$$

In the planar polar coordinates, we have $dx dy = r dr d\theta$ and $x^2 + y^2 = r^2$. Then Eq.(C.2) becomes

$$
\begin{aligned}
I_0^2(a) &= \frac{1}{4} \int_0^\infty r dr \int_0^{2\pi} d\theta e^{-ar^2} \\
&= \frac{\pi}{2} \int_0^\infty e^{-ar^2} r dr \\
&= \frac{\pi}{4a},
\end{aligned}
\tag{C.3}
$$

which gives

$$I_0(a) = \frac{1}{2}\sqrt{\frac{\pi}{a}}. \tag{C.4}$$

Generally, one can calculate the integrals

$$I_n(a) = \int_0^\infty e^{-ax^2} x^n dx \quad (a > 0). \tag{C.5}$$

When $n = 1$, we have

$$I_1(a) = \int_0^\infty e^{-ax^2} x \, dx = \frac{1}{2a} \int_0^\infty e^{-ax^2} d(ax^2) = \frac{1}{2a}. \qquad (C.6)$$

When $n > 1$, we differentiate Eq.(C.5) with respect to a.

$$\begin{aligned}
\frac{dI_n(a)}{da} &= \int_0^\infty e^{-ax^2} x^n (-x^2) dx \\
&= -\int_0^\infty e^{-ax^2} x^{n+2} dx \\
&= -I_{n+2}(a).
\end{aligned} \qquad (C.7)$$

Using Eq.(C.7), we have

$$I_n(a) = \int_0^\infty e^{-ax^2} x^n \, dx = \begin{cases} \dfrac{1 \times 2 \cdots (n-1)}{(2a)^{\frac{n}{2}}} \dfrac{1}{2} \left(\dfrac{\pi}{a}\right)^{\frac{1}{2}}, & n = 2, 4, \cdots \\ \dfrac{2 \times 4 \cdots (n-1)}{(2a)^{\frac{(n+1)}{2}}}, & n = 3, 5, \cdots. \end{cases} \qquad (C.8)$$

C.2 $\Gamma(n)$ functions

The Gaussian integrals are related to the $\Gamma(n)$ functions. The $\Gamma(n)$ functions are defined as

$$\Gamma(n) = \int_0^\infty e^{-x} x^{n-1} dx \quad (n > 0). \qquad (C.9)$$

Integrating by parts, we have

$$\begin{aligned}
\Gamma(n) &= -\int_0^\infty x^{n-1} d(e^{-x}) \\
&= -x^{n-1} d(e^{-x})\big|_0^\infty + \int_0^\infty e^{-x} dx^{n-1} \\
&= (n-1) \int_0^\infty e^{-x} x^{n-2} dx \\
&= (n-1)\Gamma(n-1).
\end{aligned} \qquad (C.10)$$

Now we consider some special cases.

(1)

$$\Gamma(1) = \int_0^\infty e^{-x} dx = 1. \qquad (C.11)$$

(2)

$$\Gamma\left(\frac{1}{2}\right) = \int_0^\infty e^{-x} x^{-\frac{1}{2}} dx$$

$$= 2 \int_0^\infty e^{-\sqrt{x}^2} d\sqrt{x} \tag{C.12}$$

$$= \sqrt{\pi}.$$

(3) For $n = 1, 2, 3, \cdots$,

$$\Gamma(n) = \int_0^\infty e^{-x} x^{n-1} dx = (n-1)!. \tag{C.13}$$

(4) For $n = \frac{1}{2}, \frac{3}{2}, \cdots \frac{2m+1}{2}, \cdots$,

$$\Gamma(n) = \Gamma\left(\frac{1}{2}\right) \cdot \frac{1}{2} \cdot \frac{3}{2} \cdots (n-1)$$

$$= \frac{1}{2} \cdot \frac{3}{2} \cdots (n-1)\sqrt{\pi}. \tag{C.14}$$

In terms of $\Gamma(n)$ functions, we can express the Gaussian integrals as

$$I_n(a) = \int_0^\infty e^{-ax^2} x^n dx$$

$$= \frac{1}{a^{\frac{n+1}{2}}} \int_0^\infty e^{-u^2} u^n du$$

$$= \frac{1}{2a^{\frac{n+1}{2}}} \int_0^\infty e^{-y} y^{\frac{n-1}{2}} dy \tag{C.15}$$

$$= \frac{\Gamma\left(\frac{n+1}{2}\right)}{2a^{\frac{n+1}{2}}}.$$

C.3 Gaussian integrations with source

Let us consider the Gaussian integral with source term.

$$\int_{-\infty}^\infty e^{-\frac{1}{2}ax^2 + Jx} dx = \int_{-\infty}^\infty e^{-\frac{1}{2}a\left(x - \frac{J}{a}\right)^2 + \frac{J^2}{2a}} dx$$

$$= \int_{-\infty}^\infty e^{-\frac{1}{2}ax'^2} e^{\frac{J^2}{2a}} dx' \tag{C.16}$$

$$= \left(\frac{2\pi}{a}\right)^{\frac{1}{2}} e^{\frac{J^2}{2a}}.$$

We can calculate the following integrals similarly.

$$\int_{-\infty}^\infty e^{-\frac{1}{2}ax^2 + iJx} dx = \left(\frac{2\pi}{a}\right)^{\frac{1}{2}} e^{-\frac{J^2}{2a}}. \tag{C.17}$$

$$\int_{-\infty}^{\infty} e^{\frac{1}{2}iax^2+iJx}dx = \left(\frac{2\pi i}{a}\right)^{\frac{1}{2}} e^{-i\frac{J^2}{2a}}. \tag{C.18}$$

$$\int_{-\infty}^{\infty}\int_{-\infty}^{\infty}\cdots\int_{-\infty}^{\infty} dx_1 dx_2 \cdots dx_N e^{-\frac{1}{2}x\cdot A\cdot x+J\cdot x}$$

$$= \left[\frac{(2\pi)^N}{\det(A)}\right]^{\frac{1}{2}} e^{\frac{1}{2}J\cdot A^{-1}\cdot J}. \tag{C.19}$$

For the functional Gaussian integrals, we have

$$\int D\varphi e^{-\int d^d x\left(\frac{1}{2}\varphi K\varphi - J\varphi\right)} = \mathcal{N}e^{\int d^d x\left(\frac{1}{2}J\cdot K^{-1}\cdot J\right)}, \tag{C.20}$$

$$\int D\varphi e^{i\int d^d x\left(\frac{1}{2}\varphi K\varphi + J\varphi\right)} = \mathcal{N}e^{i\int d^d x\left(-\frac{1}{2}J\cdot K^{-1}\cdot J\right)}, \tag{C.21}$$

where $\mathcal{N}$ is the normalized parameter related to the definition of functional integration. For complex φ with hermitian K, we have

$$\int D\varphi^\dagger D\varphi e^{-\int d^d x[\varphi^\dagger\cdot K\cdot\varphi + J^\dagger\cdot\varphi + \varphi^\dagger\cdot J]} = \mathcal{N}e^{\int d^d x[J^\dagger\cdot K^{-1}\cdot J]}. \tag{C.22}$$

Using the Taylor expansion

$$F[\varphi] = \sum_{m=0}^{\infty} \frac{1}{m!} \int dx_1 \cdots dx_m F^{(m)}(x_1, \cdots, x_m)\varphi(x_1)\cdots\varphi(x_m), \tag{C.23}$$

we have

$$\int D\varphi F[\varphi]e^{\int d^d x\left(-\frac{1}{2}\varphi\cdot K\cdot\varphi + J\cdot\varphi\right)}$$

$$= F\left[\frac{\delta}{\delta J}\right] \int D\varphi e^{\int d^d x\left(-\frac{1}{2}\varphi\cdot K\cdot\varphi + J\cdot\varphi\right)} \tag{C.24}$$

$$= F\left[\frac{\delta}{\delta J}\right] \mathcal{N}e^{\int d^d x\left(\frac{1}{2}J\cdot K^{-1}\cdot J\right)}.$$

Appendix D

Grassmann algebra

The fermion field operators obey the anti-commutation relations. Correspondingly, the fermion field functions can not be simply described by the ordinary number. The anti-commuting numbers have to be introduced. Since the anti-commuting numbers were first introduced by Hermann Grassmann, we call these numbers Grassmann variables.

A Grassmann algebra (also called exterior algebra) is an algebra constructed from a set of generators θ_i obeying the anti-commutation relation

$$\{\theta_i, \theta_j\} = 0. \tag{D.1}$$

The index i of θ_i runs from 1 to n. n is called the dimension of the algebra. Later we will generalize θ_i to $\theta(x)$ for an infinite dimension.

From Eq.(D.1), we have

$$\theta_i^2 = 0. \tag{D.2}$$

Thus the square and all higher powers of a generator vanish. When we expand an element of the Grassmann algebra, we have only finite sum of the following terms due to Eq.(D.2).

$$f(\theta_i) = f^{(0)} + \sum_i f_i^{(1)} \theta_i + \sum_{i_1 < i_2} f_{i_1,i_2}^{(2)} \theta_{i_1} \theta_{i_2} + \cdots + f^{(n)} \theta_{i_1} \theta_{i_2} \cdots \theta_{i_n}, \tag{D.3}$$

where the coefficients $f^{(i)}$ are ordinary numbers. We define differentiation with respect to the generators by

$$\frac{d\theta_i}{d\theta_j} = \delta_{ij} \tag{D.4}$$

and

$$\frac{d}{d\theta_i} \theta_{i_1} \theta_{i_2} \cdots \theta_{i_m} = \delta_{ii_1} \theta_{i_2} \cdots \theta_{i_m} - \delta_{ii_2} \theta_{i_1} \theta_{i_3} \cdots \theta_{i_m}$$
$$+ \cdots + (-1)^{m-1} \delta_{ii_m} \theta_{i_1} \theta_{i_2} \cdots \theta_{i_{m-1}}. \tag{D.5}$$

The minus sign comes from the anti-commutation relation when the factor θ_{i_k} is anti-commuted to the left so that the derivative operator can be applied directly. From Eq.(D.4), we have

$$\left\{ \frac{d}{d\theta_i}, \theta_j \right\} = \delta_{ij}, \tag{D.6a}$$

$$\left\{ \frac{d}{d\theta_i}, \frac{d}{d\theta_j} \right\} = 0. \tag{D.6b}$$

All the higher derivatives with respect to the same generator θ_i are zero.

The integration of the generators of Grassmann algebra is defined by

$$\int d\theta_i = 0, \tag{D.7a}$$

$$\int d\theta_i \theta_i = 1. \tag{D.7b}$$

The definition Eq.(D.7) is made to guarantee the translation invariance of the integration, which is an important property of the ordinary integration. For any function $f(\theta)$, its expanded form is

$$f = f_1 + f_2\theta. \tag{D.8}$$

Then

$$\begin{aligned}
\int d\theta f(\theta + \eta) &= \int d\theta[f_1 + f_2(\theta + \eta)] \\
&= \int d\theta(f_1 + f_2\theta) + \int d\theta f_2\eta \\
&= \int d\theta f(\theta).
\end{aligned} \tag{D.9}$$

Comparing the definitions of the differentiation and integration (Eqs.(D.4) and (D.7)), we can see that the operators of differentiation and integration for Grassmann variables are the same. Thus the differentials $d\theta_i$ obey the same anti-commutation relations as $\frac{d}{d\theta_i}$.

$$\{d\theta_i, d\theta_j\} = 0, \tag{D.10a}$$

$$\{d\theta_i, \theta_j\} = \delta_{ij}. \tag{D.10b}$$

Now we consider the variable transformations in the integral involving Grassmann algebra. For a linear transformation in one-dimension such as

$$\theta' = \eta + a\theta, \tag{D.11}$$

where η is an anti-commuting number and a is an ordinary number, we have

$$\int d\theta f(\theta) = \int d\theta' f(\theta')$$

$$= \int d\theta' f(a\theta + \eta)$$

$$= a \int d\theta' f(\theta) \tag{D.12}$$

$$= \int d\theta' \left(\frac{d\theta}{d\theta'}\right)^{-1} f(\theta).$$

It should be noted that Eq.(D.12) is different from the transformation formula for ordinary integration

$$\int dx f(x) = \int dx' \frac{dx}{dx'} f(x). \tag{D.13}$$

The Grassmann integral exhibits the opposite behavior as compared to the ordinary integral. In general, under a linear transformation for an n-dimensional Grassmann algebra

$$\theta'_i = \sum_j^n a_{ij}\theta_j + \eta_i, \tag{D.14}$$

we have

$$\int d\theta_n \cdots \theta_1 f(\theta) = \int d\theta'_n \cdots \theta'_1 \left[\det\left(\frac{d\theta}{d\theta'}\right)\right]^{-1} f(\theta). \tag{D.15}$$

We have a factor of the inverse of the Jacobian determinant instead of the Jacobian determinant.

Now we evaluate the Gaussian integrals for Grassmann algebra. Since the Dirac fermion field is complex, we introduce the complex Grassmann variables. θ_i and θ_i^* obeying the anti-commutation relations

$$\{\theta_i, \theta_j\} = \{\theta_i^*, \theta_j^*\} = \{\theta_i, \theta_j^*\} = 0, \tag{D.16}$$

where θ_i^* is the conjugate generator defined by

$$\left(\theta_i\right)^* = \theta_i^*, \tag{D.17a}$$

$$\left(\theta_i^*\right)^* = \theta_i, \tag{D.17b}$$

$$\left(\theta_{i_1}\theta_{i_2}\cdots\theta_{i_n}\right)^* = \theta_{i_n}^* \cdots \theta_{i_2}^* \theta_{i_1}^*, \tag{D.17c}$$

$$\left(\lambda\theta_i\right)^* = \lambda^* \theta_i^*. \tag{D.17d}$$

For one dimensional case, we have

$$\int d\theta d\theta^* e^{\theta^* a \theta} = \int d\theta d\theta^* (1 + a\theta^* \theta)$$

$$= a \tag{D.18}$$

$$= e^{+\ln a}.$$

In general, we have

$$\int d\theta_1 \cdots \theta_n d\theta_1^* \cdots \theta_n^* e^{\theta^\dagger A \theta} = \det(A) \tag{D.19}$$

and

$$\int d\theta_1 \cdots \theta_n d\theta_1^* \cdots \theta_n^* e^{(\theta^\dagger A \theta + \theta^\dagger \rho + \rho^\dagger \theta)}$$

$$= \det(A) \exp(-\rho^\dagger A^{-1} \rho). \tag{D.20}$$

Generalization from the variable to the continuum limit $\theta_i \rightarrow \theta(x)$ for applications in anti-commuting fields is straightforward. The anti-commutation relation of the variable $\theta(x)$ is given by

$$\{\theta(x), \theta(y)\} = 0. \tag{D.21}$$

The ordinary differentiations in Eqs.(D.4) and (D.6) are replaced by the functional derivative.

$$\frac{\delta \theta(x)}{\delta \theta(y)} = \delta^4(x - y) \tag{D.22}$$

and

$$\left\{ \frac{\delta}{\delta \theta(x)}, \theta(y) \right\} = \delta^4(x - y), \tag{D.23a}$$

$$\left\{ \frac{\delta}{\delta \theta(x)}, \frac{\delta}{\delta \theta(y)} \right\} = 0. \tag{D.23b}$$

A functional of $\theta(x)$ can be expanded like

$$f(\theta) = f^{(0)} + \int dx_1 f^{(1)}(x_1)\theta(x_1) + \cdots$$

$$+ \int dx_1 \cdots dx_n f^{(n)}(x_1, \cdots, x_n)\theta(x_1) \cdots \theta(x_n). \tag{D.24}$$

The integration rules for continuous Grassmann variables are

$$\int d\theta(x) \, 1 = 0, \tag{D.25a}$$

$$\int d\theta(x)\theta(x) = 1. \tag{D.25b}$$

The Gaussian integral over fermion fields is given by

$$
\int D\psi \int D\bar{\psi} \exp\Big\{ \int d^4x' d^4x\, \bar{\psi}(x') A(x',x) \psi(x)
$$

$$
+ \int d^4x\, [\bar{\psi}(x)\rho(x) + \bar{\rho}(x)\psi(x)] \Big\} \tag{D.26}
$$

$$
= \det(A) \exp\Big[- \int d^4x' d^4x\, \bar{\rho}(x') A^{-1}(x',x) \rho(x) \Big].
$$

Appendix E

Euclidean representation

The flat physical spacetime is the Minkowski spacetime. The Wick rotation makes the calculations easier because we can transform the calculations in the Minkowski spaceitme into those in the four-dimensional Euclidean space. We denote a space point in the Euclidean space by $x_E = (\mathbf{x}, x_4)$. The four-dimensional Euclidean space is obtained from the Minkowski spacetime by the transformation

$$x_i \to x_i, \quad ix_0 \to x_4. \tag{E.1}$$

Under the transformation Eq.(E.1) and the Wick rotation $t' \to -it$, x_4 becomes real. The calculations can then be performed in real Euclidean space. The transformation of volume element is given by

$$d^4 x_E = d^3 x dx_4 = d^3 x i dt = i d^4 x. \tag{E.2}$$

The distance transforms as

$$(dx_E)^2 = \sum_{i=1}^{3} (dx_i)^2 + (dx_4)^2 = -(dx)^2. \tag{E.3}$$

The kinetic term for a scalar field is given by

$$\partial_\mu \phi \partial^\mu \phi = \partial_0 \phi \partial^0 \phi + \partial_i \phi \partial^i \phi = -(\partial_0 \phi)^2 - (\nabla \phi)^2 = -(\partial_E \phi)^2. \tag{E.4}$$

The d'Alembert Operator is given by

$$\Box = \frac{\partial^2}{\partial t^2} - \nabla^2 = -\frac{\partial^2}{\partial x_4{}^2} - \nabla^2 = -(\partial_E)^2 = -\Box_E. \tag{E.5}$$

The generating functional for a free scalar field in the Euclidean representation has the form

$$W_E^0[J] = \int D\phi e^{-\int d^4 x_E \left\{ \frac{1}{2} [(\partial_E \phi)^2 + m^2 \phi^2] + J\phi \right\}}. \tag{E.6}$$

We can also define the Euclidean momentum space by the following transformation relating the momentum k in the Minkowski spacetime to the k_E in the Euclidean space.

$$k_E = (\mathbf{k}, k_4) \quad \text{with} \quad k_4 = -ik_0. \tag{E.7}$$

The volume element in the momentum space is given by

$$d^4 k_E = d^3 k dk_4 = -d^3 k i dk_0 = -i d^4 k. \tag{E.8}$$

and the distance in momentum space has the form

$$(dk_E)^2 = \sum_{i=1}^{3} (dk_i)^2 + (dk_4)^2 = -(dk)^2. \tag{E.9}$$

For the factor $k \cdot x$, we have

$$k \cdot x = k_\mu x^\mu = k_0 x^0 - \mathbf{k} \cdot \mathbf{x} = k_4 x^4 - \mathbf{k} \cdot \mathbf{x}, \tag{E.10}$$

which is not equal to $k_E \cdot x_E$. However we always have Eq.(E.10) in the integration over $d^3 k$. We can change $\mathbf{k} \to -\mathbf{k}$ and then replace $k \cdot x$ by $k_E \cdot x_E$. As an example, the Feynman propagator in the Euclidean representation has the form

$$
\begin{aligned}
\Delta_F(x) &= -i \int \frac{d^4 k_E}{(2\pi)^4} \frac{e^{-ik_E \cdot x_E}}{k_E^2 + m^2} \\
&= -i \int \frac{d^4 k_E}{(2\pi)^4} \frac{e^{-ik_E \cdot x_E}}{\mathbf{k}^2 + k_4^2 + m^2}.
\end{aligned}
\tag{E.11}
$$

Since k_4 is real, the integration in Eq.(E.11) contains no poles on its integration path and is thus well defined.

Appendix F

Some useful formulas

(1)

$$(\boldsymbol{\sigma} \cdot \mathbf{A})(\boldsymbol{\sigma} \cdot \mathbf{B}) = \mathbf{A} \cdot \mathbf{B} + i\boldsymbol{\sigma} \cdot (\mathbf{A} \times \mathbf{B}). \tag{F.1}$$

To prove Eq.(F.1), we use the commutation relations for σ_i

$$\sigma_i \sigma_j = i\epsilon^{ijk}\sigma_k + \delta_{ij}, \tag{F.2}$$

where ϵ^{ijk} is the antisymmetric Levi-Civita symbol.

$$\epsilon^{ijk} = \begin{cases} 1 & \text{even permutation of 1,2,3} \\ -1 & \text{odd permutation of 1,2,3} \\ 0 & \text{otherwise.} \end{cases} \tag{F.3}$$

From Eq.(F.2), we can easily obtain

$$\sigma_i \sigma_j - \sigma_j \sigma_i = 2i\epsilon^{ijk}\sigma_k, \tag{F.4a}$$

$$\sigma_i \sigma_j + \sigma_j \sigma_i = 2\delta_{ij}. \tag{F.4b}$$

Using the above relations, we have

$$(\boldsymbol{\sigma} \cdot \mathbf{A})(\boldsymbol{\sigma} \cdot \mathbf{B}) = (\sum_{i=1}^{3} \sigma_i A_i)(\sum_{j=1}^{3} \sigma_j B_j)$$

$$= \sum_{i=1}^{3}\sum_{j=1}^{3} A_i B_j (i\epsilon^{ijk}\sigma_k + \delta_{ij}) \tag{F.5}$$

$$= \sum_{i=1}^{3}\sum_{j=1}^{3} A_i B_j (i\epsilon^{ijk}\sigma_k) + \mathbf{A} \cdot \mathbf{B}.$$

Using the relation

$$\sum_{ijk} \epsilon^{ijk} A_i B_j \sigma_k = \sum_k (\mathbf{A} \times \mathbf{B})_k \sigma_k = \boldsymbol{\sigma} \cdot (\mathbf{A} \times \mathbf{B}), \tag{F.6}$$

we have

$$(\boldsymbol{\sigma} \cdot \mathbf{A})(\boldsymbol{\sigma} \cdot \mathbf{B}) = \mathbf{A} \cdot \mathbf{B} + i\boldsymbol{\sigma} \cdot (\mathbf{A} \times \mathbf{B}). \tag{F.7}$$

(2) The trace relations of the Gamma matrices γ^μ ($\mu = 0, 1, 2, 3$).

$$\text{Tr}(\gamma^\mu) = 0, \tag{F.8a}$$

$$\text{Tr}(\gamma^{\mu_1} \gamma^{\mu_2} \cdots \gamma^{\mu_n}) = 0 \quad \text{(n is an odd number)}, \tag{F.8b}$$

$$\text{Tr}(\gamma^\alpha \gamma^\beta) = 4\eta^{\alpha\beta}, \tag{F.8c}$$

$$\text{Tr}(\gamma^\alpha \gamma^\beta \gamma^\mu \gamma^\nu) = 4(\eta^{\alpha\beta}\eta^{\mu\nu} - \eta^{\alpha\mu}\eta^{\beta\nu} + \eta^{\alpha\nu}\eta^{\beta\mu}), \tag{F.8d}$$

$$\text{Tr}(\gamma^5) = 0, \tag{F.8e}$$

$$\text{Tr}(\gamma^5 \gamma^\alpha \gamma^\beta) = 0, \tag{F.8f}$$

$$\text{Tr}(\gamma^5 \gamma^\alpha \gamma^\beta \gamma^\mu \gamma^\nu) = -4i\epsilon^{\alpha\beta\mu\nu}, \tag{F.8g}$$

where $\epsilon^{\alpha\beta\mu\nu}$ is the totally antisymmetric symbol with $\epsilon^{0123} = 1$.

(a) Using

$$\gamma^\alpha \gamma^\beta + \gamma^\beta \gamma^\alpha = 2\eta^{\alpha\beta}, \tag{F.9}$$

we have

$$(\gamma^0)^2 = 1, \tag{F.10a}$$

$$(\gamma^i)^2 = -1 \tag{F.10b}$$

and

$$\gamma^0 = \gamma^i \gamma^0 \gamma^i, \tag{F.11a}$$

$$\gamma^i = -\gamma^0 \gamma^i \gamma^0. \tag{F.11b}$$

Taking the trace on both sides of Eq.(F.11), we obtain

$$\text{Tr}\gamma^0 = \text{Tr}(\gamma^i \gamma^0 \gamma^i) = -\text{Tr}\gamma^0, \tag{F.12a}$$

$$\text{Tr}\gamma^i = -\text{Tr}(\gamma^0 \gamma^i \gamma^0) = -\text{Tr}\gamma^i, \tag{F.12b}$$

which leads to

$$\text{Tr}\gamma^\mu = 0. \tag{F.13}$$

(b) Using Eq.(F.9), Eq.(F.12) and Eq.(F.13), it is easy to see that Eq.(F.8b) holds.

(c) Taking the trace on both sides of Eq.(F.9), we obtain

$$\text{Tr}(\gamma^\alpha \gamma^\beta) + \text{Tr}(\gamma^\beta \gamma^\alpha) = 8\eta^{\alpha\beta}. \tag{F.14}$$

Since $\text{Tr}(\gamma^\alpha \gamma^\beta) = \text{Tr}(\gamma^\beta \gamma^\alpha)$, we find

$$\text{Tr}(\gamma^\alpha \gamma^\beta) = 4\eta^{\alpha\beta}. \tag{F.15}$$

(d) Using Eqs.(F.9) and (F.15), we have

$$\mathrm{Tr}(\gamma^\alpha\gamma^\beta\gamma^\mu\gamma^\nu)$$
$$= -\mathrm{Tr}(\gamma^\beta\gamma^\alpha\gamma^\mu\gamma^\nu) + 8\eta^{\alpha\beta}\eta^{\mu\nu}$$
$$= -\mathrm{Tr}(\gamma^\alpha\gamma^\mu\gamma^\nu\gamma^\beta) + 8\eta^{\alpha\beta}\eta^{\mu\nu} \qquad (\mathrm{F.16})$$
$$= \mathrm{Tr}(\gamma^\alpha\gamma^\mu\gamma^\beta\gamma^\nu) + 8\eta^{\alpha\beta}\eta^{\mu\nu} - 8\eta^{\alpha\mu}\eta^{\beta\nu}$$
$$= -\mathrm{Tr}(\gamma^\alpha\gamma^\beta\gamma^\mu\gamma^\nu) + 8\eta^{\alpha\beta}\eta^{\mu\nu} - 8\eta^{\alpha\mu}\eta^{\beta\nu} + 8\eta^{\alpha\nu}\eta^{\beta\mu},$$

which gives

$$\mathrm{Tr}(\gamma^\alpha\gamma^\beta\gamma^\mu\gamma^\nu) = 4(\eta^{\alpha\beta}\eta^{\mu\nu} - \eta^{\alpha\mu}\eta^{\beta\nu} + \eta^{\alpha\nu}\eta^{\beta\mu}). \qquad (\mathrm{F.17})$$

(e) Since

$$\mathrm{Tr}(\gamma^5) = \mathrm{Tr}(i\gamma^0\gamma^1\gamma^2\gamma^3) = -\mathrm{Tr}(i\gamma^0\gamma^1\gamma^2\gamma^3) = -\mathrm{Tr}(\gamma^5), \qquad (\mathrm{F.18})$$

we have

$$\mathrm{Tr}(\gamma^5) = 0. \qquad (\mathrm{F.19})$$

(f) When $\alpha = \beta$, using Eqs.(F.10) and (F.19), we obtain Eq.(F.8f). When $\alpha \neq \beta$, using

$$\gamma^5 = i\gamma^0\gamma^1\gamma^2\gamma^3, \qquad (\mathrm{F.20})$$

and Eqs.(F.8c),(F.9) and (F.10), we obtain

$$\mathrm{Tr}(\gamma^5\gamma^\alpha\gamma^\beta) = 0. \qquad (\mathrm{F.21})$$

(g) When two of α, β, μ, ν are equal, using Eq.(F.21), we have

$$\mathrm{Tr}(\gamma^5\gamma^\alpha\gamma^\beta\gamma^\mu\gamma^\nu) = 0. \qquad (\mathrm{F.22})$$

When $\alpha \neq \beta \neq \mu \neq \nu$, exchanging the order of γ in $\mathrm{Tr}(\gamma^5\gamma^\alpha\gamma^\beta\gamma^\mu\gamma^\nu)$ only contributes a factor of -1. Then we have

$$\mathrm{Tr}(\gamma^5\gamma^\alpha\gamma^\beta\gamma^\mu\gamma^\nu) = \epsilon^{\alpha\beta\mu\nu}\mathrm{Tr}(\gamma^5\gamma^0\gamma^1\gamma^2\gamma^3). \qquad (\mathrm{F.23})$$

Using Eq.(F.20), we have

$$\mathrm{Tr}(\gamma^5\gamma^0\gamma^1\gamma^2\gamma^3) = i\mathrm{Tr}(\gamma^0\gamma^1\gamma^2\gamma^3\gamma^0\gamma^1\gamma^2\gamma^3) = -4i. \qquad (\mathrm{F.24})$$

Thus we have

$$\mathrm{Tr}(\gamma^5\gamma^\alpha\gamma^\beta\gamma^\mu\gamma^\nu) = -4i\epsilon^{\alpha\beta\mu\nu}. \qquad (\mathrm{F.25})$$

Similarly we can prove a more generalized form of Eq.(F.8)

$$\mathrm{Tr}(\not{a}\not{b}) = 4a \cdot b, \qquad (\mathrm{F.26a})$$
$$\mathrm{Tr}(\not{a}\not{b}\not{c}\not{d}) = 4(a \cdot b)(c \cdot d) - (a \cdot c)(b \cdot d) + (a \cdot d)(b \cdot c), \qquad (\mathrm{F.26b})$$
$$\mathrm{Tr}(\gamma^5\not{a}\not{b}) = 0, \qquad (\mathrm{F.26c})$$
$$\mathrm{Tr}(\gamma^5\not{a}\not{b}\not{c}\not{d}) = -4i\epsilon^{\alpha\beta\mu\nu}a_\alpha b_\beta c_\mu d_\nu. \qquad (\mathrm{F.26d})$$

For example,

$$\begin{aligned}
\text{Tr}(\not{a}\,\not{b}\,) &= \text{Tr}(\gamma \cdot a)(\gamma \cdot b) \\
&= \text{Tr}(\gamma \cdot b)(\gamma \cdot a) \\
&= \frac{1}{2}\text{Tr}(a_\mu b_\nu \{\gamma^\mu, \gamma^\nu\}) \\
&= 4a \cdot b.
\end{aligned} \tag{F.27}$$

(3) d-dimensional integral in polar coordinate form

$$\int d^d k F(k^2) = \frac{2\pi^{\frac{d}{2}}}{\Gamma\left(\frac{d}{2}\right)} \int_0^\infty dk k^{d-1} F(k^2), \tag{F.28}$$

where $k^2 = k_1^2 + k_2^2 + \cdots + k_d^2$.

Eq.(F.28) can be derived in the following way. We first evaluate the integral

$$I_d = \int d^d k e^{-\frac{1}{2}k^2}. \tag{F.29}$$

Eq.(F.30) is a Gaussian integral. Thus we have

$$I_d = (\sqrt{2\pi})^d. \tag{F.30}$$

On the other hand, I_d can be expressed as

$$\begin{aligned}
I_d &= C(d) \int_0^\infty dk k^{d-1} e^{-\frac{1}{2}k^2} \\
&= C(d) 2^{\frac{d}{2}-1} \int_0^\infty dx x^{\frac{d}{2}-1} e^{-x} \\
&= C(d) 2^{\frac{d}{2}-1} \Gamma\left(\frac{d}{2}\right).
\end{aligned} \tag{F.31}$$

Comparing Eq.(F.31) with Eq.(F.30), we have

$$C(d) = \frac{2\pi^{\frac{d}{2}}}{\Gamma\left(\frac{d}{2}\right)}. \tag{F.32}$$

$C(d)$ is the factor before the integral on the right hand side of Eq.(F.28), which proves Eq.(F.28).

(4) The volume and surface area of d-dimensional spheres.

Using Eq.(F.28), we can obtain the volume V_d of a d-dimensional sphere with the radius of R.

$$\begin{aligned}
V_d &= \int_{\sum x^2 < R^2} d^d x \\
&= \frac{2\pi^{\frac{d}{2}}}{\Gamma\left(\frac{d}{2}\right)} \int_0^R dr r^{d-1} \\
&= \frac{\pi^{\frac{d}{2}}}{\frac{d}{2}\Gamma\left(\frac{d}{2}\right)} R^d.
\end{aligned} \tag{F.33}$$

The corresponding sphere surface area $S_d(R)$ in d-dimension is

$$S_d(R) = \frac{dV_d}{dR} = \frac{2\pi^{\frac{d}{2}}}{\Gamma\left(\frac{d}{2}\right)} R^{d-1}. \tag{F.34}$$

(5)

$$\int \frac{d^d p}{(p^2 + 2pq - m^2)^\alpha} = (-1)^{-\alpha} i \pi^{\frac{d}{2}} \frac{\Gamma\left(\alpha - \frac{d}{2}\right)}{\Gamma(\alpha)} \frac{1}{(q^2 + m^2)^{\alpha - \frac{d}{2}}}, \tag{F.35}$$

where the integration is over a d-dimensional spacetime.

We use $I_d(q)$ to denote the integral in Eq.(F.35). Using Eq.(F.28), we have

$$I_d(q) = \frac{2\pi^{\frac{d-1}{2}}}{\Gamma\left(\frac{d-1}{2}\right)} \int_{-\infty}^{\infty} dp_0 \int_0^{\infty} \frac{r^{d-2} dr}{(p_0^2 - r^2 + 2pq - m^2)^\alpha}. \tag{F.36}$$

Since the integral is Lorentz invariant, we perform the integration in the reference frame $q_\mu = (\mu, 0)$. In this reference frame, $2p \cdot q = 2\mu p_0$. Introducing a new variable $p'_\mu = p_\mu + q_\mu$ and substituting for p_μ, we have

$$\begin{aligned}
I_d(q) &= \frac{2\pi^{\frac{d-1}{2}}}{\Gamma\left(\frac{d-1}{2}\right)} \int_{-\infty}^{\infty} dp'_0 \int_0^{\infty} \frac{r^{d-2} dr}{[(p'_0 - q_0)^2 - r^2 + 2\mu p_0 - m^2]^\alpha} \\
&= \frac{2\pi^{\frac{d-1}{2}}}{\Gamma\left(\frac{d-1}{2}\right)} \int_{-\infty}^{\infty} dp'_0 \int_0^{\infty} \frac{r^{d-2} dr}{[p_0'^2 - r^2 - (q^2 + m^2)]^\alpha}.
\end{aligned} \tag{F.37}$$

We rotate the Minkowski space to the Euclidean space.

$$I_d(q) = (-1)^{-\alpha} i \frac{2\pi^{\frac{d-1}{2}}}{\Gamma\left(\frac{d-1}{2}\right)} \int_{-\infty}^{\infty} dp'_0 \int_0^{\infty} \frac{r^{d-2} dr}{[p_0'^2 + r^2 + (q^2 + m^2)]^\alpha}. \tag{F.38}$$

We use the Euler-Beta function

$$B(x, y) = \frac{\Gamma(x)\Gamma(y)}{\Gamma(x + y)} = 2 \int_0^{\infty} dt\, t^{2x-1} (1 + t^2)^{-x-y} \tag{F.39}$$

to reformulate the integral. Eq.(F.39) is valid for Re $x > 0$ and Re $y > 0$. Introducing new variables

$$x = \frac{1 + \beta}{2}, \quad y = \alpha - \frac{1 + \beta}{2}, \quad t = \frac{s}{M}, \tag{F.40}$$

we have

$$\int_0^{\infty} ds \frac{s^\beta}{(s^2 + M^2)^\alpha} = \frac{\Gamma\left(\frac{1+\beta}{2}\right) \Gamma\left(\alpha - \frac{1+\beta}{2}\right)}{2(M^2)^{\alpha - \frac{1+\beta}{2}} \Gamma(\alpha)}. \tag{F.41}$$

Introducing $M^2 = q^2 + m^2 + p_0'^2$ and using Eq.(F.41), we obtain

$$I_d(q) = (-1)^{-\alpha} i \frac{2\pi^{\frac{d-1}{2}}}{\Gamma\left(\frac{d-1}{2}\right)} \frac{\Gamma\left(\frac{d-1}{2}\right) \Gamma\left(\alpha - \frac{d-1}{2}\right)}{2\Gamma(\alpha)} \int_{-\infty}^{\infty} \frac{dp_0'}{(M^2)^{\alpha - \frac{d-1}{2}}}$$

$$= (-1)^{-\alpha} i \pi^{\frac{d-1}{2}} \frac{\Gamma\left(\alpha - \frac{d-1}{2}\right)}{\Gamma(\alpha)} \int_{-\infty}^{\infty} \frac{dp_0'}{(q^2 + m^2 + p_0'^2)^{\alpha - \frac{d-1}{2}}}.$$

(F.42)

Using Eq.(F.41) one more time gives

$$I_d(q) = (-1)^{-\alpha} i \pi^{\frac{d}{2}} \frac{\Gamma\left(\alpha - \frac{d}{2}\right)}{\Gamma(\alpha)} \frac{1}{(q^2 + m^2)^{\alpha - \frac{d}{2}}},$$

(F.43)

which is Eq.(F.35).

(6)

$$\int d^d p \frac{p^2}{(p^2 + 2pq - m^2)^\alpha} = -\frac{i\pi^{\frac{d}{2}}}{\Gamma(\alpha)} \frac{(-1)^{-\alpha}}{(q^2 + m^2)^{\alpha - \frac{d}{2}}} \left[-q^2 \Gamma\left(\alpha - \frac{d}{2}\right) \right.$$
$$\left. + \frac{d}{2}(q^2 + m^2)\Gamma\left(\alpha - 1 - \frac{d}{2}\right) \right],$$

(F.44)

where the integration is over a d-dimensional spacetime.

Differentiating Eq.(F.43) with respect to q^μ gives

$$-\alpha \int d^d p \frac{2p^\mu}{(p^2 + 2pq - m^2)^{\alpha+1}}$$
$$= (-1)^{-\alpha} i \pi^{\frac{d}{2}} \frac{\Gamma\left(\alpha - \frac{d}{2}\right)}{\Gamma(\alpha)} \left(-\alpha + \frac{d}{2}\right) \frac{2q^\mu}{(q^2 + m^2)^{\alpha - \frac{d}{2}+1}}.$$

(F.45)

Making a variable transformation $\alpha + 1 \to \alpha$ and using $\beta\Gamma(\beta) = \Gamma(\beta + 1)$ yields

$$\int \frac{d^d p\, p^\mu}{(p^2 + 2pq - m^2)^\alpha} = (-1)^{-\alpha} i \pi^{\frac{d}{2}} \frac{\Gamma\left(\alpha - \frac{d}{2}\right)}{\Gamma(\alpha)} \frac{q^\mu}{(q^2 + m^2)^{\alpha - \frac{d}{2}}}.$$

(F.46)

Differentiating Eq.(F.43) again with respect to q^ν gives

$$\int d^d p \frac{p^\mu p^\nu}{(p^2 + 2pq - m^2)^\alpha} = -\frac{i\pi^{\frac{d}{2}}}{\Gamma(\alpha)} \frac{(-1)^{-\alpha}}{(q^2 + m^2)^{\alpha - \frac{d}{2}}} \left[-q^\mu q^\nu \Gamma\left(\alpha - \frac{d}{2}\right) \right.$$
$$\left. + \frac{1}{2}\eta^{\mu\nu}(q^2 + m^2)\Gamma\left(\alpha - 1 - \frac{d}{2}\right) \right].$$

(F.47)

Contracting μ with ν, we obtain Eq.(F.44).

Appendix G

Jacobian

We consider functions of two variables: $u(x, y)$ and $v(x, y)$. The Jacobian determinant is defined by

$$
\frac{\partial(u, v)}{\partial(x, y)} \equiv \begin{vmatrix} \left(\dfrac{\partial u}{\partial x}\right)_y & \left(\dfrac{\partial u}{\partial y}\right)_x \\ \left(\dfrac{\partial v}{\partial x}\right)_y & \left(\dfrac{\partial v}{\partial y}\right)_x \end{vmatrix} \tag{G.1}
$$
$$
= \left(\frac{\partial u}{\partial x}\right)_y \left(\frac{\partial v}{\partial y}\right)_x - \left(\frac{\partial u}{\partial y}\right)_x \left(\frac{\partial v}{\partial x}\right)_y.
$$

There are several relations for Jacobian that are useful in the calculations of the thermodynamic derivatives. When $f = f(u, v)$ and $g = g(u, v)$ are two functions of u and v, we can prove the following chain rule.

$$
\frac{\partial(f, g)}{\partial(x, y)} = \frac{\partial(f, g)\partial(u, v)}{\partial(u, v)\partial(x, y)}. \tag{G.2}
$$

Interchanging two columns of the determinant, we have

$$
\frac{\partial(u, v)}{\partial(x, y)} = -\frac{\partial(v, u)}{\partial(x, y)}. \tag{G.3}
$$

Setting $v = y$, Eq.(G.1) becomes

$$
\frac{\partial(u, y)}{\partial(x, y)} = \left(\frac{\partial u}{\partial x}\right)_y. \tag{G.4}
$$

Setting $f = x$ and $g = y$ in Eq.(G.2), we have

$$
\frac{\partial(x, y)\partial(u, v)}{\partial(u, v)\partial(x, y)} = 1. \tag{G.5}
$$

Using Eq.(G.4) and the chain rule, we obtain

$$
\begin{aligned}
\left(\frac{\partial u}{\partial x}\right)_y &= \frac{\partial(u,y)}{\partial(x,y)} \\
&= \frac{\partial(u,y)\partial(u,x)}{\partial(u,x)\partial(x,y)} \\
&= -\frac{\left(\frac{\partial u}{\partial y}\right)_x}{\left(\frac{\partial x}{\partial y}\right)_u},
\end{aligned}
\tag{G.6}
$$

which can be rewritten as

$$
\left(\frac{\partial u}{\partial x}\right)_y \left(\frac{\partial x}{\partial y}\right)_u \left(\frac{\partial y}{\partial u}\right)_x = -1.
\tag{G.7}
$$

Appendix H

Geodesic equation

In the Euclidean space or Minkowski spacetime, a line without changing its direction is a straight line. The shortest line in the Euclidean space or Minkowski spacetime is a straight line. Thus we call them the flat space or flat spacetime. In the curved spacetime such as the Riemann spacetime, there is no more a straight line. However, one can extend the concept of straight line to the curved space. A straight line in the Euclidean space is a line without changing the direction characterized by the tangent of the line. Similarly we define the 'straight line' or precisely so-called geodesics in the Riemann spacetime a line in which the tangents of nearby points are parallel. We denote $u^\mu = \frac{dx^\mu}{d\lambda}$ as the tangent of the curve $x(\lambda)$ with λ as the curve parameter. Then a geodesic is a line determined by the following equation

$$\nabla_u u = 0. \tag{H.1}$$

In the component notation, we have

$$u^\alpha u^\mu_{;\alpha} = u^\alpha u^\mu_{,\alpha} + \Gamma^\mu_{\alpha\beta} u^\alpha u^\beta = 0. \tag{H.2}$$

Inserting $u^\mu = \frac{dx^\mu}{d\lambda}$ and $u^\alpha \frac{\partial}{\partial x^\alpha} = \frac{d}{d\lambda}$ into Eq.(H.2), we have

$$\frac{d^2 x^\mu}{d\lambda^2} + \Gamma^\mu_{\alpha\beta} \frac{dx^\alpha}{d\lambda} \frac{dx^\beta}{d\lambda} = 0. \tag{H.3}$$

Eq.(H.3) is called the *geodesic equation*. It describes a line drawn in such a way that keeps its tangent as parallel as possible.

A geodesic is also a curve with minimal distance between any two points. The shortest line in the Euclidean space or Minkowski spacetime is a straight line. In the curved spacetime such as the Riemann spacetime, we will prove that the shortest line is a geodesic line. We can use the variation principle to derive the equation that describes the shortest line in the

Riemann spacetime. The distance S of a line connecting two points A and B in the Riemann spacetime is defined as

$$S = \int_B^A ds. \tag{H.4}$$

The line has the minimum distance S is the shortest line. Therefore we determine the minimum of S by variation.

$$\delta S = \delta \int_B^A ds = 0. \tag{H.5}$$

The line element ds is given by

$$dS = (g_{\alpha\beta} dx^\alpha dx^\beta)^{\frac{1}{2}}. \tag{H.6}$$

For a line described by a parameter λ, i.e. $x^\mu = x^\mu(\lambda)$, Eq.(H.6) reads

$$dS = (g_{\alpha\beta} \dot{x}^\alpha \dot{x}^\beta)^{\frac{1}{2}} d\lambda \tag{H.7}$$

with

$$\dot{x}^\alpha = \frac{dx^\alpha}{d\lambda}. \tag{H.8}$$

Eq.(H.5) becomes

$$\delta \int_B^A (g_{\alpha\beta} \dot{x}^\alpha \dot{x}^\beta)^{\frac{1}{2}} d\lambda = 0. \tag{H.9}$$

Due to the mathematical similarity, one can define $\mathcal{L} = (g_{\alpha\beta} \dot{x}^\alpha \dot{x}^\beta)^{\frac{1}{2}}$ as the Lagrangian and consider S in Eq.(H.4) as the action for particle motion in the Riemann spacetime.

The variation equation Eq.(H.9) leads to the Euler-Lagrange equation

$$\frac{\partial (g_{\alpha\beta} \dot{x}^\alpha \dot{x}^\beta)^{\frac{1}{2}}}{\partial x^\nu} - \frac{d}{d\lambda} \frac{\partial (g_{\alpha\beta} \dot{x}^\alpha \dot{x}^\beta)^{\frac{1}{2}}}{\partial \dot{x}^\nu} = 0. \tag{H.10}$$

Eq.(H.10) can be rewritten as

$$\begin{aligned}
&\frac{\partial (g_{\alpha\beta} \dot{x}^\alpha \dot{x}^\beta)^{\frac{1}{2}}}{\partial x^\nu} - \frac{d}{d\lambda} \frac{\partial (g_{\alpha\beta} \dot{x}^\alpha \dot{x}^\beta)^{\frac{1}{2}}}{\partial \dot{x}^\nu} \\
&= \frac{1}{2(g_{\alpha\beta} \dot{x}^\alpha \dot{x}^\beta)^{\frac{1}{2}}} \frac{\partial g_{\alpha\beta}}{\partial x^\nu} \dot{x}^\alpha \dot{x}^\beta - \frac{d}{d\lambda} \frac{(g_{\alpha\nu} \dot{x}^\alpha + g_{\beta\nu} \dot{x}^\beta)}{2(g_{\alpha\beta} \dot{x}^\alpha \dot{x}^\beta)^{\frac{1}{2}}}.
\end{aligned} \tag{H.11}$$

We can take the parameter λ as the distance of s, then

$$g_{\alpha\beta} \dot{x}^\alpha \dot{x}^\beta = g_{\alpha\beta} \frac{dx^\alpha}{d\lambda} \frac{dx^\beta}{d\lambda} = g_{\alpha\beta} \frac{dx^\alpha}{ds} \frac{dx^\beta}{ds} = 1. \tag{H.12}$$

Using Eq.(H.12), we have

$$\frac{1}{2}g_{\alpha\beta,\nu}\dot{x}^{\alpha}\dot{x}^{\beta} - \frac{d}{ds}(g_{\alpha\nu}\dot{x}^{\alpha}) = 0 \qquad (H.13)$$

or

$$g_{\alpha\nu}\frac{d^2x^{\alpha}}{ds^2} + (g_{\alpha\nu,\beta} - \frac{1}{2}g_{\alpha\beta,\nu})\frac{dx^{\alpha}}{ds}\frac{dx^{\beta}}{ds} = 0. \qquad (H.14)$$

Using the relations

$$g_{\alpha\nu,\beta}\frac{dx^{\alpha}}{ds}\frac{dx^{\beta}}{ds} = g_{\beta\nu,\alpha}\frac{dx^{\beta}}{ds}\frac{dx^{\alpha}}{ds} = g_{\beta\nu,\alpha}\frac{dx^{\alpha}}{ds}\frac{dx^{\beta}}{ds}, \qquad (H.15)$$

Eq.(H.14) becomes

$$\frac{d^2x^{\mu}}{ds^2} + \frac{1}{2}g^{\mu\nu}(g_{\alpha\nu,\beta} + g_{\beta\nu,\alpha} - g_{\alpha\beta,\nu})\frac{dx^{\alpha}}{ds}\frac{dx^{\beta}}{ds} = 0. \qquad (H.16)$$

Using the relation between the connection $\Gamma^{\alpha}_{\mu\nu}$ with the metric $g_{\mu\nu}$

$$\Gamma^{\alpha}_{\mu\nu} = \frac{1}{2}g^{\alpha\lambda}(g_{\mu\lambda,\nu} + g_{\nu\lambda,\mu} - g_{\mu\nu,\lambda}), \qquad (H.17)$$

we have

$$\frac{d^2x^{\mu}}{ds^2} + \Gamma^{\mu}_{\alpha\beta}\frac{dx^{\alpha}}{ds}\frac{dx^{\beta}}{ds} = 0. \qquad (H.18)$$

Eq.(H.18) is just the geodesic equation Eq.(H.3).

Bibliography

Abrikosov, A. A., Gorkov, L. and Dzyaloshinski, A., *Methods of Quantum Field Theory in Statistical Physics*, Prentice Hall, Englewood Cliffs, New Jersey, 1963.

Balian, R. and Zinn-Justin, J. (eds.), *Methods in Field Theory*, North Holland Publishing, Amsterdam, and World Scientific, Singapore, 1981.

Bellac, M. Le, Mortessagne, F. and Batrouni, G. G., *Equilibrium and Non-Equilibrium Statistical Thermodynamics*, Cambridge University Press, Cambridge, 2004.

Bjorken, J. D. and Drell, S. D., *Relativistic Quantum Mechanics*, McGraw-Hill, New York, 1964.

Cheng, T. P. and Li, L. F., *Gauge Theory of Elementary Particle Physics*, Clarendon Press, Oxford, 1984.

Coleman, S., *Aspects of Symmetry*, Cambridge University Press, Cambridge, 1985.

Dirac, P. A. M., *The Principles of Quantum Mechanics*, Oxford University Press, Oxford, 1935.

Dirac, P. A. M., *General Theory of Relativity*, Princeton Landmarks in Physics Series, Princeton University Press, Princeton, New Jersey, 1996.

Einstein, A., *The Principle of Relativity (A Collection of Original Memoirs on the Special and General Theory of Relativity)*, Dover Publications Inc., New York, 1952.

Fetter, A. L. and Walecka, J. D., *Quantum Theory of Many-Particle System*, McGraw-Hill, New York, 1971.

Feynman, R. P. and Hibbs, A. R., *Quantum Mechanics and Path Integrals*, McGraw-Hill, New York, 1965.

Feynman, R. P., Leighton, R. B., and Sands, M. L., *The Feynman Lecture on Physics*, Vols. 1, 2 and 3, Addison-Wesley, New York, 1965.

Greiner, W., *Quantum Mechanics — An introduction*, Springer-Verlag, Berlin Heidelberg, 1994.

Greiner, W., *Relativistic Quantum Mechanics — Wave Equations*, Springer-Verlag, Berlin Heidelberg, 2000.

Greiner, W. and Müller, B., *Gauge Theory of Weak Interactions*, Springer-Verlag, Berlin Heidelberg, 2000.

Greiner, W. and Müller, B., *Quantum Mechanics — Symmetries*, Springer-Verlag, Berlin Heidelberg, 1994.

Greiner, W., Neise, L. and Stöcker, H. *Thermodynamics and Statistical Mechanics*, Springer-Verlag, Berlin Heidelberg New York, 1995.

Greiner, W. and Reinhardt, J. *Field Quantization*, Springer-Verlag, Berlin Heidelberg, 1996.

Greiner, W. and Reinhardt, J. *Quantum Electrodynamics*, Springer-Verlag, Berlin Heidelberg, 1996.

Griffiths, P. G., *Introduction to Quantum Mechanics*, Prentice-Hall Inc., New Jersey, 2005.

Hao, B. L., et al, (eds.), *Progress in Statistical Physics* (in Chinese), Science Press, Beijing, 1981.

Hartle, J. B., *Gravity: An Introduction to Einstern's General Relativity*, Addison-Wesley, New York, 2002.

Ho-Kim, Q. and Pham, X. Y., *Elementary Particles and Their Interactions — Concepts and Phenomena*, Springer-Verlag, Berlin Heidelberg, 1998.

Huang, K., *Quantum Field Theory*, John Wiley & Sons, New York, 1998.

Huang, K., *Statistical Mechanics*, John Wiley & Sons, New York, 1987.

Itzykson, C. and Zuber, J. B., *Quantum Field Theory*, McGraw-Hill, New York, 1980.

Jackson, J. D., *Classical Electrodynamics*, John Wiley & Sons, Toronto, 1999.

Kadanoff, L. P., *Statistical Physics*, World Scientific, Singapore, 2000.

Kadanoff, L. P. and Baym, G., *Quantum Statistical Mechanics*, Addison-Wesley Publishing Co., Inc., 1962.

Kardar, M., *Statistical Physics of Particles*, Cambridge University Press, Cambridge, 2007.

Kardar, M., *Statistical Physics of Fields*, Cambridge University Press, Cambridge, 2007.

Landau, L. D. and Lifschitz, E. M., *Mechanics*, Elsevier (Singapore) Pte Ltd, 2007.

Landau, L. D. and Lifschitz, E. M., *The Classical Theory of Fields*, Elsevier (Singapore) Pte Ltd, 2007.

Landau, L. D. and Lifschitz, E. M., *Quantum Mechanics — Nonrelativistic Theory*, Pergamon Press Ltd, Oxford, 1977.

Landau, L. D. and Lifschitz, E. M., *Statistical Physics*, Addison-Wesley, Reading, MA, 1974.

Leader, E. and Predazzi, E., *An Introduction to Gauge Theories and Modern Particle Physics*, Vols. 1 and 2, Cambridge University Press, Cambridge, 1996.

Lee, T. D., *Particle Physics and Introduction to Field Theory*, Taylor and Francis, New York, 1981.

Liu, L. and Zhao, Z., *General Relativity* (in Chinese), Advanced Education Publishing, Beijing, 2006.

Ma, S. K., *Modern Theory of Critical Phenomena*, Benjamin/Cummings, Reading, MA, 1976.

Mandl, F., *Introduction to Quantum Field Theory*, Interscience, New York, 1959.

Mosel, U., *Path Integrals in Field Theory — An introduction*, Springer-Verlag, Berlin Heidelberg, 2004.

Negele, J. W. and Orland, H., *Quantum Many-Particle Systems*, Addison-Wesley Publishing Co., Reading, Massachusetts, 1988.

Pathria, R. K., *Statistical Mechanics*, Elsevier (Singapore) Pte Ltd, 2003.

Peskin, M. E. and Schroeder, D. V., *An Introduction to Quantum Field Theory*, Addison-Wesley Publishing Co., Reading, Massachusetts, 1995.

Reichl, L. E., *A Modern Course in Statistical Physics*, 2nd Ed., John-Wiley & Sons, Inc., New York, 1998.

Ryder, L. H., *Quantum Field Theory*, 2nd Ed., Cambridge University Press, New York, 1996.

Sakurai, J. J., *Modern Quantum Mechanics*, Addison-Wesley Publishing Co., Inc., New York, 1994.

Sakurai, J. J., *Advanced Quantum Mechanics*, Addison-Wesley Publishing Co., Inc., New York, 1967.

Schulman, L., *Techniques and Applications of Path Integrals*, John Wiley and Sons, New York, 1981.

Schutz, B., *A First Course in General Relativity*, Cambridge University Press, Cambridge, 2009.

Schwabl, F., *Statistical Mechanics*, Springer-Verlag, Berlin Heidelberg, 2006.

Stednicki, M., *Quantum Field Theory*, Cambridge University Press, New York, 2007.

Wang, C. T. *Statistical Physics* (in Chinese), Tsinghua University Press, Beijing 1991.

Wang, Z. C., *Thermodynamics · Statistical Physics* (in Chinese), Advanced Education Publishing, Beijing, 2003.

Weinberg, S., *Gravitation and Cosmology*, Wiley, New York, 1972.

Weinberg, S., *Quantum Theory of Fields*, Vols. 1 and 2, Cambridge University Press, New York, 1996.

Wen, X. G., *Quantum Field Theory of Many-Body Systems*, Oxford University Press, New York, 2007.

Yang, C. N., *Selected Papers 1945–1980 with Commentary*, W. H. Freeman, San Francisco, 1983.

Zee, A., *Quantum Field Theory in a Nutshell*, Princeton University Press, Princeton, New Jersey, 2003.

Zheng, J. Y., *Quantum Mechanics* (in Chinese), Science Press, Beijing, 2000.

Zhou, S. X., *A Course on Quantum Mechanics* (in Chinese), Advanced Education Press, Beijing, 2004.

Index

S matrix, 180, 181, 191, 194, 195, 198, 200, 313
$SU(N)$ symmetry, 42
$SU(N)$ symmetry transformation, 98
g factor, 283
1-particle-irreducible graph, 160
1-particle-reducible graph, 160
1PI, 158
1PI four-point function, 169
1PI graph, 160
1PI vertex Feynman diagram, 167
1PI vertex function, 217
2-parameter Feynman formula, 226

abelian, 98
abelian gauge boson field, 130
abelian gauge symmetry, 114
abelian gauge transformation, 128
abelian group, 97
absolute fluctuation, 386
absolute temperature scale, 376
acceleration, 299, 306
action, 23, 53, 73, 83, 90, 91, 102, 105–110, 113, 119, 120, 122, 143, 166, 178, 263, 298, 465, 467
action in quantum mechanics, 306
action operator, 18, 27, 31, 32, 35, 107, 109
active rotation, 247
adiabatic compressibility, 368
adiabatic heat capacity, 368
adiabatic process, 365, 368

adjoint field operator, 45
adjoint spinor, 43
advanced propagator, 139
Ampere's law, 265
amplitude probability, 139
amputated Green's function, 159, 175, 195
angular coordinate, 330
angular frequency, 418
angular momentum, 54, 55, 83, 293–295
angular momentum operator, 293, 294, 332, 335
angular momentum tensor, 83
annihilation operator, 4, 6, 10, 12, 29, 31, 37, 39, 41, 63, 64, 70, 80, 85, 94, 134, 182, 189, 196, 200, 249–252, 273, 274, 326–328
anomalous dimension, 236
anomalous magnetic moment, 284, 288
anti-causal propagator, 139
anti-commutation relation, 5, 44, 68, 74, 111, 112, 197, 202, 495–498
anti-commutator, 66
anti-commuting number, 495, 497
anti-hermitian, 46
anti-symmetrization operator, 7, 253
anti-symmetrized state, 7–9
antiparticle, 41, 63, 66, 70, 71, 79, 124, 139, 239, 243
antisymmetric, 7, 48, 49, 54, 55, 58,

Printed in the United States
By Bookmasters